Water Supply

Third Edition

A C Twort
BSc, FICE, FIWES

F M Law
BSc, Dip. Hydrol., MICE, FIWES

F W Crowley
OBE, BSc, FICE, FIWES

Edward Arnold

First published in Great Britain 1963 by
Edward Arnold (Publishers) Ltd, 41 Bedford Square, London WC1B 3DQ

Edward Arnold, 300 North Charles Street, Baltimore, Maryland 21201, USA

Edward Arnold (Australia) Pty Ltd, 80 Waverley Road, Caulfield East, Victoria 3145, Australia

Second Edition 1974
Reprinted 1975, 1979, 1982
Third Edition 1985

British Library Cataloguing in Publication Data

Twort, A.C.
Water supply.—3rd ed.
1. Water-supply engineering
I. Title II. Law, F.M. III. Crowley, F.W.
628.1 TC405

ISBN 0-7131-3513-1

Printed and bound at
Thomson Litho Ltd, East Kilbride, Scotland

Preface

The third edition of this book has had to deal with substantial changes in water supply technology during the past ten years. Much more data are now available about the demand for water. The organization of water supply has changed towards multipurpose functioning and total control of the hydrological environment on a river catchment basis, in the face of a growing need to exercise control over the pollution of natural waters. Hydrological and hydrogeological techniques have increased in sophistication, a new approach has been made to flood estimation, and yields are now increasingly based upon a comprehensive assessment of all water movements within a catchment. Treatment processes have diversified: some new methods of treatment have been pursued, whilst previous methods have come under closer control arising from increasing experience. The past ten years have seen an enormous increase in knowledge concerning many hundreds of complex organic compounds existing at low concentrations in all waters. Save for a few cases, their significance in relation to human health is for the most part unknown, but knowledge of their existence is bringing about substantial changes in the philosophies of approach to water treatment, water disinfection, and pollution control. New piping materials for the distribution of water have come into common use, creating a need to exercise greater discrimination in the choice of materials for particular circumstances. More knowledge has become available concerning the life of distribution systems, and the organizational and technical measures necessary to preserve them.

In all these matters there are both forward movements into new techniques and a re-emphasis on old techniques of proven worth and economy, now better appreciated.

The authors acknowledge with gratitude information and illustrations made freely available by Binnie & Partners and by many manufacturers, suppliers, and consulting engineers.

The chapters on water treatment (Chapters 6–9 inclusive) still follow and include much of the material provided for the previous edition by Dr R C Hoather and Dr C Wilcocks, formerly of the Counties Public Health Laboratories (now Bostock Rigby), but with additions and changes made by the present authors. Ms C R Jackson and Mr D D Ratnayaka also substantially assisted in the rewriting of water treatment material, Dr F C Wood providing the material on seawater desalination.

Mr R V C Loadsman updated notes on the design of reinforced concrete service reservoirs. Mr D E Foddy provided advice on pumps, Mr A R George advice on ultrasonic meters, and Mr C D Walters advice on certain financial aspects of water policy.

To all who have assisted us we express our grateful thanks, but must make clear that responsibility for the statements and opinions expressed in this book is solely our own.

1984

A C Twort
F M Law
F W Crowley

Contents

Abbreviations used in bibliographies

ASCE	American Society of Civil Engineers
AWWA	American Water Works Association
BSI	British Standards Institution
BWA	British Waterworks Association
DoE	Department of the Environment
EEC	European Economic Community
FAO	Food and Agriculture Organization
HMSO	Her Majesty's Stationery Office
IASH	International Association of Scientific Hydrology
ICE	Institution of Civil Engineers
I Chem. E	Institution of Chemical Engineers
IMTA	Institution of Municipal Treasurers and Accountants
IPHE	Institution of Public Health Engineers
ISE	Institution of Sanitary Engineers
ISO	International Standards Organization
IWE	Institution of Water Engineers
IWES	Institution of Water Engineers and Scientists
IWSA	International Water Supply Association
NWC	National Water Council
SCI	Society of Chemical Industry
SWTE	Society for Water Treatment and Examination
USGS	United States Geological Survey
WHO	World Health Organization
WMO	World Meteorological Office
WRA	Water Research Association
WRB	Water Resources Board
WRC	Water Research Centre

1

How much water is needed

1.1 Categories of consumption

It is convenient to divide water consumption into the following categories.

(1) *Domestic*
in-house—drinking, cooking, ablution, sanitation, house cleaning, car washing, clothes washing;
sprinkling—garden watering, lawn sprinkling;
standpipe—from public standpipes and fountains.

(2) *Trade*
industrial—factories, industries, power stations, docks;
commercial—shops, offices, restaurants, small trades etc.;
institutional—schools, hospitals, universities, government offices etc.

(3) *Agricultural*
crops, livestock, horticulture, greenhouses, dairies, farmsteads.

(4) *Public*
street watering, public parks, sewer flushing, fire-fighting.

(5) *Losses*
consumer wastage—leakages and wastages from consumers' premises, misuse or unnecessarily wasteful use of water by consumers;
distribution losses—leakages and overflows from service reservoirs, leakages from mains, service pipes, and service pipe connections, leakages from valves, hydrants, and washouts;
metering and other losses—source meter errors, supply meter errors, unrecorded consumptions.

In most countries all supplies are metered except standpipe and public supplies. In the UK only trade and agriculture supplies are metered, domestic supplies being unmetered. In countries where full metering is adopted, losses are derived by deducting total metered supplies and estimated standpipe and public consumption from the total put into the system. In the UK the unmetered domestic demand must be estimated before a figure for losses can be derived, or losses are estimated directly so that the domestic demand can be derived. The latter practice was frequently adopted in the past and, until recently, gave rise to an underestimate of losses and an overestimate of domestic demand in many published figures.

1.2 Validity of consumption figures

Published figures of consumption vary widely because of widely varying standards of supply, for example: losses may vary from 5% to 55% of the total supply; meters may not register correctly; supplies may be intermittent or inadequate; consumers may waste much water if there is no efficient waste prevention system in force, or if they are not billed efficiently for water supplied by meter. Comparison of consumptions by different water undertakings is therefore difficult when the factors mentioned above are not defined. Also comparison of total consumptions may be frequently misleading since trade consumption may be large or small irrespective of the size of population served. Figure 1.1 shows the total consumption per capita for 130 cities and indicates how figures can vary from 100 to 700 litres per capita per day (lcd) irrespective of the size of the city[1]. Even when trade consumption is

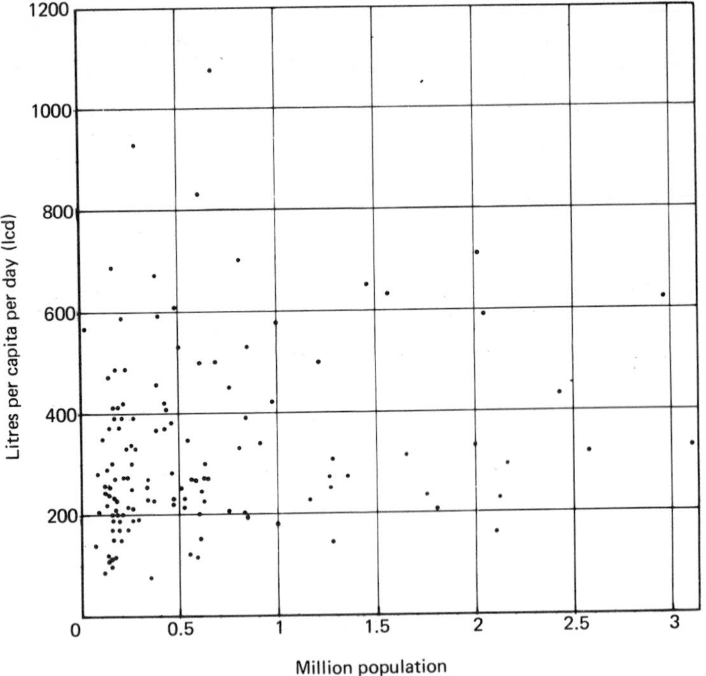

Fig. 1.1 Average daily consumptions from 130 municipal systems in the Americas, South Africa, Europe, Asia, and Russia in 1969. (Janczewski, *Proc. IWSA Congress*, 1974.)

eliminated, and losses are realistically estimated, comparison of domestic consumptions should exclude sprinkler and standpipe consumption. For example, in the USA the apparently high domestic consumption figures often quoted are in fact comprised of 20% to 60% sprinkler consumption[2] which, in that country, is very high. In many other countries, e.g. the UK, sprinkler consumption (or garden watering) is a negligible amount when expressed as an average over the year.

The accuracy of consumptions expressed per capita is also dependent upon the accuracy to which the population served can be estimated. Frequently population estimates are inexact because water supply boundaries may not coincide with census boundaries, census data may be out of date or suspect, and it may be difficult to estimate numbers of migrant, commuting, or holiday visitor populations.

For valid comparisons figures of consumption should relate only to 24-hour supply systems at reasonable pressure, with reasonably efficient waste and leak prevention measures applied.

1.3 Consumption surveys

A consumption survey is necessary before forecasting future demand or when investigating losses from a system. The steps involved illustrate the difficulties of measuring consumption.

(1) The source output meters must be checked for accuracy. Various means may have to be used for this: the source output to (or from) some tank may be measured, or a V-notch weir may be installed, or a pitometer may be used to measure pipe flow. Large errors can be found: up to ±40% error is not unknown*.

(2) Supply metering practices must be checked in order to assess how accurately they are measuring water taken by consumers. Some actions required are as follows.

(a) Selected consumers' meters are read by an engineer over a period and compared with billings.
(b) New meters are installed on a number of premises and their readings compared with previous billings.
(c) The percentage of meters not working at any given time is checked by inspection; the age of the meters is ascertained and how frequently they are called in for repair.
(d) Pitometer or other flow tests are made to check the accuracy of meters on all large supplies.
(e) The interior of mains is inspected for sand and sliming which could affect metering; the water quality is checked for any tendency to corrode meters unduly.

* There is often a complexity of pipes and meters at a major source and it is not unknown to find a Venturi tube connected to the wrong indicating and recording gear.

In poorly maintained undertakings a high incidence of stopped meters will be found. Where more than 15% of meters are stopped at any one time, or many meters are over ten years old, a substantial amount of under-recording by the supply meters must be suspected. It is particularly important to test all large consumers' meters.

(3) The rate of domestic consumption in typical dwellings must be checked. The steps involved are as follows.

(a) Classify dwellings into not more than five or six identifiable separate classes according to size or value and whether occupied by upper, middle, or lower income groups.
(b) From meter record books select 40 to 50 dwellings in each class in which the meter appears to be working satisfactorily; visit each dwelling and discard those not typical of their class; find the number of persons in each during the previous period of metering; calculate the per capita consumption.
(c) Fix new meters on selected properties in each class; record meter readings daily; find the number of persons in each property; calculate the per capita consumption.
(d) Survey standpipes for breakages; meter some standpipes and, by enquiry, estimate the number of persons dependent thereon, hence estimate the per capita consumption.
(e) Estimate sprinkling demand by comparing demand during a dry period with demand immediately after the onset of rain.

(4) Trade and agricultural consumptions must be investigated. Where records of these are incomplete or unreliable, a range of establishments must be visited to assess their probable consumptions so that this information can be used to estimate the total trade and agricultural consumptions. Test metering of supplies to government offices, public gardens, military establishments etc. is advisable since such supplies can be much higher than estimated.

(5) The final step is to build an estimate of the total demand using the data obtained from the foregoing investigations. The total domestic demand is usually worked out in district units: within each district the areas of different classes of housing are plotted and measured, the appropriate densities of population per hectare according to housing class are applied and multiplied by the appropriate per capita consumptions found from field tests. Standpipe consumptions are separately assessed. Trade, institutional, and agricultural demands are estimated either by analysing the meter records, or by the methods outlined under step 4 above.

The total demand estimated from the foregoing investigations is the total potential demand assuming all persons receive an adequate supply, but this may not be the case. If the sourceworks are inadequate or the distribution mains are not large enough, some people may not receive as

much water as they need, so there will be some unsatisfied demand. As a consequence the equation for supply and demand has to be written as follows:

$$\begin{aligned}\text{actual supply} = {} & \text{total potential demand} + \text{consumer wastage} \\ & + \text{distribution losses} - \text{unsatisfied demand} \\ & \pm \text{metering errors}\end{aligned}$$

If the metering error has been estimated this term can drop out of the equation. If the supply is 24-hour and adequate to meet all demands then the unsatisfied demand term can drop out also, but where the supply is inadequate or pressures are too low for consumers to obtain what they want, a high unsatisfied demand can exist at the same time as there is high consumer wastage and high distribution losses. Such a situation is not unusual: figures for consumption per capita (dividing the total supply by the total number of consumers) may appear reasonable, but may conceal high losses and waste, offset by large unsatisfied demand.

1.4 Domestic in-house demand

Domestic in-house demand varies widely from one house to another, even in identical housing. The 'in-house' demand is that used for domestic purposes, excluding garden watering and lawn sprinkling, but including washing cars and cleaning down outside yards. It is quite common to find households in which the per capita consumption is two or three times the average. Figure 1.2 illustrates the scatter of results obtained in an actual investigation; all surveys give similar results.

Practical difficulties stand in the way of accurately measuring the per capita consumption in more than about 20 to 30 households per class of dwelling at a time. All properties must have no leakage from plumbing systems or service pipes; all meters must be proved accurate by testing; the number of persons each day in each household must be found; the weather should preferably be 'average'; holiday times must be avoided and both weekday and weekend consumptions must be assessed. Attempts are sometimes made to gain a better estimate of the average consumption by measuring total flows to a large group of households, but then other inaccuracies apply. The population of a large group is not static and is difficult to assess accurately; the meter on the supply main will register all leaks downstream of it, and leakage from service pipe connections to mains is a major source of distribution losses; there will also be a degree of consumer wastage occurring which cannot be prevented.

Hence it is best to use small samples accurately undertaken. In such tests it is important not to discard results showing an unusually high per capita consumption if there is no evidence of meter inaccuracy or other error. In surveys in Istanbul of this type, for instance, one in five

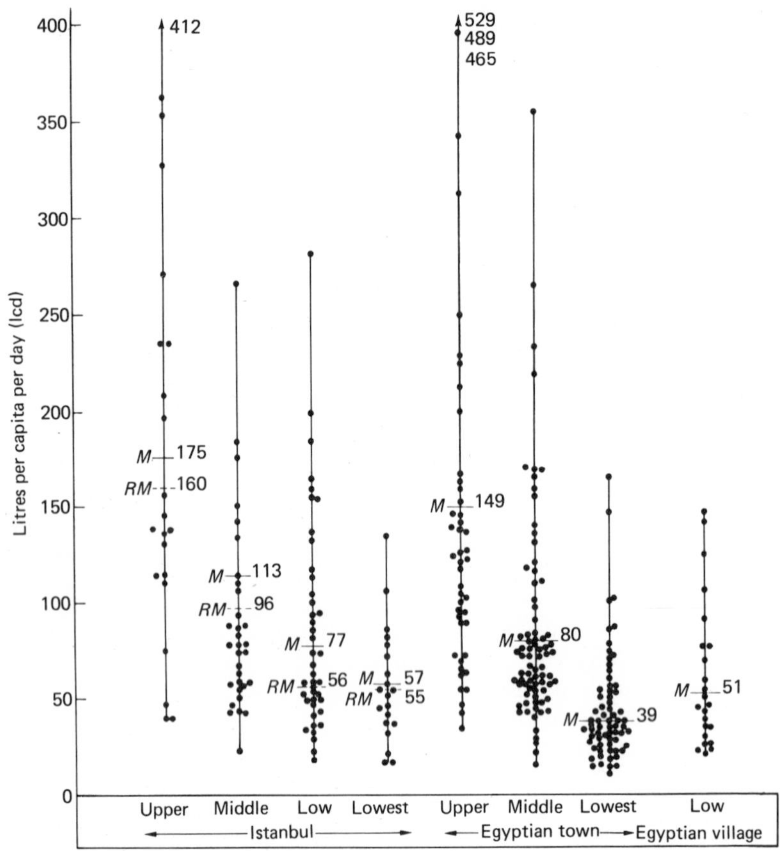

Fig. 1.2 Variation of household consumption within a given class of dwelling. *M* is the mean of all results within a class; *RM* is the reduced mean which excludes households taking more than two times the mean of the others within the class.

households were found with a per capita consumption of two or more times the average[3]. The same phenomenon has been found in English surveys[4]. These 'high consumers' must not be ignored because they raise the average consumption by a substantial amount. They can occur in households where bathing is unusually frequent, in which water is extravagantly used, or in which large families reside requiring much clothes washing. Unusually low per capita consumptions tend to occur in households occupied by elderly persons, or in which the occupants are present only in the evenings.

The mean value of per capita consumption derived from a field test should be the total consumption in all households surveyed (of a given class) divided by the total population counted in those households. It should not be the mean of the individual household values because it is unlikely that the sample size can be large enough to ensure that it contains a representative proportion of high and low consumers, nor can it be large enough to ensure that the occupancy distribution per household within the sample is the same as that applying to the whole class from which the sample is drawn. For the same reason households for test metering should not be chosen at random, but should be selected by inspection to ensure that they comprise as nearly as possible a representative sample of the class of houses to be surveyed. If random selection is adopted this may result in an unfortunate bias of the sample towards lower or higher grade properties within the class selected and test results may have to be 'adjusted', which is unsatisfactory.

Referring only to in-house consumption, some reliable test results are given in Table 1.1. The most accurate results (Grade I) come from individually test metering selected properties using new meters. Less accurate results come from inspection of meter records (Grade II) and bulk metering a group of properties (Grade III). Grade II results will include some consumer wastage as the properties will not have been inspected for waste. Grade III results will additionally include some proportion of distribution losses from the supply main downstream of the bulk meter and from the service pipes and their connections. Leaks from service pipe connections are commonly found to account for a considerable proportion of the total distribution losses.

From Table 1.1 it will be seen that domestic in-house water consumption in the typical middle class dwelling having a kitchen, a bathroom, and water-borne sewerage centres about 115 lcd. This figure appears to apply to this type of housing in most parts of the world irrespective of climate: departures from it can nearly always be explained in terms of different circumstances applying. Thus the higher consumption values reported from the USA (after a substantial amount has been deducted for sprinkler demand) relate to the generally higher standard of middle class housing that occurs in the USA than occurs elsewhere, and the substantially higher incidence of private swimming pools. In lower or upper class dwellings figures for consumption per capita are more variable than for middle class dwellings for the obvious reason that standards of housing (and of income) are more variable in these classes and therefore more difficult to subdivide accurately.

1.5 Components of domestic demand

The most reliable analysis of domestic demand is reported by Thackray *et al.*[4]. These are the results from daily diary records of water use by 850 householders in the two English towns of Malvern and Mansfield where, exceptionally for the UK, all domestic supplies are metered. However, in those towns the total in-house domestic consumption tends to be

Table 1.1 Test measurements of in-house domestic consumption.

Test type (see Section 1.4)	Number of houses in test sample	Location	Year	Consumption (lcd): House or income class: High	Upper middle	Middle	Lower middle	Low	Comments
		England							
II	461	Malvern[5]	1976	150	117	86–105	72		
II	392	Mansfield[5]	1976	126	115	93–115	89		
II	977	South West Water Authority[6]	1977			103 (av.)			Urban
						106 (av.)			Rural
						111 (av.)			Plymouth City
III	2626	East Anglia Water Authority[7]	1977		125–136	112–126			Night-time flows accounted for up to 16 lcd of the figures quoted
		Worldwide							
I	192	Istanbul[3]: Turkey	1975	175		116		56	260 lcd in 'luxury' class houses
I	48	Sakaka[8]: Saudi Arabia	1975		120	110	70		120 = compounds, 110 = new flats
II	173	Lesotho[9]: Africa	1976		123				'Houses with plumbing systems'
II	56	Cairo[10]: Egypt	1976	260		149		55–70	Consumer wastage included
II	17	Palembang[11]: Indonesia	1970	204	152				'Detached houses', some use by non-residents may be included
II	1351	Hong Kong[12]	1977	182	127	127	97	47–52	182 = luxury flats, 127 = upper and middle class flats, 97 = flats occupied by lower income groups, 47–52 = government low cost flats. (All flats individually metered.)
II	38	Alexandria[13]: Egypt	1976	180–220	160	130	100	55–70	
II	259	Melbourne[14]: Australia	1972			110–117			Metered properties
III	2300 (approx.)	USA[15]—41 areas	1965						Some distribution leakage and consumer wastage included
		eastern states—average				193 (av.)			
		eastern states—lower valued					148		
		western states—average				254 (av.)			

about 15 lcd below the national average because of the lower than average standard of housing and other factors. A summary of the figures reported by Thackray *et al.* is given in Table 1.2. In the table a suggested design value is given which endeavours to compensate for the special factors applying at Malvern and Mansfield.

Overseas, in hot climates, a different breakdown of demand applies: more water is used for ablution, food preparation, and washing clothes. Ablution in ordinary domestic premises is almost universally by means of a shower, or from under a running tap. In Muslim countries the practice is also to wash in running water before prayers. The total domestic consumption depends principally upon the liberality, or otherwise, of the supply. Most reported low figures of per capita consumption relate to supplies which are restricted in some manner.

Asian (or squatting pan) toilets are widely used, mostly pour-flushed using buckets of water or flushed by means of a hose pipe attached to a nearby tap. Occasionally flushing cisterns are installed. The Asian toilets are said to need only 6 lcd for flushing[16], but this statement is misleading since they are more often flushed with the water used for anal cleansing which may amount to 20 or 30 lcd. This form of personal cleaning after defaecation is widely adopted throughout the Middle East

Table 1.2 Breakdown of average in-house domestic demand in the UK.

Usage	Test results at Malvern and Mansfield		Suggested amended value for design purposes		Comments
	lcd	%	lcd	%	
WC toilet flushing (average cistern capacity 8.5 litres)*	32–33	33	37	32	Test figures believed low
Kitchen and washbasin use for drinking, cooking, dish-washing, cleaning, handwashing etc.	32–34	34	38	33	Allow for increase in use of dish-washing machines
Ablution					
by baths	16 }	18	20	17	Shower use small in the UK, but may increase
by showers	1 }				
Laundering of clothes by washing machine	11–13	12	16	14	Only 26% of the houses under test had automatic washing machines, but this will increase
Car washing, plant watering, and patio use	2–3	3	4	4	
Totals	97–98	100	115	100	Note increase

*Standard maximum capacity = 9.0 litres.

and Asia, toilet paper being rarely used. In higher income properties a Western-type pedestal pan toilet may be installed with a flushing cistern, but often an Asian toilet will be installed nearby in addition for anal cleansing (as a bidet is used in Continental Europe). The average consumption for sanitation may therefore reach 30 lcd, but this represents a smaller proportion of the total average domestic consumption than in the UK because of the higher amounts used for ablution, clothes washing (more frequent in a hot climate and mostly done by hand), and food preparation (such as fish and rice cleaning). Table 1.3 gives the mean actual measured consumptions by usage in 63 households drawing water from private wells, in a hot equatorial humid climate, and also gives slightly lower values for design purposes which should ensure an adequate supply.

Table 1.3 Net domestic demand in hot climates overseas.

	*Mean measured consumption in 63 households in a hot climate		Suggested mean design values for liberal supply	
Usage	lcd	%	lcd	%
(a) Sanitation use: Asian toilets plus anal cleansing	28	16	25	17
(b_1) Drinking, food preparation, and cooking	14	8	12	8
(b_2) Dish-washing, cleaning house etc.	13	8	12	8
(c) Ablution, by showers and under running taps	85	48	60	40
(d) Clothes washing	35	20	30	20
(e) Miscellaneous: yard washing, car cleaning, plant watering	(Included above)		11	7
Totals	175	100	150	100

* The figures refer to water dipped from wells, except for drinking water taken from standpipes. Each house has its own well. Where water was pumped from wells, the consumption was about 25% higher.

Tables 1.2 and 1.3 give only *net* domestic consumption figures. To them should be added a percentage for distribution system losses which, for large systems, may be in the range 20–25% of the total supply.

Drinking water demand in the UK has been assessed by the WRC[17] in 1980. Data from 3632 persons in 1320 households in 100 different locations in England, Wales, and Scotland were obtained. The mean drinking consumption of water was found to be 0.96 lcd taken from domestic taps plus 0.63 lcd water content taken in purchased beverages. The total mean consumption was therefore 1.59 lcd. Two-thirds of all consumers had a total drinking consumption in the range 0.75–2.00 lcd; those drinking less (about 9% of the sample) were believed to be mainly infants.

1.6 Sprinkler and garden watering demand

Sprinkler and garden watering demand in the UK is negligible when expressed as an average daily amount for the whole year, but on any particular day in hot dry weather of summer it can add 20% to the total domestic demand. For an average sized town in the UK this would represent an extra 50 lcd on such days. If the dry spell is prolonged the sprinkler demand will increase until it represents a 30% to 40% addition to the demand.

In the USA much higher sprinkler demand occurs and it is much more prolonged. Swimming pools are also more numerous than in the UK. This out-of-house demand, occurring throughout the summer season, is the principal reason for the total domestic demand figures in the USA being substantially higher than in the UK. Linaweaver *et al.*[2], from a study of 2300 properties throughout the USA, found the annual average sprinkler demand to be 40 to 70 lcd in eastern states and 185 lcd in western (drier) states. Maximum day sprinkler consumptions were from 10 to 30 times these figures. As a result, in the USA 25% to 50% of the total annual domestic demand is for sprinkler use, mainly for watering lawns. Kinnersley[18] reported that when Denver restricted the use of sprinklers to one in three days unless a licence was purchased, the total annual consumption fell by 20%. He also reported similar figures from South Africa where average out-of-house demand accounted for 40% of the total domestic demand, rising to 73% in some years. Weeks and Mahon[15] reported similarly in respect of South Australia where 33% to 55% of the total annual demand in three typical Australian residential areas was for sprinkler use.

1.7 Air conditioners

Air conditioners of the desert cooler type, i.e. in which air passes through a water cascade or spray, may cause a large consumption. They are principally used in hot dry climates, i.e. not in tropical monsoon climates, and each such cooler may consume 10 to 20 litres of water per hour[19].

1.8 Standpipe demand

Standpipe demand depends upon the number of standpipes provided, the distance over which consumers must fetch water, the usages permitted from the standpipe, the degree of consumption control exercised at the standpipe, and the daily hours of supply allowed. At a WHO conference in 1978[20], standpipe consumptions of 20 to 60 lcd were reported from various parts of the world.

The consumption from a standpipe may comprise the following components:

(1) water taken away for drinking and cooking,
(2) water taken away for other domestic purposes, e.g. ablution and clothes washing,
(3) water used for bathing and laundering at the standpipe,
(4) water used for watering animals at or near the standpipe,
(5) water used for cleansing vessels at the standpipe, and spillage and wastage.

Usage 4 is frequently forbidden (or notionally forbidden) and sometimes usages 2 and 3 are also discouraged. Where the supply is intended for drinking and cooking only, take-away consumption may be limited to 7 lcd or thereabouts, but inevitable spillage and wastage will cause the minimum consumption to be about 10 lcd unless strict control is exercised at the standpipe.

Where all-purpose usage (other than the watering of animals) is permitted the take-away consumption for usages 1 and 2 was found in rural Egypt to be in the range 13 to 18 lcd with a mean of about 17 lcd[21]. Spillage and unpreventable at-the-tap usage will cause the total gross consumption to rise to 25 lcd. This latter figure should be regarded as the minimum provision to be made for all domestic purposes.

Where there is little or no control exercised at the tap, and in any case where bathing and laundering are permitted at standpipes, then 45 lcd needs to be provided. If purpose built bathing and laundering facilities are provided at the standpipe then 65 lcd should be allowed. These figures are illustrated by actual experience: in India 50 lcd is the usual standpipe allowance; in Indonesia 15 to 20 lcd occurs where the water is *sold* from standpipes; in rural Egypt where all-purpose uncontrolled use tends to occur, field test figures of 46, 52, and 71 lcd were obtained.

1.9 Consumer wastage

Consumer wastage is a term used to mean all leakage, wastage, and misuse of water on consumers' premises. It is usually estimated by measuring minimum night flows (MNFs) to domestic premises during the hours 01.00 to 04.00 when legitimate demand should be least. Giles[22] advises that 1.0 l/h per household should be allowed as the lowest MNF to domestic areas, but where waste prevention measures are only moderate 2.0 l/h per household should be allowed. Some investigation results are given in Table 1.4.

For purely domestic systems an MNF of 2.0 l/h per connection is likely. Of this perhaps 1.5 l/h can be regarded as wastage and 0.5 l/h as demand. For households containing an average of three persons these figures imply about 12 lcd wastage and 4 lcd minimum nocturnal demand. The total of 16 lcd, or its equivalent of 2 l/h per connection

Table 1.4 Minimum night flow tests on domestic premises.

Reference	MNF (litre per hour per connection)	Comments
Gledhill[23] (reported by Giles)	1.2	Test repeated three times and supply mains tested before and after for leaks
Reid[24]	1.0	Tests on leak-free portions of distribution systems serving 750–1000 dwellings in Manchester
Heide[25]	0.5	Average night toilet-flushing usage on a test of sixty houses in Germany
Wijntjes[26]	0.3–0.5	Netherlands practice to allow 0.1 l/h per person for night consumption
Cook[7]	1.5–2.1	Deduced from 13 to 17 lcd night consumption tests on five systems supplying 200 to 2200 persons after 'a moderate leak detection exercise'. (Three persons per household assumed.)
Shaw Cole[27]	1.6	Quoted as usual USA experience

MNF, is the lowest likely to be experienced with moderate waste detection measures in force. Where waste prevention measures are slack or non-existent, consumer wastage will be much higher, being frequently 50 lcd or above. Multiple or 'block' metering will tend to encourage high consumer wastage since, under this system, where one meter measures the supply to a block of properties and the landlord pays the water charges (recovering the cost by means of the rents charged), individual householders do not pay a sum related to the amount of water they use. If a supply is given free of charge and no attempt is made to curb demand, very high consumptions can result in hot climates. On the Kainji hydroelectric construction project in Nigeria in 1968 where the water supply was free and unrestrained, senior staff in married quarters used 860 to 950 lcd during the wet season and over 1600 lcd in the dry season[28].

1.10 Design figures for in-house domestic demand

Design figures for in-house domestic demand are given in Table 1.5 and they include an amount for unavoidable consumer wastage where waste prevention measures are reasonably efficient. They are averages for city-block areas and represent a reasonable allowance for all in-house domestic needs exclusive of sprinkler irrigation or garden watering. They apply to a system which affords a 24-hour supply at reasonable pressure.

The *mean* overall in-house demand for large urban areas containing a typical mix of upper, middle, and lower class properties, including a reasonable allowance for unavoidable consumer waste, may be taken as

Table 1.5 Suggested design allowances for in-house domestic water demand and standpipe demand (including an allowance for unavoidable wastage).

Class	Occupancy	Type of property	Allowance (lcd) European	Allowance (lcd) Elsewhere (except USA)
A1	Highest income groups	Villas; large detached houses; large luxury flats	180	230–250*
A2	Upper range of income groups	Detached houses; large flats	160	200*
B1	Middle income groups	Houses and flats with one or two WCs and three or more taps, one bath and/or shower	140	160*
B2	Lower middle income groups	Houses and flats with only one WC, one kitchen, and one bathroom	125	
		block metered		160+
		individually metered		130
C1	Low income groups	Tenement blocks or government housing with high density occupation, with one shower, one WC or Asian toilet, and one or two taps		
		block metered or free		130+
		individually metered		70–90
		Small rural cottages (UK)	70–90	
C2	Lowest income groups above poverty line	Low grade basic tenement blocks comprising one or two roomed dwellings with high density occupation		
		with common washrooms (unmetered)		110
		with one tap and one Asian toilet per household (block metered)		90
C3	Lowest income groups	One tap dwellings with shared toilet facilities or none; dwellings with intermittent or low pressure supplies		50–55
D	Standpipes	In urban areas with no control	70+	
		In rural areas under village control	45	
		As above but with good washing and laundering facilities at the standpipe	65	
		Minimum supply for drinking, cooking, and ablution	25	
		Drinking and cooking only	8–10	

* In the USA figures run 30 lcd higher in eastern states and up to 80 lcd higher in western (hotter) states.

follows: 125–140 lcd in western European domestic urban areas and in other similar cities in similar climates; 200–250 lcd in the USA, the higher figure occurring in the hotter drier states. The higher demand in the USA is primarily due to high average standards of housing and living and the hotter summer climate. The figures exclude sprinkling demand and garden watering.

1.11 Trade and industrial demand

Trade and industrial demand may be large or small irrespective of area or population served. Four classes can be distinguished as follows.

(1) *Cooling water demand*—usually abstracted direct from rivers and estuaries and returned to the same with little loss.
(2) *Major industrial demand*—factories using 1000 to 20 000 m^3/d or more for paper making, chemicals manufacturing, production of iron and steel, oil refining etc., the supply for these frequently being obtained wholly or partly from private sources.
(3) *Large industrial demand*—factories using 100 to 500 m^3/d for food processing, vegetable washing, bottling drinks, ice making, chemical production etc., the supply being frequently taken from the public water undertaking.
(4) *Medium to small industrial demand*—factories using less than 50 m^3/d comprising many types making a wide range of products, the great majority taking their water from the public water undertaking.

Overall mean consumptions on a country-wide basis are as follows.

Usual basic consumption for trade purposes in towns with 250 000 population or more	25 lcd
Average trade consumption in England and Wales 1982 (potable water only)	82 lcd
Typical trade consumptions in large industrial cities of the UK and western Europe	100–200 lcd
Typical trade consumptions in industrial cities of the USA	500 lcd

Some more consistent figures of demand by various industries are given in Table 1.6 and figures for typical light industrial estates are given in Table 1.7.

Chapman[29] reports that the English Industrial Estates Corporation and the Welsh Development Agency allow 150 m^3/d per hectare of factory floor area. In their experience this is generally adequate for production, domestic, and fire-fighting purposes for normal industrial estates. Based upon a current average of 325 workers per hectare of factory *floor* area that gives 0.46 m^3/d per worker.

Table 1.6 Figures representing typical magnitudes of demand by various manufacturing industries. (Note that individual factory consumptions vary widely and recycling of water can reduce consumption substantially.)

Process or product	Information source	Consumptions quoted
Automobiles	(a)	12 m^3 per vehicle
Bakery	(d)	2 m^3/t of product
Canning		
generally	(a), (b)	25 m^3/t, 20 m^3/t
meat and vegetables	(c)	30–35 m^3/t
fish	(c)	60 m^3/t
Confectionery	(d), (c)	12 m^3/t, 35 m^3/t
Chemicals: plastics	(a), (b)	9–23 m^3/t, 30–80 m^3/t
Concrete products	(d)	1 m^3/t
Concrete blocks	(d)	1 m^3 per 200 blocks
Food-processing		
biscuits, pet foods, cereals, and pasta	(c)	8–15 m^3/t
jams, chocolate, cheese, and cane sugar	(c)	20 m^3/t
frozen vegetables, poultry, and beet sugar	(c)	45–50 m^3/t
Iron castings (small production)	(d)	0.4 m^3/t used in sand moulds
Laundry	(d)	20 m^3/t of laundry
Leather production	(a), (e)	70 m^3/t, 80 m^3/t
Meat production and slaughtering	(a), (c)	5 m^3/t of livestock, 40 m^3/t of fresh meat
Paper production (from wood pulp)	(f), (g), (b), (h)	135–150 m^3/t, 90 m^3/t average UK, 240 m^3/t average USA
fine quality paper	(h)	800 m^3/t or more
Rubber, synthetic	(a), (i)	12–13 m^3/t
Steel	(f), (j)	10 m^3/t or more
Terrazo tiles	(d)	1 m^3 per 10 to 20 m^2 of tiles

Information sources:

(a) Weeks C R and Mahon T A, A Comparison of Urban Water Use in Australia and US, *JAWWA*, April 1973, p. 232.

(b) Rees J, *The Industrial Demand for Water in SE England*, Weidenfeld, 1969.

(c) Whitman W E and Holdsworth S D, *Water Use in the Food Industry*, Leatherhead Food Research Association, 1975.

(d) Binnie & Partners, *Survey of Water Use in Riyadh Factories*, 1977.

*(e) Powell S T and Bacon H E, Magnitude of Industrial Demand for Process Water, *JAWWA*, August 1950.

*(f) Ronalds A F, *The Industrial Use of Water*, Melbourne University Press, 1963.

*(g) Walter J W, Water Quality Requirements for the Paper Industry, *JAWWA*, March 1971.

(h) Van der Leeden F, *Water Resources of the World*, Water Information Centre Inc., 1975.

*(i) Callinan B R, *Primary Production in Australia*, 1965.

*(j) Moffat G, *Industrial Requirements and Water Problems in Australia*, Report 19, Water Research Foundation of Australia, 1964.

*These references are quoted in reference (a).

Table 1.7 Light industrial estate water consumption.

Usage	Consumption allowance
Basic factory requirement for cleaning and sanitation	0.05 m^3/d per worker*
Average consumptions in light industrial estates with no large water-consuming factories	0.25 to 0.50 m^3/d per worker*
Average consumptions in light industrial estates which include a proportion of factories engaged in food-processing, ice-making, and soft drink manufacture	0.9 to 1.1 m^3/d per worker*
Typical factory consumptions in SE England†	
clothing and textiles	90% under 6 m^3/d per factory
leather, fur, furniture, timber products, printing, metalworking, and precision engineering	70% to 80% under 25 m^3/d per factory
plastics, rubber, chemical products, mechanical engineering, and non-metallic products	70% to 85% under 125 m^3/d per factory

* Average number of workers per hectare of developed factory plot areas (excluding access roads) is typically 60 to 100.
† *Source*: Rees J, *The Industrial Demand for Water in SE England*, Weidenfeld, 1969.

1.12 Commercial and institutional demand

In the UK supplies to many small shops and offices are not metered, the charge for water being on a flat rate basis according to the rateable value of the premises. In other countries, whilst all supplies are normally metered, supplies to government offices, military establishments etc. may not be metered or, if metered, may be given a free supply of water on the basis that 'government' has no need to charge itself for its own services. In such cases very large consumptions may occur due to inattention to waste. For commercial metered supplies some figures of demand which have been used are given in Table 1.8.

1.13 Agricultural demand

Private water resources are frequently used for farming, especially for the irrigation of crops. The public supply is principally used in the UK for dairies, cattle troughs, farmhouse purposes, horticulture, and greenhouse cultivation. Estimated consumptions for these purposes are given in Table 1.9, assuming all water for the purpose stated is obtained from the public supply.

Potential crop irrigation demand (by sprinkling) is best estimated by considering the rainfall deficiency which can, with profit, be made up by water during the growing season. In the opinion of some specialists the rainfall of the drier part of England in four years out of five is some 125 mm per annum short of the precipitation required for the full development of crops. On one hectare 125 mm precipitation would represent 1250 m^3. If this were required over a three-month growing

Table 1.8 Allowances frequently adopted for water consumption in commercial and institutional establishments.

Usage	Consumption allowance
For small trades and small lock-up shops in urban areas, small offices etc.	In the UK—10 to 12 lcd Elsewhere—up to 25 lcd (applied as per capita allowance to the whole urban population)
Offices	65 l/d per employee, but actual consumption may be three to four times this if waste is not attended to
Departmental stores	100 to 135 l/d per employee
Hospitals	350 to 500 l/d per bed
Hotels	250 l/d per bed
Schools	25 l/d per pupil and staff for small schools, rising to 75 l/d per pupil and staff in large educational establishments

Note: The figures for offices, stores, and schools apply to the days when those establishments are open. The figures generally are supported by data published by Thackray and Archibald[30].

Table 1.9 Agricultural demand in the UK.

Usage	Consumption allowance
Intensive dairy farming areas, including water used for milk cooling	80 l/d per hectare of grazing fully utilized, or 1350 l/d per cow
Average agricultural demands for mixed farming	57 l/d per hectare of farmland generally[31]
Glasshouses	12 400 l/d per hectare in winter rising to three times this or more in summer

season the average daily demand during this period would be 14 m^3 per day per hectare cultivated.

1.14 Public usage of water

Public usage of water in the UK is not measured separately from trade and industrial consumption. Overseas public usage may comprise water (supplied mostly free of charge) for fountains, parks, governmental buildings and their grounds etc. The consumption for the watering of green areas in hot climates may be particularly large. Figures quoted for public use vary so widely that no general rule can be given, but a provision of 45 lcd for these purposes would not be unusual in hot dry climates where there is an adequate supply of water to meet such a demand. Reed[32] reports figures ranging from nil to 104 lcd for 'public services and other known uses' in some western European cities, but the nature of this consumption is not defined. The public demand for fire-fighting, routine fire-hydrant testing, temporary building supplies, and sewer flushing is normally so small when expressed as an average daily supply per capita during the year that it seldom need be separately estimated.

1.15 Metering errors

The ordinary small domestic water meter tends to under-record consumption by at least 2% when meters are up to five years old, and by at least 3% when meters are over five years old. Phillips[33] found 3.2% mean under-registration by a random sample of 50 meters of age 10 to 20 years. Some of this under-registration is due to failure of the meter to register the low night flows caused by dripping taps, leaking ball valves etc. which comprise the major portion of unavoidable consumer wastage mentioned in Section 1.9 above. Stopped meters will account for a further measure of under-registration unless a proper allowance is made for this by substituting estimated consumptions during the period the meter was not working. Large under-registrations are quite commonly found. Gebhardt[19] found 16.7% mean under-registration to six blocks of flats in Johannesburg. Hong Kong is reported to have found 17% mean under-registration of recorded flow on 900 meters of age 1 to 10 years taken at random from their system; this is equivalent to 14.5% under-registration of the true flow[34].

Bulk supply meters (whether on the supply lines to major consumers or at the sourceworks) can be subject to large errors if not properly positioned, calibrated, and maintained. Despite the best attention, they will seldom record to better than ±5% accuracy (often the indicator scale cannot be read more accurately than this). Most engineers never 'meter the same water twice' since it is a common experience to find 5% to 10% discrepancy between two meters on the same line. Thackray *et al.*[4] mention two cases of this sort where the total supply measured by aggregating the recordings of house meters to houses in a cul-de-sac was 6% and 7% more than the total amount of water measured by a bulk flow meter on the supply main to such houses. This kind of experience is common, even when—as in the case quoted above—tested meters are used: hence the possible inaccuracy of all meter readings must be constantly borne in mind when analysing demand data.

1.16 Total losses

Where all supplies are metered the apparent loss, or 'unaccounted-for water', is the difference between the total metered input to the system and the total supplied as measured by consumers' meters. If the metering system is accurate, unaccounted-for water in this case excludes consumer wastage and leakage (which will be measured and sold via consumers' meters), but in practice the unaccounted-for water will include under-registration of consumers' meters and water taken through unmetered connections. Where, as in the UK, domestic supplies are not metered and therefore have to be estimated, the estimated losses will include consumer leakage and wastage.

In practice total losses may vary from 5% to 55% or more of the total supply and, assuming the source input meters are correct, the three elements of total loss are as follows:

(1) service reservoir leakages and overflows,
(2) leakages from distribution mains and service pipe connections,
(3) leakage and wastage on consumers' premises, or (in the case where domestic supplies are metered) all under-recordings of water supplied due to faulty meters or meter reading, illicit or unmetered connections, and other water usage that has not been allowed for, such as wastage from public standpipes and unmetered consumptions for watering public gardens, public offices etc.

Leakages from reservoirs do not normally form a significant part of total losses. The Water Research Centre (WRC) in a test on 123 service reservoirs in the UK[35] found 3 per 100 had gross leaks, 5 per 100 leaked 2% to 4% of their capacity per day, and 10 per 100 leaked 0.5% to 1.5% of their capacity per day. Except where gross leaks and overflows occur, reservoir leakage is therefore usually only a small quantity in relation to the total daily supply.

Leakages from mains can be divided into two parts: (a) trunk main leakages and (b) other losses from distribution mains and service pipe connections. With respect to the former, on a test of 107 trunk mains up to 150 years old, the WRC found 18% leaking 310–620 l/h per km, 21% leaking 620–1860 l/h per km, and 7% leaking 1860–6200 l/h per km. No relationship was found between diameter of main and rate of leakage. These results imply a minimum average leakage rate of 300 l/h per km, but this must be regarded as high and applying only to systems which include very old mains. With trunk mains under 50 years old which are inspected regularly, losses are usually low if all visible bursts have been repaired. With respect to the other losses, the major part of these arises from the smaller mains and from service pipe connections. The WRC estimated these distribution losses as ranging from nil to 12.5 l/h per connection over a typical sample of mains, with a median value of 1.5 l/h per connection. If there are 120–150 connections per km of main (see Table 14.2) this implies typical leakage rates of 180–225 l/h per km from urban distribution mains in average condition; older mains will probably have 50% greater leakage and newer mains up to 50% less. A high proportion of such leakage is found to occur from service pipe connections to mains.

Consumer wastage is best estimated by measuring minimum night flows to purely domestic areas and, as shown in Section 1.9 above, is not less than 1.0 l/h per connection, and probably averages 2.0 l/h per connection in large systems. These low flows are mostly not registered on consumers' meters and are therefore included in the 'metering errors' of a fully metered system. For valid minimum night flow assessment there must be no service reservoirs on the system nor large storages on consumers' premises which fill at night. Allowance must also be made for legitimate night demand by factories on night shift, and by hotels, all-night cafes, hospitals, power stations, airports etc. which remain active throughout the 24 hours. Ingham[36], in a report on a Sutton undertaking serving a population of 280 000, estimated 34% of the MNF was legitimate night consumption by metered trade supplies (of which 26% was actually read from meters on the night of test and the balance of 8% estimated). The usual period of measurement is 01.00 to 04.00 hours, but Heide[25] has shown that, in a domestic supply area, it is more accurate to measure the night flow during the period 02.30 to 03.30 hours when domestic use of toilets is at a minimum. This finding may not apply to cities outside Europe.

Minimum total losses may be assessed in the first instance by considering the minimum night flow likely to occur. Consumer wastage will be not less than 1.0 l/h per connection, largely unregistered by meters, and distribution leakage is likely to be a minimum of 1.5 l/h. The total of these two, 2.5 l/h per connection, represents a theoretical loss rate of 20 lcd with an average of three persons per connection. On supplies which total 180–250 lcd this loss would represent 11% to 8% of the supply. The order of magnitude of this figure is supported by data quoted by Reid[24] and Cook[7] who measured MNFs of 6.5–8.7% and 10.4–13.6% respectively for domestic areas. Since these flows are largely unregistered by meters, they are also found with fully metered systems. Keller[37] found unaccounted-for water in the USA ran at a mean of 9.5% in 476 undertakings in 1965 and at 10.9% for 354 undertakings in 1970. Other data support these figures for fully metered systems where metering is at maximum practicable efficiency. The figures can also apply to small domestic systems where domestic supplies are unmetered and there is good maintenance of the distribution system.

For large cities it is difficult to keep total apparent losses below 16% of the total supply, whether this is measured by the MNF applying in the case where domestic supplies are not metered, or measured as the total unaccounted-for water which is the difference between the source meter readings and the total of the supply meters. The reason is partly that it is more difficult to keep a large distribution system leak-free than a small system, partly that supply metering cannot be so accurate, and partly that there may be considerable legitimate nocturnal demand included in the MNF. For whole cities the WRC found an average flow of 6 l/h per connection (80% of results in the range 3 to 11 l/h per connection): this would represent 16% night flow line assuming a gross consumption of

300 lcd and three persons per connection. Mostly figures of about 18% MNF are the best that can be achieved for whole town supplies, and figures in the range 23% to 26% are by no means unusual*. Such figures will contain a substantial proportion of legitimate nocturnal consumption. In major industrial areas MNFs of 40% or more of the daily average are not unusual, but in these cases a large part of the MNF may arise from 24-hour working in some industrial plants. In general, total losses or total unaccounted-for water, when based upon the domestic consumption given in Table 1.5, should not exceed 25% of the total supply. Loss rates in excess of this indicate that more attention should be paid to leakage and waste reduction. The fact that some large, fully metered water authorities show losses very much less than this, down to 5% or even 3%, is scarcely explicable except on the grounds that some other basis for definition of losses has been taken†.

An alternative method of estimating losses is to express them as per capita daily amounts after assuming some reasonable figure per capita for legitimate domestic demand and knowing the population served. On this basis Herrington[38] reported losses were in the range 49–55 lcd for three English water authorities and were 87 and 106 lcd for two other authorities where special conditions, e.g. old mains and poor ground conditions, applied. Shaw Cole[27] considered 31 lcd represented 'uncontrolled leakage in plumbing fixtures' and to this would need to be added distribution leakage. Keller's figures reported above[37] for fully metered systems give distribution losses in the range 38 to 46 lcd, i.e. Keller's estimate added to Shaw Cole's would give 69 to 77 lcd for total losses including both distribution losses and consumer wastage. In comparison 25% losses on a total supply of 250 to 300 lcd would give 62.5 to 75.0 lcd total losses and this is probably usual, provided that the demands are assessed at their true level.

Summary with regard to losses—many different loss figures could be quoted for different undertakings, but they are of little design value since they virtually comprise only a description of the state of a particular undertaking. Generally, for undertakings supplying populations of 100 000 or more, it is commonly found difficult to get night flow lines or unaccounted-for water below 16–17% of the total average daily supply. Usually such a low figure is only obtained immediately after some leak detection and repair exercise. If only routine waste detection measures

* For example, NW Leicester 18% (1970), Tehran 23% (1976), Damascus 26% (1976), W Suffolk 26.9% (1970)—all after intensive leak detection and repair. (*Sources:* Annual Reports and IBRD.) Phillips[33] reporting in 1983 on the fully metered town of Malvern in the UK showed that unaccounted-for water varied annually from 21% to 29% in the years 1972–1981. He surmised that the variation was closely related to annual climate, mains bursts increasing in cold winters.

† For example, one UK water undertaking consistently maintained that its losses were 7% of the total supply—when questioned it turned out that 11% was regarded as 'unavoidable losses' and the 7% was in excess of that.

are applied, losses or night flow lines will usually be in the range of 22–25% of the total supply. Where figures higher than this are experienced, they indicate a need for more attention to be paid to leakage and waste reduction. In small systems losses should not be above 20% of the total supply and, for individual housing estates, losses of 10–15% should be possible.

1.17 Price effect on consumption

When water is sold by meter the economist's price–demand relationship applies:

$$Q = kP^e$$

where Q is the consumption at price P per unit of consumption, k is a constant, and e is a coefficient which measures the 'elasticity of the demand'. Since an increase of price will tend to reduce consumption, e is negative, i.e. Q is proportional to $1/P^e$. When $e = 0$ then $P^e = 1.0$, implying that changes of price have no effect on consumption: the elasticity of the demand is then said to be nil. This is the case where the price is so low that it has no effect on consumption, or the need for water is so great that it must be 'had at any price'. When $e = -1.0$ then Q is proportional to $1/P$, in which case small changes in P cause almost the same proportionate change in Q.

Hanke[39] has collected some thirty estimates of the value of e, most of which apply to USA water supplies. The ranges of values quoted are:

Water usage type		*Estimated e value*
USA:	domestic in-house use	−0.6 to −0.2
	sprinkler use	−1.5 to −0.7
	industrial use	−0.9 to −0.3
	total use	−1.1 to 0.0
UK:	industrial use	−6.7 to −1.7

It should be noted that lower e values in the range −0.5 to −0.2 must be difficult to prove. At $e = -0.2$ a 69% increase of price would be necessary to achieve a 10% reduction in Q; at $e = -0.4$ a price rise of 30% is required to give the same effect. It is, however, seldom possible to obtain values of Q for large samples which are more accurate than ±10%, and sudden price rises of 30% or more are very seldom experienced. Furthermore a price rise may only temporarily reduce consumption. To measure the permanent effect of a price rise the consumption some time after the price rise has to be compared with the consumption before, but due to the time lag that must elapse, it is almost impossible to obtain the same conditions for the test after the rise as occurred in the test before the rise. Not only will differences of weather, season of year, and incidence of public holidays influence demand, but there will also be one-way changes due to inflation of prices and incomes, and improvement of standards of housing. However, for sprinkler and industrial demand there is no question that a price–demand relationship exists, sprinkler demand being readily curtailed by a price rise. Hence any total demand which includes sprinkler usage (as in the USA) must also have some elasticity. The elasticity of demand by industry will, however, vary from one factory to another, being dependent upon the importance of water in the manufacturer's plant processes.

Generally, where in-house consumptions for domestic purposes are similar to those given in Table 1.5 and reasonably effective systems for waste prevention are in force, the *e* value applying is too small to result in any significant permanent change of demand following a price change. As Linaweaver *et al.*[2] report: 'whether consumers are metered or on a flat rate basis appears to have little influence on domestic [in-house] use in USA'. Primarily in-house domestic demand relates to the class of dwelling in which the consumer lives and his income, and it is for this reason that the classification by type of property is adopted in Table 1.5.

Where metering of domestic supplies occurs, its principal effect is to curb consumer wastage in the absence of other means of waste prevention: it poses the threat that, if consumption is not controlled, the consumer will have to pay a financial penalty. In hot climates where the 'comfort value' of water is great, and where inspection and waste prevention measures may be weak, metering is essential if water demand is to be kept down to reasonable amounts—but it can only do so if the metering, billing, and income collection procedures are efficient. In temperate or cold climates, at the rates of charge for water commonly applied, metering is unlikely to reduce in-house domestic consumption below that which occurs with unmetered supplies accompanied by efficient inspection and waste prevention measures. If this were not so, the average domestic consumption in the UK where no domestic meters are installed ought to be higher than consumptions in many similar countries where all domestic supplies are metered, but there is no clear evidence of any such difference.

1.18 Total water consumption

Total water consumption for all purposes varies widely, as indicated in Fig. 1.1, and there is no discernible relationship between size of population served and consumption per capita. A classification of the range of demand is given in Table 1.10.

1.19 Increase of consumption with time

The trend of increase of water consumption in England and Wales during the period 1961–82 is given in Fig. 1.3[40]. If losses are taken as 25% of the total supply, the unmetered domestic and small trade

Table 1.10 Range of total water demand.

Location	Typical total demands per capita reported (lcd)
England and Wales 1982	
(a) Average unmetered domestic and small trade demand	157 } total 319
(b) Average metered trade demand (potable water)	82 } total 319
(c) Unaccounted-for water taken as 25% of the total	80 } total 319
Worldwide	
(a) Highest consumptions in highly industrialized areas with high standards of living; large cities in the USA (e.g. Philadelphia 741, Hamilton 667, Rome 651, San Francisco 608)	600–700
(b) Major cities with substantial industry and/or high average standards of housing (e.g. Perth 503, Glasgow 500, Belfast 495, Sydney 491, Marseilles 486, Antwerp 456, Turin 425, Wellington 410, Milan 400)	400–500
(c) Cities in temperate or oceanic climates; whole country means (e.g. Bangkok 355, Bologna 350, Liverpool 349, Denmark average 340, Durban 332, Stockholm 328, Plymouth 326, London 314, Frankfurt 312, Copenhagen 311, Japan average 303)	300–350
(d) Mixed urban and rural areas; city suburbs; towns with moderate industry (e.g. Jerusalem 291, Tel Aviv 281, Southampton 278, Birmingham 273, Bordeaux 270, Berlin 268, Barcelona 267, Dublin 251, Paris 249, Sheffield 235, Hamburg 229)	200–300
(e) Mixed urban and rural areas with low proportion of industry; towns with moderate to low standards of housing; towns with water supply under strict control or limited in quantity (e.g. Brussels 178, Colombo 135, Italian small towns 100–200, Swedish non-industrial towns 161, USSR municipal supplies 164)	150–200
(f) Small towns with little industrial demand; areas with low standards of housing; settlement areas	90–150
(g) Low standard housing and a high proportion of standpipe supplies	70–90
(h) Standpipe supplies: rural areas with mainly standpipe supplies	25–60

Sources: Various, circa 1975–78.

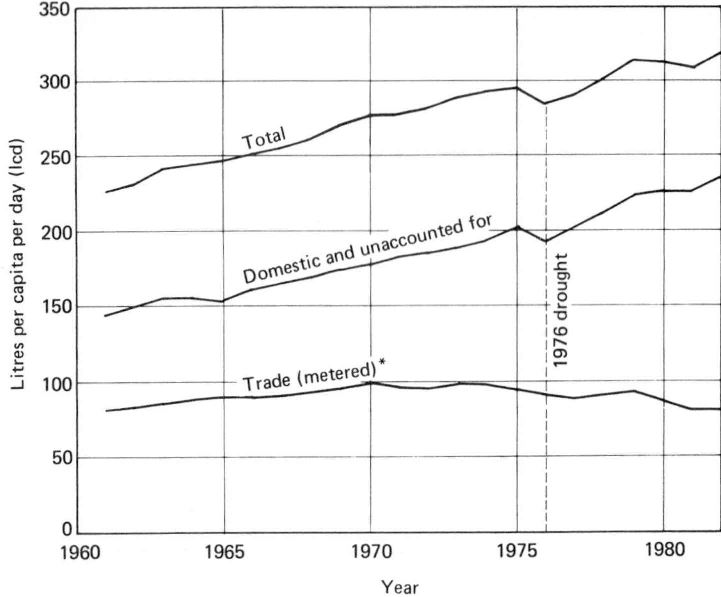

Fig. 1.3 Consumption per capita in England and Wales[40].
*Trade excludes non-potable supplies.

demand can be estimated as increasing by an average of 3.2 lcd per annum between 1961 and 1982. During the same period, trade and agricultural demand increased by about 2.0 lcd per annum from 1961 to 1970, and thereafter fell by 1.5 lcd per annum from 1971 to 1982.

It should be noted that an unknown amount of the increase in the unmetered (domestic) demand in England and Wales during the period 1961–75 may be due to the policy of removing meters from small shops and trades which was then pursued, transferring this demand from 'trade' to 'domestic'. Prior to 1961 the average annual increase of domestic demand in major towns for the period 1900–49 varied from 0.3 to 0.7 lcd per annum, and for the period 1950–59 ranged from 1.0 to 1.8 lcd per annum[41].

1.20 Estimates of future demand

Estimates of future demand are difficult to make and many have proved wrong. For short periods ahead of five to ten years straight-line extrapolation of previous growth over the past ten to twenty years may prove right, but for longer periods ahead the possible growth of population served, the possible increase of per capita consumption, and the possible increase of trade and agricultural demand must all be separately assessed. If both population and consumption are expected to increase, a geometric demand curve for the future consumption will be produced, showing ever-increasing rates of consumption. It then becomes important to assess whether total consumption per capita will continue to increase or will tend to level out. Considerable debate occurs on this matter.

Any attempt to base forecasts on the growth of component consumptions for toilet flushing, ablution, use of laundry- and dish-washing machines etc. runs into difficulty because of the need to add together components each of which has a wide possible range of error. Statistics relating to the present and future incidence of washing machines etc. are difficult to obtain and seldom reliable. Far more reliably domestic consumption per capita is related to class of housing, since a better class of house not only possesses more water-using facilities which the householder is then tempted to use, but also expresses his higher standard of living. Hence plans for rehousing and for new housing, which are usually available, together with forecasts of population growth form the best basis for forward estimating of consumption. Income growth can also be a relevant factor, but in most cases it is sufficiently well expressed in the standards of housing which are proposed.

There are also signs that, once housing standards have reached the level at which consumers have a reasonable range of facilities for water-borne sewerage, bathing, laundering and dish-washing, the per capita demand will tend to stabilize at about the figures given in Table 1.5. The main factors influencing per capita consumption in households with an adequate range of water facilities appear to be climate and social habits. Some societies have the habit of bathing more frequently and using water more liberally than others: obviously climate affects these habits. In Sweden domestic demand is reported to have 'levelled off' in the 1970s to about 200 lcd; in the USA it appears to have levelled off to a slightly higher figure, dependent again upon the class of housing. In the UK domestic usage of water was traditionally put at about 90 lcd during the 1920s and in the 1980s is now believed to be about 115 lcd. The discrepancy between the 0.4 lcd per annum average increase this implies from 1920 to 1980 and the figures reported in Section 1.19 above suggests that the main portion of the actual increase which has occurred in unmetered (mainly domestic) supplies in the UK is due to improvement in the mean standards of housing: it is not caused by people under static housing conditions substantially increasing their take. Also some of the increase in unmetered demand in the UK can be due to an increasing incidence of leakages as distribution systems grow older, there having been a definite tendency to underestimate leakages in the UK until quite recently.

For the reasons stated in Section 1.17 the price of water at the usual rates ruling is unlikely to have a permanent quantifiable effect upon in-house consumption, even if domestic metering is gradually brought into the UK. It does, however, have a substantial effect in those areas (particularly the USA and Australia) where much domestic water is taken for out-of-house consumption for sprinkling lawns and filling swimming pools.

In the respect of trade demand the rate of increase will be primarily dependent upon the level of economic activity, as the decline in UK consumption in the later 1970s shows. A period of substantial growth can be followed by one of little growth or actual decline. Also the pricing policy for water supplied to industry can have an effect, and so can any shortages imposed during a drought situation. Both can result in industrialists adopting water-recovery measures, causing a permanent drop in consumption until further expansion takes place. Notwithstanding these comments it is prudent for an undertaking to have a margin of water in hand at all times, sufficient to meet an unexpected large new trade demand: if this is not done, potential new industry may be lost to the area served by the water undertaking. The margin to be adopted depends upon the nature of the area served and the types of industry that might be expected to come to the area, but a margin of about 3% on the total supply would not appear extravagant.

Demand constraint measures are being promulgated in the UK and by the World Bank for overseas projects. Within the UK dual flush or low volume (6-litre) flushing cisterns for use with WCs could reduce consumption by about 10 lcd, but the process will tend to be slow. Any savings achieved may be offset by increased installation of bidets, now tending to come into favour. Introduction of more showers is unlikely to reduce ablution consumption because of their greater frequency of use than baths. Domestic metering is unlikely because of its high cost and a lack of evidence that it would reduce consumption. Probably the best contribution to reduction of UK consumption will come from reduction of distribution leakage and consumer wastage. Automatic control of distribution pressures to the minimum necessary for an adequate supply may have a significant part to play in this matter.

In overseas countries efficient metering coupled with an appropriate tariff structure properly enforced exerts the most effective control over demand and wastage. Insistence upon fine spray showers may have some effect, but the use of low volume pour-flushed Asian-type pans can have little effect where the principal sanitation use for water is for anal cleansing (see Section 1.5). Where supplies are very limited restricted hours of supply may have to be adopted but, unless these are very short, e.g. of the order of two periods of two hours per day, the reduction in demand is likely to be small.

1.21 Maximum day's consumption

The maximum consumption for a day is usually expressed as a percentage of the average annual daily supply. Some figures are given in Table 1.11: the highest percentages are primarily caused by a hot summer climate and a large amount of sprinkler demand.

Table 1.11 Maximum day's consumption.

Location	Ratio: maximum day to average annual daily consumption
UK	
Rural areas in which water is used for spray irrigation of crops	140%–150%
Seaside and holiday resorts	130%–140%
Residential towns: rural areas	122%–125%
Industrial towns	117%–122%
Exceptional peak due to garden watering in prolonged hot dry weather	150%–170%
Worldwide	
Typical peak domestic demands in the USA due to sprinkler demand	
western states[2]	215%–340%
eastern states[2]	195%–295%
Some large cities in the USA, Canada, and Europe	150%–200%
Peak USA in-house domestic demand (excluding sprinklers)	
western states[2]	180%–185%
eastern states[2]	130%–140%
Cities with hot dry summers	135%–145%
Cities in equable climates	125%–135%
Large cities with substantial industrial demand in western Europe, the Middle East, and Asia (e.g. Singapore 123%, Hong Kong 122%, Penang 116%, Damascus and Toulon 115%, Marseilles 111%, Barcelona 109%)	110%–125%

1.22 Maximum week's consumption

The maximum consumption for a week is of considerable design importance to a water undertaking, because the overdraw above normal for this length of time cannot usually be met from service reservoir storage and must therefore be matched by source output capacity. Maximum week demands are usually only a few per cent less than maximum day demands, i.e. not more than 5% less. In areas with hot dry summers this maximum week demand may be prolonged for several weeks during a 'hot spell'.

1.23 Maximum hourly rate of consumption

The maximum hourly demand depends upon the size of area served and the nature of the demand. For domestic areas, excluding sprinkler demand, Adams[42] arrived at the peak flow factors shown in Fig. 1.4

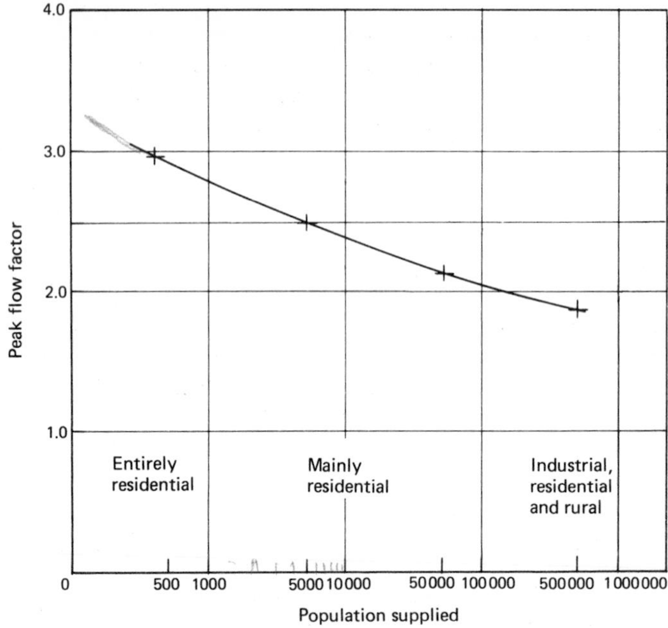

Fig. 1.4 Ratio of peak hourly flow to annual average flow in an undertaking. (Adams, *JIWE*, 1955.)

which are applicable to the UK and were derived from an investigation of areas in Liverpool. The peak flow period is usually between 08.30 and 10.00 hours for residential areas, and about noon for mixed industrial and residential areas. However, sprinkler demand can create an evening peak value of 3.0 or more. Tighe[31], investigating rural areas in Ireland, found peak flow factors of 1.4 to 2.0, but for individual or groups of dairy farms he found it was 4.0 to 4.5 occurring at morning and evening times due to milk cooling demand. In the USA Linaweaver *et al.*[2] found the peak hourly factor for in-house domestic demand in sample housing areas was about 2.6–2.8 for eastern states and 4.3–4.9 for western states. Sprinkler or garden watering increased the factor to 6.0. All these factors apply to the annual average daily demand.

At the other end of the scale Tessendorff[43] reports peak flows to a 'one-family' household measured 83 l/min for a 20-second duration, and were estimated as 30–35 l/min for a 10-minute duration. These figures are thought to be applicable to all households with up to twenty inhabitants (but presumably not to one- or two-occupant households). A total flow of 30–35 l/min for a 10-minute duration is equivalent to the demand from two cold taps open and a WC cistern filling and would not be unusual, but in practice the usual 13 mm service pipe could not supply this amount unless the distribution pressure were excessive.

'Diversity factors' must apply to water demand as they do to electrical demand, and the peaking factor to be adopted must relate to the number of properties under consideration. In practice the peak hourly demand exerted by small groups of properties within an urban distribution network is not an important design factor, partly because the minimum size of distribution mains laid is usually 100 mm and all such mains are interconnected at both ends, and partly because simultaneous drawoff lowers local pressures so that flow rates to taps are reduced and consumers' demands are met in rotation according to the way in which pressures are restored as some consumers are satisfied (see Section 15.10). The most important factor influencing the design of small mains within large networks is their capacity to meet local fire demand from a single hydrant or from a group of hydrants in one area.

References

(1) Janczewski H, Demand for Water in Towns 1969–1990, *Proc. IWSA Congress*, 1974.
(2) Linaweaver F P, Geyer J C and Wolff J B, *A Study of Residential Water Use*, John Hopkins University, Baltimore, Maryland, USA, 1967. See Tables 2, 3 and 4.
(3) Binnie & Partners, Technical Report No. 1 on Leakage Control Studies: Istanbul, February 1976.
(4) Thackray J E, Cocker V and Archibald G, The Malvern and Mansfield Studies of Domestic Water Usage, *Proc. ICE*, February 1978, and Discussion, August 1978, paragraphs 85 and 185, pp. 481 and 501.
(5) Reference 4, Table 5.
(6) McClure A J C, Reference 4, Discussion, Table 14.
(7) Cook R G, Reference 4, Discussion, Table 15.
(8) Binnie & Partners, Field Survey: Sakaka, 1975.
(9) Binnie & Partners, Field Survey: Lesotho, 1976.
(10) Taylor-Binnie, Report on Greater Cairo Sewerage, Annexe 4.3.1, 1978.
(11) Binnie & Partners, Field Survey: Palembang, 1970.
(12) Tomkinson T H, Hong Kong Water Authority, *private communication*, 1978.
(13) Camp, Dresser, & McKee, Field Survey: Alexandria, 1976.
(14) Weeks C R and Mahon T A, A Comparison of Urban Water Use in Australia and US, *JAWWA*, April 1973, p. 232.
(15) Reference 2, Table 3 and p. 20.
(16) *Appropriate Sanitation Alternatives*, The World Bank, 1978, p. 56.
(17) Water Research Centre, *Drinking Water Consumption in Great Britain*, Technical Report 137, November 1980.
(18) Kinnersley D J, Water Use and Consumption, *Proc. IWSA Congress*, 1980.
(19) Gebhardt D S, *Urban Water Requirements in Johannesburg*, 1966.

(20) *Public Standposts for Developing Countries*, Bulletin No. 11, WHO, International Reference Centre, The Hague, May 1978.
(21) Binnie-Taylor, Report on Egyptian Provincial Water Supplies, October 1979.
(22) Giles H J, *Proc. Symposium on Waste Control*, IWES, 1974, p. 13.
(23) Gledhill E G B, An Investigation of the Incidence of Underground Leakage, *JIWE*, 1957, p. 117.
(24) Reid J, *Proc. Symposium on Waste Control*, IWES, 1974, p. 105.
(25) Heide G F, Energy Saving Through Reduction of Leakage, Open Forum Paper No. 7, *Proc. IWSA Congress*, 1980.
(26) Wijntjes W C, Reference 20, p. 34.
(27) Shaw Cole E, Water Losses and Leakage Control, *Proc. IWSA Congress*, 1978.
(28) MacKichen R W A, Water Supply for Construction of Town of Kainji etc., *JIWE*, November 1970, p. 461.
(29) Chapman M L, Discussion on Reference 30, p. 436.
(30) Thackray J E and Archibald G G, The Severn–Trent Studies of Industrial Water Use, *Proc. ICE*, August 1981, p. 403.
(31) Tighe B J, Rural Water Supply in the Republic of Ireland, *JIWE*, August 1967, p. 479.
(32) Reed E C, Report on Water Losses, *Aqua.: JIWSA*, **8,** 1980.
(33) Phillips J H, Domestic Metering: An Interim Review, *JIWE*, August 1972, p. 337, also Water Usage and the Quantification of Unaccounted-for Water, *JIWES*, August 1983.
(34) Anon., Hong Kong: 30% of Supply is Unaccounted For, *World Water*, November 1983, p. 28.
(35) *Waste Control and Leak Detection in Water Distribution Systems*, WRC Regional Meeting, Autumn 1978.
(36) Ingham G L, Water Losses and the Consumer, *Proc. IWES Symposium*, December 1981.
(37) Keller C W, Analysis of Unaccounted-for Water, *JAWWA*, March 1976, p. 159.
(38) Herrington P, Evidence to the Public Enquiry on the Broad Oak (Canterbury) Reservoir Proposal, July 1979.
(39) Hanke S H, A Method for Integrating Engineering and Economic Planning, *JAWWA*, September 1978, p. 487.
(40) *Public Water Supply in 1975 and Trends in Consumption*, Central Water Planning Unit Technical Note No. 19, 1977, and data from the NWC, 1983.
(41) *Manual of British Water Engineering Practice*, 3rd Ed., IWE, 1961, p. 69. Figures for Birmingham, Liverpool, London and Manchester.
(42) Adams R W, The Analysis of Distribution Systems, *JIWE*, November 1955, p. 540.
(43) Tessendorff H, Peak Demands: Results of German Research Programme, *Proc. IWSA Congress*, 1980.

2

Waterworks administration and finance

2.1 Types of water supply organization

Most water undertakings come under some form of public control. The control can be direct, as when the undertakings are run by government itself or by the local municipal authorities, or it can be indirect, as when a water undertaking is run by some form of national water authority, or by a joint board or a water company. The control exercised by government is usually of three kinds: financial, qualitative, and administrative.

Financial control usually comprises control of capital expenditure and rates of charge for water. As a result of this control the government may need to enquire into the efficiency of the undertaking, the manner in which it operates in detail, and its proposals for additional capital expenditure on new works. Government requirements in these matters may have to be met before sanction is given to raise capital or increase charges to consumers.

Quality control is also frequently exercised by the central government. Adherence to published standards of quality for water supplied will almost certainly be required. The rights of consumers and the manner in which water services are to be provided may be laid down. It is usual for this type of control to be exercised through some Act or Acts passed by Parliament but, in addition, any of these Acts may permit the appropriate Minister of the government to issue Regulations specifying in more detail how the undertakings shall meet the requirements of the Acts. In the UK the Third Schedule to the Water Act 1945 sets out a code of practice for waterworks day-to-day operations and permits the Minister to issue further regulations from time to time, giving him powers to sanction certain actions by the water undertakings provided that the procedures laid down are complied with.

It is in the area of administrative control that water undertaking vary most. Those which come under direct government control usually have to comply with most civil service practices, whilst those which come under direct local government control must comply with local government regulations. The other undertakings which come under indirect control usually have much greater administrative freedom, such as the Water Authorities and Water Companies in the UK. The water companies must, however, comply with the various Companies Acts, as well as complying with all the same water regulations as apply to the other undertakings. Their charges to consumers must be authorized by government, and their profits and consequent dividends on share capital must also not exceed certain limits laid down by regulations[1]. These limits are such that the return on capital invested in water companies is no greater than the return on money loaned to local authorities and other public bodies. The single real enjoyment a company has is that, subject only to the provisions of the Companies Acts, it generally has a measure of freedom to organize itself internally as it pleases.

In some countries there may be no codified or legislated form of control of the water undertakings, and the personnel of the undertaking may only form a part of a department of central government. Hence the practice of these water undertakings will simply be to follow 'acknowledged procedures' which have been found by experience to be acceptable to the government Ministries involved, e.g. Ministry of Finance, Ministry of Health, Ministry of Planning etc. Often such systems can work well and will comprise practices which are very little different from those adopted under full legislated systems.

Whatever system of administration is adopted it must be borne in mind that water undertakings basically comprise a 'manufacturing' industry which obtains, processes, distributes, and sells water on a 24-hour-a-day basis. Executive powers must be delegated in a precise manner so that an efficient round-the-clock technical service can be given to consumers and under which prompt and decisive actions can be taken whenever necessary to maintain adequate and wholesome supplies.

2.2 Internal organization of a water undertaking

The tasks to be performed by a water supply undertaking are relatively simple. Water has to be procured, treated, and distributed, and money has to be collected for it. These four tasks are the same in any part of the world and, as a consequence, the organization of most water undertakings is very similar.

The typical essentials of organization are shown in diagrammatic form in Fig. 2.1 and they reflect the four principal tasks outlined above. There is a Managing Director or Chief Engineer and Manager in charge and directly below him are the Heads of the four chief executive departments: Supplies or Sourceworks, Distribution, Chemical and Bacteriological or Quality Control, and on the administrative side Finance and Administration. The Heads of these departments are the key personnel in any undertaking who must individually and collectively keep the undertaking running efficiently, and who must advise the Managing Director on all relevant matters having a bearing on the overall policy of the undertaking.

Whilst four main departments can be identified as shown in Fig. 2.1, there can be a number of other departments according to the size of the

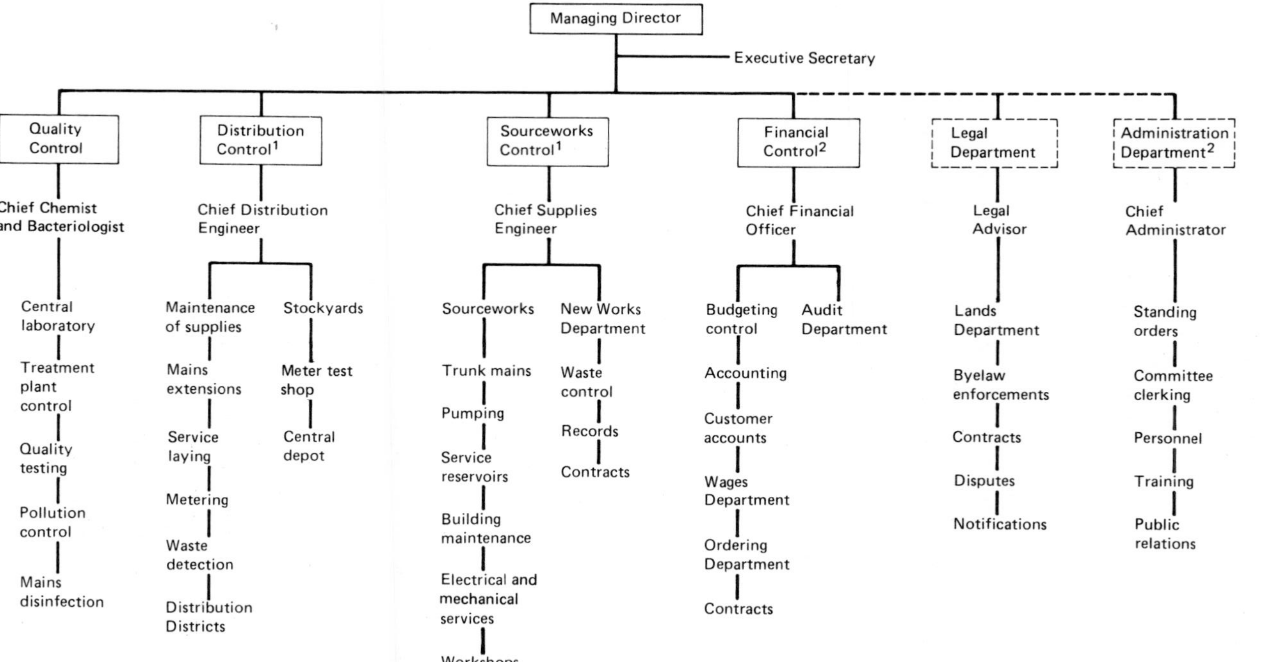

Alternatives:
(1) A Chief Engineer may be in charge of the departments, with a Deputy Engineer (Distribution) and a Deputy Engineer (Sourceworks and Transmission) to assist him.
(2) A Chief Financial and Administrative Officer may be appointed, with a Chief Accountant and a Chief Administrator to assist him. The post of Chief Treasurer may be created to handle all income collection.

Fig. 2.1 Organization of a large waterworks undertaking.

undertaking. Quite frequently Finance and Administration will form separate departments. In the larger undertakings there may be a need for a separate Legal Department which handles the enforcement of byelaws, problems of non-payment and other disputes with consumers, and the drawing up of contracts for the purchase of materials or the execution of work.

Sometimes Distribution and Sourceworks are put together under one Chief Engineer who has two Chief Assistants under him to run these two departments, or there may be other more complex arrangements, such as a Chief Engineer, a Deputy Engineer, and Senior Engineers taking various responsibilities between them. In this case the Chief Engineer will often be concerned with overall policy whilst the Deputy Engineer runs the works on a day-to-day basis.

A separate Training Department with a Training School is essential in the larger undertakings if the problem of reducing wastage of water is to be tackled efficiently. Sometimes two or more undertakings find it economic to set up a Joint Training School. A Measurement and Waste Control Department is advisable in the larger undertakings in order to ensure that it is one engineer's responsibility to make waste control effective.

2.3 Staffing levels

The number of staff and operatives employed to run a water undertaking varies more with the legal framework under which the undertaking operates than with the nature of the technical functions to be performed. By and large the field functions of an undertaking do not vary greatly from one undertaking to another, provided that the undertakings are of the same size and are direct suppliers of water to consumers and not bulk suppliers. The internal functions, however, vary according to the format of the undertaking. Numbers employed tend to increase when the undertaking is only a part of the manifold activities of government and numerous internal routines must then be followed; numbers tend to decrease as the undertaking gains autonomy. Where a reasonable degree of autonomy pertains, as in the case of the water companies and as usually occurs with the municipal or joint board authorities, the total number of employees is normally in the range of 60 to 100 per 100 000 consumers served for undertakings supplying above 100 000 consumers. The median value is about 85 employees per 100 000 consumers served[2]. Of this total number of employees, 50 to 60 per cent would be manual workers as distinct from the technical, financial, and administrative staff working in offices.

Another measure of staffing levels is the number of employees per 1000 m^3/day of water supplied. This appears to range from about 2.5 to 5.0 for the water companies in the UK, with a median value of about four[3]. The higher ratio tends to occur in the more rural undertakings, but not exclusively so.

2.4 Effect of type of organization on finance

Undertakings which come directly under the control of central government or local municipal authorities normally have to prepare annual forecasts of expenditure which are submitted to the government or local authority for authorization. The authority will fix the necessary water charges to raise the income needed to meet the estimated expenditure.

Unfortunately both central and local government authorities may be under considerable pressure to expend money on all kinds of public services, and may also be under pressure not to raise water charges for political reasons. Hence it can happen that the water charges fixed are insufficient to cover the forecasted expenditure. One possibility is for the government or local government authority to use money from other sources, e.g. from general rates and taxes, to make up the deficit on the water account: this is called 'deficit funding'. Quite as frequently, however, the officer in charge of the water undertaking may be told that he must cut the undertaking's expenditure so that it does not exceed the expected income: either because there is no deficit funding or because the amount of it is insufficient. The inevitable consequence is that money for maintenance of the undertaking is cut, and this leads to increasing losses of water through unrepaired pipes and through the inability to employ sufficient staff to curb consumer wastage. These misfortunes tend to result in a lowering of the income of the undertaking as more water is wasted, making it more difficult and more expensive to bring the undertaking back into a good condition.

Where only indirect control is exercised over a water undertaking it is usually under some statutory obligation to keep income in line with expenditure, and it therefore sets water charges at the necessary level to achieve this. Even so government may, for political reasons, refuse to sanction the necessary levels of charge, in which case either deficit funding or cutting of services becomes necessary.

It should not be thought that any government wilfully ignores the obvious need to finance water supplies: the problem is that public services of all kinds make immense demands on public funds and governments sometimes find themselves unwilling or unable to sanction the levels of direct and indirect taxation that all such services would necessitate. The special problem applying to water undertakings is that, once not enough money is provided for proper maintenance, the situation goes from bad to worse and becomes increasingly expensive to rectify. The poorer a water undertaking the more essential it is that it should be efficient: but the reverse is more often the case through lack of adequate staff.

2.5 Capital financing

When a new works has to be constructed its capital cost may be borrowed or it can be generated by savings from past income. In the former case the burden of capital repayment falls on future consumers; in the latter case past consumers pay more in order to save money for future expenditure on capital projects. Hence a mixture of both borrowing and setting aside savings from revenue can spread the cost of capital works over past and future consumers.

Loan capital can be repaid in annual instalments over a fixed period by the 'capital-plus-interest' method or by the 'annuity' method*. Alternatively a 'sinking fund' may be set up which comprises annual savings invested on a regular basis so that the amount accumulated with interest will be sufficient to redeem the loan in one single payment when the loan period ends[4]. Savings from revenue, i.e. excesses of income over expenditure after all loan charges have been met, can additionally be accumulated for use as capital for future works or for the replacement of worn-out assets. These savings are invested in just the same manner as a sinking fund, except they may be called a 'capital savings account' or 'reserve' or 'renewals' fund.

Provided that sufficient profits are made by an undertaking, depreciation of an asset value achieves the same effect[5], but it uses a different accounting method. Instead of showing the savings from revenue as a separate reserve or renewals fund, the amount taken from profit and saved is applied to reducing the book value of an asset†. Thus it may be decided to 'write off' the cost of some mechanical plant in ten years, so its value is depreciated by 10% of the original cost per annum ('straight-line' depreciation). It can be seen that if the loan for the purchase of the plant initially is also repaid over a term of years, both the loan is repaid and savings are made to purchase new plant when the old plant wears out. As Gilliland[6] has put it:

> 'Depreciation is not just an accounting convention. It is the mechanism by which the cost of the economic use of assets is charged in a revenue account before a surplus is declared. It is that amount of expenditure which should reflect the use made, that is the physical consumption of assets provided during an accounting period.'

In practice varying amounts of capital will need to be raised each year by an expanding waterworks as it extends its distribution system, replaces worn-out plant, and builds new works. Loan periods will vary

* In the capital-plus-interest method the loan is paid off by equal annual instalments, interest being paid annually on the balance of loan remaining, hence the size of the annual payments reduces year by year. In the annuity method the annual payments are made equal, hence the ratio of capital repayment to interest increases with each instalment.

† The depreciation is debited, i.e. charged to, the revenue account, and credited to the asset value, thus reducing it. The actual money amount of the depreciation is reflected in the balance sheet as an increase of the other assets held, such as the cash or investments.

and the actual useful lives of assets may depart substantially from the period set for repayment of the loan. When numerous loans have been raised they may be treated as a whole in one 'consolidated loans fund' and, where accumulated savings are large enough, 'internal borrowing' may take place whereby funds accumulated for long-term use are used temporarily to finance short-term projects.

In an inflationary situation the money cost of replacing an asset will be greater than its original cost. Hence if the savings made from revenue are only sufficient to pay off the original loan, more money will have to be found to replace the asset. This can be countered by making additional savings from revenue by any of the methods described above. For example the amount set aside annually to a renewals fund may be based upon the expected inflated cost of the asset when it has to be renewed. Similarly, if profits are high enough, depreciation may be set high enough to generate extra money to meet inflation of costs. By such means it is possible to ensure that consumers pay enough to ensure that savings are in line with the rising price of replacing an asset. If this is not done, an undertaking may be confronted with the need to increase charges steeply when renewal of a major asset unexpectedly becomes necessary. This situation has tended to arise in the UK where it is evident that large parts of the distribution systems, constructed piecemeal during the late nineteenth and early twentieth centuries, have begun to approach the end of their trouble-free service life and now need renewal. Their original cost has long ago been paid off, but not enough savings have been generated from revenue to replace them at today's prices. Consequently steep increases of water charges will become necessary if large sections of these distribution systems have to be replaced at one time. This sort of situation needs to be avoided.

2.6 Water pricing policies

As mentioned in Section 1.17 a variation in the price charged for water mainly affects the amount used for sprinkling, trade, and industrial uses. It does not affect supplies which are given free, such as supplies by standpipe, and it has little effect on in-house demand, whether this is metered or paid for on a 'flat rate' basis as in the UK. Only where the domestic supply is metered and an exceptionally high charge is made for the water, such as when desalinated water is supplied, is there a significant relationship between price and in-house demand. Nevertheless, provided that domestic supplies are metered, penalty charges can be applied once a reasonable amount of water has been taken, so that consumers can be induced not to waste water and not to use it extravagantly.

A pricing difficulty is that a large proportion of the costs which have to be met by a water undertaking are fixed charges, such as capital repayment charges, administration, and operational costs. These do not change much with increasing demand for water, until new capital works have to be built. The costs which relate to the amount of water consumed, such as the costs of pumping and treatment, represent only a small proportion of the total price charged. Hence the cost per unit of additional water supplied from existing sources, i.e. the 'short run marginal cost' (SRMC) of water*, tends to reduce as the output from existing sources is increased up to the maximum possible. If a further increase of demand then takes place, a relatively steep rise in charges is likely to follow because of the large capital expenditure required on new works. When the consumer experiences this price increase he may reduce his consumption but this will be 'after the event', and his action will have virtually no effect on the total costs faced by the undertaking.

To ameliorate this situation the charge for water can be based upon the 'long run marginal cost' (LRMC)[7]. This price reflects the increased cost which would occur if further capital projects have to be built to augment supplies as demand increases†. Such a price permits an effective price–demand relationship to apply and, if consumers so wish, they can reduce their consumption at this higher price, thus postponing the time when new works must be constructed and thereby achieving an economy.

Whilst the calculation of an appropriate LRMC may be somewhat inexact[8] (because future schemes to procure more water may not be fully investigated and chosen), the adoption of a pricing policy based upon the LRMC has three beneficial financial effects for an undertaking as follows. Firstly, it extracts from consumers a financial contribution towards future capital expenditure (see Section 2.5). Secondly, it foreshadows to consumers the rising prices that will accompany rising consumption and gives them an opportunity to reduce their demand if they so wish. Thirdly, it results in the smoothing out of price variations to consumers, preventing sudden large increases occurring whenever a new scheme must be financed.

2.7 Appropriate technology

Many water undertakings throughout the world have very limited financial resources. They may have to supply water to low income populations and consequently have a low revenue, and the government may be unable to fund waterworks deficits (Section 2.4) and have only limited offshore earnings to pay for imported items. The designer of new works for such undertakings must be careful to take these financial constraints into consideration, and must use the 'appropriate technology' which best matches the undertaking's resources. A first advisable

* The short run marginal cost is the cost of supplying another unit of water, e.g. 1 m^3/d, extra to the present demand.

† The price will therefore generate some savings from revenue to meet future capital expenditure.

step is that the financial and administrative abilities of the undertaking should be examined by specialists experienced in the financing and management of public utility undertakings. Many constructive recommendations may follow from such an examination, and the engineer-designer will be made aware of the physical constraints applying, whether these refer to lack of trained operatives and an overriding need for more training, or to actual financial limitations to year-by-year expenditure in terms of both offshore and onshore currency. The need to avoid large lump-sum outlays of capital expenditure may challenge the engineer to devise schemes which can be constructed in phases, so that expenditure can be kept more closely in line with income.

The term 'appropriate technology' has unfortunately tended to imply the use of processes which are primarily cheap, irrespective of their efficiency. This is a simplistic view that can be erroneous. It is not necessarily the case that use of a great deal of hand-labour, for instance, will be both cheap and efficient. Often it requires a great deal of organizational ability to achieve good results with extensive use of unskilled labour and, if this ability is missing, the result will be neither cheap nor efficient. Similarly to assume that labour-saving equipment should not be used when local labour is cheap may be to commit another error of judgement. Often appropriate technology means 'appropriate plant': not the sort of plant which is so complex and automated that, as soon as it develops a fault, no one knows how to repair it locally, but the sort of plant which local people can understand and can repair themselves when it has a breakdown. Right through the design of any works the designer must ask himself whether he is providing something which the local personnel will be able to operate efficiently and to mend when it develops a fault. The processes of water treatment, disinfection, and pumping can all be made complicated or relatively simple. Plant does not have to be the latest in automated design to be 'appropriately efficient', but the plant must not be so crude and so dependent upon a lot of hand-labour that the chances are that it will never be run efficiently. Often what is most required is reliable, easily repairable plant which is sufficiently labour-saving to keep labour requirements within the ability of the water undertaking to provide and train.

The standards set for works quality and for quality of the water produced do not have to be 'the best' that can be procured. Works quality must be satisfactory, durable, and fit for the purpose; water quality has to be safe and acceptably palatable. To achieve consistent water quality the control systems must be simple and effective, and within the ability of local personnel to apply successfully. Palatability is important because it is the means whereby a public water supply undertaking can gain the confidence of consumers and, in consequence, can call upon them not to waste water and to pay for the supply.

Financial and economic constraints will often require that new projects are devised for only a limited period ahead in steps, rather than for a full 20- or 30-year term. This means that sourceworks, trunk mains, pumping stations etc. have to be under-designed rather than over-designed to meet every possible contingency. The peak hourly demand on a day of maximum demand may be a design figure which is difficult to meet except at excessive extra cost. The designer may decide that this is too costly a provision, with regard to the rarity of the event, and may design for a lesser figure. If the risk materializes and consumers are short of water for an hour or two, no great harm will be done, but the financial benefits of the policy may be large. By such means and many others the engineer can ensure that the schemes are the most appropriate to the resources and needs of the undertaking.

2.8 Discounting and economic appraisal of capital projects

The total cost of a project over the whole of its useful life consists of the initial capital outlay, any subsequent phased outlays of capital, plus all the annual operating and maintenance costs to the end of its use. It therefore consists of 'lumpy' outlays of capital spread over time, plus varying amounts of yearly expenditure. Some schemes may have a high initial cost and a low annual cost, and others the reverse: thus the sum total expenditure in either case may be much the same, but the incidence of the payments is different.

To elucidate the difference in timing of expenditure all future costs can be brought to their 'present value' using an annual 'discount rate' which represents the benefit obtained by deferring payments. For example if £100 is held now and 5% interest is given, its value at the end of Year 1 will be £105. In the reverse manner if £100 has to be paid at the end of one year, we need hold only £95.24 now because invested at 5% it will amount to £100 at the end of Year 1. The £95.24 represents the present value of £100 payable in one year's time 'discounted at 5% per annum'.

The formulae for working out present values are[9] as follows.

(1) For a single payment of £x after n years:

$$\text{present value (£)} = \frac{x}{(1+r)^n}$$

(2) For repeated annual payments of £y per annum from years 1 to n:

$$\text{present value (£)} = \frac{y}{r}\left(1 - \frac{1}{(1+r)^n}\right)$$

where r is the discount rate.

The discount rate chosen can be used to represent not necessarily the rate of interest that will have to be paid on capital money borrowed, but the interest or 'return on capital' which the money could bring were it invested in something quite different (the 'social opportunity cost' of capital). When capital is scarce, an appropriately high discount rate can express this. A more esoteric conception is that the discount rate should represent the 'social time preference', i.e. to what extent people prefer a benefit now rather than the same benefit in the future—an idea which may seem valid in principle but is difficult to pin down in value[10,11].

The same kind of assessment (albeit with more attendant difficulties) can be applied to the 'benefits' derived from some project, provided that the benefits can be expressed in money terms. In water supply this is difficult because charges for water usually express the cost of production and not the 'benefit' to the consumer, but there are other benefits that can also be taken into account, such as the recreational use of reservoirs and their catchments, fishing rights, amelioration of flood damage etc. If these can be expressed in money terms the whole flow of benefits from a project can be summed according to their present value and compared with the present value of the 'costs' of the project[12,13]. By this and various other means projects can be scrutinized and compared under various assumptions concerning the rate of return on capital as expressed by the discount rate. Although schemes are seldom chosen on their present value alone, it is usual to calculate the present values over a range of discounts to reveal the influence, if any, of varying the assumptions. Unfortunately, even the range of discounts that might reasonably be applied is a matter of debate among economists, and figures currently used are quite different from those used five years ago, and may change again. The comments made in Section 14.4 are therefore appropriate.

2.9 Recent developments in water supply organization

During the past two decades the chief development in water supply organization has been an appreciation of the need to control water on a river catchment basis. The whole cycle of abstraction and use of water and its return as wastewater has to be controlled on a catchment basis if the environment is to be preserved. This first became evident in the small highly industrialized countries, such as England and Holland, and in the industrialized regions of the USA. The uncontrolled discharge of effluents to relatively small catchments or short rivers brought about a sudden and potentially alarming decline of purity in natural waters.

The development of many new chemical processes, coupled with more sophisticated methods of analysis, revealed that many new and unnatural substances could be entering water supplies as a result of the wastewater discharges to rivers. These could arise from the use of pesticides, insecticides, herbicides, and fertilizers in agriculture as well as from new factory effluents. Analysis showed that the various contaminants represented a large range of complex substances[14], some of which might be harmful to humans[15]—although proof of the toxicity, if any, of very minute quantities of these in drinking water presents problems which do not yet appear to be solvable*.

In England and Wales the first step to control and license all abstractions within a river catchment was taken in 1963 when the River Authorities were set up. It quickly became apparent that the control of water quality on a river catchment basis was just as important as the control of quantity. Hence ten years later the Water Act of 1973 set up large catchment-based regional Water Authorities in England and Wales which controlled not only all abstractions from a river basin but also all effluent discharges to it, whether from municipal sewage disposal plants or from industrial plants. The new Water Authorities were additionally required to promote the environmental and recreational uses of water. This 1973 Act has been one of the most constructive pieces of environmental protective legislation passed in recent years. Great benefits have followed and many rivers have been restored to a condition better than they have been for a century, as illustrated by the improved condition of the Thames. The single handicap of the new Authorities has been their regional size, making them somewhat remote from the public and only capable of coming under indirect control via the various representatives on them, or via Members of Parliament who may question the Minister responsible for their performance. They have also been faced with much backlog work rectifying the poor state of many previous municipal sewage plants, and consequently have had to embark on a number of costly renewal schemes with a consequent increase of charges to the consumer. However, any reversion to smaller sized authorities would be unthinkable since they would be unable to exercise control on a river catchment basis: water supply would again become divorced from wastewater disposal and no proper planning for the use and disposal of water would be possible.

Holland and to a lesser extent Belgium illustrate the problems and costs arising when control on a river catchment basis is not possible. The River Rhine and its tributaries form the largest source of freshwater available to either country, but the Rhine also carries the effluents from many large industries along its banks in West Germany and France. Forced to use the Rhine water because other freshwater sources are limited, some of the newer treatment plants in Holland and Belgium contain as many as six stages of treatment in order to make the water fit for public consumption[16]. In the UK, where catchment control is properly exercised, no more than three stages of treatment are the normal requirement.

In the USA similar occurrences have been experienced and these led

* Substances can be detected at the low level of one part per thousand million parts of water. Whilst some of these substances are known to be toxic at levels ten thousand times greater, they are possibly not toxic at the low levels found—the difficulty is to prove this.

to 'blanket edits' under the Clean Water Act 1972 and the Safe Drinking Water Act 1974. The former Act required all municipal authority sewage effluents to receive the 'best practicable treatment' by 1977, whilst the latter Act set uniform quality standards for public water supply. The time targets set were remarkably ambitious and resulted in immense sums of money being spent in haste on the rehabilitation of many municipal sewage plants. Richardson[17] has reported the consequences in terms of inappropriate expenditure, promotion of schemes which were unnecessarily complex and expensive, and the solution of one problem only leading to another. In some cases effluents were turned out of a quality much superior to that of the receiving body of water, small communities 'were saddled with large, sophisticated treatment facilities with excess capacity, resulting in extremely high operation and maintenance costs', and advanced waste treatment 'usually doubled or tripled the volume of sludge, giving rise to solids waste disposal problems'.

Administratively the USA experience has not been a happy one. To control the estimated \$30 000 million expenditure on new works 1973–81, two-tier and sometimes three-tier organizations were set up. This resulted in long delays before schemes could be sanctioned because of the many officials involved at the various levels of authorization and their tendency to 'keep an unnecessary amount of paperwork in orbit'. Severe delays to schemes occurred and the whole programme has had to be rescheduled.

These experiences indicate that wide but simple control is the most effective and, for maximum return on outlay, discrimination through planned use of water has to be exercised. Standards set for effluents need to be related to the quality of the receiving water and the planned use for that water. All kinds of solid and liquid waste disposal, to both overground and underground, need to come under a single planning control. By this means some rivers can be scheduled primarily for use in the carriage of wastewaters, and others can intentionally be preserved at higher quality for use as sources of public water supply. For maximum efficiency multipurpose direct-acting authorities are best, their power and obligation being clearly set out by statute. Control can be exercised by relevant inspectorate systems and by requiring all capital programmes to be submitted for prior approval by government.

2.10 Current water charges

In 1983 a survey by the National Utility Services[18] revealed that commercial prices charged for water in mid-1983 were in the range 20–29 pence/m^3 in England and Wales. Higher charges up to 45 pence/m^3 were reported from Belgium and Germany, and up to 75 pence/m^3 in Australia. Lower charges in the range 10–18 pence/m^3 were reported from Scotland, Eire and Canada, and in some cities of the

USA. Wide variations occur within these ranges and the survey covered only some typical cities.

References

(1) *The Structure and Management of the British Water Industry*, IWES, 1979, p. 38.
(2) IMTA Water Statistics 1970–71, also certain Annual Reports of Undertakings.
(3) NWC Manpower Statistics 1973–74.
(4) *An Introduction to Engineering Economics*, ICE, 1969, pp. 117–122.
(5) Reference 1, pp. 185–188.
(6) Gilliland E J, Financial Problems, Paper No. 2, p. 2-9, *Proc. Symposium on Water Services: Financial, Engineering and Scientific Planning*, IWES, 1977.
(7) Reference 1, pp. 171–173.
(8) Walker D L, Possible Policies for the Future, Paper No. 7, p. 7-5, Symposium referred to in Reference 6.
(9) Reference 4, Appendix F, p. 169.
(10) Reference 4, Chapter 4.
(11) Herrington P R, Choices within the Water Industry: Does Economics Help?, Paper No. 6, pp. 6-5 and 6-20, Symposium referred to in Reference 6.
(12) Reference 4, Chapter 5.
(13) Kuiper E, *Water Resources Project Economics*, Butterworth, 1971.
(14) Fielding M and Packham R F, Organic Compounds in Drinking Water and Public Health, *JIWES*, September 1977, p. 353.
(15) Burke T, Hyde R A and Zakel T F, The Performance and Cost of Activated Carbon for Control of Organics, *JIWES*, July 1981, p. 329.
(16) Schalenkamp M, New Experiences in Removal of Organics in Surface Water Treatment, *Aqua.: JIWSA*, **4**, 1981, p. 279.
(17) Richardson W H, Experience Gained with the US Pollution Control Policies, *Aqua.: JIWSA*, **2**, 1981, p. 245.
(18) Anon., NUS Monitors Water Costs, *Water Services*, October 1983, pp. 445–446.

3

Hydrology

3.1 Water resource surveys

In assessing the water resources of an area the modern approach is to consider all possible means of development, and to examine comprehensively the hydrology of the catchment involved. Although traditional developments may frequently be adopted, the full range of possible developments needs to be considered as listed below.

(1) *Surface water*
(a) river intake,
(b) reservoir for direct supply,
(c) reservoir depending upon indirect gravity or pumped inflow,
(d) river abstraction guaranteed by releases from upstream storage or by inter-river transfers,
(e) tanks fed by collected rainfall.

(2) *Groundwater*
(a) springs,
(b) wells and boreholes,
(c) adits and galleries,
(d) river abstraction guaranteed by groundwater regulation pumping,
(e) reservoir fed by groundwater pumping,
(f) riverside collector wells,
(g) artificial recharge,
(h) storage in river bed sands.

(3) *Water reclamation*
(a) desalination and blending,
(b) reuse of treated effluent.

(4) *Bulk supply from adjacent undertaking*

Whereas formerly it was thought sufficient to evaluate the output of a source on the basis of rainfall, losses, and catchment area, nowadays many catchments are already partially developed so that it becomes necessary to adopt an inventory of all water available. To do this a hydrological survey is conducted in which all flows into and out of the catchment are quantified and balanced, so ensuring that all have been accounted for. The flows to be measured for a particular catchment will be as follows.

(1) *Inflows (or gains)*
(a) precipitation,
(b) surface runoff into area,
(c) groundwater movement into area,
(d) irrigation returns,
(e) sewage and industrial effluent returns,
(f) imports by water distribution systems,
(g) canal transfers and leakages into area.

(2) *Outflows (or losses)*
(a) evaporation and transpiration,
(b) surface runoff out of area,
(c) groundwater movement out of area,
(d) irrigation abstractions,
(e) water supplies taken to other areas,
(f) power station evaporative losses,
(g) sewer infiltration taken out of area.

(3) *Storage*
(a) soil moisture changes,
(b) change in contents of impounding reservoirs,
(c) aquifer storage changes.

Most of the above terms except groundwater flows and leakages can be measured directly. Table 3.1 shows a typical simple water balance trial and it will be noted that a column for residual discrepancy is given. These will frequently arise, and where they show repeated seasonal variation this may indicate that not all storage effects have been properly accounted for. However, over a few years the errors should cancel out: if not, some miscalculation or misunderstanding of the factors involved may be indicated. Frequently evapotranspiration is made the residual term so that errors are compounded in it.

To sum up, a water balance survey provides a means of:

(1) understanding water use in a catchment,
(2) checking that catchment data are adequate and accurate,
(3) quantifying average resources.

3.2 Catchment areas

In theory the topographic catchment to a source can be readily found by examining a contour map, but in practice difficulties occur. Many parts of the world are still unmapped at an adequate scale; even where maps exist, the contours on them may be largely extrapolated or may be out of date because of subsequent constructional work. Personal knowledge of a catchment is therefore important to the engineer responsible for defining it. To gain this knowledge a walking survey of key sections of the topographic divide may be desirable, particularly for an impounding

Table 3.1 A typical water balance trial.

Month	Rainfall +	Evaporation of intercepted rainfall −	Evapotranspiration −	Runoff[a] −	Net boundary transfer by groundwater flow +	Soil moisture storage change[b] ±	Groundwater storage change ±	Change of water in transit[c] ±	Residual discrepancy ±
January	60.8	3.3	4.2	30.3	14.2	−14.5	−6.5	−0.1	16.1
February	63.2	3.3	8.4	42.6	14.2	−11.9	7.8	−2.0	17.0
March	44.6	5.1	28.2	50.2	16.0	3.3	−0.9	20.2	−0.3
April	61.7	7.4	48.7	48.3	16.2	3.8	4.8	1.4	−16.5
May	15.1	2.4	78.7	35.6	16.8	58.5	3.9	0.0	−22.4
June	11.9	2.2	59.7	28.6	16.6	39.7	5.2	0.0	−17.1
July	35.8	7.2	44.8	25.4	17.7	11.6	6.5	0.0	−5.8
August	25.5	3.5	30.7	23.8	16.9	4.0	6.5	0.0	−5.1
September	35.7	4.4	23.2	23.6	17.8	−9.2	3.9	0.0	−3.0
October	40.3	3.8	15.1	18.7	17.0	−19.8	2.6	0.0	2.5
November	147.6	5.9	10.7	11.1	14.7	−69.0	−15.6	−22.3	27.7
December	57.6	3.5	−2.7	14.8	12.8	−12.0	−14.3	4.5	33.0
Total	599.8	52.0	349.7	353.0	190.9	−15.5	3.9	1.7	26.1

Notes:
Year—1970
Units—mm over area
Area—71.5 km^2
(a) Runoff artificially high in months when well field test pumping is discharged to river (February to December).
(b) Surface retention has been included with soil moisture storage change.
(c) Water in transit is falling from the root zone to the water table.

reservoir. When larger scale maps cannot be obtained, those at 500 000 scale by the USA Air Force in their world 'Tactical Pilotage'* series are recommended. Where possible an aerial survey followed by detailed photogrammetric contouring is the best policy. Associated ground control surveys will be needed as well, but a rapid mapping programme measured in weeks rather than months can be completed for a modest cost. Direct examination of stereo pairs of air photographs may resolve some uncertainties. Large inaccessible basins can sometimes be examined from satellite photographs, libraries of which are now available internationally.

Having located the boundary by visualizing downhill flow at right-angles to the contour lines, one may nevertheless come across water-courses marked on maps that appear to cross the boundary divide. Where such a watercourse is not the top pound for some canal, it may well be a contour leat. A leat is an open channel gravity catchwater, often built years ago for bringing water to a mill or mine or to augment the flow into some reservoir. Leats may either collect all the drainage from land on their higher side or take only the flows of the major streams which are intercepted: a survey is necessary to assess this. The capacity of the leat relative to the local runoff has also to be assessed: it may be large enough to contain any flood or it may be so small that it is overtopped many times a year.

A reservoir bywash channel is a contour catchwater starting at the main inlet stream at the head of a reservoir and passing down the rim of the reservoir to discharge to the spillway. These channels were often built in Britain in the late Victorian era in an attempt to minimize reservoir siltation by diverting turbid spate water, thus improving the quality of water taken directly into supply without treatment. A bywash normally intercepts all tributaries en route and has side weirs to permit flood spates to pass into the reservoir without damage. To evaluate the effect of such bywashes it is necessary not only to calculate the uncontrolled and the diverted drainage areas but also to assess the effect of rules by which any control sluices have been, or still are, operated. Bywashes are only of value where a reasonable proportion of average runoff comes as baseflow of good clarity.

The groundwater catchment to a source is not so readily defined, even where contoured maps of the water table are available. In the majority of cases they represent a reduced version of the surface contours, but there are classic cases showing groundwater reaching an outlet from a recharge point many kilometres beyond the surface topographic drainage limit. Natural water table levels tend to rise and fall together in any one region but, particularly where pumping abstraction is taking place, the groundwater boundary may migrate seasonally. This illustrates the

* Obtainable in the UK from the Director of Military Survey, Ministry of Defence.

fact that there is a continual balancing effect proceeding in an aquifer where steady abstraction is coupled with intermittent replenishment. Satisfactory confirmation of such conditions may require as many as one observation well per km^2 once away from flat terrain.

3.3 Rain gauges

Precipitation is measured with a rain gauge, which is hardly more than a standard cylindrical can, so designed that rainfall is trapped within it and does not evaporate before it can be measured[1]. It should be placed on level ground with no ground falling away steeply on the windward side. Obstructions affecting local wind flow should be a distance away from the gauge of at least twice their height above the gauge. Ideally rainfall should be measured at ground level, but there is danger of rain splashing into the gauge if placed so low. However, the higher the rim is placed, the more will some rain be blown away from the can and go unrecorded. For instance many international gauges are set one metre high: these can be expected to read 3% lower than the 0.3 m tall British gauge unless the local climate is characterized by light airs.

A rain gauge is almost infinitely small compared with the area it will be taken to represent, especially when taking into consideration the rapidity with which rainfall intensity can vary both in time and space. However, by judiciously siting gauges in typical catchment settings, they will be found surprisingly effective and consistent as time goes by. Where a site is regularly exposed to very windy conditions it is advisable to install a turf wall around it (Fig. 3.1). In exposed conditions this can be effective in producing ground-level readings and will prevent a loss of 5% to 20% of the catch.

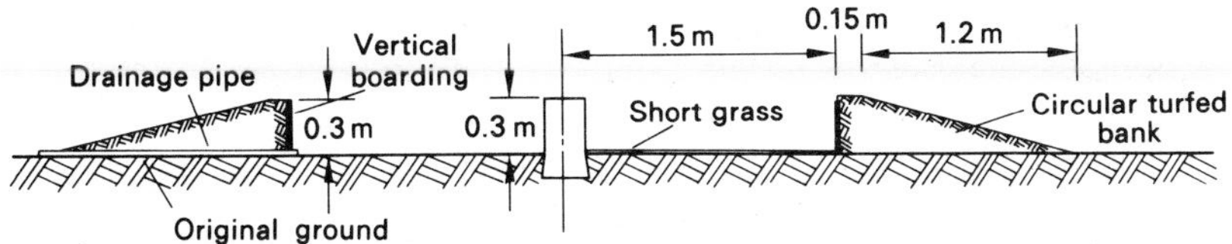

Fig. 3.1 Rain gauge with turf wall protection.

When new gauges have to be sited, and it is desired that the resulting records should be published, the advice of an experienced meteorologist should be sought. Many national meteorological institutions provide pamphlets designed to ensure good standard observation practice and the World Meteorological Organization plays an effective coordinating role.

It is particularly important that gauges should not overflow during significant storms. In the past gauges have frequently done this, either because of irregular emptying or because of inadequate capacity. The inner can should hold the once-in-fifty-years daily fall that can be expected at the site. If and when the inner can overflows the surplus should be held in the outer casing and measured. It should be noted that conventionally rainfall is measured at a fixed hour each morning (normally 9.00 hours GMT) and the quantity collected is attributed to the previous day's date. Where fresh snow is found lying at the time of measurement a core through it should be taken on level ground and melted to find the equivalent rainfall. Specialized problems occur in particularly snowy areas and snow-course surveys are then required[1].

Continuously recording rain gauges are invaluable for flood studies. The usual type gives a daily chart record of the accumulating contents of a rain-filled container: this empties by a tilting siphon principle each time 5 mm has collected. A more recent development is the tilting bucket gauge linked to a magnetic tape recorder which will run for at least one month: each time the bucket tilts to discharge 2 mm the event is recorded for subsequent computer totalling.

3.4 Stream gauges

Flow records at an intake or dam site, however short, are invaluable for revealing the potential yield of a source. Proven techniques for gauging which are in common use are:

(1) current meter gauging in a natural river section of known dimensions[2],
(2) sharp edged weir plates[3],
(3) broad crested[3] and triangular profile weirs[4],
(4) critical depth flumes[5],
(5) calibration of an existing set of sluices[6].

All but (5) require measurement of stream level in a section with a unique head–discharge relationship. It is preferable to install, in a common stilling well, a chart recorder with an unlimited natural level scale together with a punched or magnetic tape level recorder. The latter makes processing more accurate, but the chart records, generally on a weekly scale, give the necessary immediate visual check on conditions. Flow integrators, and charts reading flow directly, are to be avoided because they are a deterrent to continued improvement of the rating curve for the site.

Choice of gauging method involves many factors including cost and staff availability[6]. As a result of the difficulty of getting to a river at a time of peak flood flow, and of arranging for the necessary measurements by current meter, it may take a long time to obtain a complete rating curve for a river section. Despite the low capital cost of this method of gauging, the need for repeated attendance at the site and for frequent analysis of the results makes heavy subsequent demands on staff time. Each time a major flood is thought to have shifted the river bed profile there is an uncertainty about the rating curve until this is rechecked: therefore the alternative of a standard gauging structure is more attractive to the engineer, particularly so for catchments of less than 500 km^2.

Sharp edged plate weirs, of rectangular or vee shape depending upon the sensitivity required, are generally only suitable for springflows or for debris-free small streams. The need to keep the weir nappe aerated at all stages limits their use, but they are frequently used for low flow surveys as they can be rapidly placed in small channels. Limited measurements of this kind are best used on a temporary basis because only an all-range gauging station can adequately provide the answers to the range of problems a water engineer faces.

Of all the weirs that designers have tried, perhaps the most successful has been that of Crump[4]. It has a simple and efficient flow characteristic $Q = 1.966\, H^{3/2}\, \text{m}^3/\text{s}$, whilst operating up to the total head at which tailwater reaches 75% of that upstream (relative to crest height). Using the crest tapping designed by Crump it is possible to attain reasonable results at up to 90% submergence, but this versatility is marred by a tendency for the crest tappings to block in floods carrying a high sediment load. A minimum head of 6 cm on the weir is necessary for accuracy, but with compounding of the weir crest this can be achieved. The flat-vee variant is another possibility[5].

Critical depth flumes are appropriate on smaller catchments of, say, under 100 km^2 which have a wide flow variation and where sensitive results are required. Essentially a contraction of the channel forces flow to attain critical depth over a fixed section whatever the upstream head. A unique upstream head–discharge relationship is then created such that for a rectangular throat

$$Q = 1.70\, C_v b h^{3/2}\, \text{m}^3/\text{s}$$

where b is the effective width (m), h is the effective head (m), and C_v is the approach velocity coefficient.

Introduction of a shape factor is possible for trapezoidal or U-shaped throats.

Calibration of an existing sluice structure may be achieved using formulae derived from laboratory model tests, but these should be checked by current meter measurements wherever possible. Sometimes long, relatively homogeneous sluice keeper's records may exist, and the calibration of such sites can then produce records of flow of several decades in length for a modest cost. What these records may lack in hydraulic accuracy may be offset by their statistical value as a long record.

New gauges should be installed to the standards set by the responsible regional or national hydrometric agency. Early processing and publication of the assembled flow measurements is eminently desirable, and it is important that the original level charts should be carefully preserved. This may necessitate microfilming.

3.5 Evaporation and percolation gauges

It is difficult to measure evaporation with any certainty by means of a gauge because of the edge effects associated with such an instrument. The standard approaches to the problem are either direct or indirect[7]:

(1) direct—raised pan or sunken tank, lysimeter or percolation gauge,
(2) indirect—evapotranspiration formulae based upon meteorological data.

For measuring evaporation the USA Class A pan[7] is of galvanized iron or monel metal, 1.21 m in diameter and 25.5 cm deep. It is set on a standard wooden framework 10 cm above ground level, thus allowing air to circulate all round it. As a result measured evaporation is higher than that of a natural water surface and a reduction factor must be applied. This is generally taken to be 0.7, but it can vary between 0.35 in areas of low humidity and very strong winds, surrounded by bare soil, and 0.85 where high humidity and light winds prevail[8]. The British Symons sunken tank is 1.83 m square and 61 cm deep, with the rim 7.5 cm above ground level. It is more nearly a model of reservoir evaporation but suffers from inconsistent results if it is not in tight contact with the surrounding ground. The heat storage of a small tank is correspondingly small, whereas a large lake takes time to warm up or cool down. As a consequence tank results do not quite match the evaporation of a nearby lake in regions with strong seasonal temperature variations. Peak lake evaporation rates in the Kempton Park experiment[9] occurred up to a month after peak tank measurements and this was explained by heat storage theory.

Annual open water evaporation ranges from 600 mm in Northern Europe, through 1500 mm in much of the tropics, to more than 2500 mm in hot arid zones.

Unfortunately no standard percolation gauge has yet been devised. Most are formed by a large diameter pipe sunk about one metre into the ground and carefully refilled with the original soil layers. A surface cover of short grass is irrigated by a trickled hose if potential evapotranspiration is to be measured. Any water percolating through the soil layers during rainfall is drained from the bottom of the pipe to a nearby access manhole where it is measured[10]. Deduction of the measured percolate from rainfall measured by an adjacent rain gauge gives the evapotranspiration loss from a non-irrigated gauge, but only between dates of equal soil moisture in the gauge will this figure be meaningful. A lysimeter measures water balance elements but on a larger scale. The term normally covers a small catchment plot underlain by impermeable geology, with a single type of vegetation, and from which all overland flow and shallow drainage can be measured. Rainfall is measured above the canopy of vegetation on the plot so that the total evaporative losses from the canopy and at ground level can be inferred. The technique is particularly useful for showing the relative water consumption characteristics of moorland, forest, arable crops, and pasture.

The best technique for predicting water losses indirectly is that developed by Penman[11]. His equation depends upon measurements of radiation (or sunshine duration), wind run, vapour pressure, and air temperature, all of which should be taken at the same site. At altitudes above, say, 1000 m McCulloch's[12] fuller version of Penman's equation should be used as it makes express allowance for the corresponding pressure drop as well as adjusting the radiation term for latitude. Either open water losses or potential catchment transpiration losses can be calculated according to the reflectivity and roughness coefficients used. A basic assumption is that the ground vegetation cover will be short (akin to grazed grass) and that sufficient soil moisture will be available to keep up potential loss rates.

3.6 Effects of storage

The temporary storage of rainfall in the soil and aquifer layers of a catchment can be significant, but is only quantified with difficulty. Soil moisture variations occur predominantly in the first 0.5–1.0 m below the surface. From the driest (Wilting Point) condition to the wettest drained state (Field Capacity) may mean a rise from 3% to 10% water content in a sandy soil or from 20% to 40% in a clay soil. Thus the maximum range of water storage in one metre of soil may be as much as 20 cm. Additional water may be held under waterlogged conditions whenever the drainage rate is lower than the rainfall intensity.

The total quantity of water stored in an underlying aquifer is often much greater, but its fluctuations may be smaller. Observed groundwater levels can be converted into storage estimates, either by correlating their fall over a dry season with springflows from their catchment or by multiplying by the storage coefficient of the aquifer if this is known (see Section 4.17).

Measuring soil moisture storage can be done by drying augured soil samples, but a destructive process of this kind has obvious limitations. Where regular sampling is required, say at monthly intervals, a neutron measurement probe, lowered down 45 mm diameter access tubes[13], can be used. In general groundwater pumping will not affect soil moisture storage, which is more likely to be influenced by local ditch levels, tile drainage, and farming practice. Groundwater development will alter the pattern of natural groundwater storage changes and consequently springflows locally. Where the water pumped is returned to the local stream, then the overall effect will be to reduce wet season flows and raise dry season ones. Surface reservoir impoundment modifies river flows so directly that the effect is normally reviewed at least monthly so that river flows measured downstream can be adjusted back to natural conditions.

3.7 Consumptive use

Water consumed is usually lost by evaporation or incorporation in some product. For an urban catchment in the UK it is likely that about 90% of domestic water supplies are returned to local rivers as effluent, either directly by sewerage systems or indirectly via soakaways; returns from industrial use are slightly higher[14]. Where garden watering is more dominant, as in much of the USA and Australia, this figure can fall significantly. In attempting to check figures it is often difficult to quantify mains leakage which, although a temporary loss to resources, must return to the surface eventually. Precise checks are complicated by sewer infiltration where pipes are below the water table.

Losses in the personal use of water are small, probably only 1% of consumption, and more significant sources of loss can be expected in industrial and horticultural activities. Fire-fighting use is negligible when averaged over the year. Generally the major loss will be in garden watering, which may fluctuate widely in quantity from year to year. It can be quantified by careful examination of a water undertaking's daily demands in differing temperature and soil conditions, or it may be computed theoretically by irrigation consumptive use methods providing the watered area is known.

River flows are likely to be directly affected wherever power stations[15] are built. A modern 2000 megawatt station on full load can lose about 65 Ml/d by cooling tower evaporation, but this reduces with lower load factors.

Consumption by irrigation has seen considerable growth in recent years because of the potential for enhancing crop yields and securing their reliability. Where, as in England and Wales, spray irrigation has to be separately licensed, the seasonal total and potential daily maximum rate can be found from local water authority records. However, gravity diversions for subsurface irrigation are much more difficult to evaluate because they are unlikely to be controlled: they may be made through a variety of diversion canals or by sluices ('slackers') from embanked main rivers. The water is used to ensure that local ditch levels are kept high enough to permit crop roots to draw water from the associated water table, but low enough to encourage good root growths. Obtaining an exact estimate of irrigation consumption of this type on a major basin while it is happening is rarely possible. Some indication of the quantities involved in English conditions are shown in Table 3.2.

3.8 Data collection and preservation

Symons, the great Victorian who initiated the British Rainfall Organization as long ago as 1861, had a realistic attitude to the collection of

Table 3.2 Subsurface irrigation quantities in England.

Area	Average seasonal rainfall (mm)	Consumptive water use for subsurface irrigation in the growing season (Ml/d per km^2)						
		M	A	M	J	J	A	S
South level fens Cambridgeshire (arable)	225	—	0.3	0.8	1.3	1.2	0.6	—
Somerset moors (pasture)	290	—	—	0.3	0.2	0.4	0.4	0.1

data. He once stated: 'It is one of my fixed rules never to ask for data unless I see my way quite clearly to utilizing it.'. This standard is still relevant for any engineer approaching a hydrological task. The corollary holds also: when faced with a problem one must search for the data needed. If the data do not exist, one must set about measuring these and, once measured, the original records should be carefully filed for safe preservation. Records are needed for those parts of the hydrological cycle upon which an existing source depends in order that the severity of current conditions is known and any deterioration in performance is spotted. An initial need at almost every source is the measurement of rainfall on the catchment concerned, but streamflow figures are more important for any river source, as are aquifer levels for groundwater supplies. Both must be supplemented by careful notes of the quantities recharged or released, and of associated changes in storage. Careful annotation is required if the distinction between measured and natural river flow is to be preserved. Information relating to wells should embrace both pumping and rest water levels together with power consumption and pump output figures.

The resulting hydrological data are of permanent value and should be treated accordingly. Original records should be bound or microfilmed, indexed, and archived for future reference. Computer-filed data may need frequent transposition to the latest type of installed system so that it does not suffer gradual inaccessibility. It is disappointing to find few autographic rain recorder charts going back for more than twenty years. When original data must be dispensed with, one solution is to offer them to a regional or national hydrometric agency for safe keeping.

3.9 Catchment rainfall

Although attempts have been made with varying success to estimate the precipitation over a region from formulae (using factors such as distance from the sea, altitude, and aspect), the only safe means of obtaining a convincing catchment rainfall figure is to depend upon rain gauges in the area concerned. (Most of these would be needed to confirm the formula applied anyway.) Rainfall (and snowfall) is best computed monthly and summed each year. Weekly figures are to be avoided because they neither correspond to the duration of significant hydrological events nor precisely fit an annual calendar.

The most accurate method of computation is laborious. The usual practice is to take the monthly rainfall total recording by each gauge and plot this on a map of the catchment. Readings from around the outside of the catchment are also required. Isohyetal lines, i.e. lines connecting positions of equal rainfall, are then drawn on the map. If any of these lines shows an unusual feature, the gauge readings must be checked and the plottings be critically reviewed. It is sometimes more revealing to plot isopercental lines instead of isohyetal lines, but this depends upon long average rainfall figures being known for most sites. Isopercental lines are lines connecting together locations on a map which have the same percentage of the average annual rainfall or of the average monthly rainfall—the choice being one of convenience. Once defined they are a reliable means of estimating missing individual gauge readings at sites where the average is known.

When the isohyetals have been adjusted to give the most likely picture of the rainfall distribution, the total precipitation on the area for the period considered is obtained by planimetering areas between isohyetals and multiplying each area by the rainfall upon it. However, this method is time-consuming and subjective. The most popular method of weighting gauge readings objectively by area is that of Thiessen. An area around each gauge is obtained by drawing a bisecting perpendicular to the lines joining gauges, as shown on Fig. 3.2. The portion of each polygon so formed lying within the catchment boundary is measured and the rainfall upon each is assumed to equal the gauge reading. The total precipitation is the weighted average of these values. One drawback is that, if the gauges are altered in number or location, major alterations to the polygonal pattern ensue. The gauges must also be reasonably evenly distributed if the results are to lie within a few per cent of the isohyetal method.

The number of rain gauges required to give a reliable estimate of catchment rainfall increases where rainfall gradients are marked. A minimum density of 1 per 25 km^2 should be the target, bearing in mind that significant thunderstorm systems may be only about 20 km^2 in size. In hilly country, where orographic effects may lead to large and consistent rainfall variations in short distances, it can be necessary for the first few years to adopt the high densities suggested in Table 3.3.

In large areas of the tropics there is great variation in rainfall from place to place on any one day, but only a small gradient in annual totals: in such areas the rain gauge densities of Table 3.3 will be excessive and more attention should be given to obtaining homogeneous records of long duration at a few reliable sites.

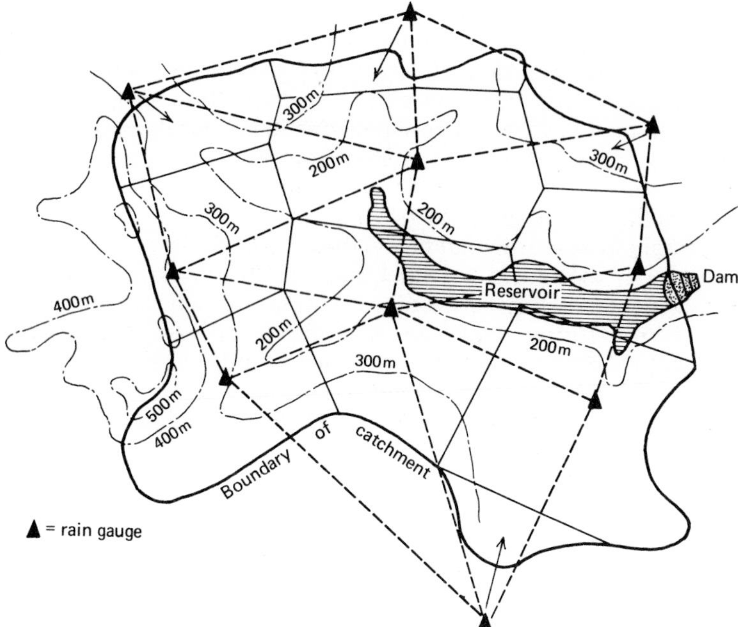

Fig. 3.2 Thiessen's method of estimating general rainfall over an area.

Table 3.3 Rain gauges required in a hill area[16].

Catchment area (km^2)	Number of gauges
4	6
20	10
80	20
160	30

When quoting average rainfall it is important to specify the period of years concerned. World practice now tends towards the use of the three most recent decades, i.e. 1951–80. Such periods can still be too short for completely stable averages to be found, for example the period 1916–50 was 5% to 10% wetter than the period 1881–1915 over most of western Britain. This raises the whole question of climatic trends, but as no marked shift of average rainfall has occurred in the last century which cannot readily be explained as an expected statistical variation, this can be ignored by the practical engineer. Care must be observed in semi-arid zones where occasional runs of five years or more of either extreme dry or wet conditions are known to occur.

3.10 Losses

A significant proportion of rainfall is lost by immediate evaporation or by the later transpiration of vegetation growing on water-absorbing soils. In some cases there will also be deep infiltration that eventually emerges in the sea without ever appearing in surface drainage channels. Catchment losses are best estimated from a water balance conducted over a number of years on the catchment concerned, or on one of similar rainfall, geology, and land use.

In Britain losses rarely fall below those estimated by such a method[17] for the rocky catchment of Allt Uaine at 330 to 880 m altitude in the Scottish highlands. These averaged 242 mm per annum, with a range from 117 to 456 mm. However, a more typical loss rate in England would be 450 mm per annum, with significantly higher figures above 500 mm per annum generally only occurring where afforestation predominates. A review of average gauged loss estimates in temperate Britain[18] is given in Table 3.4.

Table 3.4 British catchment losses.

	Loss (mm)				
	<380	381 to 440	441 to 500	>500	Total number of catchments
England					
Number of catchments	20	51	93	41	205
Average rainfall (mm/annum)	1150	940	850	880	
Wales					
Number of catchments	4	10	9	11	34
Average rainfall (mm/annum)	1510	1520	1440	1900	
Scotland					
Number of catchments	12	19	12	4	47
Average rainfall (mm/annum)	1320	1280	1310	1920	

Table 3.5 gives some idea of the variation in loss in different regions of the world. Although it might be thought that losses would be higher in years hotter than average, this is often more than offset by the concurrent dryness of the weather which leads to a deficit of moisture in the soil which in turn leads to a limit on the transpiration of water by the growing vegetation. As soil dries out and approaches wilting point it has been shown that evapotranspiration rates can drop to only about one-tenth of those to be expected from weather data[21]. Some plants are much more successful at their control of water use in drought conditions than others, for instance pine trees have a marked ability to conserve water in this way.

Table 3.5 Typical losses in various parts of the world.

Country	Location	Catchment cover	Annual rainfall (mm)	Annual loss (mm)	Marked seasonal variation
Nigeria	Ibadan	Rain forest	2500	2350	No
Iraq	Kirkuk	Irrigated pasture	250–400	1800	Yes
Malaysia[19]	Kuantan	Forest	2950	1450	No
Sri Lanka	Kirindi Oya	Mixed forest	1650	1230	Yes
Malaysia[19]	Kuala Lumpur	Forest	2410	1210	Yes
Sierra Leone[20]	Freetown	Forest	5795	1145	Yes
Hong Kong	Islands	Grass	2100	1050	No
Zaire	Fimi	Rain forest	1700	1040	No
Japan	Ota	Conifer forest	1615	890	Yes
Australia	Perth	Mixed grass/forest	875	760	Yes
South Africa	Transvaal	Mixed grass/forest	870	760	Yes
Kenya	Tana	Forest/savannah	1100	730	Yes
India	Bombay	Rain forest	2550	700	Yes
Lesotho	Maseru	Grassland	600	530	Yes
Holland	Castricum	Low vegetation	830	450	Yes
Britain	South England	Pasture/arable	600–900	450–530	Yes
	Midlands	Pasture/arable	650–850	440	
	Central Wales	Moorland/forest	1500–2300	480–530	
	Pennines	Moorland/forest	1150–1800	410–460	
	NE England	Moorland/pasture	700–1250	380	
	South Scotland	Moorland/pasture	600–1800	360–410	
	North Scotland	Moorland/pasture	1250–2500	330–380	
Algeria	Hamman Grouz	Scrub	420	400	Yes
Russia	Moscow	Agricultural	525–600	375	Yes
South Korea	Han basin	Forest	1180	320	Yes
Oman	Oman	Rock	160	130	Yes
Iran	Khatunabad	Bare ground	150–550	50–200	Yes

Losses increase with the density and height of natural growth and crops: this is particularly so with mature coniferous forests. However, predicting the additional loss compared with the alternative of a grazed pasture cover is still not easy. It has been shown that the loss is due to intercepted raindrops being evaporated back into the atmosphere at rates of up to five times normal transpiration values for short grass. This is because water laid out in thin films on vegetation can take up available heat in the atmosphere more readily: the sight of a forest steaming gently in a short spell of sunshine between showers is not uncommon. To quantify the extra loss to be expected requires[22] an idea of the depth of water the canopy of vegetation can hold during a shower, the frequency of showers, and their magnitude relative to local grass transpiration rates. One point to note is that whilst rain is being evaporated off the outsides of leaves, water will not be transpired through them. As a result perhaps only 90% of interception losses will be an addition to the catchment losses that would prevail anyway.

3.11 Long average runoff

This can be obtained in at least three ways for those sites where runoff data are brief or absent.

(1) Deduct loss estimates from catchment rainfall figures.
(2) Correlate brief runoff records with those of a nearby long-period station.
(3) Adjust a short-term flow duration curve by a knowledge of the relationship between short- and long-period duration curves elsewhere.

Correlations are best done graphically (rather than by uncritical best-fit techniques) so that runoff values can be examined for odd features which are prevalent in periods of snow or scattered thunderstorms. Daily data will produce too much scatter and annual values too few points; monthly figures are a good compromise.

A flow duration curve is obtained by putting the daily flows of a stated period (usually a year) into a ranked list of decreasing size. They can then be plotted in dimensionless form, as in Fig. 4.13, p. 114, to show the percentage of time any particular flow is exceeded. Where the number of daily flows for analysis is large, the figures are first tallied off into about twenty different flow ranges. The number of events in each range is found and the cumulative total of events above the lowest flow in each range then follows; the curve of flow duration is constructed by plotting these last two factors. The area below such a curve equals the runoff volume of the period concerned. These curves are particularly useful for intake and catchwater yield studies (Section 4.9) or for run-of-river hydropower investigations.

If the long average flow duration curve is compared with that for a single year it will generally lie completely above or below. At each 10% point along the time axis an adjustment factor can be found which will transform the single year value into the average conditions value at that point. These factors can then be applied to a flow duration curve for the same year at a new or temporary site nearby in order to estimate its own average characteristics. However good this technique is at describing the range of flows to be expected in future, it is not always easy to take the average runoff volume from under the curve because of the large quantity of flow under the acutely shaped flood limit of the graph.

3.12 Drought rainfall

Although it is now more normal to calculate water supply yields by direct use of local flow records rather than by recourse to complex rainfall analysis, it is important to know the severity of past drought events which have strained water resources. Definitions of drought based upon the number of consecutive days without measurable rainfall are only relevant in those areas (mostly outside the UK) where reliance has to be placed on small storages using roof catchments. In the UK the longest known spell without any recorded rain at all[23] was for 73 days from 4 March 1893 at Mile End in London; Calama in Chile is believed to have had virtually no rain for 400 years until a sudden storm fell in 1972.

Since 1861, when Symons first circulated *British Rainfall*, the most notable droughts for water engineers in Britain have been 1887–89, 1921, 1933–34, 1943–44, 1949, 1959, 1972–73, and 1975–76. Where undertakings relied upon river intakes, or had a minimal amount of storage, 1949, 1959, and 1976 were critical because of the exceptionally low summer rainfall and high temperatures. Spring sources were hard hit in 1921 as autumn rainfall was insufficient to prevent the flow recession, which began in a dry spring, continuing until January 1922 in many parts. Annual rainfall in 1921 was the lowest in over 100 years in South East England.

Many of Britain's major cities are supplied from large upland reservoirs that are only really taxed in two-year droughts. Such a drought occurred in 1933–34 when there were two dryish summers with a remarkably dry winter intervening. Conditions were worst over Wales and the Midlands. 1943–44 followed a similar pattern in southern England and, because preceding years had also been dryish, very low flows were experienced in spring-fed rivers. Many records were broken in 1975–76 and an emergency Drought Act was passed to conserve dwindling supplies.

It should be clear from these few examples that the engineer must examine his data carefully to locate the drought with a duration and magnitude which is most likely to deplete his undertaking's resources. Sometimes source yields are quoted as 'safe' or 'reliable' because they could have been taken during a quoted notable drought event. The responsible engineer seeks to anticipate what may occur in the future in order to be prepared. Ever since rainfall has been measured people have sought to infer expected averages and extremes. Some prefer to plan against a recurrence of a minimum already on record. Others, claiming that history never quite repeats itself, consider statistical inference is more appropriate. Either approach requires many years of data before it can be really convincing. To use less than fifteen years' data is to court trouble.

To forecast drought rainfall either one can use a knowledge of recorded minimum rainfall for a region, expressed as a percentage of the average[24], or one can carry out a statistical analysis of available rainfall measurements. A useful study of the latter sort was carried out by Law[25] for areas where the coefficient of variation of annual rainfall is between 8% and 20%. (Coefficient of variation = the standard deviation expressed as a percentage of the mean value.) Figure 3.3 summarizes his results for the worst conditions that are likely in a century. Tabony has produced a fuller analysis on the same topic[26].

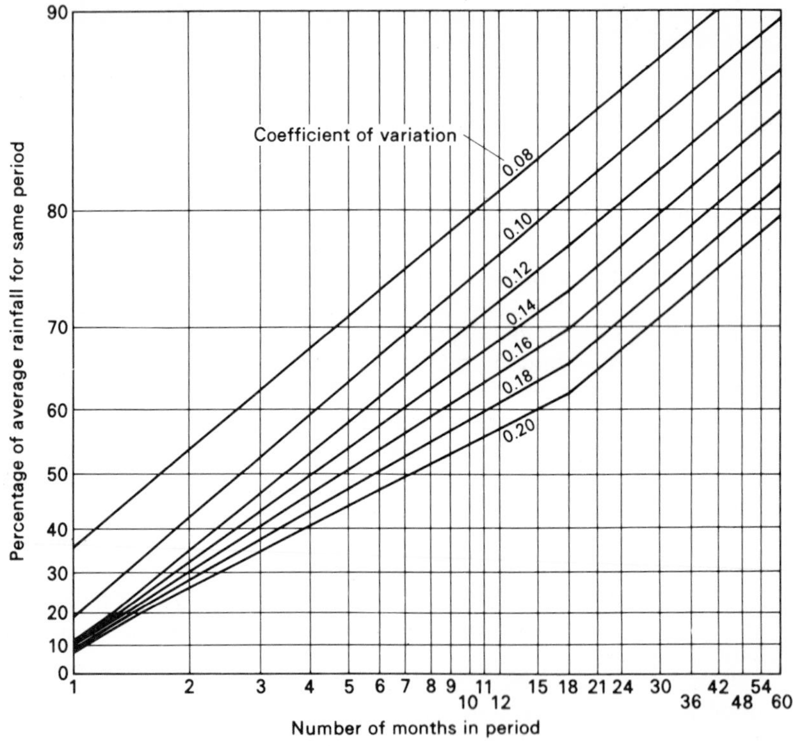

Fig. 3.3 Minimum rainfalls corresponding to 1% probability for given coefficients of variation of annual rainfall.

3.13 Drought runoff

The prediction of the minimum daily flow that will be reached during some future drought is not a light task. When recorded runoffs are expressed as rates per unit catchment area it can often be seen that geology and topography are the major influences unless man has interfered. It must be obvious to the reader from his knowledge of the countryside that the dry weather flow of many small catchments is zero. However, large rivers in temperate zones do not dry up. In these circumstances no information can be as valuable as that given by a gauging station on the river at the point under consideration.

Several studies have attempted to predict minimum flows of specified severity after regional analysis of flow records. Notable examples include Malaysia[27], South Africa[28], New York State[29], and the United Kingdom[30]. Estimation at an ungauged site is fraught with difficulties. Transposition of results even between natural catchments requires climatic and geological similarities. The Institute of Hydrology equations for UK regions are of the following general form, although the coefficients in this example are specific to northern England and southern Scotland:

$$\begin{aligned}&\text{10-day mean flow exceeded 95\% of the time}\\&= (7.6BFI^{0.5} + 0.0263SAAR^{0.5} - 1.84)^2\end{aligned}$$

where *BFI* is a baseflow index (see Table 3.6) and *SAAR* is the standard (1941–70) average annual rainfall in mm. The result is a percentage of the long-term average daily flow.

Table 3.6 Typical baseflow indices for various rock types[30].

Dominant permeability characteristics	Dominant storage characteristics	Example of rock type	Typical *BFI* range
Fissure	High storage	Chalk	0.90–0.98
		Oolitic limestones	0.85–0.95
	Low storage	Carboniferous limestone	0.20–0.75
		Millstone Grit	0.35–0.45
Intergranular	High storage	Permo-Triassic sandstones	0.70–0.80
	Low storage	Coal measures	0.40–0.55
		Hastings Beds	0.35–0.50
Impermeable	Low storage at shallow depth	Lias	0.40–0.70
		Old Red Sandstone	0.46–0.54
		Metamorphic-Igneous	0.30–0.50
	No storage	Oxford and London Clay	0.14–0.45

Qualifications are required in certain terrains. Bournes are streams which run strongly when the water table is high, but which dry out gradually from their headwaters as the water table falls away from the stream bed. As a result a chalk or limestone stream cannot be expected to have a good dry weather flow unless the water table ensures perennial flow.

As a general guide it may be said that minimum flows are likely to be between 5% and 10% of the long-term average daily flow (ADF) unless there is a notable aquifer in the catchment. Even then dry weather flows are unlikely to be better than 25% ADF, although higher values can occur in the tropics where there are two brief dry seasons each year. Minimum natural daily flows are about 80% and rarely less than 70% of the minimum mean monthly flow in all but the smallest of catchments.

3.14 Sediment estimates

Water can carry a surprisingly large amount of solid material during its passage to the sea. Over different reaches of a river there will be at one time stability, erosion, or deposition. Although most sediment is carried in suspension, the bed load of a stream should not be ignored. This term refers to the material that rolls and bumps along the bed generally in intermittent surges during flood periods. The quantity of suspended solids in the runoff from a natural catchment rises and falls with the stream hydrograph, but the relationship is neither direct nor precise. Ideally sampling should enable one to estimate a sediment rating curve[31] giving the relationship between flow and solids carried. However, it is well known that a flow reached on a rising flood frequently carries a larger proportion of sediment than when a river is falling. This is particularly the case when the catchment is being 'washed down' by a rainstorm at the end of a long dusty drought.

Figure 3.4 shows a comprehensive range of suspended sediment loads experienced in rivers of different size. To these must be added perhaps 10% to 20% to cover bed load which cannot normally be measured.

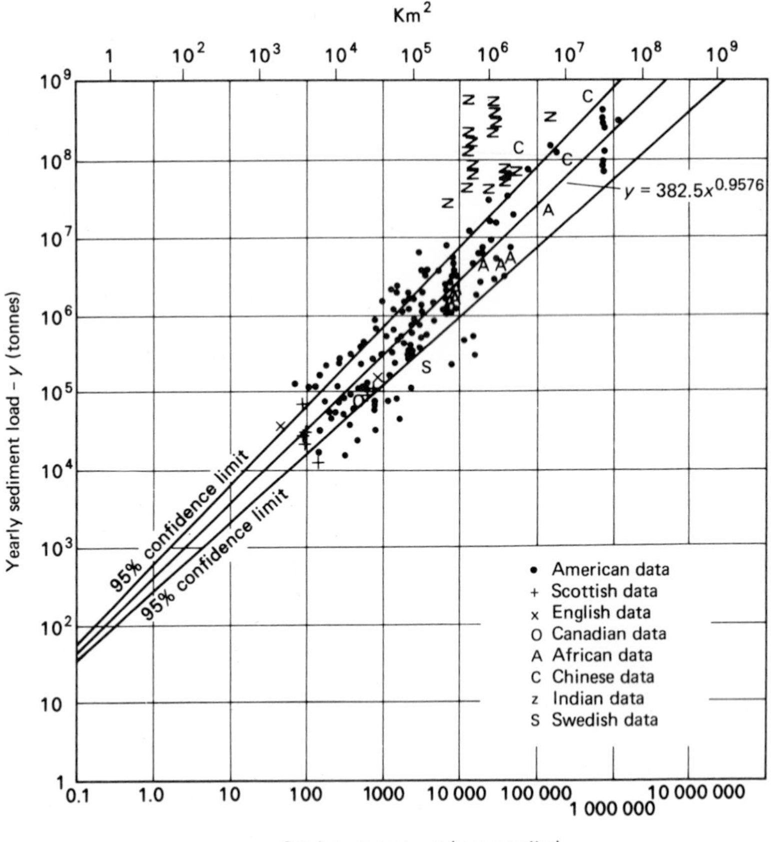

Fig. 3.4 Average suspended sediment concentrations. (After Fleming[32].)

Sometimes bed load is dredged from a river on a commercial basis at a spot where a flattening gradient favours its deposition and, in such a case, reasonable estimates can be made. However, it is at reservoirs that such deposition most concerns the water engineer. Dams in areas of active soil erosion have been known to fill rapidly with sediment in only a few years. Algeria has provided some of the best known cases of severe sediment regimes. One review[33] examined thirteen basins of between 105 and 5488 km^2 in area with mean runoffs in the range 75 to 1100 mm/y: suspended sediment averaged over the years of record varied between 60 and 8880 t/km^2. Eight basins of various sizes produced less than 700 t/km^2, but the other five gave more than 1100 t/km^2. At any one site the highest annual sediment load recorded was up to five times the yearly mean; the load in the lowest year was often below 10% of the mean.

Sediment load is understandably biassed upwards in wetter basins. However, where there is good cover of vegetation that is not overgrazed alarming situations need not be expected. Surveys of a few drained reservoirs in England[34] showed a trapped sediment volume which can be expressed as being between 40 and 90 parts per million of inflow (by volume) over a life of about 100 years. Another review[35] expressed reduction of reservoir volume in SE Australian reservoirs in terms of unit loss per year per catchment area; in metric terms figures ranged from 0.01 to 0.20 Ml/y per km^2.

The coarsest material becomes deposited in a delta at the head of the lake, whilst the finer material spreads more evenly over the lake bed. It is advisable to operate a low scour drawoff occasionally to ensure that it does not become covered with silt; however, this action will not normally clear sediment from more than a small circle around the drawoff point.

The multiplication of a sediment rating curve by a flow duration curve (such as Fig. 4.13) will give an estimate of the expected average sediment flow to a site. However, such estimates are frequently liable to be underestimates due to the difficulty of sampling sediment at exactly the right time to catch peak concentration. Meters are available which continuously measure turbidity by variations in light transmission between two photocells immersed in the river.

A catastrophic flood, rare though it is, may move more material in a few hours than the total carried away for several years previously. Alternatively an apparently stable channel can undergo a great deal of scouring during a flood, only for the hole formed to be filled by deposition as flows subside. This and the channel cross-section should be considered when siting an intake: water will be deepest on the outside of a bend in the channel (unless a rock sill is present), whilst a shifting shoal of deposited material can often build up at the inside of the bend. Bank intakes are generally sited on the deeper water side and may well require erosion protection upstream and downstream.

3.15 Flood calculations

Hydrology for the water engineer is mainly concerned with the magnitude and risk of drought, but familiarity with low flow needs tempering with a healthy respect for floods and the potential they have for damage. (The Tamar, in Cornwall, experienced a flow minimum of 59 Ml/d on 5 October 1959; hardly more than a month later came the record flood of 35 000 Ml/d.) When locating any works in or by a river, it is necessary to estimate the spectrum of floods that will occur during the scheme's lifetime and the consequential effects.

Problems fall broadly into the following groups:

(1) sizing gauging weirs and intakes,
(2) estimating floor levels for treatment plant, pump houses etc.,
(3) designing and reviewing reservoir spillways,
(4) planning for construction floods.

Each country tackles flood estimation in its own way but there tend to be:

(1) road and urban drainage engineers concerned with the rapid response of paved areas to short, frequently recurring storms,
(2) land drainage and river engineers concerned with relatively rare floods which the riparian community has decided it can no longer tolerate,
(3) dam engineers concerned with the possible overtopping and failure of a structure in a very rare but conceivable event, which could lead to loss of life[36].

As a result, different organizations may publish widely varying techniques, and published information may take some unearthing. It normally takes the following forms:

(1) an empirical flood-peak formula, based upon the size of area within a given region—Creager's is a well known version[37],
(2) a regional flood probability graph[38],
(3) a rainfall of known probability converted into runoff by an impermeability factor or by a unit hydrograph (see Section 3.19),
(4) a meteorologist's estimate of Probable Maximum Precipitation[39] (PMP), converted into the Probable Maximum Flood by using the worst reasonable assumptions about catchment response.

Following the Institution of Civil Engineers' recommendations in 1967, and the spur given by the disastrous Bristol floods in July 1968, a UK Flood Studies Team was set up. Its Report[40], probably the most detailed of any in the world, was published in 1975 and included a parallel study for the Republic of Ireland. This was complemented by an Institution of Civil Engineers' guide[36] applying the results to reservoir spillway designs and reviews in the UK.

3.16 Maximum rainfalls

Whereas droughts are often widespread, maximum rainfall intensities can be very localized. When considering the latter, two additional factors, area and time, have to be brought in. Figure 3.5 is a plot of the maximum measured rainfalls at individual points in the world. It must be stressed that such colossal quantities are precipitated only in the most unusual hill areas in certain climatic zones once the duration exceeds a day. In addition the line does not represent a continuous event: this would only be possible for a storm lasting up to, say, four hours.

The maximum figures for Britain are seen to be about 20–30% of world maxima, with the lowland easterly part of the country suffering less severely in long duration storms. The greatest is the Martinstown, Dorset, storm on 18 July 1955 in which 280 mm (11 in) of rain were officially noted in 18.5 hours, with an unofficial measurement at the heart of the storm claiming 350 mm (14 in). Such maximum records are gradually rising as time progresses, although towards some physical upper limit. It is quite possible to go for many years without any outstanding event and then to have several clustered close together. For example Singapore experienced no higher daily fall than 328 mm in a day in the 76 years up to 1969; however, the peak fall at the Lower Pierce Reservoir in 1969 was 369 mm and this was exceeded elsewhere on the island in 1978 when 537 mm fell on the second and third days of December. Many countries have had compilations of extreme events produced, e.g. Indonesia[43], and these should never be overlooked.

Individual long rainfall measurements can be analysed statistically to provide duration–intensity–frequency estimates and many national meteorological agencies make these available[44]. These are often adequate for storms that can be expected more frequently than once every fifty years, but rare storms need thorough regional studies for their evaluation. The UK Meteorological Office contribution to the Flood Studies Report[40] gives rainfall intensities for any time duration up to one month for any point in the United Kingdom for probabilities ranging up to 1 in 10 000 years. Tables are also given relating maximum point rainfall to that appropriate to the catchment area during a widespread storm. Thus whereas a 10-minute rainfall on a 10 km^2 catchment may only be 83% of the worst point of precipitation within it, this would rise to 96% over a six-hour storm. The corresponding figures for a 1000 km^2 catchment in the temperate English climate are 32% and 83%. In the tropics there are indications that the very localized nature of frequent convective storms gives way to more consistently widespread rain in the rarest of storms.

Maximum rainfalls vary with the season of the year because thunderstorm intensities are associated with high sea and air temperatures. However, high rainfalls over a day or so may occur at any time of year

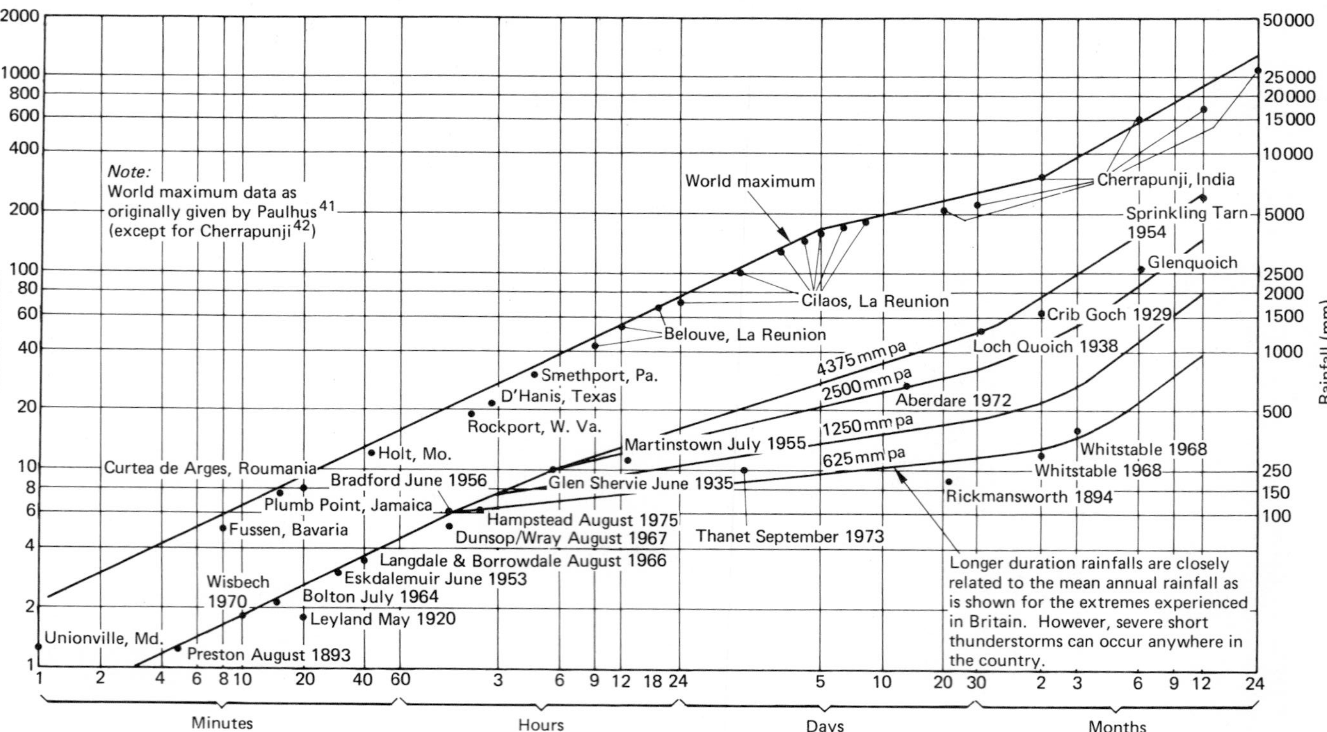

Fig. 3.5 Maximum recorded world and British rainfalls.

wherever the weather system can bring in moist air steadily and the conditions (often orographic) exist to cause its precipitation.

3.17 Maximum runoffs

Peak runoff rates are notoriously difficult to estimate because of the damage, debris, and sediment associated with major floods. However, most countries have endeavoured to relate maximum flow measurements to catchment area, and sometimes to catchment slope, geology, and climatic type as well[45]. Knowledge of the possibilities of nature is gradually increasing as Fig. 3.6 demonstrates, consequently it becomes necessary to introduce probability techniques to measure the degree of rarity of a flood.

In lowland areas peak runoff rates are normally much lower than Fig. 3.6 suggests because of the greater likelihood of drier soil conditions before a storm and the manifold possibilities for detention of runoff away from drainage channels. In flat areas flood levels and volumes can be of much greater importance because these may determine pumping station levels and freeboard provision for lowland reservoirs. Although in East Anglian fenland entire agricultural catchments may be drained quite adequately to the sea through large pumping stations capable of pumping no more than 13 mm of runoff on their catchment per day, more severe conditions can occur. After a wet summer in 1968 at Bough Beech Reservoir, Kent[46], a September storm giving over 150 mm of rain in 12 hours led to a runoff of about 135 mm at a time when the reservoir was beginning to fill for the first time. The reservoir level rose 7.6 m in 14 hours despite an outlet pipe being open. However, neither of these figures is large on the world scale. For instance in Hong Kong runoffs of over 300 mm have been experienced during typhoons. Empirical flood equations derived in one country or one area are rarely of use anywhere else until their coefficients have been calibrated with local data.

It has been found widely possible to form regional flood probability plots by first finding the mean annual flood for a catchment (Fig. 3.7) and then the variability about that mean (Fig. 3.8). The mean annual flood is the arithmetic average of a long series of maximum instantaneous flows, taking only the highest in each year. The Flood Studies Report treats Ireland as one region and divides Britain into seven areas when relating the mean annual flood to catchment characteristics. For all but the Thames–Lee–Essex area:

$$\text{mean annual flood } (\mathrm{m^3/s}) = aA^{0.94}\,STMFRQ^{0.27}\,S1085^{0.16}\,RSMD^{1.23}\,SOIL^{1.03}(1 + LAKE)^{-0.85}$$

where a is a regional factor (see Table 3.7), A is the catchment area (km^2), *STMFRQ* is the number of stream confluences/km^2 as shown on a 25 000 scale map, *S1085* is the main stream slope between two points 10% and 85% of the stream length from the site (m/km), *RSMD* is the

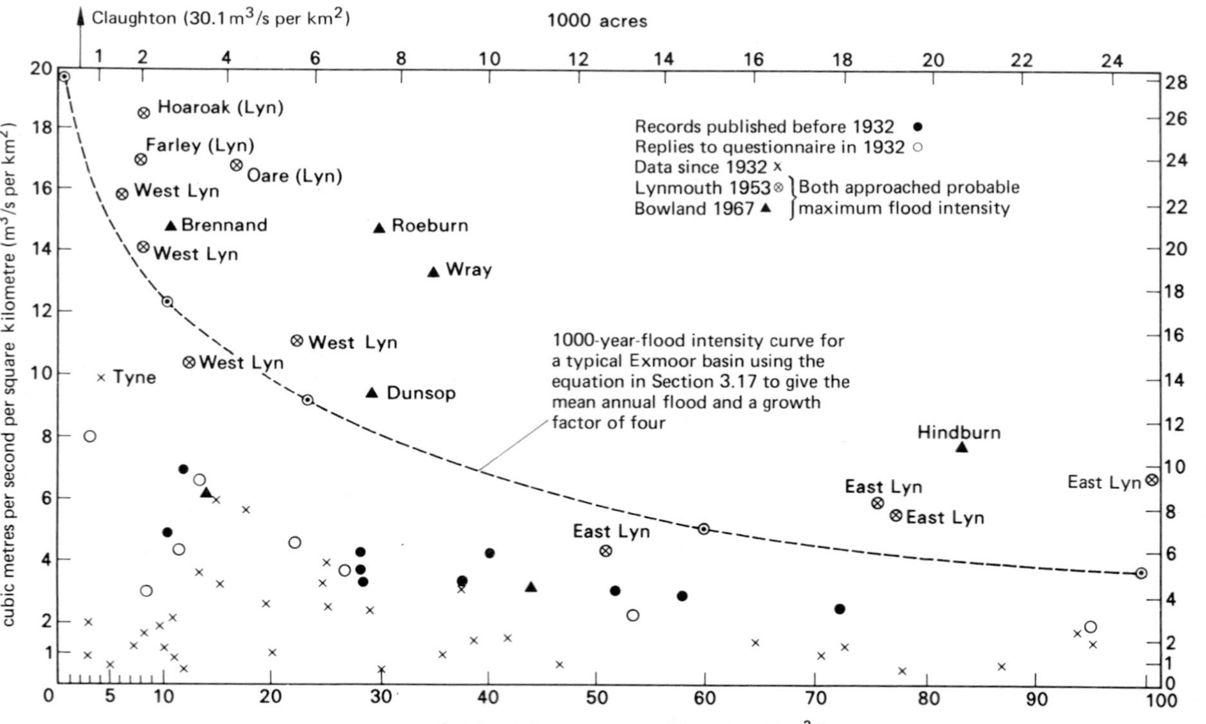

Fig. 3.6 Flood data in Britain in and since the ICE 1933 Report. (*Proc. ICE*, 1960.)

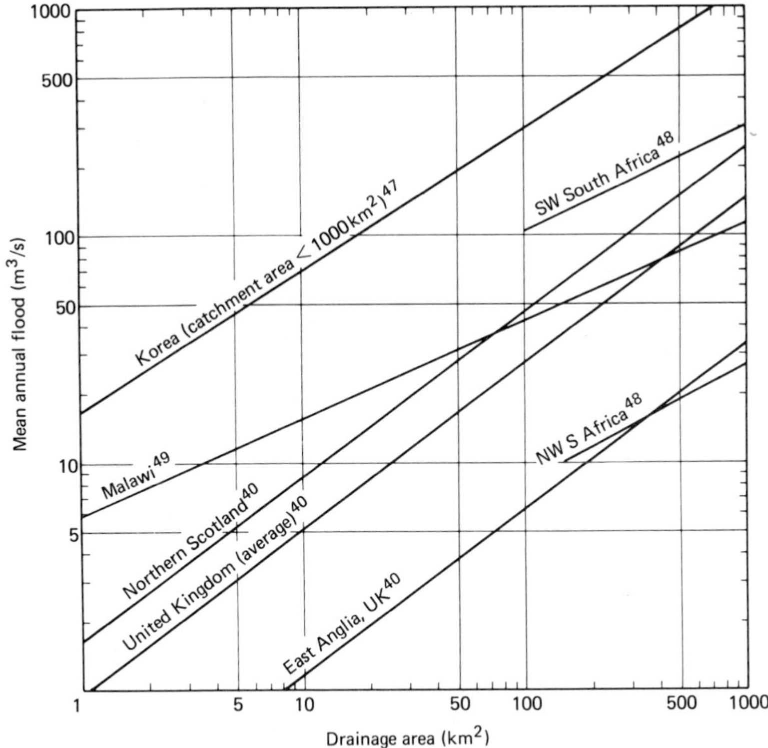

Fig. 3.7 Variation of mean annual flood with drainage area.

5-year return period one-day rainfall minus an expected effective soil moisture deficit, *SOIL* is the mapped index of infiltration capacity of catchment soils (range 0.15 to 0.50), and *LAKE* is the fraction of *A* that drains through major lakes.

Providing the regional characteristics can be checked with flood records over one or two years at the site concerned, then reliable results will ensue. Once the mean annual flood can be found, it can be multiplied by the factors given in Fig. 3.8 to find all rarer events.

3.18 Flood estimates for weir and intake design

It is not economic to gauge the complete range of flows at any but a few completely spring-fed catchments: the size of the structure required to contain the rarest flood would be so large as to be unrealistic in most valley settings. However, a small weir which is undersized and frequently bypassed by floods is of dubious value so that one must seek to design a sensibly sized structure.

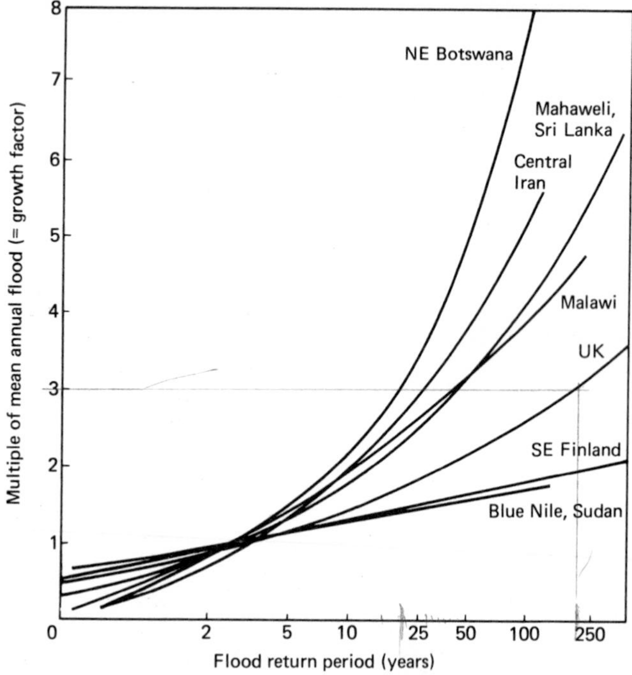

Fig. 3.8 Regional dimensionless flood frequency curves.

Table 3.7 Regional factors for mean annual floods.

Region	Factor '*a*'
North Scotland	0.0186
Central Britain	0.0213
East Anglia	0.0530
Southern England	0.0234
South West England	0.0315
Lowland Ireland	0.0172

Often the best approach is to ensure that the stilling well and recorder are capable of withstanding and recording the highest possible water level, whilst the structure itself is designed to contain only, say, the 10-year flood. To achieve this may well require a site where the normal flood plain is missing and the channel flow is kept between banks over a very wide range. In the event of a worse flood being recorded, every attempt should be made to gauge the flows at or near their peak, whether by float or current meter. If the stage–discharge curve cannot be extrapolated in that way, it will be necessary to resort to theoretical extension of the rating or to model testing. Rating extension can be done by drawing up separate graphs of river level versus mean velocity and cross-sectional area: when extrapolated independently and multiplied a realistic rating is formed.

Where gravity intakes are built, frequently combined with measurement weirs, similar principles apply. Such intakes need to be screened to prevent the entry of debris into an aqueduct, but such screens should only be placed parallel to flood flow lines. If they are set directly across a stream the build-up of debris has been known to so increase the upstream head that bypass flows occur: it may also lead to the collapse of the screens if the loading is unexpected. Catchments which are wooded, steep, or urban are particularly liable to cause this trouble.

In the case of pumped intakes the prime requirement is to prevent the entry of flood water to any part of the pump house containing sensitive equipment. This normally means setting the pump-house floor at least as high as the maximum known flood level at the site, almost regardless of the severity of that event: should the event recur no one will be thanked for saving a small proportion of the cost by keeping the floor lower. Where study shows that even higher flood levels are possible, the designer must exercise a nice judgement between the value of continuous operation of the intake and the cost of raising the pump house. Conditions vary widely, but designing for a 100-year flood with a small amount of freeboard would not be unreasonable. Alternatives exist: the intake can frequently be sited remote from the pump house, which may then be set in the valley side, away from flood influences and frequently with less detriment to amenities, or resort may be made to flood insurance to cover against the chance of a really extreme flood damaging the works.

Two techniques for obtaining probable flood levels can be summarized as follows.

(1) Rank all flood levels at the site in order of decreasing severity using the peak level for each complete flood. The highest level in n years is assumed* to be of that return period, the second level of $n/2$ years return period etc. Plot these values on level versus log return period paper and interpret with a straight line up to the bankfull level and a line of lower slope thereafter.

(2) Obtain a level–discharge rating for the site of the river intake and pump house, extrapolated as necessary by the Manning formula. Estimate the peak flood discharge of the required probability from published regional figures (Fig. 3.7) or, preferably, by direct analysis of local flood flows. Convert the resultant discharge to level using the rating.

* If such assumptions are to be refined, refer to p. 71 of Reference 40 for an appropriate, unbiassed plotting position (in the statistical sense).

Method (1) is of dubious value if too few levels over the flood plain are known and (2) requires data that are often unavailable. So reliance on local knowledge, with a liberal measure of freeboard above the worst remembered event, is likely to prevail in many designs. The growing availability of official flood plain planning maps for the 100-year flood in the USA, Australia, and the UK can be of considerable help. The dramatic change to levels by accretion, erosion, and canalization should not be overlooked.

3.19 Design flood hydrographs

In constructing a design flood hydrograph for sizing a reservoir spillway there are two main techniques, the first of which is the more usual:

(1) converting a design rainstorm, after losses, into flow by use of a unit hydrograph,
(2) choosing a historic flood hydrograph that can be scaled up to the peak flow (or volume) suggested by probability analysis.

It is not always immediately apparent which sort of rainstorm may most threaten to overtop a dam. A small lake inundating little of its catchment might suffer most in a short thunderstorm, whereas a major reservoir may cope with a large volume fall from a frontal or monsoonal system before its freeboard is swamped. Moreover, the combination of possible rain intensities in different seasons of the year with losses varying with a wide range of soil moisture states can make design complex. Unless clear information is available to the contrary it is usual to compound pessimistic assumptions in order to give a margin of safety.

Rainfall figures for storms of varying duration and given return periods (right up to the bounding possibility of probable maximum precipitation) are often available from the appropriate national meteorological agency. These must be formed into a realistic series of consecutive rainfall increments, perhaps ten in total, to make up a design storm profile. Possible profiles can be termed 'cloudburst', 'bellshaped', 'crescendo', and 'historic'. In the first the rainfall increments are in order of descending magnitude, whilst in the second they rise up to the midpoint of the storm and then fall again. The crescendo profile is the worst conceivable, with the rainfall increments getting gradually worse until the storm suddenly ceases. A historic profile means the general form of the severest known storm. The symmetrical bellshaped storm is favoured where it is desired to keep close to an equality of return periods for the storm and the resulting runoff.

Estimates of infiltration and surface detention losses are then made and deducted from these design rainfalls to find the effective rainfall. As losses may well range[50] from 2 to 25 mm/hour, local experience of the prevailing soil and rock strata is important. The effective rainfall increments are multiplied by the ordinates of a unit hydrograph and the resulting increments of runoff are summed to obtain the flood hydrograph. The arithmetic is best carried out by computer.

A unit hydrograph (UH) is intended to show the runoff response of a catchment to unit effective rainfall in unit time, e.g. 10 mm in one hour. It presupposes that rainfall over the catchment is uniform and that runoff will increase linearly with effective rainfall. Thus the runoff from 20 mm of effective rainfall in one hour is taken to be exactly twice that due to 10 mm, and so on. Despite the simplifications it is the best available method, especially on small uniform catchments: the only proviso is that the unit hydrograph should be derived from major floods rather than from small spates.

Derivation of a unit hydrograph[40] from concurrent rainfall and flow measurements is beyond the scope of this book. However, a method of obtaining an adequately severe UH synthetically is that suggested by the US Bureau of Reclamation[51]. This states, using the original units:

(1) Time of concentration

$$T_c = \left(\frac{11.9\,L^3}{H}\right)^{0.385} \text{ hours}$$

where L is the length of the longest tributary (miles) and H is the fall along the stream (feet).

(2) Time to peak for the hydrograph (1 inch of runoff over the catchment)

$$T_p = \tfrac{1}{2}(\text{rainfall duration}) + 0.6T_c \text{ hours}$$

Thus, for a 1 inch $\frac{1}{2}$ hour unit hydrograph

$$T_p = 0.25 + 0.6T_c$$

(3) Peak runoff rate $= \dfrac{484\,AQ}{T_p}$ cusecs

where A is the catchment area (square miles) and Q is the unit rainfall (e.g. 1 inch).

(4) The time base of the hydrograph is effectively five times T_p. Table 3.8 gives the coordinates along the unit hydrograph: each point is given as a multiple of T_p and a proportion of the peak runoff.

However, it must be stressed that other than very rare floods will be overestimated by these equations in temperate, flat, and permeable regions: in such areas a simple triangular UH with a base of 2.5 times the time to its peak should suffice. Its peak flow will equal $2.2/T_p\,\text{m}^3/\text{s}$ for each km^2 of catchment for every 10 mm of effective rainfall. T_p often lies in the range from $1.3(\text{area})^{1/4}$ to $2.2(\text{area})^{1/4}$ hours where the area is in km^2.

Table 3.8 Synthetic unit hydrograph coordinates.

Time ratio (T/T_p)	Discharge ratio (q/q_p)	Time ratio (T/T_p)	Discharge ratio (q/q_p)	Time ratio (T/T_p)	Discharge ratio (q/q_p)
0	0	1.0	1.00	2.4	0.18
0.1	0.015	1.1	0.98	2.6	0.13
0.2	0.075	1.2	0.92	2.8	0.098
0.3	0.16	1.3	0.84	3.0	0.075
0.4	0.28	1.4	0.75	3.5	0.036
0.5	0.43	1.5	0.66	4.0	0.018
0.6	0.60	1.6	0.56	4.5	0.009
0.7	0.77	1.8	0.42	5.0	0.004
0.8	0.89	2.0	0.32	∞	0
0.9	0.97	2.2	0.24		

3.20 Reservoir flood control

The storage of water above natural ground level is a potential hazard to life and property downstream in times of flood. Every impounding reservoir, or artificially raised lake, therefore needs proper provision for the passage of floods. This is made mandatory by law in many countries. There are two objectives: to protect the dam from costly damage or failure and to ensure that downstream interests are at no greater risk from natural (or artificial) floods than before the dam was built.

The best modern practice[36] is to ensure that any new dam can pass:

(1) the probable maximum flood without overtopping where failure would lead to loss of life in a community in the flood zone below the dam, or
(2) in situations of lower risk, a design flood of appropriate specified severity.

Table 3.9 sets out the guideline standards for the UK which, with local modification, can provide a framework for design choice in other countries. Particularly important is the need to set concurrently the initial reservoir level prior to the design flood and the accompanying wind/wave run-up margin that completes the determination of freeboard.

Probable maximum precipitation (PMP) estimates (Table 3.10) should be sought from authoritative sources and converted, with pessimistic assumptions about catchment response, into a probable maximum flood estimate. Such a flood is the upper bound to all possible floods which, although lying well beyond most historical experiences in temperate areas, may nevertheless be approached relatively frequently in those countries that suffer effective rain producing systems, such as monsoons and typhoons.

Where, as in many cases in the past, a spillway has been designed to

Table 3.9 Reservoir flood and wave standards by dam category[36].

Category	Initial reservoir condition	Dam design flood inflow		Concurrent wind speed and minimum wave surcharge allowance
		General standard	Minimum standard if rare overtopping is tolerable	
A Reservoirs where a breach will endanger lives in a community	Spilling long-term average daily inflow	PMF	0.5 PMF or 10 000-year flood (take larger)	Winter: maximum hourly wind once in 10 years Summer: average annual maximum hourly wind Wave surcharge allowance not less than 0.6 m (applies to A and B)
B Reservoirs where a breach (1) may endanger lives not in a community (2) will result in extensive damage	Just full, i.e. no spill	0.5 PMF or 10 000-year flood (take larger)	0.3 PMF or 1000-year flood (take larger)	
C Reservoirs where a breach will pose negligible risk to life and cause limited damage	Just full, i.e. no spill	0.3 PMF or 1000-year flood (take larger)	0.2 PMF or 150-year flood (take larger)	Average annual maximum hourly wind Wave surcharge allowance not less than 0.4 m
D Special cases where no loss of life can be foreseen as a result of a breach and very limited additional flood damage will be caused	Spilling long-term average daily inflow	0.2 PMF or 150-year flood (take larger)	Not applicable	Average annual maximum hourly wind Wave surcharge allowance not less than 0.3 m

Notes:
Where reservoir control procedure requires, and discharge capacities permit, operation at or below specified levels defined throughout the year, these may be adopted providing they are specified in the certificates or reports for the dam. Where a proportion of PMF is specified it is intended that the PMF hydrograph should be computed and then all ordinates be multiplied by 0.5, 0.3, or 0.2 as indicated.

Table 3.10 Probable maximum precipitation (PMP) estimates.

Location	Country	PMP (mm)				
		20 min	1 hour	6 hours	15 hours	24 hours
South Ontario	Canada	—	—	410	—	445
Guma	Sierra Leone	81	183	—	580	630
Batang Padang	Malaysia	76	157	—	292	297
Shek Pik	Hong Kong	101	220	—	915	1200
Garinono	Sabah	81	162	—	620	675
Brenig	Wales					
	May to September	74	109	183	—	254
	October to April	38	72	165	—	272
Tigris (50 200 km^2)	Iraq	—	—	60	—	167
Jhelum, Mangla	West Pakistan					
(2500 km^2	December to May	—	—	185	—	295
sub-area)	January to November	—	—	365	—	575

Note: Except where area is stated, all values refer to small catchments.

pass the worst flood intensity suggested by regional records, a slim risk of failure must have been tacitly accepted although perhaps not explicitly stated. Fortunately, however, there are many cases where reservoirs are drawn down at the onset of a flood and an additional factor of safety therefore exists.

3.21 Spillway flood routing

Once the area of a catchment that is inundated exceeds 2%, a distinct reduction of the flood peak inflow can be expected because of the storage effect. For instance, with barely any overflow from a small spillway, 2.5 m freeboard in such a situation would be sufficient to contain a 50 mm catchment runoff. Flood reduction by storage is termed 'routing' and calculation of its effect is nowadays accomplished by computer. However, to examine the mechanics more closely it is worth detailing a convenient manual method, known in the USA as the 'Modified Puls' method[52].

It is necessary to determine beforehand:

(1) the initial reservoir level and spillway overflow, if any, at the onset of the design flood,
(2) the elevation–capacity curve for the storage,
(3) the spillway head–discharge characteristic,
(4) any major drawoff from or transfer into the lake that would be occurring during the flood.

Information on (2) and (3) should be carried far enough to cover a moderate amount of overtopping where this is conceivable.

Reservoir routing then involves solving the equation:

inflow (I) = outflow (O) + change in storage (S)

for successive units of time during a flood, i.e.

$$\frac{(I_2 + I_1)}{2} = \frac{(O_2 + O_1)}{2} + (S_2 - S_1) \qquad (1)$$

where I_1 is the inflow rate at the start of a time unit and I_2 is the inflow rate at the end of the same time unit and the start of the next one.

This equation is solved by rewriting it as:

$$\frac{(I_1 + I_2)}{2} + \left(S_1 - \frac{O_1}{2}\right) = S_2 + \frac{O_2}{2} \qquad (2)$$

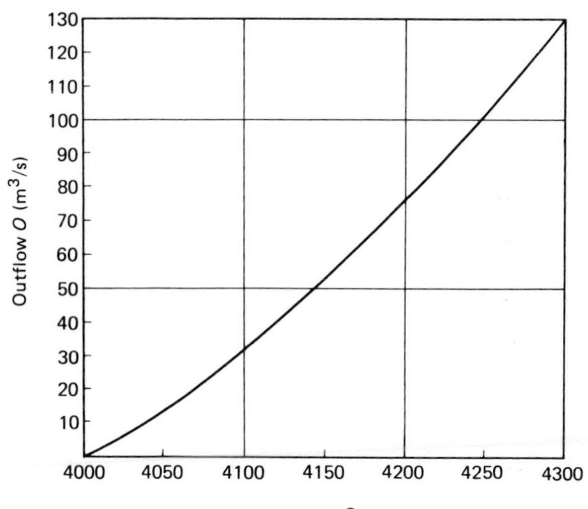

Fig. 3.9 Flood routing storage indication curve.

A storage indication curve (as Fig. 3.9) is then drawn up which shows outflow against $(S + O/2)$ in appropriate units. Table 3.11 sets out an example of a routing calculation. The steps required are as follows.

(1) Compute a numerical value for the left hand side of equation (2) for given values of I_1, I_2, S_1, and O_1 for the first time increment.
(2) This value equals $(S_2 + O_2/2)$ and therefore the storage indication curve can be entered and the corresponding outflow O_2 can be read off.
(3) The left hand side of equation (2) can be computed most rapidly for the next time increment by deducting this value of O_2 from the

Table 3.11 Flood routing calculations.

Period (hours)	Average inflow, $\frac{1}{2}(I_1 + I_2)$ (cumecs)	Elevation, spillway crest— 200.60 (metres)	Storage indication, $S + O/2$ (cumecs $\times \frac{1}{2}$ hour)	Instantaneous outflow (cumecs)
0– ½	1.9	200.74	4012.0	1.9
½– 1	1.9	200.74	4012.0	1.9
1 – 1½	6.0	200.76	4016.1	2.8
1½– 2	18.1	200.91	4031.4	6.7
2 – 2½	40.0	201.19	4064.7	18.2
2½– 3	69.2	201.55	4115.7	36.5
3 – 3½	99.2	201.99	4178.4	65.1
3½– 4	117.6	202.32	4230.9	90.7
4 – 4½	130.6	202.59	4270.8	112.5
4½– 5	125.5	202.67	4283.8	119.0
5 – 5½	114.7	202.64	4279.5	116.8
5½– 6	101.1	202.55	4263.8	108.5
6 – 6½	85.2	202.39	4240.5	96.0
6½– 7	68.2	202.27	4212.7	81.3
7 – 7½	54.9	202.06	4186.3	68.6
7½– 8	46.6	201.90	4164.3	58.5
8 – 8½	40.6	201.77	4146.4	50.1
8½– 9	36.0	201.67	4132.3	44.0
9 – 9½	32.4	201.59	4120.7	38.8
9½–10	29.1	201.52	4111.0	34.7
10 –10½	26.2	201.46	4102.5	31.5
10½–11	23.6	201.41	4094.6	28.8
11 –11½	21.5	201.35	4087.3	26.2
11½–12	19.6	201.31	4080.7	23.9

Notes:
Total inflow = 1309.7 cumecs × ½ hour.
Gain of storage = 57 cumecs × ½ hour.
Total outflow = 1263.0 cumecs × ½ hour.

corresponding storage indication value. The resulting figure $(S_2 - O_2/2)$ becomes $(S_1 - O_1/2)$ for the following increment. The procedure is then repeated until the peak of the outflow hydrograph is passed (Fig. 3.10).
(4) A useful final check is to ensure by adding the columns that the total inflow is sufficiently close to the total outflow plus gain in storage.

3.22 Reservoir safety reviews

Where reservoirs must be regularly inspected by a competent and approved engineer, reviews of spillway adequacy will be normal. Experience shows that the spillway condition at an old dam can change markedly over a period of, say, ten years. There may be additional storm or flood data to take into account, manifestly different freeboard conditions due to crest settlement, and better techniques emerging for

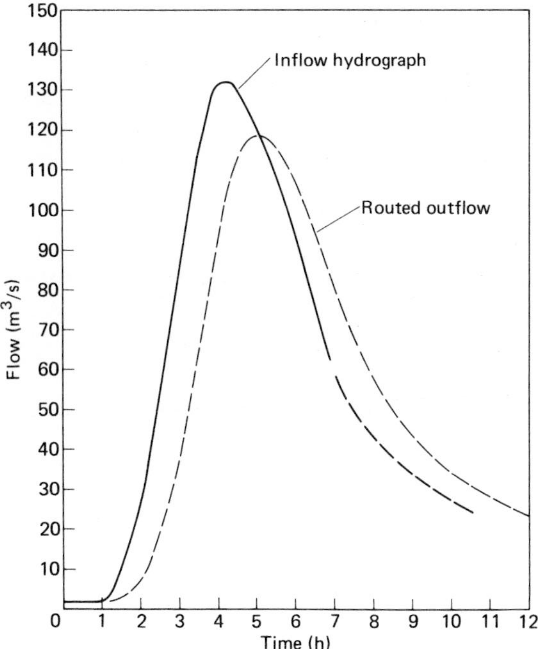

Fig. 3.10 Routed flood hydrograph.

flood synthesis and routing. Old dams are often found to have inadequate ability to pass the expected flood over the spillweir or safely past the toe of the dam because of over-simplification of complex spillway hydraulics.

The questions listed below are an attempt to highlight those criteria or data which must pass under review. Not all the topics will be involved in each spillway design or inspection. The importance of a thorough check is reinforced by a growing awareness of the scale of dam failure floods[36]. For dams storing substantial volumes the flood failure peak, Q_F, can be approximated very roughly by:

$$Q_F\ (m^3/s) = 1.3(\text{dam height})^{2.5}$$

where the dam height is in metres.

Mathematical assessment of dambreak flood waves is now possible[53], although the pace of failure of a dam must always remains conjectural.

(1) *The catchment*
(a) What is the soil type and underlying geology?
(b) What is the topographic catchment area?
(c) What are the main stream length and fall to the reservoir inlet?

(d) What is the average annual rainfall?
(e) What is the land use in broad terms (% forest, % urban)?
(f) What local flood records exist?
(g) Is there a natural drainage system in the valley or is it 'dry'?
(h) Is there a valley obstruction, e.g. embankment with small culvert, which could give way or conversely assist routing in a major flood?
(i) What are the implications of the design flood within the catchment even before it reaches reservoir storage, e.g. floating debris?
(j) What method is appropriate for obtaining the design flood(s) for the catchment, and what are the dominant factors in the estimation?

(2) *The reservoir and dam*
(a) What is the water area at spillway crest level?
(b) What is the normal retention level if different from that used for 2(a)? Can it be guaranteed with acceptably high probability at the commencement of the design storm?
(c) Is an elevation–capacity table available up to flood levels?
(d) What artificial inflow could occur during times of flood, e.g. from a catchwater or pumps?
(e) What significant drawoff or scour release could be taking place?
(f) Is the reservoir site manned?
(g) Do statutory reservoir level records exist as a complete series? What do they indicate about past flood probability?
(h) What freeboard to dam crest exists (minimum and maximum)?
(i) What is the wave wall height, if any? Could it bear a sustained flood level against it? Is it continuous?
(j) Are there any wave or wind records for the lake?
(k) If a bigger flood is to be passed, should the dam be raised, the spillway widened, or the spillweir lowered?

(3) *The spillway*
(a) What is the spillway rating? Is it accurate?
(b) Does it cover constrictions in the channel below the spillweir?
(c) Is the spillweir in the condition the rating implies?
(d) Can the spillweir block with floating debris?
(e) What was the spillway originally designed to pass?
(f) Has its capacity ever been seriously tested?
(g) If the spillway rating is partly dependent upon gates how would it be changed in the event of their remaining closed?

(4) *The channel downstream*
(a) Is there a desirable limit to the flood peak to be passed downstream?
(b) Has the stream, or a similar local river, ever experienced a catastrophic flood?
(c) Can flood damage to the spillway outlet channel be accepted?

(5) *The consequence of overtopping and failure*
(a) If a flood overtops the dam, is this acceptable *vis-à-vis* its stability, and what would the discharge rating then be?
(b) Is there a limit to permissible overtopping to avoid failure of the dam, e.g. 10 cm for 15 minutes?
(c) Is there risk to life if the dam fails and empties completely over a period of a few hours?
(d) What value of property might be lost if the dam fails?
(e) Does the reservoir storage represent more than 10% of the flood volume or peak if it fails? (If not, as with many park lakes, it will generally be unnecessary to design for a very rare flood.)
(f) Would the loss of water supply, power, or irrigation by failure be such as to require maximum dam protection?
(g) Would failure cause overtopping of a downstream dam?

3.23 Dam diversion floods

Impounding dams require the diversion of the river at the site at an early stage in their construction, whether this is by tunnel, culvert, or flume. In sizing diversion works which have only temporary value, higher risks are taken than is usual on flood control works. Typically a 1 in 10 chance of flooding the foundations works might be accepted, subject to a review of the financial implications of a contract set-back, but it is important to note also that the longer the construction period during which the incomplete dam might be overtopped, the larger the design period must be. Flood estimates for different probabilities can be estimated by the regional technique covered in Section 3.17. It is a good precaution to confirm these estimates by measuring dam site flood levels at the earliest possible opportunity so that at least the mean annual flood can be checked at an early stage.

A diversion requires a lead-in channel with a stank or coffer dam to protect the main dam. This can often be raised quite economically in order to surcharge the diversion in a major flood, and will also give a modest amount of flood routing. As soon as the main dam exceeds the coffer dam height, then additional flood water may usually be coped with and the risks generally lessen unless the catchment is a very large one. Flood routing studies should be carried out to discover the potential for rapid and critical rises in the impounded level either just before or just after placing the 'plug' in the diversion route. Schemes seem unerringly to attract a major flood at about this time: Llyn Celyn in North Wales, while first filling, rose 5 m in three days due to 183 mm runoff; the Kariba Dam on the Zimbabwe–Zambia border experienced a succession of unexpectedly large floods during construction and filling—several reviews of the catchment flood potential resulted[54].

References

(1) *Guide to Hydro-meteorological Practices*, 2nd Ed., WMO, Geneva, 1970.

(2) United States Geological Survey, *Techniques of Water Resources Investigations of the USGS*, Book 3, Chapters A6–A8 on stream gauging procedure, US Government Printing Office, Washington, 1968.

(3) BS 3860: *Methods of Measurement of Liquid Flow in Open Channels*, Part 4A: *Thin Plate Weirs*, BSI, 1981.

(4) *Crump Weir Design,* Technical Note No. 8, WRB, Reading, 1967.

(5) Ackers P, White W R, Perkins J and Harrison A J M, *Weirs and Flumes for Flow Measurement*, John Wiley & Sons, 1978.

(6) Charlton F G, *A Review of Methods of Measuring Flow in Open Channels*, Report No. 75, Construction Industry Research and Information Association, London, 1978.

(7) *Measurement and Estimation of Evaporation and Evapotranspiration*, Technical Note No. 83, WMO, Geneva, 1966.

(8) Doorenbos J and Pruitt W O, *Crop Water Requirements*, Irrigation and Drainage Paper 24, FAO, 1975.

(9) Lapworth C F, Evaporation from a Reservoir near London, *JIWE*, **19,** No. 2, 1965, pp. 163–181.

(10) Rodda J, Downing R A and Law F M, *Systematic Hydrology*, Newnes-Butterworth, 1976, pp. 88–89.

(11) Penman H L, *Vegetation and Hydrology*, Technical Communication No. 53, Commonwealth Bureau of Soils, Harpenden, 1963.

(12) McCulloch J S G, Tables for the Rapid Computation of the Penman Estimate of Evaporation, *East African Agricultural and Forestry Journal*, **XXX,** No. 3, 1965.

(13) Gardner C M U, *The Soil Moisture Databank*, Institute of Hydrology Report No. 76, 1981.

(14) Thackray J E and Archibald C C, The Severn–Trent Studies of Industrial Water Use, *Proc. ICE*, August 1981, p. 428.

(15) Ege H D, Management of Water Quality in Evaporation Systems and Residual Blowdown, in: *Water Management by the Electric Power Industry*, University of Texas at Austin, Texas, 1975.

(16) Smith C C (Chairman), Report of the Joint Committee to Consider Methods of Determining the General Rainfall over any Area, *Trans. IWE*, **XLII,** 1937, p. 231.

(17) Reynolds G, Rainfall, Runoff and Evaporation on a Catchment in West Scotland, *Weather*, **24,** No. 3, 1969, pp. 90–98.

(18) Water Data Unit, DoE, *Surface Water: United Kingdom 1971–73*, HMSO.

(19) Enex/Drainage and Irrigation Department, *Average Annual Surface Water Resources of Peninsular Malaysia*, Water Resources Publication No. 6, 1976.

(20) Ledger D C, The Water Balance of an Exceptionally Wet Catchment Area in West Africa, *J Hydrology*, **24,** 1975, pp. 207–214.

(21) Penman H L, The Dependence of Transpiration on Weather and Soil Conditions, *J Soil Science*, **1,** 1949, pp. 74–89.

(22) Rutter A J, Studies in Water Relation of *Pinus Sylvestris* in Plantation Conditions: Part IV—Direct Observations and Evaporation of Intercepted

Water and Evaporation from the Soil Surface, *J Applied Ecology*, **3,** 1966, p. 393.

(23) Holford I, *The Guinness Book of Weather Facts and Feats*, Guinness Superlatives Ltd, 1977.

(24) Glasspoole J, The Reliability of Rainfall over the British Isles, *Trans. IWE*, **35,** 1930, pp. 174–199.

(25) Law F, Estimation of the Yield of Reservoired Catchments, *JIWE*, **9,** No. 6, 1955, p. 476.

(26) Tabony R C, *The Variability of Long Duration Rainfall over Great Britain*, Scientific Paper 37, Meteorological Office, 1977.

(27) Enex/Drainage and Irrigation Department, *Magnitude and Frequency of Low Flows in Peninsula Malaysia*, Hydrological Procedure No. 12, Malaysia, 1976.

(28) Midgley D C and Scheurenberg R J, Aid to the Appraisal of Water Resources: Sequences of Deficient River Flow in Some Regions in South Africa, *The Civil Engineer*, South Africa, May 1967.

(29) Darmer K I, *A Proposed Streamflow Data Program for New York*, USGS Water Resources Division Open File Reports, Albany, New York, 1970.

(30) *Low Flow Studies Report*, Institute of Hydrology, Wallingford, 1980.

(31) United States Geological Survey, *Techniques of Water Resources Investigations of the USGS*, Book 3, Chapters C1–C3 on fluvial sediments, US Government Printing Office, Washington, 1970.

(32) Fleming G, Design Curves for Suspended Load Estimation, *Proc. ICE*, **43,** 1965, pp. 1–9.

(33) Hydrotechnic Corporation, Report on *Resources en Eau de Surface et Possibilités de leur Amenagement*, Region d'Algerie Orientale, April 1970.

(34) Reference 10, Table 6.13.

(35) Technical Committee on Surface Water, *Sediment Sampling in Australia*, Hydrology Series No. 3, Australian Water Resources Council, 1969.

(36) *Floods and Reservoir Safety: An Engineering Guide*, ICE, 1978.

(37) Creager W P, Justine J D and Hinds J, *Engineering for Dams*, John Wiley & Sons, 1945.

(38) Enex/Drainage and Irrigation Department, *Magnitude and Frequency of Floods in Peninsula Malaysia*, Hydrological Procedure No. 4, Malaysia, 1974.

(39) *Manual for Estimation of Probable Maximum Precipitation*, Operational Hydrology Report No. 1, WMO, Geneva, 1973.

(40) *Flood Studies Report*, Five volumes, NERC, 1975 (available through the Institute of Hydrology).

(41) Paulhus J L H, Indian Ocean and Taiwan Rainfalls set New Records, *Monthly Water Review*, **93,** No. 5, 1965, pp. 331–335.

(42) Dhar O N and Farooqui S M T, A Study of Rainfalls Recorded at Cherrapunji Observatory, *Int. Assoc. Hydrol. Sciences*, **XVIIII,** No. 4, 1973, pp. 441–450.

(43) Tadjudin H and Kaul F J, *Maximum Recorded Floods in Indonesia*, Guideline BP 5B, Directorate General of Water Resources Development, Indonesia, 1983.

(44) Bell G J and Chin P C, *The Probable Maximum Rainfall in Hong Kong*, Royal Observatory, Hong Kong. See Table 7.3 (as revised 1980).

(45) *Flood Design Manual for Java and Sumatra*, Direktorate Penyelidikan Masalah Air, Indonesia, 1983 (available in the UK through the Institute of Hydrology).
(46) Hallas P S and Titford A R, The Design and Construction of Bough Beech Reservoir, *JIWE*, **25,** 1971, p. 297.
(47) Binnie & Partners/Hyundai Engineering Co., *Hydrologic Services, Rural Infrastructure Project Report*, Volume 4, Appendix L, *Flood Frequencies* (by the UK Institute of Hydrology), Ministry of Construction, 1978.
(48) *Design Flood Determination in South Africa*, Report 1/72, Hydrological Research Unit, University of Witwatersand, 1972.
(49) Drayton R S *et al., A Regional Analysis of River Floods and Low Flows in Malawi*, Report No. 72, Institute of Hydrology, Wallingford, 1980.
(50) *Australian Rainfall and Runoff: Flood Analysis and Design*, 2nd Ed., Institution of Engineers, Australia, 1977.
(51) US Bureau of Reclamation, *Design of Small Dams*, US Government Printing Office, Washington, 1977.
(52) *US Bureau of Reclamation Manual*, Volume IV—*Water Studies*, Part VI—*Flood Hydrology*, US Government Printing Office, Washington, 1951.
(53) Chen C L and Armstrong J T, Dambreak Wave Model: Formulation and Verification, *JASCE*, **106,** HY5, 1980, p. 147.
(54) Reeve W T N and Edmunds D Y, Zambesi River Flood Hydrology and its Effect on Design and Operation of Kariba Dam, *Proc. River Flood Hydrology Symposium*, ICE, 1965, p. 169.

4

Estimation of yield

Part I
Yield estimation for surface sources

4.1 Definition of yield

The term 'yield' has an imprecise meaning and the following terms will therefore be used:

(1) 'yield from experience'—the abstraction taken from a source over a number of years,
(2) 'historic yield'—the steady supply that could just be maintained through a repetition of the worst drought on record,
(3) 'probability yield'—the steady supply that could just be maintained through a drought of specified severity,
(4) 'operating yield'—the supply that could be given under a fixed set of operational rules,
(5) 'failure yield'—the steady supply that could be maintained for a given percentage of time (or would fail to be achieved for a given percentage of time due to reservoir emptiness etc.).

Gross yield is the total available from a source. *Net* yield is the water remaining for supply after any compensation water or residual flow has been left for other riparian interests. Sometimes deductions also have to be made for losses, notably for reservoir surface evaporation.

Current waterworks practice in the United Kingdom is to adopt definition (3) for planning purposes, but to evaluate (4) for emergency drought operation. Major sources, whose failure would seriously disrupt industry or which are irreplaceable within any undertaking, are generally assigned yields for a once-in-100-years drought (1% yield). Minor sources may well be run at 3% or 4% yields. The term 'safe' or 'reliable' yield has no precise definition, but it is commonly adopted for the yield maintainable through a 2% (once-in-50-years) drought which is a frequently adopted standard.

Standards vary widely, but they generally rise with the economic development of a country. Developing nations may improve intermittent supplies until they are reliable through a 20- or 30-year drought, but the advent of industrialization normally leads to the need for even greater reliability, e.g. Malaysia and Singapore have accepted the need to provide city water supplies that are adequate through a 50-year drought. The American approach was to adopt historic yields, but the unusual severity of recent recorded droughts has led to a shift towards taking a probability yield instead.

Choice of risk must depend upon:

(1) the social and economic consequences of supply failures of different magnitudes,
(2) the extent to which demand might be lowered (voluntarily or by legislation) during a drought crisis,
(3) the available reserves elsewhere in the undertaking or the possibilities of temporarily reducing residual river flows,
(4) the uncertainty about population and unit consumption growth.

No source can be said to have a fixed yield because catchment conditions and consumer requirements change with time. (Examples are given in Section 4.13.) It may be necessary to compute, not simply the yield of an individual source, but also the amount (frequently different) by which a source raises the yield of the water supply system of which it becomes a part. In fact it is essential that a water engineer should be able to appraise the 'system yield' of an undertaking prior to planning any new development (see Section 4.11).

4.2 Types of surface source

Once the natural streamflows available at an intake are inadequate, a storage scheme of some sort is necessary. This chapter covers the estimation of yield for the basic alternatives shown in Fig. 4.1.

Direct supply storage (whether artificial or natural) is characterized by the impounding of gravity inflow and the piping of the majority of the stored water directly into supply. The yield may be augmented by a 'catchwater', which diverts water from an adjacent area directly by tunnel or round the contour by canal or pipeline. Lake Vyrnwy in Wales is a classic Victorian example of such a scheme. Plover Cove and High Island storages in Hong Kong are similar hydrologically, but are formed by damming off the sea.

Pumped storage reservoirs differ from direct supply reservoirs by the fact that the major part of their inflow is obtained by pumping. The reservoirs supplying London are of this type, being filled from the Thames and the Lee. A pumped storage reservoir may be formed by damming a side valley to the main stream or by raising embankments to enclose a flat area in a river valley. A more novel form of such storage is to create bunded storage areas in sea estuaries, as has been proposed for the Wash and Dee estuaries in England. These would be filled by

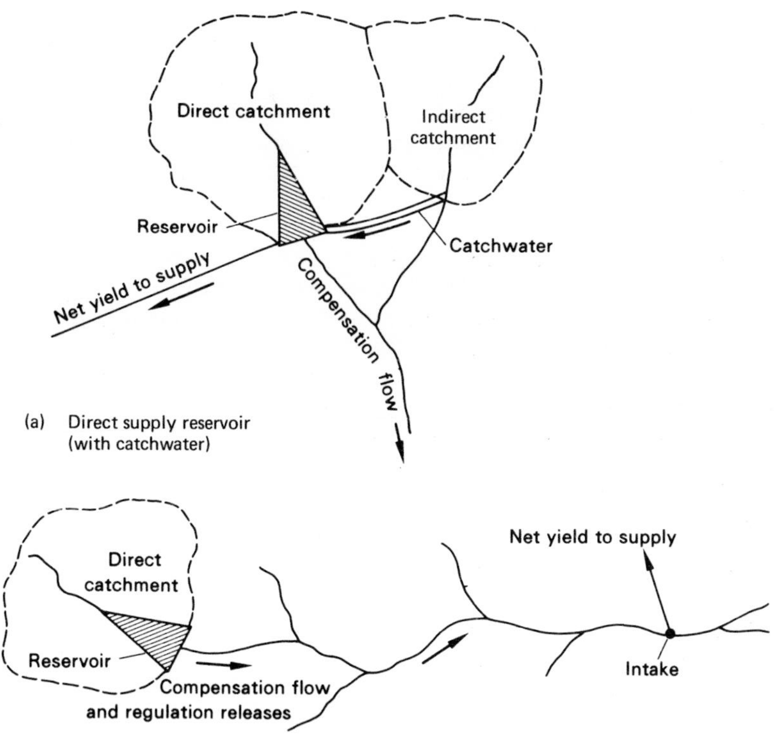

(b) Regulating reservoir

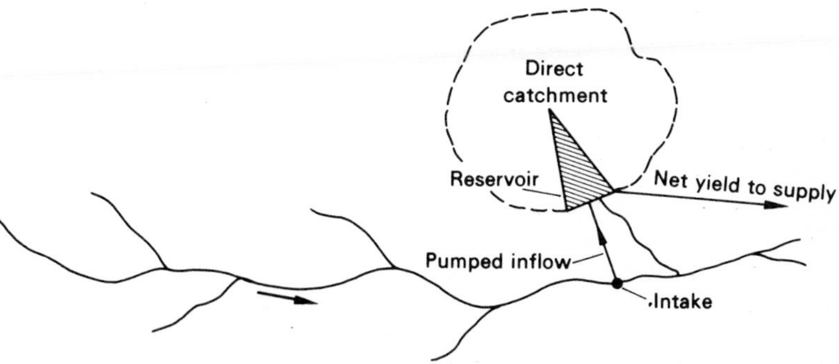

(c) Pumped storage reservoir

Fig. 4.1 Types of reservoir.

pumping from just upstream of tidal sluice gates on rivers flowing into the estuary. A pilot storage was successfully completed on the Wash estuary[1], but the full scheme has not yet been proceeded with owing to the lessening of water demand in the early 1970s.

Regulating reservoirs primarily impound water for its later release to a river when flows at some downstream abstraction point would otherwise become too low. Storage used in this way usually produces a higher yield than would a direct supply reservoir because the runoff from the catchment below the reservoir assists in maintaining flow at the abstraction point. If the demand centre is downriver there can be a large saving in aqueduct costs because of the shorter length required. The major Bhatsai dam project for the supply of water to Bombay is a recent example.

4.3 River intake yields

Supplies which amount to only a small proportion of the average flow of a river, say less than 5% ADF*, can often be abstracted straight from a river intake. However, raw water storage may still be required as a safety precaution against accidental pollution of the river upstream. Toxic and taste-producing pollutants can swiftly contaminate a river from factory plant and chemical transport road accidents, so that monitoring of storage may be essential[2].

To calculate yield it is rarely satisfactory to 'create' a flow record from rainfall data, although HYSIM[3], Sacramento[4], and Stanford[5] models are steadily improving the art. In any case all these methods involve calibration, so there is no substitute for a long, reliable series of runoff measurements at the location concerned. Self-evidently the historic yield for a direct abstraction is the minimum flow on record at the point of abstraction, but corrections may have to be made for irreversible catchment developments that have occurred since then. The net yield is simply the amount remaining after the deduction of any prescribed residual flow from this gross figure (see Section 4.10). The main adjustments for catchment development tend to be:

(1) deductions for increasing irrigation consumptions,
(2) deductions for growth in the 'export' of water out of the catchment by pipe or canal etc.,
(3) additions for increased effluent returns, providing these originate from water 'imported' from outside the catchment or from the use of storage.

In assessing the above a water resources survey (see Section 3.1) is invaluable. However, the timing of the events considered must be taken

* ADF = average daily flow, measured over a period of years.

into account, e.g. if the extra irrigation ends in July and the lowest flows are in September, the former is not significant.

To estimate an intake yield on the basis of risk of failure the following procedure is recommended.

(1) Locate a flow record of adequate length (preferably twenty years or more) for the river concerned, or for one hydrologically similar.
(2) Adjust this record back to natural conditions, where necessary by adding back 'exported' water and deducting 'imported' water etc.
(3) Find the lowest weekly flow in each year and list these in order of size, i.e. 'rank' them, Rank 1 being the smallest, and so on.
(4) Plot flow, Q, against percentage ranking on an appropriate probability paper which will give straight-line interpretation (for an example of this see Fig. 4.2) where

$$\% \text{ ranking} = \frac{(m-0.44)}{(n+0.12)} \times 100\%$$

n is the total number of years of record and m is the rank number for a given value of Q.

This method of plotting is preferred because it is unbiassed in a strict statistical sense. When drawing a line through the plotted points the engineer should try to infer the future population of annual low flow events rather than obtain a perfect fit to the plotted points. Convincing extrapolation of the line requires that it should be drawn far enough to give a reasonable explanation of any 'outlying' event, but care must be taken not to choose a line which infers that some relatively recent event must be one of extreme rarity, i.e. a 'once-in-1000-years' happening, since later investigations may disclose that something similar has been experienced 'within living memory' although records of it are missing. Also different types of probability paper may have to be tried out (see Section 4.5).

It is normally best to take weekly flow rather than daily flow, since the latter can be affected by coincidental interferences with flow, such as the operation of mill ponds, sudden coincidental changes of demand or effluent return etc. Also a guide can be obtained by plotting the typical recession of flow in dry weather: this will indicate the likely range of flow in the driest week.
(5) From the chosen flow probability line read off the flows for varying risks and transpose these to the intake site being studied, either by correlation or in the ratio of the estimated mean flows of the sites considered.
(6) The deduced figures for the intake site under (5) are then plotted on a gross-yield–return-period graph (Fig. 4.3), together with such adjustments as are necessary to give the net yield to supply. Where water

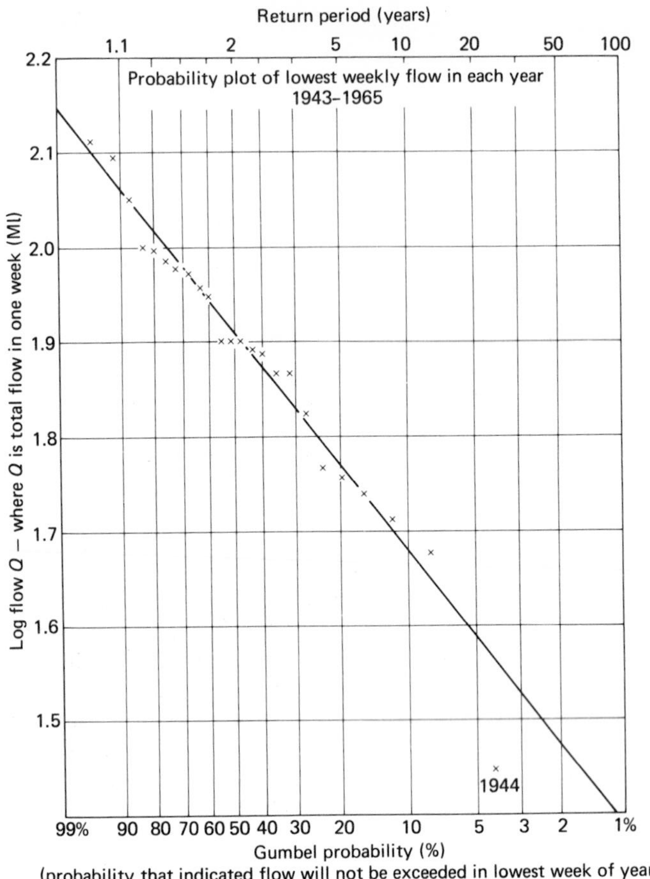

Fig. 4.2 Dry weather flow probability plot.

supply demands regularly peak at a time of year when flows are not at their lowest, a similar probability study is required for this peak demand period.

4.4 Storage estimation

For yield calculations it is important to estimate storage volumes accurately. Aerial surveys can be used, being relatively cheap and sufficiently accurate provided that proper ground control is used. Instrument surveys should provide at least four accurately surveyed contours through the site, as a smooth area–level curve cannot otherwise be drawn. Capacities are computed by planimetering areas at each contour level, plotting them, and from the resulting curve calculating

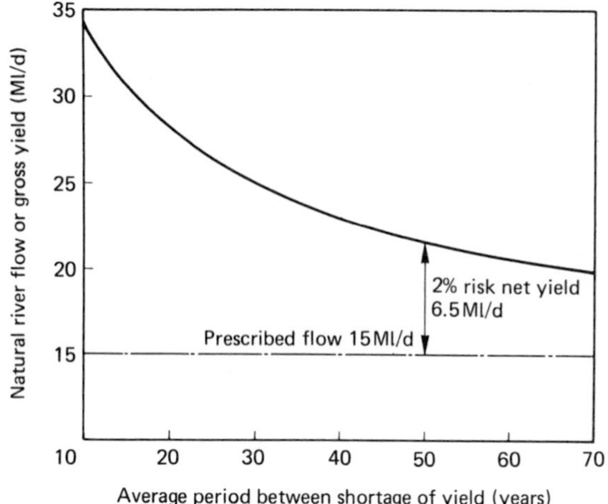

Fig. 4.3 Yield–return-period graph.

mean volumes for each unit rise of level. Areas are usually taken to the centreline of any dam or embankment, so that a deduction must be made for the volume occupied by the upstream half of the embankment unless the material was obtained from a borrowpit inside the reservoir.

The capacity–elevation graph should be extended to the maximum possible level of storage, including surcharge volumes above spillway crest as this will be required for flood routing. The limiting factor for maximum storage is usually the need to avoid flooding some nearby community or some low point in the catchment rim.

Allowance must be made for unusable bottom water or 'dead' storage in order to obtain the effective or 'live' capacity. The outlet level to the reservoir may have to be set above valley bottom for supply reasons, to avoid sediments, or to give sufficient minimum depth to prevent bed vegetation growth. Operational control rules may also reduce live storage, particularly if the river is to be used also for flood control, in which case it will be kept in a partly drawn down state at certain times of the year. If the flood fails to come, the reservoir may not be full at the commencement of a severe drought period (Fig. 4.16, p. 119, shows an example of this). However, in practice flood storage provisions seldom reduce live storage by more than 10%. Special storage allowances are occasionally necessary: in the USSR allowance is sometimes made for ice floes which beach in cold weather; in South Wales, in the cold winter of 1962–63, a small direct supply reservoir was thought to be full until the snow-covered ice sheet across the lake cracked and dropped substantially—the water left was found to be so small that supplies had to be restricted.

4.5 Yield of direct supply reservoirs

A reservoir supplements inflow during some critical low flow period so that a larger drawoff can be maintained than could be derived from the inflow alone. Thus

$$\text{yield} = \frac{\text{inflow volume over critical period} + \text{effective storage}}{\text{length of critical period from full to empty}} \qquad (1)$$

In the past this problem was usually approached by graphical techniques using a cumulative (or mass) runoff plot as Fig. 4.4. It will be seen that

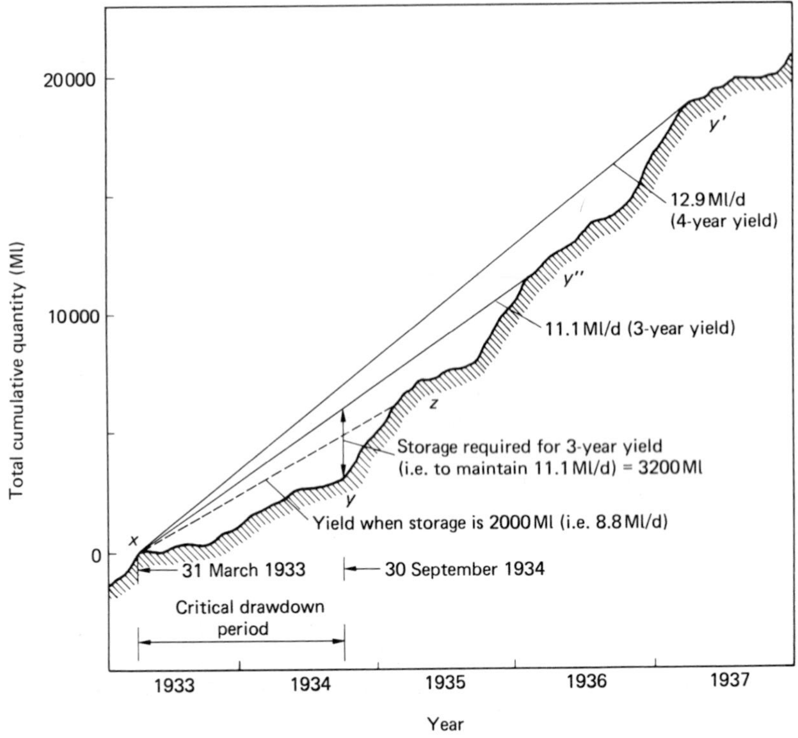

Fig. 4.4 5-year mass-flow diagram.

the reservoir is bound to be full at the beginning of the drought because previously inflow has exceeded yield. Although such a graphical method is useful to the novice, it will be found more precise and rapid to use equation (1). To do this different possible critical periods must be examined (normally in monthly units) until the minimum yield is located for a given storage capacity.

The way in which the critical period lengthens as storage is increased is seen by reference to a cumulative minimum runoff diagram (Fig. 4.5). This simply compresses on to one diagram the driest segments of the mass curve. Each point represents the minimum runoff historically

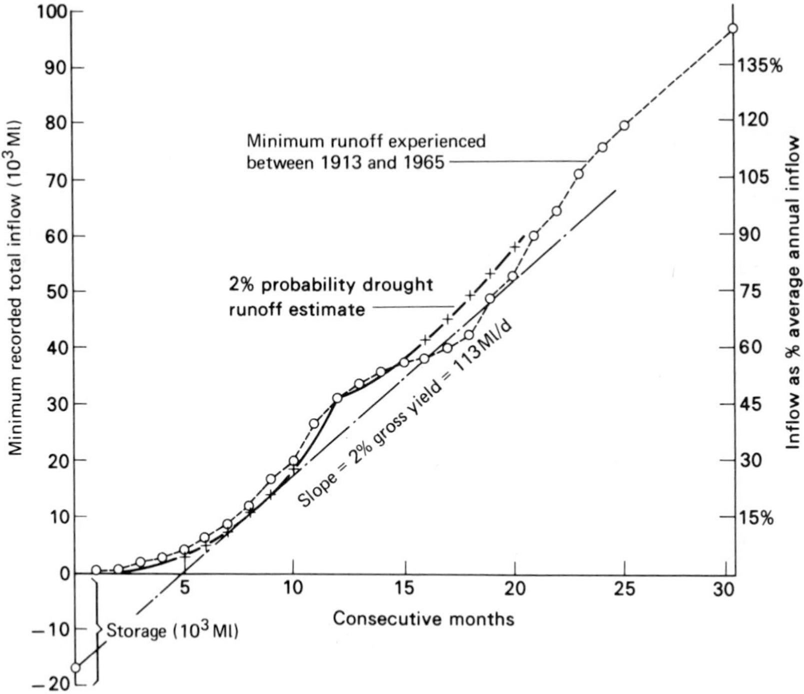

Fig. 4.5 Minimum runoff diagram.

recorded for a consecutive period of months and, as a result, adjacent points may be entirely different drought events. By drawing the storage on the negative ordinate and striking a tangent to the curve, the minimum yield can be found from the slope of the line because it expresses equation (1). Adjustment to this equation is needed where lake evaporation exceeds the evapotranspiration loss from inundated land. The deduction from inflow volume is minimal for most UK reservoirs, but can be dramatic in the monsoonal climate of India or the arid zones of Australia.

Historic minimum runoff diagrams, such as Fig. 4.5, normally have an irregular shape. This is because they portray events which are likely to differ in incidence. For example if a 30-year flow record is available there will be 30 summer low flows recorded, and one of these might be a '50-year summer drought' (called a 2% drought for convenience because, over a long period, there will be, on average, two such events of equal or worse severity per century*). In the same record, however, there will be only 15 independent values for the lowest 2-year drought, and the worst recorded may represent only, say, a 20-year (5%) event. Consequently it is desirable to supplement knowledge of historic drought yields with those that can be estimated for the future by synthetic means. Synthetic minimum runoff diagrams formed by flow probability analysis first attained prominence in the hydrological surveys of various English basins in the 1960s[6]. They can be formed as follows.

(1) Locate a long monthly flow record for a hydrologically similar catchment to that forming the subject of study. This requires average runoff, seasonal rainfall pattern, rainfall variability, hydrogeology, and catchment cover to be similar. (If flow records are non-existent refer to the end of this section.) The record used must be checked for accuracy.

(2) Tabulate the flow record for as long a period of complete years as possible and produce the monthly average flows. Check that they represent a realistic water balance.

(3) Subject the flow volumes over every time period of interest to probability analysis. Much could be written about this, but basically it has to be remembered that short-time-period events are markedly skew, whilst flow volumes for, say, 10-year periods stay very close to their mean, usually being Normally distributed. Thus the use of a wide range of probability functions may be appropriate and to simplify the problem it is reasonable to adopt certain approximations according to the following procedure.

(4) For any period short enough to exclude regularly any wet season (e.g. 1, 2, 3 . . . 11 months) rank all independent events starting with the driest and proceeding until almost one event per year of record is listed. (By independent is meant that no monthly flow can appear in more than one drought total.) Table 4.1 is an example using a 50-year flow record. The partial duration series so formed is in effect a search for the 'tail' of an exponential distribution[7] and should be plotted on appropriate paper. Figure 4.6 is an example of probability paper that is normally satisfactory over a wide range of rivers. Interpretation of the plots is made by drawing the best family of lines that will account for both outliers and any flat patches that arise from chance coincidence of flow totals. Reading off, say, the 50-year values to form the lower part of the minimum runoff diagram should show steadily rising increments of flow as time progresses into the first wet season.

(5) For longer time periods above, say, one year the number of independent events that can be located grows fewer, the values less truly 'extreme', and the plot less realistic to interpret. It is therefore preferable to sum and rank the complete flow series beginning on a fixed

* For any single century, less or more than two such events may take place.

Table 4.1 Ranking of drought flows (duration four months).

Rank	Runoff in megalitres (Ml)	Starting month	Year	Return period (years)
1	28.40	June	1934	50.00
2	41.50	July	1959	25.00
3	50.50	June	1949	16.66
4	56.30	July	1947	12.50
5	60.90	June	1933	10.00
6	64.00	June	1921	8.33
7	66.90	June	1929	7.14
8	72.20	July	1955	6.25
9	72.80	April	1938	5.55
10	74.70	May	1935	5.00
11	76.20	June	1961	4.54
12	78.00	August	1937	4.16
13	80.30	June	1940	3.84
14	85.10	April	1928	3.57
15	92.90	April	1957	3.33
16	96.40	May	1963	3.12

date each year. This date is chosen by examining the average monthly flows to find the lowest appropriate period. Thus the 18-month series beginning where the driest 6-month period normally commences will lie below the other 11 lines that could be produced by starting in every other calendar month in turn. Such a series embraces events from wet to dry and can appropriately be plotted on Normal probability paper (Fig. 4.7) using the unbiassed plotting position formula given in Section 4.3 above. The engineer should check that the driest event of any chosen duration features in the ranked series of data; if not, an alternative rank list should be formed beginning with the month in which this driest event took place. The principle is to locate as rapidly as possible the event probability line which gives the lowest runoff volume at the desired return period. Where average flows are very low, e.g. less than the equivalent of 250 mm per annum over the catchment, even long-time-period droughts may be skewed and it may be necessary to plot the logarithmic values of runoff on Gumbel (Weibull) or even Normal distribution paper to obtain a family of straight lines. Hardison[8] shows American examples of these distributions and suggests rules that work well in the authors' experience. Shortened these say:

(a) Use a log–Normal distribution if the coefficient of skew of the logs is algebraically greater than 0.2.
(b) Use a Normal distribution if the coefficient of variation of the flows is less than 0.25.
(c) Use a Weibull distribution if neither a log–Normal nor a Normal distribution is selected by the criteria given.

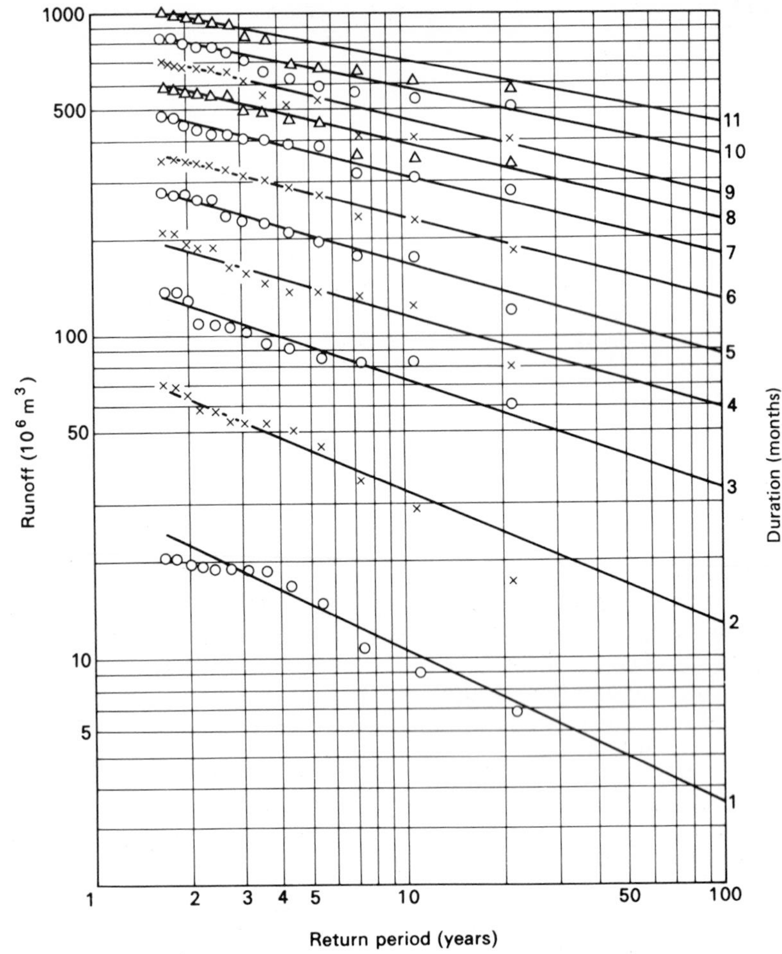

Fig. 4.6 Independent events probability analysis (one–eleven month periods).

(6) From the lines on Figs 4.6 and 4.7 the necessary figures can be read off to complete the synthetic runoff diagram (Fig. 4.5). The usual calculation then follows.

(7) The results of yield estimates should be presented in the form of a yield–storage relationship, as Fig. 4.8, so that it is possible to see whether further storage would bring a good return or whether more inflow is required to make use of the available capacity. If the relationship is made dimensionless by expressing yield and storage as proportions of the average flow, it is readily possible to transpose the relationship from the gauged site to the reservoir site being studied.

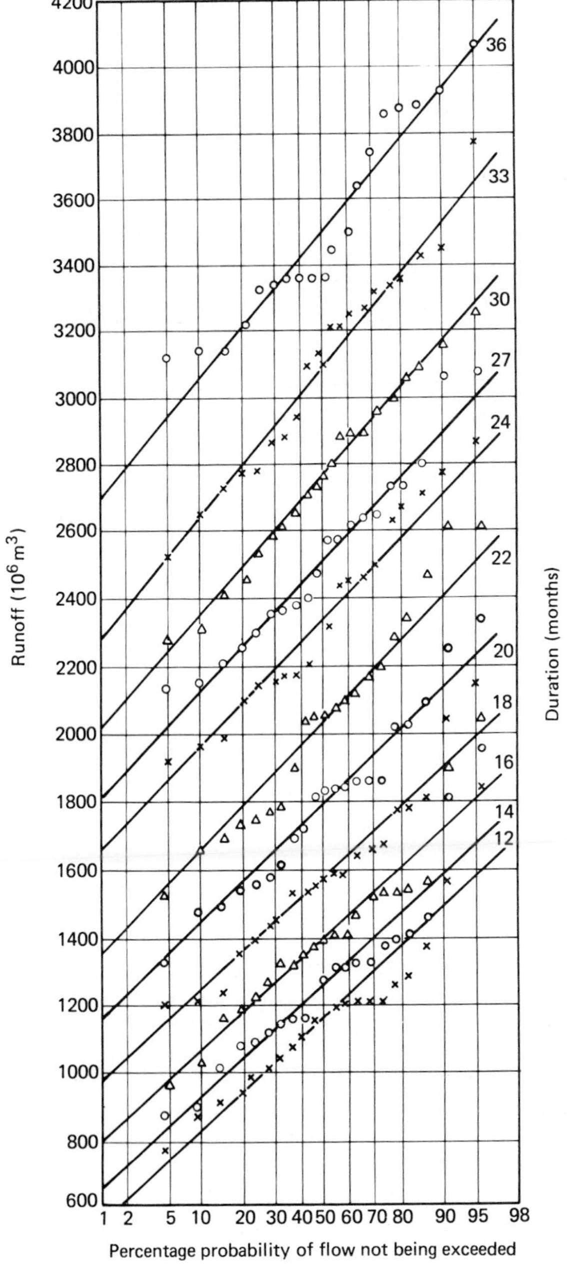

Fig. 4.7 Overlapping events probability analysis (twelve–thirty-six month periods).

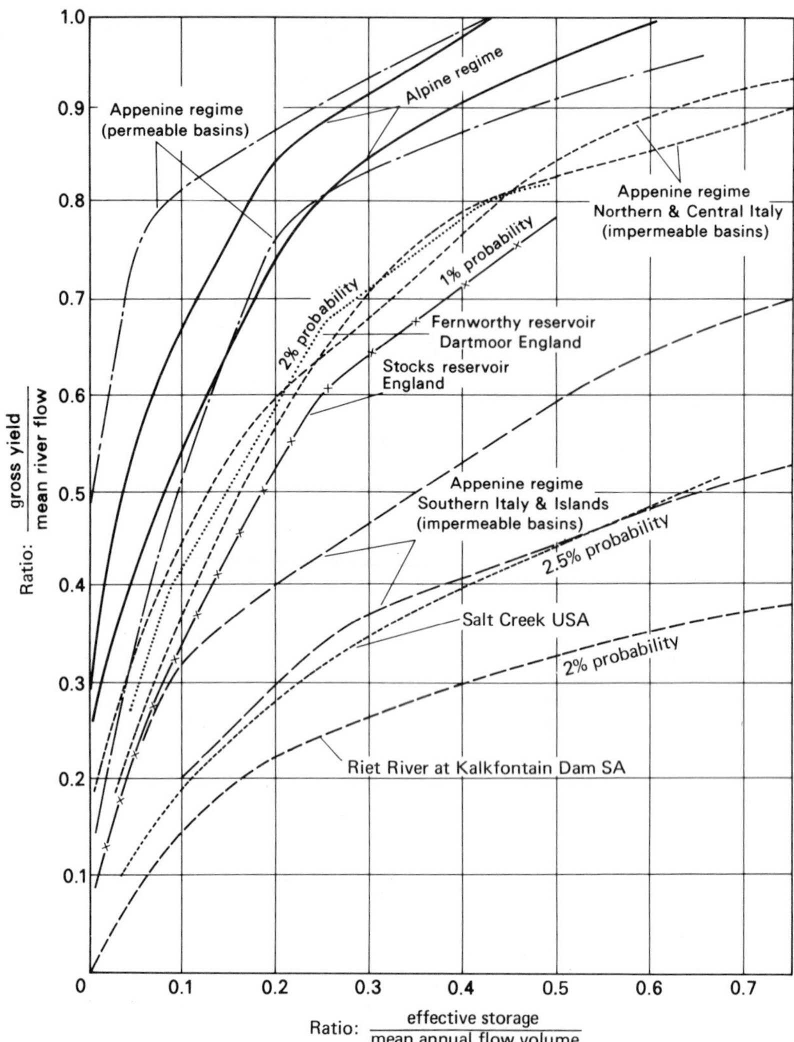

Fig. 4.8 Typical yield–storage relationships for direct supply reservoirs.

Since equation (1), p. 96, is linear in form, points on the yield–storage curve can be joined by a straight line if they involve the same critical period. If yield analyses are carried out in weekly rather than monthly units, the yield curve slope changes so smoothly that it appears to form a curve. However, smoothing by curves obscures the abrupt intersections which occur when, with increasing storage, the critical period moves from a single dry season to a period including two dry seasons, and so on.

If there should be a short period of runoff record concurrent with a longer rainfall record[9], it can be possible, particularly in high rainfall areas, to correlate the records seasonally. From this a long runoff record can be created for analysis as described above.

Alternatively, where rainfall and potential evapotranspiration data exist, it may be possible to synthesize a flow record by a set of equations which form a physical model of how the catchment soils and aquifers are successively wetted and drained[3,10]. However, unless the catchment concerned is either totally impermeable or totally permeable this method can become complex and is best left to the hydrological specialist. Average flows are adjusted by changes in the size of the soil moisture content stored within the range of the vegetation roots; minimum flows depend strongly upon the chosen recession constant. However, neither value can be determined reliably without calibration against a period of flows that includes a drought. Computer routines improve the speed of modelling, but optimizing the constants employed is always a matter of judgement and compromise. Such a model is more often used in groundwater studies, but is then less amenable to calibration.

4.6 Yield of pumped storage reservoirs

The yield of a pumped storage reservoir is dependent both upon the incidence and range of daily flows that pass the intake and upon the ability of the pumps to abstract water from such flows efficiently, with a realistic pump capacity, whilst leaving any statutory residual flow to pass downstream. In order not to be over-optimistic by 10% to 20% on yield, it is essential to obtain, or generate, daily flows for the intake site.

Where a long record exists it is possible to carry out a day-to-day operation study to follow reservoir performance over such a period. This is simply a daily balance of runoff, pumped inflow, drawoff, and storage change, given any rule that may exist for, say, residual flows. Table 4.2 illustrates how such calculations may be set out.

Table 4.2 Pumped storage reservoir calculation.

(1) Date	(2) Flow to intake (less impounded flow if any)	(3) Potential river abstraction	(4) Reservoir inflow	(5) Compensation release and loss	(6) Drawoff	(7) Drawdown at end of day = previous value+6+5−4−3	(8) Remarks
15/8	40	10	2	2	8	600	
16/8	34	4	1	2	8	605	
17/8	30	0	0	2	8	615	Pumping ceases

By taking from such an operation study the maximum drawdown occurring each year, it is then possible to rank the values of maximum drawdown in order of severity and plot them on probability paper to find the drawdown likely once in, say, 50 years. This is then the effective storage required for the stated drawoff. Alternatively, if the reservoir already exists, it is possible by repeating the calculations for a range of drawoffs to produce a yield–storage relationship with an associated risk of emptiness: the yield for the known storage can then be read off for any degree of risk. This process is difficult to carry through successfully unless the flow record available is long enough to make the probability plot one for interpolation only. Insufficient points from an unnatural distribution of this type do not permit confident extrapolation.

The above accounting, or simulation, technique is normally done by computer, but hand checks of the routines involved are essential, because one may easily fail to foresee situations that can arise in the different subroutines required for complex operation rules. A full program write-up becomes vital where it is to be used frequently for different studies.

It must be realized that to define a yield in this manner provides only a nominal rating for a source, to help indicate when a further resource must be brought in to meet rising demand. Where actual operational forecasts of drawoff from an existing system are required, then supply restriction rules which will occur in practice during a drought must be written into the study: results are then presented in a different form to show the quantitative shortfall of supply that is expected.

The yield of a pumped storage reservoir can often be defined more satisfactorily by calculating the yield–drawdown relationship for a synthetic 1% or 2% drought of the form described in Section 4.7 below. Yield will vary primarily with storage and secondarily with pump capacity (Fig. 4.9). Sometimes, where flows in a drought fall rapidly below the prescribed residual flow and critical drawdown periods are short, it can occur that raising intake capacity has little influence on yield. However, further capacity is essential for refilling storage after a drought whilst still maintaining supply. In temperate climates this is often found to require pumping capacity of twice the yield to supply, but standards vary. Generally to be able to refill a reservoir from empty through a single average wet season is the most secure provision. However, a desire for economy has sometimes led to the acceptance of refilling over as long a period as six wet seasons, i.e. $5\frac{1}{2}$ years. In such cases difficulty can occur over first filling if the reservoir must meet an early rising demand.

Calculating potential pumped abstractions by reference to a mean daily flow record can be misleading if flows at the intake vary rapidly within the abstraction range. Examination of hydrograph charts helps here, but it will often be satisfactory to assume that only 90% of

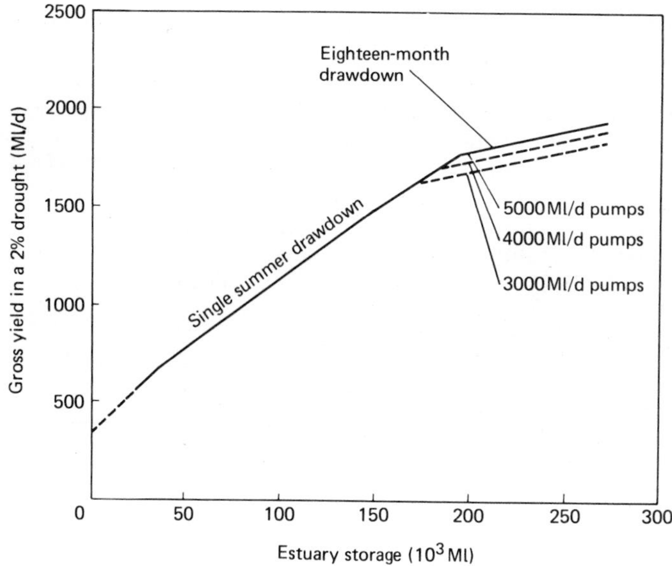

Fig. 4.9 Yield–storage diagram for a pumped storage scheme.

potential abstraction is achieved on those days when the available water falls short of the pumping capacity; 100% can be taken when there is ample water for all purposes. Different considerations apply if the intake can abstract a large proportion of the average flow past it (see Section 4.9).

Reservoir losses by evaporation are only those over and above the evapotranspiration loss of the inundated land. They can be estimated reasonably by deducting percolation gauge losses from tank evaporation figures, or by subtracting known catchment drought losses from Penman open water evaporation estimates. Where the flooded area does not form part of the intake catchment, as occurs with bunded estuary storage, then a different logic applies. An absolute balance of drought rainfall and loss is then needed and this may in some cases turn the total net reservoir inflow over the critical drought into a negative figure.

4.7 Design drought hydrographs

The creation of a 'design hydrograph' can have two purposes:

(1) to forecast a future critical event in order to prepare contingency plans for an existing source,
(2) to provide a convenient tool for the rapid and consistent testing of a variety of schemes to find their likely critical drawdown period and associated yield.

In regard to the first use, the one certainty is that history will not exactly repeat itself. Statistical techniques exist for generating long sequences of daily flows[11], but it has yet to be shown that they can be made to produce as large storage requirements as the originals. An endless variety of future possibilities exists and yield calculations on this basis may be similarly endless, unless the more limited approach, (2), is adopted.

A method for compiling a 2% design drought which has proved satisfactory in terms of giving results consistent with experience, coupled with flexibility in use, is set out below. The essential stages for compilation of a 2% design drought are as follows.

(1) Take the 2% minimum runoffs from a diagram such as Fig. 4.5, express them as percentages of the average annual flow (AAF), and show the incremental percentages per month, thus

Period	*Cumulative (per cent AAF)*	*Incremental (per cent AAF)*
driest 1 month	0.9	0.9
driest 2 months	1.9	1.0
driest 3 months	3.1	1.2
driest 4 months	4.4	1.3
driest 5 months	6.9	2.5
driest 6 months	11.5	4.6
driest 7 months	16.7	5.2
etc.		

(2) Rearrange the monthly incremental values to simulate a realistic calendar order, starting with the month in which the flow is not quite deficient of the required abstraction and maintaining the 2% totals for the lowest 1, 2, 3 . . . etc. consecutive months, thus

month (a) as seventh	5.2%	(April)
month (b) as fifth	2.5%	(May)
month (c) as fourth	1.3%	(June)
month (d) as second	1.0%	(July)
month (e) as first	0.9%	(August)
month (f) as third	1.2%	(September)
month (g) as sixth	4.6%	(October)
etc.		

This is called a 'stacked drought' because every possible 2% drought is stacked within it. Alternatively it can be made to reproduce the local drought character by putting the increments in an order representing dry weather recession. It will, however, be found impossible when stacking the drought to reproduce exactly all minimum runoff totals for periods which include wet seasons, e.g. a 13-month period, but this will not affect the drought yield in practice.

(3) The percentages under (2) are converted to flow by multiplying them by the long-term average annual flow at the point of interest, thus

month (a)	5200 Ml (say)
month (b)	2500 Ml (say)
month (c)	1300 Ml (say)
etc.	

(4) The historic record of daily flows (however short) is next consulted and a month of daily records for the same time of the year as month (a) is found which most nearly has a total runoff equal to that for month (a). The historic daily flows are then scale adjusted (see (5) and (6) below) so that their total equals the total for month (a) and these are then taken as the 'daily flows' for month (a). The method is repeated to produce daily flows for months (b), (c), (d) . . . etc.

(5) We now have a 'daily design drought hydrograph' of flows containing all the information expressed in Fig. 4.5 but extended to give daily flows. The value of this is that it can be used quickly and consistently to assess storage requirements for alternative reservoir sites, pump capacities, residual flows etc. However, before finally putting it to use in this manner, it may need some final adjustments and checks as follows.

(6) When adjusting the historic daily flows to match the ideal it is important to see that flow transitions from one month to the next are realistic. In particular no sudden drop of flow should take place.

(7) Instead of scaling down all the historic flows in a month to make the total equal the 2% total, it may be best, particularly in the absence of any really dry months of record, to apply a straight-line recession of flows below the historic record to gain the equivalent total flow for the month.

(8) The 2% daily hydrograph may need checking to see that it has a proper balance of spates and recessions for the period following the first summer. It is also important to check the deficiency probability, as well as the runoff probability. To do this the historic daily record is examined to find the daily deficiencies below, say, 40% of the average daily flow (ADF). These deficiencies are summed and tabulated monthly, and then plotted as Fig. 4.10, a curve being drawn through them. If the months chosen to form the 2% hydrograph all fall below the curve then the hydrograph is not severe enough. Each season's total deficiency is summed, ranked, and plotted on log-Normal probability paper (Fig. 4.11): curves drawn through the points should show an upper limit, the consequence of flows never being negative. The cumulative 2% deficiencies are now read off, the second summer deficiency being obtained by subtracting the first season's deficiency from the two-season's deficiency, and so on. The 2% hydrograph is then checked for its deficiency below the same 40% ADF level. If the second season's runoff

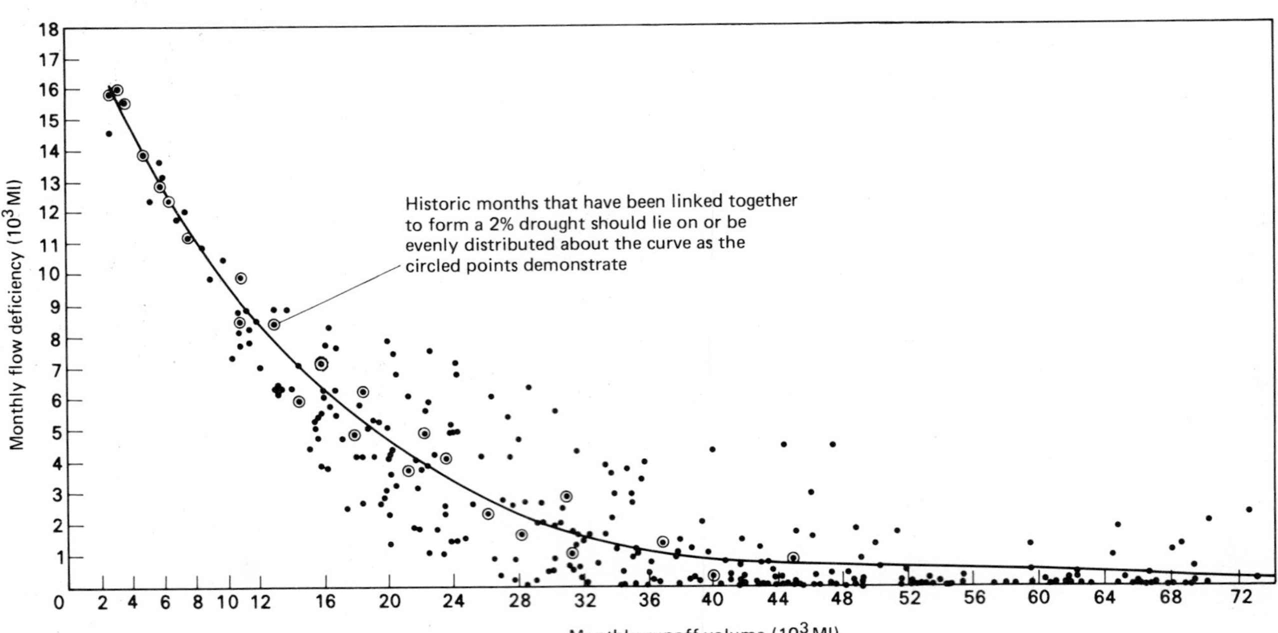

Fig. 4.10 Runoff–deficiency relationship. A check of the daily deficiencies of flow below a given flow to ensure a 2% daily design hydrograph realistically models the historic record. Note that the flow deficiency is that amount by which the daily flow hydrograph falls short of some arbitrary flow of, say, 40% ADF and is calculated monthly.

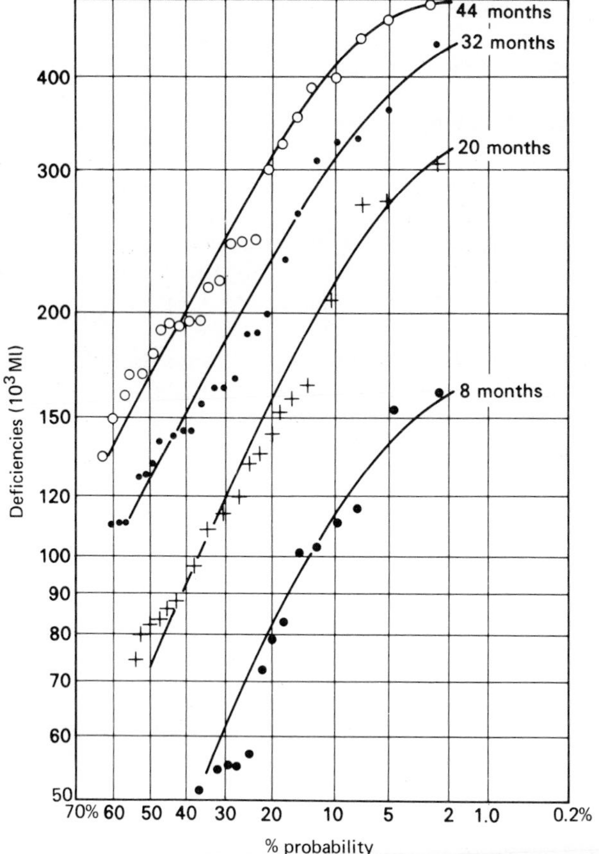

Fig. 4.11 Hydrograph deficiency probability. Seasonal deficiencies are plotted on log–Normal probability paper to check a 2% daily design hydrograph (Regulation level 596 Ml/d).

total is too low, as can be the case, an alternative selection of matching historic months is chosen to ensure that the finally adopted 2% design drought is satisfactory in terms of both runoff and cumulative release requirement.

Although the foregoing procedure requires care, and may seem complex, it nevertheless permits subsequent work to proceed much more rapidly without the need for reiteration. It does not require foresight of, or any assumption about, the critical drawdown period that finally emerges as likely for a chosen reservoir site; nor is the amount of data generated as unwieldy as that for a 500-year generated sequence.

Hydrographs of this kind have already been produced for many rivers in the United Kingdom, India, and Malaysia and are particularly useful for regulating reservoir yield studies.

4.8 Yield of regulating reservoirs

The storage required in a regulating reservoir is dictated primarily by the deficiencies of natural river flows at the downstream abstraction point during a drought. A secondary influence is the amount of water that can be impounded in the reservoir during the drawdown period on those days when regulation releases from storage are not required. The design hydrograph method given in Section 4.7 permits an assessment of both these criteria.

Simulation of the drawdown during a design drought can be carried out in the same way as for a pumped storage reservoir. Additional features are the rules governing regulation releases, and the extent to which release losses occur. An example is given in Table 4.3.

Table 4.3 Regulating reservoir calculation.

Day	(1) Natural flow at intake	(2) Dam site flow	(3) Unregulated flow at intake	(4) Flow deficiency	(5) Reservoir release	(6) Reservoir drawdown	Conditions (All units Ml)
1	48.0	4.0	44.0	Nil	2.0	Nil	Yield to supply = 30 Ml/d
2	42.0	3.5	38.5	Nil	2.0	Nil	Minimum release = 2 Ml/d
3	36.0	3.0	33.0	5.0	7.5	4.5	Maintained residual flow past intake = 10 Ml/d
4	30.0	2.5	27.5	10.5	13.5	15.5	
5	etc.						

Notes:
Column (1) = flow if there were no reservoir.
Column (2) = natural inflow to reservoir.
Column (3) = col. (1) – col. (2).
Column (4) = yield to supply + maintained residual river flow – col. (3) – minimum reservoir release.
Column (5) = col. (4) × 110% + minimum reservoir release.
Column (6) = previous reservoir drawdown + col. (5) – col. (2).

Release rules normally specify a minimum release to cover channel requirements immediately below the dam and conditions at the regulated intake downriver. These latter conditions fall into two main groups, the first of which involves maintaining a steady flow continuously below the intake, thus generally ensuring an improvement in residual drought flows in the river. The second group basically aims at preserving natural low flows below the supply intake, in which case the river is simply used as an aqueduct between reservoir and intake. As flows rise, regulation releases are cut back until the basin runoff can sustain the whole of the supply. The latter type, sometimes known as 'hands-off' schemes, are relatively simple to operate as little or no prediction of flow is necessary when calculating drought releases and no loss allowance need be made. However, more precise regulation control is required at reservoirs which must achieve a statutory requirement that flows should be maintained at or above a prescribed figure below the intake. It may take one or more days for release changes at a reservoir outlet to make themselves felt at a downriver intake (at an apparent release velocity of, say, 1 to 2 km per hour down the channel length). Thus releases may be made one day and then rain may occur on the next, so making the release, with hindsight, wasted.

Allowance for such mistiming of releases can be made by adding 10% or 15% to the theoretical total release requirement, depending upon the distance of the storage from the intake concerned. However, a better approach for operational studies is to mimic the typical decision process. If the river's recession in completely dry weather is plotted, and given mathematical expression (e.g. Fig. 4.23, p. 138), then a rule can be written making the regulation release on any day dependent upon the expected deficiency of the river flow below the desired yield on, say, the next day. The deficiency is simply calculated with the pessimistic view that dry weather will occur. Such rules, applied to rivers in North Wales and North West England, have realistically predicted waste releases of 20% or more in wet years, but only 2% or 3% of the total release needed in a major drought season.

A characteristic of the operation of a regulating reservoir is that it tends to be full for a larger percentage of time than its similarly sized direct supply counterpart. Conversely such a reservoir tends to empty with dramatic rapidity in a rare drought.

4.9 Catchwater yield

A catchwater is a stream intake that enables runoff to be impounded from a larger catchment than that which drains naturally to a dam site. Its useful yield, therefore, is the amount by which that reservoir's yield is raised. This can be much less than the average transfer of which the catchwater is capable, but will be greater than its dry weather transfer.

The stages for assessing catchwater yield are:

(1) definition of the drought flow at the catchwater intake,
(2) calculation of the proportion of this flow that can be abstracted,
(3) modification of the reservoir yield calculation by the addition of these abstractions to storage.

Drought flow at the intake site is generally taken as some proportion, or some function, of the design drought for the reservoir's direct catchment. Calculation of the flow that can be abstracted is aided by a graph of the form of Fig. 4.12. This shows, in dimensionless form, the amount that can be taken from a stream given the relative size of the catchwater (or intake pumping capacity). Figure 4.12 was derived for small British catchments with average rainfalls of 1500 mm per annum or more, but checks equally well against data from a tropical catchment in Singapore. It is based upon the integration of the area under flow duration curves (Fig. 4.13) drawn up directly from continuous flow hydrographs. To use mean daily flows for such analysis would lead to unwarranted optimism except for groundwater-dominated streams. Similar pairs of figures can readily be produced for gauged areas with less frequent rainfall.

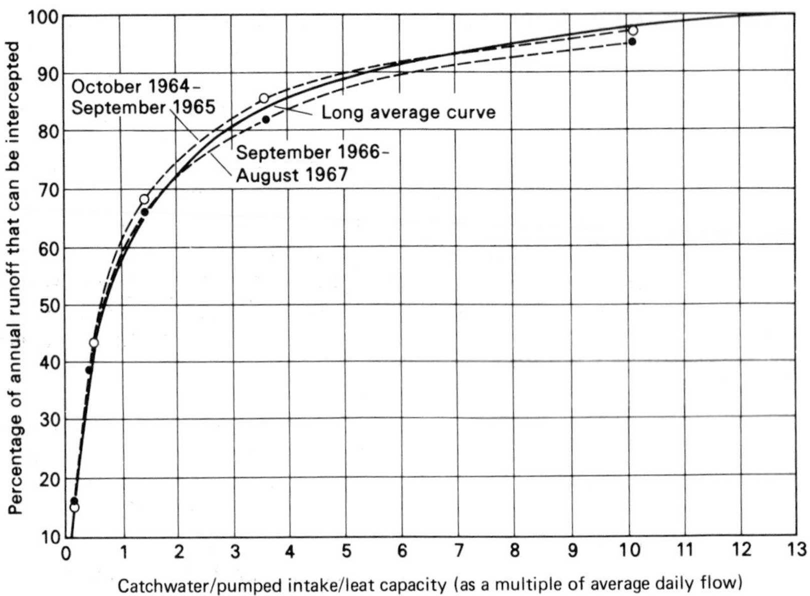

Fig. 4.12 Catchwater transfer curve. The design curve is based upon hourly flow duration data from four small mountain catchments in England and Wales.

Figure 4.12 refers strictly to average annual takes, but can be used for seasonal assessments in individual years without too much error providing the appropriate average daily flow (ADF) for the period concerned is used. One can also apply to it the intake prescribed flow (expressed as a percentage of the ADF) to see to what extent residual flow will be met.

To use the graph in typical yield calculations one should first find the average intake flow over each possible critical drawdown period. Then the catchwater capacity plus the residual flow, and the residual flow alone, are expressed as multiples of such averages. From Fig. 4.12 the

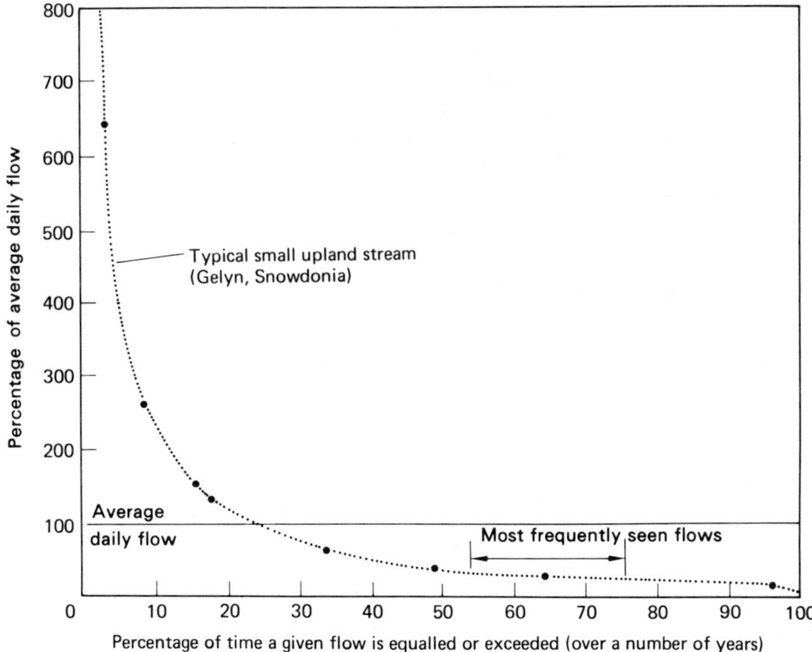

Fig. 4.13 Instantaneous flow duration curve. The curve is derived by measuring the number of hours the stream hydrographs exceed given stated levels of flow.

corresponding total availability of water is read, from which the average residual flow is deducted. The net figures, representing the potential transfers, are then added into the direct supply yield calculations of the type described in Section 4.5. The minimum yield and likely critical period are located and plotted, as in Fig. 4.14. The value of adding a catchwater to a major storage and the consequent shortening of the drought period most likely to cause failure are clearly seen.

4.10 Compensation and residual flow rules

Compensation water is the flow that must be discharged below a direct supply reservoir to compensate riparian interests for the water taken away to supply. Each country has its own water law built up over the centuries to preserve local rights[12], and the setting of a compensation flow can involve much legal dispute. In Britain, Parliamentary committees used to specify one-third of the gross yield as compensation water, but this proportion gradually fell to one-quarter as water became more expensive and mill interests became less dominant. In arid countries, where streams only flow seasonally it is normal to find that no compensation flow is required below a dam.

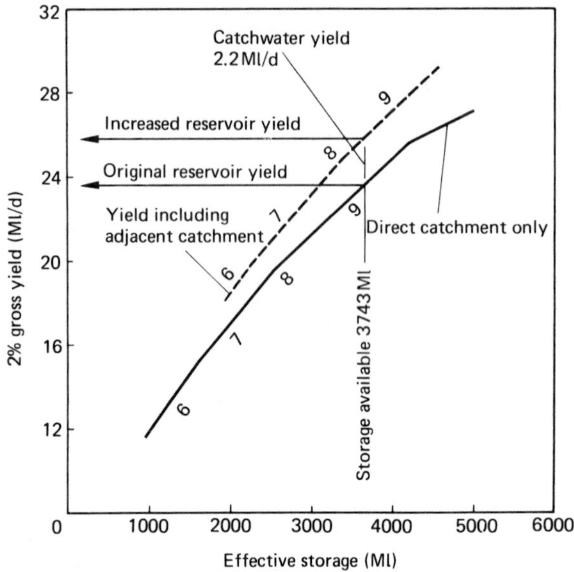

Fig. 4.14 Catchwater yield–storage diagram. Note that the figure by each straight line denotes the length of the critical period in months.

The current concern is to provide no more than is necessary for the health of the river in terms of effluent dilution, visual amenity, fish life, or navigation level. Each case requires individual scrutiny and some assessment of the economic implications. However, it is often realistic to assume that compensation water need not be greater than the flow which is normally exceeded for 90% of the time.

Fixing a 'prescribed' flow below which a river intake must not cause diminution of flow is especially important where there is no regulation of the river. This prescribed flow should never be decided cursorily: it needs careful prior investigation. Once studied and agreed it should not be capriciously altered at some later stage of negotiation. Even though the decision ultimately taken may rest upon philosophy or politics, the design engineer must bear the initial responsibility for choosing a justifiable prescribed flow, if only because he will be one of the few to realize its economic significance. To avoid misunderstandings it is a salutary exercise for the engineer to draft both the legal phraseology covering the residual flow rule (and any associated release conditions) and to translate these into corresponding 'plain language' instructions for the operatives concerned. Table 4.4 shows a variety of rules in use in Britain's temperate conditions where streams are perennial. Residual flows should be set to protect downstream interests properly[13] rather

Table 4.4 Some British residual flow conditions.

River	Location	Type of condition	Residual flow (Ml/d)	Average daily flow (%)	Remarks
Hull	Hempholme*	PF[(a)]	45	11	54 Ml/d source
Trent	Fossdyke Canal*	—	Nil	Nil	Occasional support of Witham–Ancholme Scheme
			(No abstraction 90 min either side of high tide)		
Nene	Orton	PF	136	19	Empingham Reservoir Intake
Ely Ouse	Denver*	PF	113	8	Ely Ouse–Essex Scheme
			(March to August)		
			318	23	
			(Rest of year)		
Thames	Teddington*	PF	773	13	Reduced if storage low
Rother	Robertsbridge	PF	4	3	Darwell reservoir intake about 19 km above tidal limit
			(No pumping April to September)		
Test	Testwood	—	Nil	Nil	81 Ml/d intake
Exe	Thorverton	HOF[(b)]	273	20	Wimbleball Reservoir
Usk	Llandegfedd Reservoir Intake	PF	568	23	Abstraction reaches 340 Ml/d at flows exceeding 1360 Ml/d
			(Proportion of excess abstractable)		
Towy	Nantgaredig	MMF[(c)] and	136	4	Brianne Reservoir
		HOF	682	20	
Leven	Newby Bridge	PF	273	21	Windermere Intake
			(May to September)		
			136	10.5	
			(October to April)		
Blackwater	Langford*	PF	1	0.9	Hanningfield Reservoir

Notes:
(a) PF—Prescribed Flow—the river flow which must not be diminished by abstraction. Described as the minimum residual flow (MRF) until it becomes prescribed by statute.
(b) HOF—Hands-Off Flow—the flow below which all abstraction must be supported by an equivalent upstream reservoir release. 'Natural' low flows are thereby preserved.
(c) MMF—Minimum Maintained Flow—the flow maintained at all times at a control point downstream of a supply intake on a regulated river.
* Indicates tidal limit.

than as a *quid pro quo* for getting agreement to the building of an intake. Furthermore, if every effort is made to site the intake at the tidal limit[14] to maximize natural resources, downstream interests are often non-existent and a zero prescribed flow may be countenanced.

Residual flows should not be used in an attempt to maintain navigation channels or to cure sediment deposition unless:

(1) tests show a positive relationship between freshwater flow and channel conditions,

(2) the value of the water which it is proposed to abstract is markedly less than the alternative of dredging or other mitigation works.

If water is used for these purposes it should only be taken in spate periods. However, if tidal gates require the release of summer water to clear any silt against them prior to winter floods, this probably need not exceed 1% of the annual average flow. It is also a retrograde step to use a large prescribed flow to dilute a poor quality effluent: treatment of the latter is preferable and often more economic. The visual impact of a prolonged residual flow is best judged by considering the implied river level. In some cases existing river levels can be maintained at lower flows by the use of check weirs.

The water requirements of migratory fish are known in broad terms and the seasonal cycle of the spawning and return of fish to the ocean may demand varying residual flows. To maintain fish life may require a total volume of 10% of the annual average flow to pass into the estuary each year. To satisfy the angler, as opposed to the fish, may be much more difficult.

It is rarely possible or economic to regulate a river by storage and pumping to remove more than half of the average flow of a river at its tidal limit. Thus in the foreseeable future there will always be more water reaching each estuary than a study of residual flows alone suggests.

It must be realized that as yields rise and residual flows shrink the magnitude of potential drought deficiencies will escalate. Thus it may be necessary to plan in future for rarer droughts than has hitherto been the practice. The importance of effluent returns in diminishing this risk is now widely appreciated; an alternative way of reducing risk is to relate residual flow specifically to reservoir contents, as is done on the lower Thames.

4.11 Conjunctive use and operation studies

Individually, the various resources of an undertaking may have a variety of critical drought durations to which they are most susceptible. Thus when one source fails another will still have something in hand, and if the sources are linked in a system this reserve can be turned into effective yield. Conjunctive or integrated use of resources is therefore the means of improving a system's total yield; alternatively it may lead to the discovery of operational rules that reduce costs[15].

Ideally the demand area common to a set of sources should be identified first. The basic supply system as far as the treatment works or service reservoirs should then be sketched, with all the associated constraints of capacity, drawoff, and storage. An existing or 'first-shot' set of operational rules is required. Analysis must then proceed on a trial-and-error basis unless the problem and objective are so clear and well known that one of the iterative forms of programming can be invoked.

In essence the operation of a conjunctive use system must be simulated against an assumed demand, generally day-to-day through a known flow record, with gradually varied rules until a maximum yield or minimum cost emerges. Often it becomes clear that an additional link main or extended treatment works will more than pay for itself. Normally a specified priority order for source use is defined with cheapest gravity water being used first and expensive pumped water being left unused until really necessary; typically decisions to change source outputs are made weekly. Storages with short critical periods should be kept in continuous use to avoid spills and to maximize supply. In many years it may be found that only reduced cost, off-peak electricity for pumping is all that is needed to keep the supply going[16,17].

Table 4.5 sets out a simplified system yield calculation, which shows that the essential feature is bringing the storages to a common critical drawdown period in the design drought. This is usually done for straightforward cases by trial-and-error adjustment of the total yield. In the example it can be seen that the system yield of 34 Ml/d is 4 Ml/d greater than the sum of the sources if operated separately. A gain of about 10% is not unusual, but even this may not be achieved unless adequate treatment capacity exists.

Plotting the simulated historic reservoir levels (Fig. 4.15) that result from an operation study illustrates how fully the storage is likely to be used. Rules based upon reservoir level control curves (Fig. 4.16) can then be produced. They are the means of safeguarding a basic yield whilst permitting much larger drawoffs in wetter periods. To produce a control curve requires calculating backwards in time from the zero 'live' storage level at the end of each month of the year. By examining droughts of every duration it is possible to locate the maximum storage required at any time of year which will ensure that the reservoir never quite empties while giving the chosen supply. Each supply rate requires a different control curve which can be defined either by the severest droughts of a historic record or by a design drought of specified probability. Rule derivation is best left to specialists.

To determine the relative chance of storage being at different levels, a 'probability of emptiness' technique may be used[18]. Such matrix techniques can be cumbersome and imprecise but, where they are used, Moran's method is preferred to Gould's because it avoids the latter's heavy dependence upon chance droughts in the historic record.

4.12 Shortage rules and contingency plans

The resources of some water undertakings fall so consistently short of demand that regular shortage rules must be built in to any operation

Table 4.5 System yield calculation.

	Jan	Feb	Mar	Apr	May	June	July	Aug	Sept	Oct	Nov	Dec	Individual source yield (Ml/d)	Treatment capacity (Ml/d)
Stream intake														
Pumpable flow	620	560	620	600	620	480	310	155	240	620	600	620		
Actual take	Nil	Nil	Nil	60	155	300	310	155	240	620	180	Nil	5	5–20
Minor reservoir														
Inflow	434	336	310	240	186	180	124	62	90	279	372	496		
Gravity drawoff	124	112	124	240	155	180	155	155	150	217	300	124	5	5–10
Storage (month end)	184*	184	184	184	184	184	153	60	Nil	62	134	184		
Major reservoir														
Inflow	1302	1008	930	720	837	540	372	184	270	558	1116	1488		
Gravity drawoff	930	840	930	720	744	540	589	784	629	217	540	930	20	18–30
Storage (month end)	1136*	1136	1136	1136	1136	1136	919	359	Nil	341	917	1136		
													30	
Total supply	1054	952	1054	1020	1054	1020	1054	1054	1019	1054	1020	1054 =	34	

Assumed operating rules:
(1) Cheapest water first, i.e. major reservoir, then minor reservoir, then intake.
(2) Pump intake to keep storage as full as possible, but maintain minimum reservoir treated water output.
(3) Keep reservoirs same % full, i.e. aim at emptying simultaneously.
(4) Keep within normal treatment capacity range except when reservoirs nearly empty.
* Full capacity.

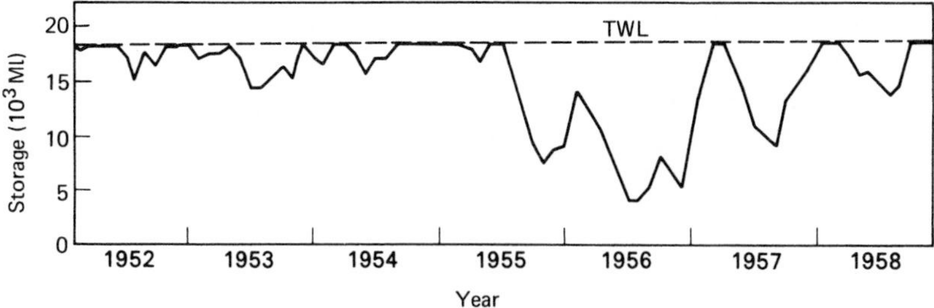

Fig. 4.15 Seven years of reservoir drawdown levels.

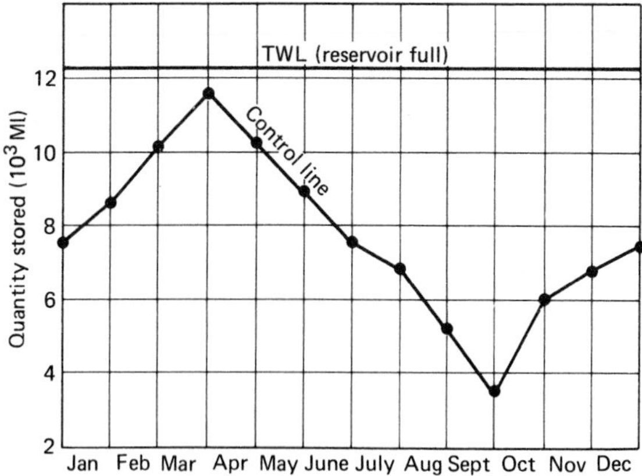

Fig. 4.16 Reservoir control curve. The control line represents the level to which the reservoir could have been drawn down during any month in the period 1910–64 (55 years) and still maintain a total outflow of 46 Ml/d to supply and 13 Ml/d to compensation. When the contents are above the control line the drawoff may exceed 46 Ml/d to supply (up to the limit of the treatment works capacity). If the contents fall to or below the control line the drawoff must be limited to 46 Ml/d to supply.

study. Thus in Hong Kong for several years the standard supply was water for four hours every other day. Where potable water is a rare commodity, rules restricting use may be the only way of ensuring economic supplies. Attempting to restrain demand in a crisis rarely cuts consumption by more than 10% to 15% if only exhortation is used accompanied by a hosepipe ban. If standpipes in the streets are the only distribution points, demand can drop to perhaps 25% of the average. Greater cuts can be achieved where garden watering dominates the supply pattern as in Australia and California.

Having obtained a nominal yield for an undertaking in a future rare drought of some specified severity, say once-in-50-years, contingency plans can be drawn up for worse occasions. These may include:

(1) demand restraint,
(2) cuts in residual flows and compensation waters,
(3) bulk supplies from neighbouring undertakings,
(4) emergency development of a new source (often a river intake with elementary treatment).

Where yields are calculated in operational terms, using such shortage rules, they are normally much higher than those found from design drought analysis of the type discussed previously (Section 4.7). Shortage indices and reliability percentages[19] have been investigated, but it is likely that valid reasons exist for each substantial undertaking determining its own approach to suit local conditions.

4.13 Increase of yield with time

A past fallacy has been the assumption that a yield, once calculated, is unchangeable. In fact many kinds of events can change a yield:

(1) a reservoir can silt up appreciably,
(2) other abstractors may consume catchment water, e.g. irrigation interests,
(3) water may be 'imported' or diverted into a catchment by pipeline and be discharged to the local stream after use,
(4) some of the original yield may be consumed and returned as effluent upstream of a main river supply intake.

In case (4) yield can become largely dependent upon the population living in the catchment concerned. In a study of the yield of Grafham Water[20] it was found that the yield would rise from 170 to 325 Ml/d between 1965 and 2000 if population and water consumption predictions held good. However, in the 1975–76 drought the general publicity restraining water use caused a consequential drop in effluent returns to the River Ouse above the intake: yield growth has been held back by this phenomenon. It is clearly important to review yield calculations regularly.

Part II
Yield of underground sources

4.14 Hydraulic and hydrological well yields

There are two distinct ways in which well or borehole yields have been defined. The first concentrates on well hydraulics and installed pumping plant. The second attempts to predict yield from the hydrogeology of the borehole site and the contributing catchment. Frequently the two results differ widely, either because the pumping plant does not properly match the borehole potential or because overpumping is adopted, until it becomes clear that the water table is dropping, never to recover. Thus to one engineer borehole yield is the lowest output which has been obtained from a source in some dry period with the currently installed pump, whereas to another engineer it is the average percolation over a source's catchment.

It will be assumed that the aim is to attain the ideal yield, whereby the source output is safely maximized, with no more than is necessary in the way of pumps. It is only possible to make a rough estimate from experience of what an individual site will be capable of delivering until such time as a borehole is sunk and test pumped for the first time. In fissured aquifers it is a matter almost of chance whether or not large quantities of water are struck, even though theoretically the general site area may be plentifully endowed with recharge and aquifer storage. Even in sand aquifers a change in the local clay fraction can drastically reduce the effectiveness of a hole and this can only be discovered by test boring and pumping. It is this chance element that makes the drilling of a small diameter pilot bore at a major development site an unnecessary distraction. If successful the pilot bore may still not guarantee the success of the main borehole; if it fails it may deter the sinking of what might still prove to be a good source. The chance element must be taken into account at every stage and no great reliance should be placed upon any prediction, however expert, that a hole at a given position will yield '*x*' Ml/d until after it is drilled. Until a prolonged pumping period is completed it is wise to have some reservation about the value of any site.

The two forms of yield calculation are of different value in the various possible forms of groundwater development. These include:

(1) river bank collector wells, including those laid in river bed sands,
(2) overflowing artesian wells,
(3) shallow shaft wells,
(4) deep boreholes,
(5) adits, including mine drainage levels or 'soughs',
(6) recharge pits and boreholes,
(7) groundwater 'mining',
(8) well fields for direct supply or river regulation.

Whichever technique is used, there is always a basic need for properly annotated information at a source. Table 4.6 shows a useful format for summarizing this in a regional study when a wide variety of sites are examined. Such data are often held by a national geological agency whose records are normally on open file. Water quality information may

Table 4.6 Water well data summary sheet.

Location .... Date ....
Borehole/Well Ref. No. ....
Original data held at ....

Boring data
Drilled by .... Date ....
Status (observation hole/production hole/abandoned/other) ....
....

Ground level .... m OD		Depth .... metres	
Casing diameter and types	....	.... to	m OD
	....	.... to	m OD
	....	.... to	m OD
	....	.... to	m OD
Lithological succession	....	.... to	m OD
	....	.... to	m OD
	....	.... to	m OD
	....	.... to	m OD
	....	.... to	m OD
Water first struck at .... m OD			
Standing water level on completion .... m OD			
Aquifer surface level .... m OD		Base	m OD

Pumping data

Pump depth	.... m OD
Date of commencing test	....
Test duration (hours) and type	....
Standing water level at start	.... m OD
Max. production/av. production	....
Max. drawdown level	.... m OD

Water quality	*During test*	*In production*
Appearance	....	....
pH	....	....
Conductivity/TDS	....	....
Iron	....	....
	....	....

Remarks
e.g. results of pump test analysis for aquifer characteristics
....
....
....

occasionally indicate that saline water has been struck or that some other characteristic has been observed which makes the yield, however large, useless for the usual purposes. Care must therefore be taken when estimating groundwater yield that the result does not imply a steady encroachment of coastal seawater into the aquifer or that a poor quality water zone of the aquifer will be drawn upon.

4.15 Pumping tests

Once a borehole has been sunk, cleared, and developed by pumping, surging, and, where appropriate, acidizing[21], a thorough pumping test should be carried out. This is normally part of the drilling contract and a temporary pump and power supply arrangement is used. The normal options[22] are to use a suction pump, an air lift arrangement, or a submersible pump. The first of these presupposes a very shallow well.

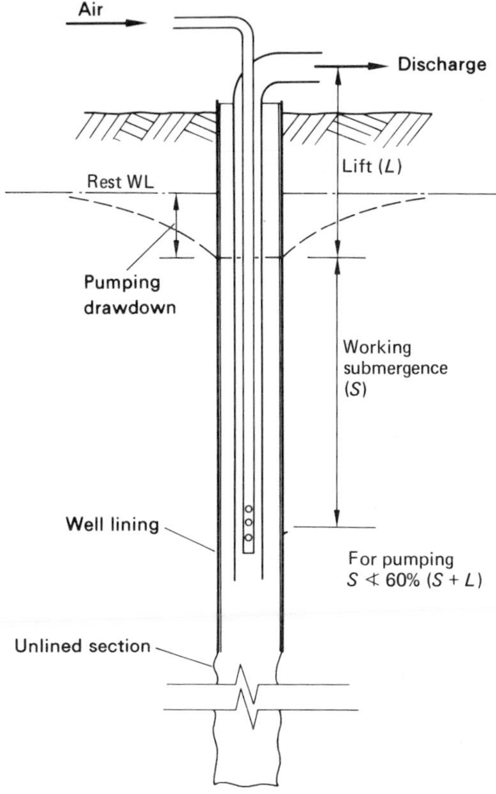

Fig. 4.17 Air lift pump.

The second, although cheap, requires certain critical conditions to be fulfilled (see Fig. 4.17) namely:

(1) the borehole depth must permit a submergence of the air line of more than 60% of its total length to the surface,
(2) the borehole diameter must be great enough to take the air line and eductor pipe,

(3) sufficient air must be supplied to create an adequate air–water mixture that will rise to the surface.

The submersible pump is most frequently used for test pumping because of the wide range of outputs possible, but marked impeller wear should be allowed for in early test pumping where abrasive sands are likely to be drawn from a hole.

A pumping test has several objectives:

(1) to find the abstraction limit of the hole and the rate at which the water level falls with time,
(2) to define the discharge–pumping-level relationship in order to choose an efficient permanent pump,
(3) to monitor the effect of the use of the source on the local environment,
(4) to determine the aquifer's permeability and storage characteristics.

Sometimes it is suggested that pumping, once initiated, should be continued until an 'equilibrium' in output and pumping level is reached. However, only rarely can the hydrogeological conditions permit this because it presupposes that an aquifer is continuously and evenly fed with natural recharge, whereas the latter will often have a marked seasonal variation. Resort must therefore be made to analyses of drawdown relative to rest water level in the non-steady state of the aquifer. Where the water in an aquifer is confined under pressure by an impermeable layer above it, and where the outcrop area is several kilometres away, steadier states will prevail, as they also do in river bank aquifers. A recommended test sequence for investigations, particularly where an aquifer is unconfined, is set out below:

(1) pump for one day at a lowish output,
(2) raise the pumping rate for the second day to a medium output,
(3) raise the pumping output for the third day to a maximum output,
(4) let the hole recover for at least as long as the time occupied by stages (1) to (3),
(5) pump for 14 days continuously at a steady rate just below the rate for stage (3),
(6) let the hole recover for 14 days.

Pumping levels and outputs from stages (1), (2) and (3) of the step-drawdown test can be plotted, as Fig. 4.18, to produce a 'type curve' to assist in locating an appropriate pump duty. Measuring the water level in a pumped well accurately can be difficult and needs a special dip tube located inside the well lining. The remaining part of the test really needs to be carried out in dry weather whilst groundwater levels are low if the closest possible approximation is to be made to a drought yield condition. However, information can be gained if wet

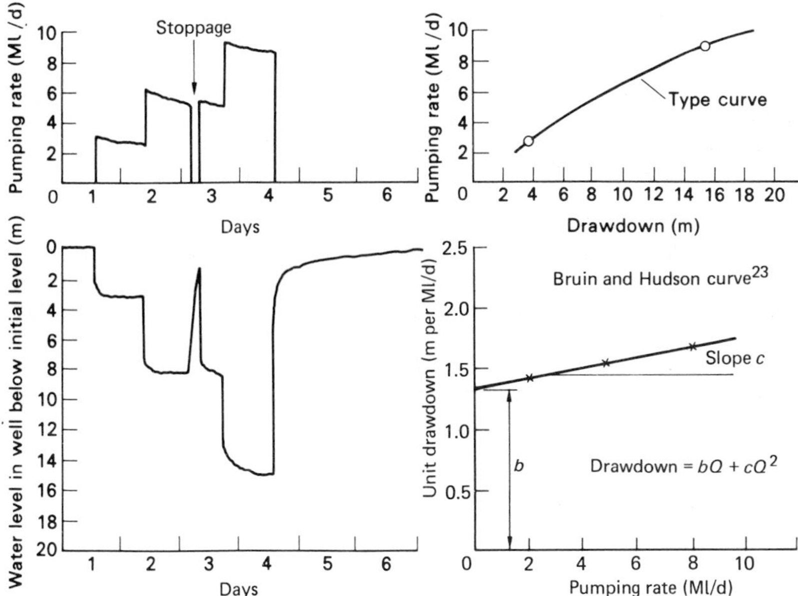

Fig. 4.18 Step-drawdown test results.

weather prevails, but this demands closer attention to additional observation well readings just outside the radius of influence of pumping; corrections for regional water table movement during the test can then be applied in order to make the test results conform to the constant water table assumption of classic well hydraulics.

To determine aquifer characteristics it is preferable to have two observation wells at suitable radii from the pumped bore, at one of which a chart recorder should be installed. The radial distance can vary widely[24], but is governed by three factors:

(1) large enough drawdown during the test to ensure measurement accuracy,
(2) small drawdown relative to the saturated aquifer thickness to ensure that flow to the well is essentially horizontal,
(3) convenient access, e.g. a roadside verge.

More often than not a distance of between 50 and 200 m is appropriate. The larger the distance the greater is the slice of aquifer that is being observed between the pumped well and the measured borehole. A careful measurement of the radial distance should be made. During the testing period, water levels are required for analysis at about equal logarithmic increments of time after each change of pumping rate. This stems from the Theis equation where drawdown is shown to be proportional to the logarithm of time since pumping commenced—thus readings, initially frequent, can be gradually spaced out to become about twice daily. They should be taken in the pumped well and the observation bores, although the former will be less precise and less useful.

Pumping rates should be adjusted in the first few minutes to roughly the desired output by, say, throttling with a valve on the discharge main. There is nothing to be gained by continual small valve adjustments in an attempt to pump precisely some target figure: this will only upset water level readings. Output will fall away more significantly with air lift pumping. Discharge should be stilled within a baffled tank and measured over a vee-notch weir before being discharged via a pipeline to a point distant enough to prevent recirculation. Alternatively discharge taken into supply during a test can be metered with an orifice plate.

Monitoring the surrounding area can be time-consuming, but is essential if the permanent use of the source is to be sanctioned by some licensing authority. Measurements may be needed within the conceivable limit of influence of the source of:

(1) springflows,
(2) main river upstream and downstream of the 'cone of depression' expected in the water table during pumping (Fig. 4.19),
(3) borehole and pond levels, especially private water supplies,
(4) river bank tube wells to check for any reversal of the natural hydraulic gradient to the river,

and even of:

(5) the soil moisture in limited areas where the groundwater level might be drawn down away from the roots of water-sensitive crops.

All these measurements must begin a sufficient time before the start of the test to allow an understanding to be gained of natural water table fluctuations and of normal gain in river flow along each major reach. Only then will the true impact of the test pumping be clear. Readings should also continue for a considerable while after pumping ceases in order to observe recovery until the original conditions are restored. Recovery measurements complement drawdown observations and make conclusions more certain.

4.16 Borehole characteristics

In addition to the type curve of the borehole, careful consideration must be given to the characteristic curve of the permanent pumping installation because, as Fig. 5.25 (p. 197) shows, the combination of the two dictates future performance and yield. The step-drawdown test results depend upon the seasonal rise and fall of the water table and thus the

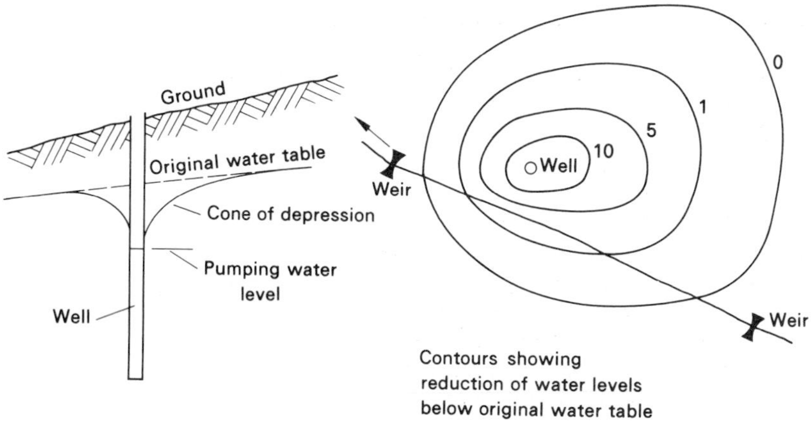

Fig. 4.19 Cone of depression about a pumped well.

type curve can have a range of positions. Each manufacturer offers a wide variety of pump characteristics and thus any particular motor power rating may imply a wide range of output and lift. The choice of a pump with a 'flat' characteristic curve can be unfortunate if the seasonal fall in water level is greater than anticipated because the output of the pump may fall significantly at a crucial time (see Section 12.4). Where a low risk of loss of output is desired, a 'steep' characteristic curve for the pump should be selected, albeit at higher cost. If electricity rates favour off-peak operation to a service reservoir as a daily routine, it is worth establishing, by a period of time-controlled pumping, the proportionate drop in daily output.

The level at which a pump is slung should ensure that it will remain submerged at the lowest water table level and highest concurrent output likely to occur. It will rarely happen that the initial pumping test can be carried out at the height of a rare drought, and therefore predictions are necessary of the output, i.e. the hydraulic yield, to be expected in, say, a 50-year drought. It is not sufficient in such a case to extend the drawdown–time diagram to the length of a single dry season, ending when the aquifer natural recharge commences, because it is frequently the cumulative depletion of some years of below-average recharge that most severely taxes a source. Estimates must be made, by extrapolation, of the rest water level and drawdown values at the end of all likely critical drawdown periods and the characteristic curve must be adjusted accordingly. As an example, this type of estimation for the chalk of Thetford in Norfolk gave initial pumped yields averaging 4.3 Ml/d, which declined to 3.4 Ml/d in a notable drought in 1970, and which are expected to fall to 2.8 Ml/d during a 50-year drought.

It is useful to replot step-drawdown results as a Bruin and Hudson[23] curve (see Fig. 4.18). This reveals in what proportion pumped drawdown is due to the nature of the aquifer or the nature of the well. Examination of the turbulent loss coefficient, *c*, can sometimes suggest high well entry losses due to poor screen or pack design; the laminar loss coefficient, *b*, describes the relative permeability of the aquifer(s) feeding the well. Good, i.e. low, figures for *c* are below 0.5 min^2/m^5.

4.17 Aquifer characteristics

The two essential characteristics of an aquifer are horizontal transmissivity, *T*, which is the product of the permeability and the wetted aquifer depth, and the storage coefficient, *S*.

Transmissivity, *T*, is the flow through unit width of the aquifer under unit hydraulic gradient. Its units are therefore $m^3/(m/d)$, often abbreviated to m^2/d. Previously gallons per day per foot were used, although this led to confusion because of the difference between the American and British gallon.

The storage coefficient, *S*, is defined as the amount of water released from the aquifer when unit fall in the water table occurs. Where free water table conditions occur it is the volume of water released from a unit volume of an aquifer (expressed as a percentage of the latter) that will drain by gravity with a unit water table fall. When the aquifer is confined under pressure it is the percentage of unit aquifer volume that must be drained off to reduce the piezometric head by unit depth. The difference between these two meanings, although subtle, is vital: whereas the former may be in the range 0.1% to 10.0%, the latter may be 1000 times smaller (demonstrating the incompressibility of water). Thus to achieve the same water table fall in unconfined and confined conditions takes 1000 times less abstraction in the latter case, all else being equal.

By considering a well as a mathematical 'sink', it has been shown by Theis[25] that drawdown in a homogeneous aquifer due to a constant discharge, *Q*, initiated at time $t=0$ is

$$h_0 - h = \frac{Q}{4\pi T}\left[-0.5772 - \ln u + u - \frac{u^2}{2(2!)} + \frac{u^3}{3(3!)} - \frac{u^4}{4(4!)} + \ldots\right]$$

$$= \frac{Q}{4\pi T} W(u)$$

where $u = (r^2 S/4Tt)$, h_0 is the initial level, and *h* is the level after time *t* in a well, distance *r* from the pumped well.

Any consistent set of units can be used. For example, if *Q* is in m^3/d and *T* is in m^2/d then h_0 and *h* must be given in metres. *S* is a fraction. $W(u)$, the 'well function' of the Theis equation, can be obtained from tables[26].

In the Theis method of solution a type curve of $W(u)$ against u is overlaid on a plot of the pump test drawdowns versus values of $\ln(r^2/t)$. Where a portion of the type curve matches the observed curve, coordinates of a point on this curve are recorded. With these match-point values the equations can be solved for S and T. A graphical expression of the Theis equation is given in Fig. 4.25 (p. 141).

However, Jacob's less exact method[27] is easier to apply and meets most situations. He pointed out that if time t is large, as in most major pumping tests, then u is small (say less than 0.01), and so the series in the Theis equation can be shortened to

$$h_0 - h = \frac{Q}{4\pi T}\left[-0.5772 - \ln\left(\frac{r^2 S}{4Tt}\right)\right]$$

$$= \frac{Q}{4\pi T}\left[2.30\log_{10}\left(\frac{4Tt}{r^2 S}\right) - 0.5772\right]$$

$$= \frac{2.30Q}{4\pi T}\log_{10}\left(\frac{2.25Tt}{r^2 S}\right)$$

Plotting drawdown against time at an observation borehole within the cone of depression thus produces a straight line (Fig. 4.20). If the drawdown for one logarithmic cycle of time is read off then

$$T = \frac{2.30Q}{4\pi(h_0 - h)}$$

where $h_0 - h$ is in metres and Q is in m^3/d.

Then reading off the time intercept, t_0 days, for zero drawdown

$$S = \frac{2.25Tt_0}{r^2}$$

where T is in m^2/d and r is in metres.

Certain qualifications are set out below.

(1) If the regional water table varies markedly during the test then the drawdown is affected and should be adjusted by an equivalent amount throughout the test pump analysis.

(2) Early time data should not be used because there will then be substantial initial vertical flow as the storage is evacuated. Boulton suggests the necessary horizontal flow conditions exist when observations are taken at $r > 0.2d$ and

$$t > \frac{5dS}{K_z} \text{ days}$$

where d is the wetted aquifer depth and K_z is the vertical hydraulic conductivity which can be taken as T/d, the horizontal hydraulic

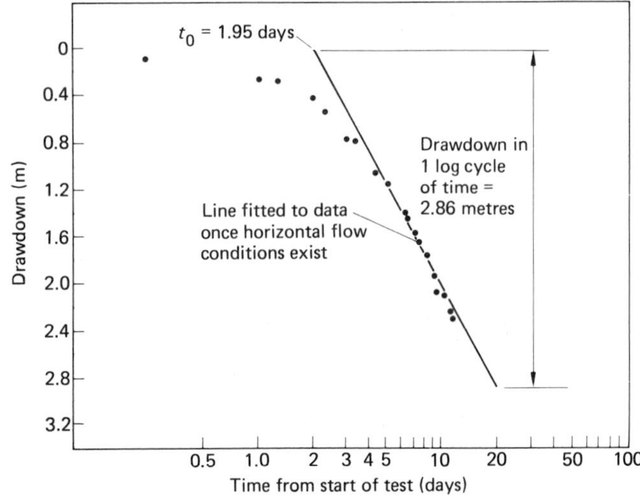

Fig. 4.20 Jacob's pump test analysis for the test pumping of a well at 4000 m³/d. Data are plotted from an observation well 200 m from the test well.

$$T = \frac{2.3 \times 4000}{4\pi \times 2.86} = 256 \text{ m}^3\text{/d per metre}$$

$$S = \frac{2.25 \times 256 \times 1.95}{200^2} = 0.028 \text{ or } 2.8\%$$

conductivity, if K_z is not otherwise known. (In horizontally layered strata K_z may be one-third or less of the horizontal conductivity.)

Some reiteration with values of S and T is required to make use of this guide.

(3) Where the pumped well only partially penetrates the aquifer[24], it will be necessary to adjust the drawdown values for the non-standard flow lines that then ensue unless the observation well is at least 1.5 times the aquifer depth away from the source.

(4) Where drawdown is large compared with the aquifer thickness, and the aquifer is unconfined, the measured drawdowns should be corrected[27] by subtracting from them

$$\frac{(\text{drawdown})^2}{2 \times \text{wetted aquifer depth}}$$

Table 4.7 shows the typical layout of pumping test results.

Analyses for S and T can also be made during the recovery part of the test at an observation bore: if only levels in the pumped well are obtainable the only reliable information that can be achieved is usually

Table 4.7 Layout of pumping test results.

Average pumping rate Q. . . . .
Results from observation well No. Radial distance r . . . .
Aquifer thickness . . . .
Regional water table change during test δh_0 per day . . . .

(1) Date	(2) Time	(3) Time from start of pumping t	(4) $\frac{r^2}{t}$	(5) Level in well h above datum. Initial value h_0	(6) Drawdown $S_1 = h_0 - h$	(7) Correction for regional water table change $\delta h_0 t$	(8) Corrected drawdown (6) ± (7)	(9) Correction for large drawdown $\frac{[\text{col. }(8)]^2}{2d}$	(10) Final drawdown (8) − (9)

an estimate of T on recovery. The methods involved are well covered in standard textbooks[22].

Once S and T are established it is then possible to predict with the above equations what drawdown below current rest water level will result at different pumping rates, different times, and other distances. Where more than one well can create a drawdown at a point of interest, the total effect can be calculated by the principle of superposition, i.e. the drawdowns due to individual well effects can simply be added. Analytical solutions[24,26] exist for many aquifer conditions, including boundary effects from impermeable faults and recharging streams.

It will be appreciated that the engineer has no control over the values of S and T found at a well site. He cannot control the way in which water drawn from the hole flows to that hole in dry periods: it may come from local aquifer storage or from a remote outcrop after a long time. However, resiting, deepening, or duplicating a bore are all options that may be called upon once S and T are known.

4.18 Yield of an aquifer

In water resource assessment it is often necessary to quantify the limiting hydrologic yield of an aquifer. This is frequently taken to equal its long average recharge wherever the quantity of water stored in the aquifer is known to be large enough to balance out almost every variation of percolation and abstraction.

First the appropriate groundwater catchment is defined from the contours of the water table. It is generally assumed that the water table reflects the ground surface to a reduced scale, but this is by no means always true. It has to be assumed that underground flow is in the direction of the major slope of the water table. This may not be true because of the influence of fissures and open bedding planes, but it is best to follow the general rule that the surface catchment is taken to be the underground catchment except where there is positive evidence to amend this. Sometimes, of course, the catchment outcrop may be remote from the point at which the wells tap the strata concerned. The average recharge may then be estimated as average rainfall minus evapotranspiration over the outcrop area, providing no surface drainage losses occur. Supplementary recharge can take place through masking clays and leaking river beds, although hydraulic gradients do not often favour the latter except in arid areas; however, its estimation can be near to guesswork until there is a long experience of successful pumping.

Aquifer percolation formulae can be of use, but need confirming evidence before being used out of the area for which they were derived. Recent research has demonstrated from tritium contents in chalk–limestone pore water and fissure water that, whereas the latter may be travelling down to the water table at more than 0.3 m/d, the pore water recharge front may move down at only 1 to 2 m/year. Determining the volumes travelling by these alternative routes is fraught with difficulty and, for the present, it seems safer to assume that the swiftest path can be adopted: any other water arriving at the water table after considerable delay is likely to enhance drought season yield by a small margin.

Wherever possible the most reliable method of analysis is the measurement of groundwater flow (baseflow) in the river draining the catchment concerned. Percolation must emerge as baseflow (except where it produces sea-bed springs) and therefore long average estimates of the latter give the former. Dry weather recessions of both flow and index well levels in the catchment can be used to help separate surface runoff and shallow drainage from true springflow. Baseflow is percolation routed through storage so, by correcting seasonal baseflow for groundwater level change, it is possible to estimate the percolation for each wet season throughout a flow record. The rise (or fall) of the catchment water table over each wet (or dry) season is multiplied by the aquifer storage coefficient to make this correction. Ranked and plotted on probability paper (Fig. 4.21) these percolation values can be interpreted to give drought estimates of aquifer replenishment.

Where a major group of wells exists, particularly if they are adit fed, it is often possible to delimit the maximum water table drawdown, averaged over their catchment, that may occur during a design drought before the wells begin to fail. Once this has been converted to an equivalent of reservoir storage by multiplying by their average storage coefficient, S, a yield calculation can be made. Table 4.8 sets out the balance of storage and recharge over possible critical drought durations to find the maximum possible abstraction rate. Column (2) is obtained from Fig. 4.21, but column (3) is an estimate from a knowledge of the catchment.

From such calculations it may be foreseen, for instance, that wells or pumps are not deep enough to average out all percolation fluctuations

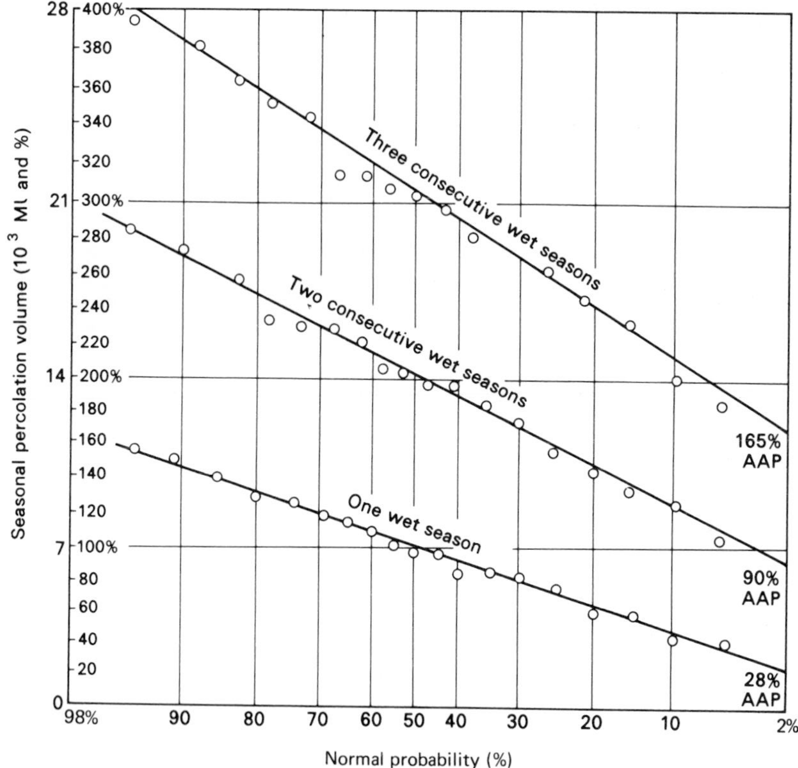

Fig. 4.21 Seasonal percolation probability plot.

to permit maximum yields. The table also illustrates why a dry period covering several years can be crucial for a well source.

Percolation gauge readings can be used for this sort of analysis, but it is difficult to be sure they represent average catchment conditions. Their rims prevent surface runoff and, being on level ground, shallow drainage does not occur so the measured percolate may be an optimistic estimate of the amount that reaches the water table under natural conditions. Practical difficulties occur when they are used for a long period: they can become moss covered or the soil may shrink away from the edge of the container.

Percolation can be estimated by a soil moisture storage balance method of the type demonstrated by Rodda[10] and Headworth[28]. Once readily available moisture in the root zone is used up by evapotranspiration in a dry period, then the loss rate drops markedly. The soil moisture deficit must be made up before excess rainfall can infiltrate

Table 4.8 Groundwater yield calculation.

(1) Drought length	(2) Recharge percolation in a 2% drought % annual average percolation	 (Ml)	(3) Total loss by springflows (Ml)	(4) Storage in aquifer that can be used during drought (Ml)	(5) Average abstraction possible [(2) – (3) + (4)]/(1) (Ml/d)
1 summer (240 days)	Nil	Nil	200	6000	19.2 (average percolation)
2 summers 1 winter (600 days)	28	1960	200	6000	13.0
3 summers 2 winters (965 days)	90	6300	200	6000	12.5 (critical)
4 summers 3 winters (1330 days)	165	11500	200	6000	13.0

once more. As Fig. 4.22 shows, the average percolation that is produced by this method depends strongly upon the amount of moisture storage within the catchment cover root depth. These storage or 'root constant' values are known for several crops, ranging from 25 mm for short-rooted grassland to 200 mm for woodland. Figures such as Fig. 4.22 require specific calculation for each climatic region.

Having found the potential yield of a groundwater catchment it is possible to compare it with the total licensed groundwater abstractions and, after some field work, with what is usually the rather lower amount of actual abstraction. This gives the remaining potential for expansion or, if too much is being taken, the extent of groundwater mining. Note that such mining can occur in one part of an aquifer while elsewhere springflows may indicate that all is well. In such cases where transmissivity is low, it is necessary to get a better spread of abstraction wells. Traditional problem spots are along coastlines where excess groundwater abstractions within towns and industrial plants have led to saline intrusion. In such situations it is necessary to reduce pumping to ensure a small net outflow to the sea to keep out salt water. This is best checked by preserving a small positive gradient of the water table to mean sea level: by drawing water from shallow depth the risk of tapping the denser water of the saline wedge below is lessened. Since seawater is 1.025 times heavier than freshwater, where the two water bodies are in equilibrium it will be found that the fresh–salt interface is $40h$ below sea level, h being the freshwater head above sea level.

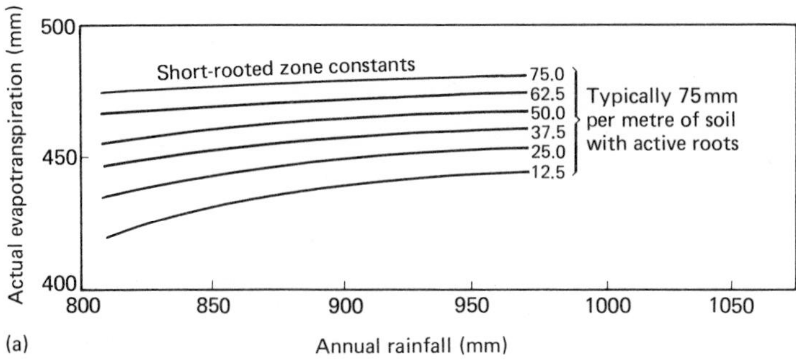

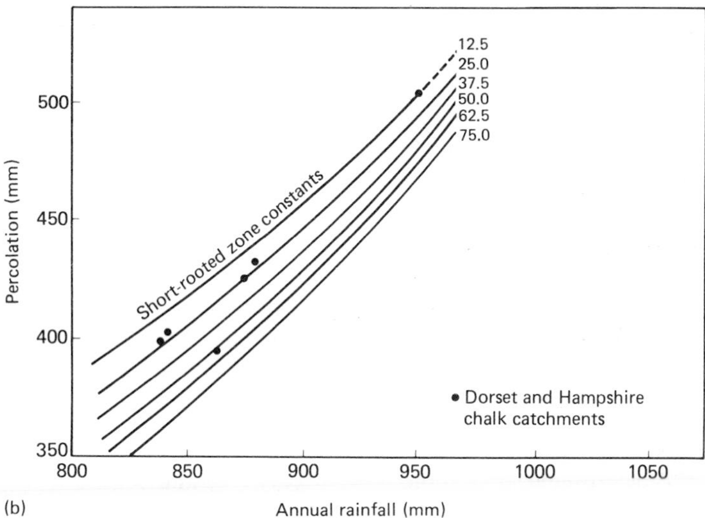

Fig. 4.22 Relationship between rainfall and (a) evapotranspiration and (b) percolation using various root constants. (After Headworth[28].)

4.19 Significant water supply aquifers

The porosity of a rock is no guide to its potential usefulness for water supply because water locked in, say, the pores of silt cannot be made to drain out sufficiently rapidly; nor will a rock sustain a yield properly if it is so free-draining that any recharge leaves through springs as rapidly as it entered the ground. Carboniferous limestone is a good example of the latter case.

Major aquifers are generally those which can be expected to have an adequate area of outcrop for recharge and a wetted depth of over 25 m with any of the following:

(1) a relatively even, dense network of small interconnecting solution fissures,
(2) many joints and bedding planes,
(3) thick sand and gravel layers.

Thus the main British aquifers[29] matching these groups are (1) the Chalk and the rather similar Oolitic Limestones, (2) the Triassic (Bunter) sandstones, and (3) the lightly cemented Greensands. To a lesser extent the Carboniferous gritstones are used, as are some river gravel deposits and occasional sheets of Pleistocene sands. Table 4.9 gives a rough indication of supply potential averaged from published results for a range of aquifers in England which has a wide variety of geological strata, but where a temperate rainfall averages only about 600 to 900 mm/year over outcrop areas. The best boreholes or wells may produce three times as much*. Data from representative aquifers in other countries are given for comparison.

Private water supplies for one or more houses can be maintained from a vast range of deposits. Boreholes from even granite bedrock are quite often successful in giving 5000 litres per day. However, with private supplies it is quality criteria, rather than quantity, which rule out strata from being considered. Some series can contain very mineralized water, as does any aquifer once it dips under a clay cover without an outlet permitting throughflow. Water at great depth remote from an outcrop is therefore frequently unusable as a potential source. However, deep sandstone aquifers in Saudi Arabia are being developed successfully as far east as that country's coastline: water is relatively fresh due to leakage losses up to salt pans or the sea.

4.20 Groundwater modelling

When it is desirable to model the behaviour of a groundwater catchment there are two basic approaches:

(1) an electrical analogue model,
(2) a mathematical model.

Both simple and sophisticated analogue models can be built, making use of the similarity between Ohm's law and Darcy's law of groundwater flow:

$$Q = Tiw$$

where T is the transmissivity and i is the hydraulic gradient through an aquifer cross-section of width w.

Electrical current is analogous to discharge, and voltage to hydraulic

* Quantities sufficient for public water supply require at least a 250 mm diameter borehole. Increasing the diameter beyond 600 mm does not often add greatly to the value of a source.

Table 4.9 Approximate average borehole capacities.

Aquifer	Nominal borehole diameter (mm)	Region	Output at large drawdown (Ml/d)	Reference
England				
Chalk				
Upper	455	East Anglia	3	
Middle	610	Southern England	8	
Bunter	760	Midlands (Pebble beds)	6	
Sandstone	455	Fylde (Lancs)	4	16
	380	Devon	2	30
Lower Greensand	610	Kent	2.5	31
Magnesian Limestone	455	Durham	5	32
Gritstones	455	Pennines	1–3	31
Worldwide				
Metamorphic	150	West, Central and South Africa	<0.1	
Dolomites	200	West, Central and South Africa	<0.5	
Sandstones	—	West and South Africa	0.5	
		Colombia	10.0	
Sands	—	Sahara	0.25	33
		Surinam	1.0	
River alluvium	—	Congo/Nile/Niger	2.5	
		Peru/Venezuela	5.0	
Volcanic	—	Central America	2.5	

Note: Where possible median output values are quoted. However, it would be dangerous to use this table other than as a starting point in planning local investigations.

gradient. Resistors form the transmissivity and capacitors the storage[34]. Such models are suitable for illustrating flows in a known aquifer to a non-technical audience, but are unwieldy if substantial alterations to S and T are desired. The decreasing cost of digital computing has made analogues relatively rare.

Mathematical models commence with simple equations representing aquifer storage effects, and range up to major digital computer simulations of pumped aquifers with all the necessary equations of flow being incorporated[35]. The most common models view an aquifer as a grid of points across which flow can be represented by finite difference techniques. They require specialist staff to produce, calibrate, and run.

Much can be learned from an equation which relates springflow, Q, directly to storage volume, S, at any instant of time.

$$Q = kS$$

where k is the recession constant of springflow out of storage. It follows that

$$Q_t = Q_0 \mathrm{e}^{-kt}$$

where k has the unit day^{-1} if t is in days. (Typical values lie between 0.01 per day in a good aquifer and 0.10 per day in a relatively impermeable catchment.) Q_0 is the initial baseflow (i.e. total springflow), Q_t is the baseflow after t days, and e is the natural logarithm base.

The slope of a logarithmic plot of dry weather flow (Fig. 4.23) will indicate k, and this enables a reasoned estimate of the groundwater stored in a catchment to be made. If storage is increased by percolation, springflow rises correspondingly. Thus the seasonal rise and fall of

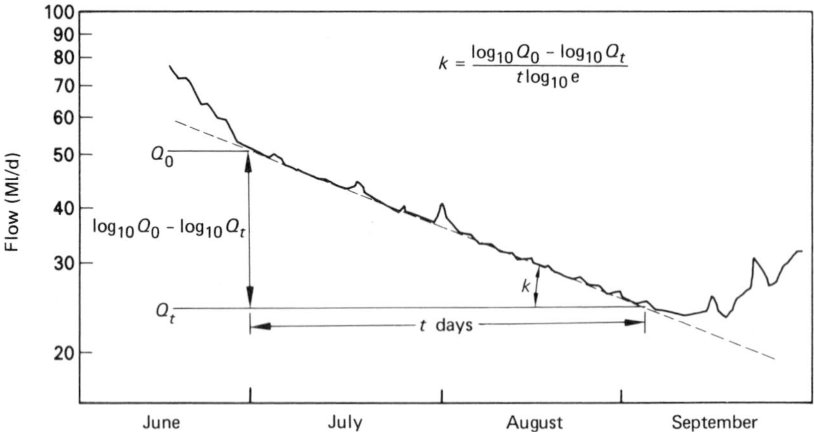

Fig. 4.23 Groundwater recession graph.

groundwater levels can be simulated once percolation estimates are available. Given a period without percolation, it is also possible to calculate the amount by which springflow will fall away. The yield of a spring source and the impact on springflows of major pumping can be quantified in this way. Pumped abstraction is the reverse of percolation and, if treated as an equal take from storage, S, each day, its seasonal effect on Q is soon seen. Figure 4.24 shows mean monthly springflows calculated only from rainfall, evaporation, and assumed figures for the recession constant. Occasional spot-flow measurements that were available for this limestone valley stream are close enough to give confidence in the use of the technique in other years.

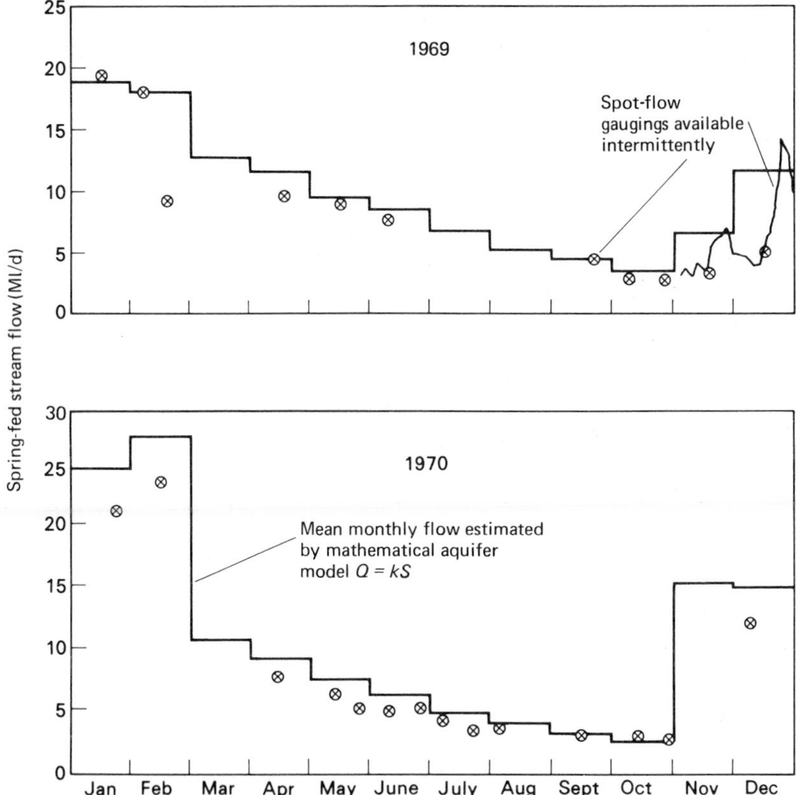

Fig. 4.24 Synthetic and measured groundwater outflow.

4.21 Impact on the environment

The abstraction of small quantities of groundwater and their return to the aquifer by soakaways in the same locality is rarely noticeable.

However, agreement to the pumping of a large part of an aquifer's potential out of a valley to some distant city may be difficult to obtain without causing a great deal of controversy. Objection may as often arise from principle as much as it does from fact, as instanced by the case where public announcement was made of the commencement of a test pumping for a certain day, leading to receipt of compensation claims in the next day's post, although the start of the test pumping was in fact delayed.

Depending upon circumstances, pumping may or may not cause:

(1) depletion and cessation of springflow (often noticed at watercress beds and village water sources),
(2) lowered groundwater levels and hence local private supply difficulties,
(3) drying out of ponds and lakes either by assisting drainage from them or reducing inflow to them,
(4) drying out of riparian land, which may greatly assist drainage but remove wetland habitat,
(5) drying out of stream lengths and lowering of bourne* heads to the detriment of fisheries and of visual appeal,
(6) damage to vegetation and crops in those limited areas where root systems could originally make use of the water table,
(7) land subsidence and soil cracking where the pumped well is set in unconsolidated strata,
(8) saline water to be pulled inland to the detriment of adjacent bores.

Measures to limit the effect of a source on people and their environment are commonsense. Whenever possible wells should be sited away from famous spring heads and wetland areas of special scientific interest, by a distance that will reduce drawdown to less than 0.5 m, even after a dry season. Quantities pumped should not be such a large proportion of resources that the seasonal character of local springflows is changed entirely in localities which could not be compensated. Predictions of the drawdown at every source within the effective radius of influence of the well should be made, so that if necessary private boreholes can be deepened, pumps lowered, and shallow wells replaced with mains supplies. Figure 4.25, which is a graphical expression of the Theis equation (Section 4.17), should assist in this. Field troughs may need to be put on the mains. Some additional compensation pumping may be required at times to keep up stream levels and scheme costs will rise accordingly.

4.22 River regulation by groundwater pumping

Regulating river flow by groundwater pumping is growing in

* A bourne is a stream fed by a spring the position of which moves up and down a valley with water table changes.

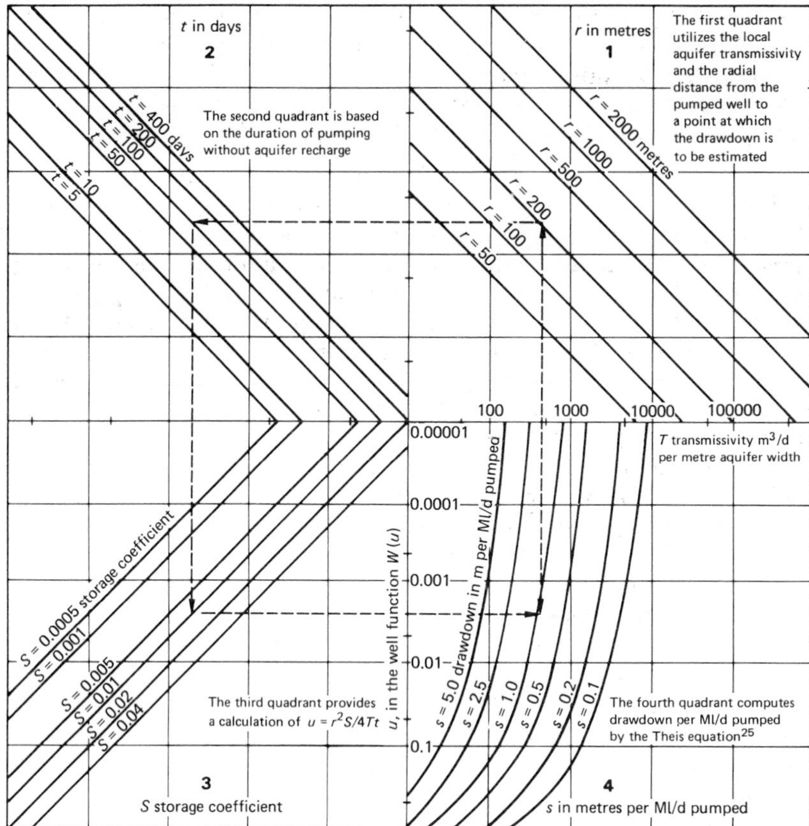

Fig. 4.25 Graphical expression of the Theis equation.

importance[36]. In essence such projects resemble those involving surface storage except that the storage is unseen, underground, and leaks. Aquifer storage exists and is free, but creating outlet capacity is expensive; with surface storage the reverse is true and it is forming the storage that consumes capital expenditure, leaving outlet costs minimal. Successful pilot schemes have been extended to major developments by the Thames[37] and Anglian[38] Water Authorities during the 1970s and at least one American example[39] exists.

Certain basic criteria need to be met:

(1) The river regulation should provide a marked increase in yield above the alternative of direct piped supplies from the same aquifer.
(2) The use of the river system as an aqueduct should bring economies in pipeline costs.

(3) The pumping capacity at the height of a design drought should exceed the consequential reduction of local springflow by the desired gain in river flow downstream.
(4) If recirculation losses through the river bed occur, it must be practicable to provide equivalent extra well capacity.

The main difficulty in yield estimation stems from the fact that, in reality, the act of regulation pumping itself causes a fall in natural river flow, thus more pumping is required and so on until, in many circumstances, springs cease completely. To avoid iterative calculations it seems best to resort to the premise that in a well field during a major drought there will be effectively no natural springflow once pumping is really under way. This means that the resource being regulated by the aquifer is the sum of runoff from areas above the regulated supply intake having no well development and uncontrollable surface drainage, if any, in the well field itself.

The maximum regulation that can be accomplished will be the amount of aquifer storage discharged to the river during the drought period which would not have drained naturally at that time.

Defining the amount of drawdown that can be created by controlled pumping in some future severe drought is hazardous. Pumping tests of the well field to achieve high regulation flows during dry periods are essential for this, but are expensive even if an off-peak electricity tariff time is used. Caution is called for until experience is built up because groundwater yields in most areas fall with time and cannot be precisely foreknown until the critical event happens. This is particularly true with regulation schemes, which may not need to be used in, say, seven years out of ten because ordinary river flow is adequate.

4.23 Artificial recharge

Artificial recharge of an aquifer is more often practised for water treatment reasons than it is to augment groundwater yields. The two basic methods adopted are the use of spreading areas or pits and the use of boreholes.

The former technique has predominated because of its simplicity and the ease with which clogging problems can be overcome. The relatively unknown performance of wells, and the need to pass only pretreated water down them, limits their usefulness except where land for recharge pits is at a premium. Untoward events can happen during recharge operations: at an Israeli site[40] the injection of water down a well caused unconsolidated sands around it to settle and, in less than two days after pumping ceased, the borehole tubes and surface pump house sank below ground level; in a British experiment[41] even the use of a city drinking water supply for recharge did not prevent the injection well lining slots from becoming constricted with growths of iron bacteria.

At times when existing surface water treatment works are working below capacity, e.g. when there is a seasonal fall in demand, it may be possible to make water available for well recharge. This is being practised by the Thames Water Authority and the Lee Valley Water Company[42]. Complete efficiency cannot be expected because of the relatively uncontrollable and 'leaky' nature of aquifer storage. However, such situations are relatively rare and for the rest of this section only recharge pits will be considered.

Although dimensions vary, a recharge pit bears resemblance to a slow sand filtration bed because replaceable filter media normally cover the base. The rating of such a pit must depend upon the rate at which raw water will pass through the filter media and the rate at which the underlying aquifer will accept water.

The former depends upon raw water quality, pretreatment of the water (if any), and the depth of water kept in the pit. Little if anything can be done to improve the latter, but the possibility of a falling off in the acceptance rate must be guarded against by careful tests to ensure the success of the filter media. Published results for infiltration rates naturally vary widely[43]. Typical figures are given in Table 4.10 for pit water depths of 1 to 2 m using settled river water. Where potable water is used and the aquifer is fissured, rates up to ten times higher may be found.

Table 4.10 Typical infiltration rates.

Underlying aquifer	Pit recharge rate (m/d)
Sand and gravel	3.0
Fine sand	0.4
Sandstone	0.3+

However, pilot tests are always required at a new site, as are initial investigation bores to ensure the pit floor will be above the water table, preferably by a margin of about 3 m after recharge 'mounding' has occurred. Iron-pan layers and other impedances to vertical flow should be avoided.

4.24 Special cases

Overflowing artesian conditions are a frequent occurrence where an aquifer is overlain by some impermeable layer in a hollow or valley. Drilling through such layers can produce some spectacular results, giving the water engineer the equivalent of an oil gusher. Such wells should be capped off as soon as possible and be equipped with a pressure recorder. If this is not done, and the water is run to waste, it is possible that regional levels may be lowered once and for all. Artesian flows may be developed in a similar fashion to springs, but sometimes they fail seasonally, bringing the necessity for occasional pumping.

Fissured aquifers are often exploited with adits which are galleries, usually about 2 m high, driven just below the water table surface from a central pumping shaft. In the lower chalk–limestone, for instance, such adits have run 300 m or more before encountering a single fissure bringing in a significant flow. The result is a well source that cannot be made to increase its hydraulic yield by a greater drawdown but which, instead, responds as a spring source since the adit will only partly fill with water. Mine drainage 'soughs' are very similar.

A patented Ranney collector well can be useful in sand and gravel deposits, especially where these are alluvial and thin. The well comprises a large diameter central shaft from which perforated collector tubes are jacked out horizontally in the most appropriate configuration, e.g. parallel to a river bank. It may, however, have a short life if the well is in reality a river intake. Suspended sediment will be pulled deeper and deeper into the river bank until the latter becomes seriously clogged and head drop increases across the deposits. Flood velocities can succeed in cleansing the bank, permitting yields to rise again, but the worst feature is the way in which yields may drop during a long dry period. Great reliance therefore should not be put on such a source for having a life greater than, say, about ten years. Its strong point is its low capital cost as it combines both resource and filtration.

Wellpoint intakes[44] can be used in sand rivers and are found particularly in areas such as South West Africa. These are like wadis in that they only flow seasonally but, in fact, there is a continuous unseen baseflow with the sand bed. The water is protected from high potential evaporation losses by this alluvial cover. In certain instances underground dams have been built to increase the output of sand river sources by creating storage within the bed. The wellpoints should be made from properly designed wire-wedge screen as low entry loss is important.

Peak-lopping reservoirs should not be overlooked. This concept is an extension of the service reservoir to the point where it will balance out the dry season peak demand to a figure closer to the annual average. The provision of such storage in a water undertaking dependent upon wells has the effect of reducing the peak resource capacity required. A smaller number of wells can meet wet season demand and refill storage prior to the onset of the peak period. The Hilfield Park Reservoir[45] north of London is the earliest known example.

References

(1) Baker A C J, Trial Banks, *Proc. ICE/CWPU Symposium on The Wash Storage Scheme*, 1976.

(2) Steering Committee on Water Quality, DoE, 1st Annual Report, Circ. 22/70, HMSO, 1972.
(3) Manley R E, Simulation of Flows in Ungauged Basins, *Hydrological Sciences Bulletin*, **3,** 1978, pp. 85–101.
(4) Codner G P and Ribeny F M J, The Application of the Sacramento Rainfall Model to a Large Arid Catchment in Western Austrália, *Proc. Inst. Eng., Australia, Hydrology Symposium*, 1976.
(5) Fleming G, *Computer Simulation Techniques in Hydrology*, Elsevier, 1975.
(6) Ministry of Housing and Local Government, *Hydrological Survey of North Lancashire Rivers*, HMSO, 1964, p. 11.
(7) Stall J B, Reservoir Mass Analysis by Low-flow Series, *JASCE San. Eng. Div.*, **SA5,** September 1962, p. 21.
(8) Hardison C H, Storage to Augment Low Flows, Paper 8, *Proc. WRA Symposium on Reservoir Yield*, 1965.
(9) Law F, The Estimation of the Reliable Yield of a Catchment by Correlation of Rainfall and Runoff, *JIWE*, **7,** 1953, pp. 272–293.
(10) Rodda J C, A Drought Study in SE England, *WWE*, August 1965, pp. 316–321.
(11) Kottegoda N T, Statistical Methods of Flow Synthesis for Water Resources Assessment, *Proc. ICE Supplement xviii*, 1970.
(12) Caponers D A, *Water Laws in Moslem Countries*, FAO I and D Paper 20/1, 1973.
(13) Law F M, Determination of Residual Flows in Rivers, *Proc. ICE*, **53,** December 1972, pp. 691–694.
(14) *Residual Flows to Estuaries*, Central Water Planning Unit (UK), Reading 1979.
(15) Cole J A, Assessment of Surface Water Sources, *Proc. ICE Conf. Engineering Hydrology Today*, 1975.
(16) Law F, Integrated Use of Diverse Resources, *JIWE*, **19,** 1965, pp. 413–461.
(17) Bolton K, Milligan C and Sankey K A, The Operation of the Water Resource System of Manchester, *Proc. Symposium Inspection, Operation and Improvement of Existing Dams*, BNCOLD, University of Newcastle Upon Tyne, 1975.
(18) McMahon T A and Mein R G, *Reservoir Capacity and Yield*, Elsevier, 1978.
(19) *Studies of the Reliability of Water Supplies*, Central Water Planning Unit, (UK), Reading, 1977.
(20) Law K K, Yield of Diddington Pumped Storage Scheme, Paper 12, *Proc. WPA Symposium on Reservoir Yield*, 1965.
(21) Stow A H and Renner L, Acidising Boreholes, *JIWE,* **19,** 1965, pp. 557–575.
(22) *Groundwater and Wells*, Edward E Johnson Inc., Saint Paul, Minnesota, 1966. See Table XXX.
(23) Bruin J and Hudson H E, *Selected Methods for Pumping Test Analysis*, Report 25, Illinois State Water Supply, 1955.
(24) Kruesman G P and de Ridder N A, *Analysis and Evaluation of Pumping Test Data*, International Institute for Land Reclamation and Improvement, 1970.

(25) Theis C V, The Relation Between Lowering of the Piezometric Surface and the Rate and Duration of Discharge of a Well using Groundwater Storage, *Trans. American Geophysical Union*, **16,** 1935, pp. 519–614.
(26) Walton W C, *Groundwater Resource Evaluation*, McGraw-Hill, 1970.
(27) Jacob C E, in: *Flow of Groundwater* (Ed. H. Rouse), John Wiley & Sons, 1950.
(28) Headworth H G, The Selection of Root Constants for the Calculation of Actual Evaporation and Infiltration for Chalk Catchments, *JIWE*, **24,** 1970, p. 431.
(29) Rodda J C, Downing R A and Law F M, *Systematic Hydrology*, Newnes-Butterworth, 1976. See Table 3.10.
(30) Sherrell F W, Some Aspects of the Triassic Aquifer in East Devon and West Somerset, *QJ Eng. Geol.*, **2,** 1970, p. 278.
(31) *Manual of British Water Engineering Practice*, 4th Ed., Volume II, IWE, 1969, p. 103.
(32) Cairney T, Investigation of Magnesian Limestone, *J Hydrology*, **XVI,** 1972, p. 330.
(33) *Groundwater in Africa*, 1973, and *Groundwater in the Western Hemisphere*, 1976, United Nations.
(34) Walton W C and Prickett T A, Hydrologic Electric Analogue Computers, *JASCE*, HY6, November 1963, p. 67.
(35) Boonstra J and de Ridder N A, *Numerical Modelling of Groundwater Basins*, International Institute for Land Reclamation and Improvement, 1981.
(36) Jackson H B and Bailey R A, Some Practical Aspects of River Regulation in England and Wales, *JIWE*, **33,** 1979, pp. 183–199.
(37) Thames Groundwater Scheme, *Proc. ICE Conf.*, 1978.
(38) Steering Committee, *Groundwater Pilot Scheme*, Final Report, Great Ouse River Authority, 1972.
(39) Todd D K, *Groundwater Resources of the Upper Great Miami River Basin and the Feasibility of their Use for Streamflow Augmentation*, Report to Miami Conservancy District, Ohio, 1969.
(40) Sternam R, Artificial Recharge of Water through Wells: Experience and Techniques, *Proc. IASH Symposium on Artificial Recharge and Management of Aquifers*, Haifa, 1967.
(41) Marshall J K, Saravanapavan A and Spiegel Z, Operation of a Recharge Borehole, *Proc. ICE*, **41,** 1968, pp. 447–473.
(42) Edworthy K J, Headworth H G and Hawnt R, Application of Artificial Recharge Techniques in the UK, in: *A Survey of British Hydrogeology*, Royal Society, 1981.
(43) *Proc. Artificial Groundwater Recharge Conf. September 1970*, WRA, 1971.
(44) Powers J P, *Construction Dewatering*, John Wiley & Sons, 1981, Chapter 17.
(45) The Hilfield Park Reservoir and Clay Lane Treatment Works of the Colne Valley Water Co., *WWE*, November 1955, p. 495.

5

Design of works to procure water

Part I
Dams

5.1 Introduction

Dams are usually made of earth, rock, concrete, or masonry. The choice of material depends firstly upon the geology of dam site and secondly upon the cost of various alternatives. Concrete and masonry dams require a hard rock foundation, but earth dams can be placed either on rock foundations or on firm clays and other sound strata, not so hard as rock. Only the most common difficulties and dangers encountered in the design and construction of dams can be mentioned here so that the reader can appreciate the complexities of the problems involved.

General principles of design

5.2 Necessary conditions to be fulfilled

For a successful construction the following conditions need to be fulfilled.

(1) The valley sides of the proposed reservoir must be adequately watertight to the intended top water level of the reservoir, and they must be stable under the raised water level.
(2) Both the dam and its foundations must be sufficiently watertight to prevent dangerous or uneconomic leakage through or under the dam.
(3) The dam and its foundations must be strong enough to resist all forces coming upon them.
(4) The dam and all its works must be durable.
(5) Provision must be made to pass all flood waters safely past the dam.
(6) Provision must be made to draw off water from the reservoir for supply purposes.

5.3 Examination of the valley

All possible sources of leakage from the proposed reservoir must be examined by a geological survey which will include borings where necessary. There must be an assurance that no fault zones or permeable strata lie through or down the valley, such that concealed leakage below ground might take place. Similarly it is important to discover any concealed river valleys that exist to make sure that they also would not cause hidden leakage. It is not sufficient to consider only the geology at the dam site. The stability of the hillsides when the water level is raised in the valley is of great importance. One of the greatest of reservoir disasters ever was the Vaiont Dam disaster of 9 October 1963 in Italy[1], when a landslide of gigantic proportions fell into the reservoir causing a 100 m high flood wave of water to pass over the crest of the 206 m high arch dam. The dam, which was then the second highest in the world, was not destroyed but the flood wave caused the deaths of 3000 people in the valley below. The reservoir has since been abandoned.

5.4 Watertightness

All dams leak to a greater or lesser extent: it is important that any such leakage must present no danger to the safety or durability of the dam and must not be of economic significance. As much as possible of all such leakage should be collected by an underdrainage system and must be measured and recorded regularly, usually monthly. When no regular inspection procedure is in force, the leakage may progress to the point where failure is sudden, as in the case of the Dolgarrog disaster[2] in Wales on 25 November 1925. There a dam to raise the level of Lake Eigiau was a low wall of concrete only 3 m high but it had, in one area, been taken only 0.5 m deep into clay foundations. The water leaked below this portion of the wall and so widened its passage that there was ultimately a sudden breakthrough of the lake waters. The wave of water destroyed another small dam below and rushed onwards to engulf the village of Dolgarrog, causing sixteen deaths.

To reduce seepage below the foundations of a dam two methods are in common use:

(1) the construction of a 'cutoff' trench across the valley below the dam,
(2) grouting the foundation beneath the dam to reduce its porosity.

A typical cutoff is illustrated in Fig. 5.1 and it is taken down sufficiently far to connect into sound rock or clay at the base. It is usually filled with concrete, the trench being about 2 m wide. The junction of the top of the cutoff with the corewall of the dam is a matter requiring the most careful design: Fig. 5.1 shows the concrete splayed out to receive the puddle clay of the corewall. In many other cases the top of the cutoff concrete was finished in the shape of a 'spearhead' so that the clay could be brought down on either side of it. It is also important to take the cutoff well into the abutments of the dam, as shown in Fig. 5.1, to reduce seepage around the ends of the dam.

A wide, shallow cutoff is shown in Fig. 5.2, and this is used where a sound foundation material exists not far below ground surface. The

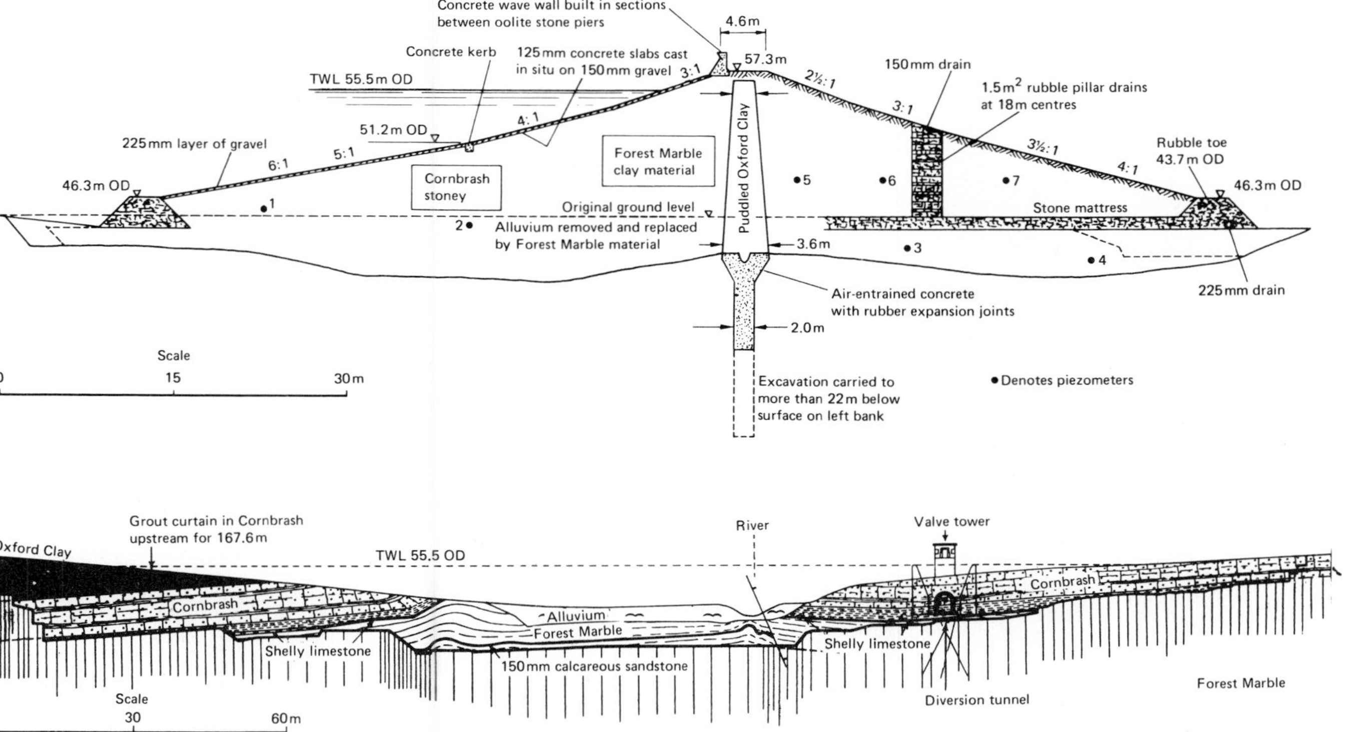

Fig. 5.1 The Sutton Bingham Dam for Yeovil and District Water Supply showing the cutoff trench. Engineers: Herbert Lapworth & Partners. (Walters *et al.*, *Proc. ICE*, 1957.)

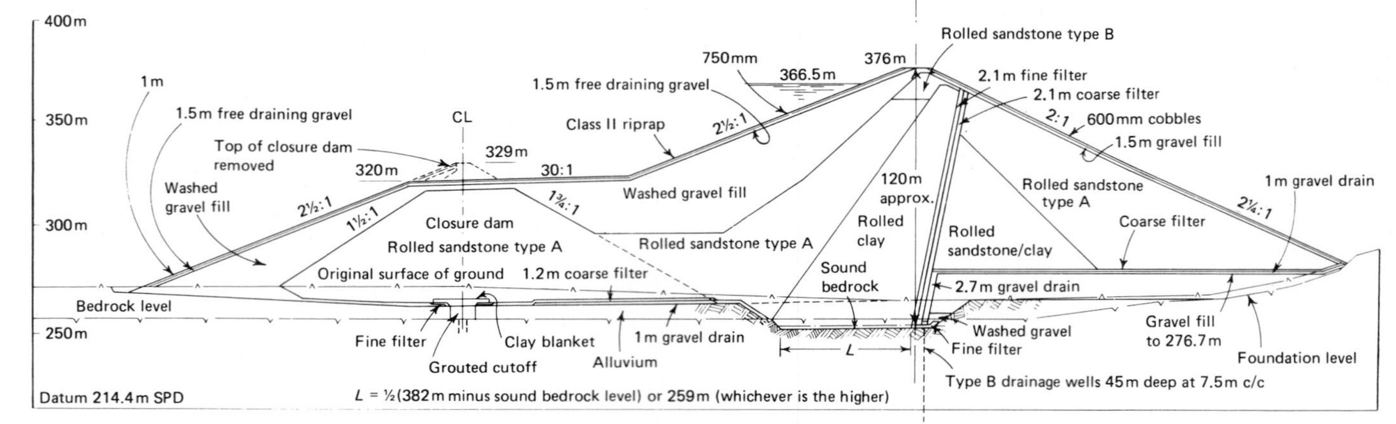

Fig. 5.2 The Mangla Dam for the West Pakistan Water and Power Development Authority, 1968. Engineers: Binnie & Partners.

corewall material, in this case 'rolled clay' which is a mixture of clay and coarser materials, is taken right down to the bottom of the cutoff.

Some cutoffs have been extraordinarily deep and therefore difficult and expensive to construct. The classic case is the cutoff for the Silent Valley Dam for Belfast[3], which reached a maximum of 84 m depth before sound rock was met. Cutoffs of up to 30 m deep are common, but very deep cutoffs are seldom necessary today because of the development of techniques of grouting which are cheaper to adopt.

A modern means of constructing a cutoff without the need for a timbered trench is to use the method known as 'diaphragm walling'. The trench is excavated by a hydraulically operated clam-shell grab fixed to the bottom of a long steel column which is lowered into the ground (see Plate 7). The trench remains open by being kept filled with bentonite clay slurry which holds the walls of the trench stable. The material is excavated from the base of the trench through the bentonite mix. When the trench has been taken deep enough to lodge into sound strata, concrete is tremied to the bottom of the trench and brought progressively upwards, thus displacing the bentonite mixture which is pumped away. Any rock encountered at the base of the trench is broken up by a heavy chisel and removed by the grab. The diaphragm is constructed in short lengths of about 5 m which are sealed together to form a continuous wall. The concrete is made plastic, i.e. capable of taking some distortion without cracking, by the addition of some bentonite to the mix so that the diaphragm can take up any necessary movement when the dam comes under water load.

Certain precautions are essential if a diaphragm wall is to be successful: the base of the trench must be carefully cleaned free of debris such as broken rock etc. and it must be taken at least 0.6 m into the bedrock, the concreting of any panel must be completed in one continuous pour of concrete, and the concrete tremie pipe must initially be placed close to the trench bottom for the first part of the pour and kept submerged below the concrete surface as it is slowly withdrawn.

Grouting consists of drilling vertical or inclined holes into the ground at intervals, and injecting into such holes mixtures of water and cement or clay, or chemical mixtures (see Fig. 5.3). This mixture penetrates into the bedding and joint planes of any permeable strata which are cut and, when the grout sets, such fissures and joints are thereby sealed off and made watertight. The holes have to be near enough for the grout to form a continuous impermeable barrier from hole to hole. It is therefore usual to grout first a series of primary holes spaced perhaps 6 or 7 m apart, then to sink secondary holes between these, test them for permeability, and grout them, and then if necessary to sink tertiary holes between them. The required spacing of holes to form a proper grout curtain varies according to the type of strata penetrated, but is

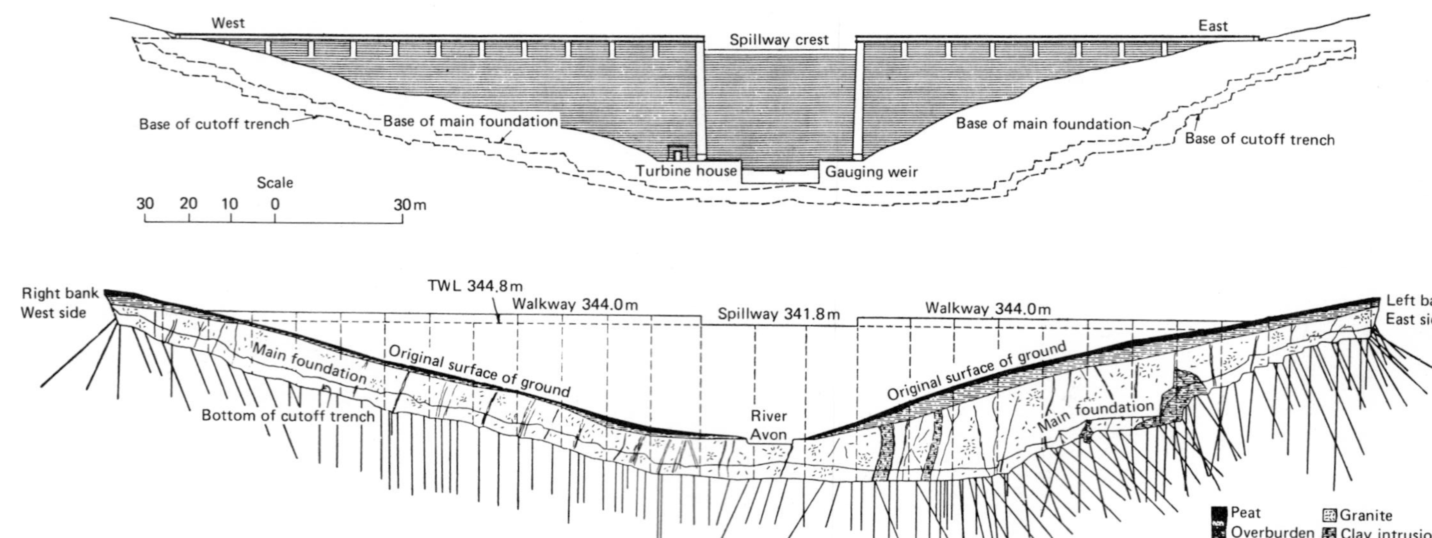

Fig. 5.3 The Avon Dam for the South Devon Water Board. Engineers: Lemon & Blizzard. (Bogle *et al.*, *Proc. ICE*, 1959.)

usually in the range 1.5 to 5.0 m. Sometimes several parallel rows of holes are grouted to form a wide band of grouting.

Usually grouting is adopted to improve the watertightness of a rock which is already basically sound, reasonably impermeable, and not liable to deteriorate or decompose even with some leakage through it. In exceptional cases it may be economically justified to attempt to grout materials other than fissured rock, such as sands, gravel, silts, clay, and mixtures of these materials[4]. These are often difficult to grout despite a wide variety of methods being adopted, so that complete success may not come to hand and additional precautions will have to be adopted in the design of the dam to allow for unavoidable leakage.

'Stage grouting' is most often used. In this process 'packers' are fixed in the boring above and below a given length or 'stage' of borehole so that water tests and grout can be applied to this section only. It is usual to carry out this procedure for progressive 10 m stages of borehole. When severely fractured rock, granular materials, or clay are encountered it is difficult or impossible to fix packers, but then the *tube-à-manchette* method can be used. The tube has perforations at intervals down its length, each perforation being covered on the outside by a thin membrane. The tube is lowered into the boring and the annular space outside it is grouted with cement grout. When the grout is set, packers are inserted inside the tube above and below a given perforation and, with a slightly raised 'breakout' pressure, the membrane and annular grout in the vicinity of the perforation can be fractured, permitting grout to be injected into the formation at that level only.

Grouting can be highly successful, but tales are legion concerning the damage that can be done and the waste of grout which occurs if the procedure does not come under the most careful control. Pushing too much grout under too high a pressure into the formation may do nothing to reduce the permeability of the ground, resulting only in the build-up of grout lenses. Sometimes the permeability of the ground can even be *increased* by inappropriate grouting techniques and, in one case on record, the whole of a partially built dam was raised 150 mm or so by unnecessary grouting. The most important point is not to use too high a pressure so that 'hydrofracturing' (or rupture of the formation) occurs, unless this process is intended for special reasons as advised by some experienced grouting specialist. Normally the maximum pressure applied should not exceed twice the overburden depth to the point of injection, i.e. at 50 m depth below ground surface not more than 100 m (or 10 bar) should be applied at that point. This holds for both grouting pressures and water test pressures. Tests for assessing the permeability of a formation should always be made on newly drilled, ungrouted holes: not on previously grouted holes. The water leakage from a boring under test (or a stage of it) is usually expressed in 'lugeons', 1 lugeon being 1 litre per minute per metre of boring under 10 bar head. The diameter of the boring does not enter this definition as it has little significant effect on test results. Permeabilities are expressed either in cm/s units or m/s units. A frequent target for acceptable permeability of the formation after grouting is 1×10^{-4} cm/s, i.e. 1×10^{-6} m/s.

5.5 Strength of the dam

Every dam must be proof against failure by sliding along its base. This applies whether the dam is of concrete, masonry, rock, or earth. Resistance to failure against shearing through any plane is another criterion also to be applied to all dams. Usually the resistance of rock and earth dams against sliding or direct shear failure is ample on account of their large width in cross-section. However, masonry and concrete dams have failed in this manner. An early failure in 1844 was that of the Bouzey Dam in France[5], a 19 m high masonry dam which moved 350 mm downstream under the static force of the water. It was cemented back on its foundations, but 11 years later, in 1895, it split horizontally about the middle. Another failure was that of the St. Francis Dam on 13 March 1928 which was a 60 m high concrete dam feeding Los Angeles[6]. It was placed on weak foundation material that cleaved so that a section of the dam broke out and a flood wave 40 m high travelled down the valley, at 65 km/hour, causing 426 lives to be lost. Concrete or masonry dams, being constructed to much steeper slopes than earth dams, have to be so designed that they are proof against overturning. Where concrete dams are arched, the abutments must be strong enough to take the end thrusts of the arch, and the foundations may also be highly stressed. The failure of the Malpasset Dam in France[7] on 2 December 1959 illustrates the catastrophic nature of an abutment failure. The dam was an arched concrete dam 60 m high only 6.5 m thick at the base. It was judged afterwards that, owing to rock weakness, the left abutment moved out ultimately as much as 2 m causing rupture of the arch in a few seconds, instantaneously releasing the whole contents of the reservoir which engulfed the Riviera town of Fréjus 4 km downstream. The dam was completely swept away.

5.6 Durability of a dam

Durability is of the utmost importance because dams are one of the few structures built to last indefinitely. In Italy the Gleno Dam[8], a multiple arch dam 43 m high, suddenly failed on 1 December 1923 when one of the buttresses cracked and burst in a matter of a few minutes. The cause was attributed to poor quality workmanship: the concrete in the arches was poor and inadequately reinforced with scrap netting used as hand grenade protection in the war and there was also evidence of lack of bond with the foundations.

The durability of an earth dam is primarily dependent upon the continuance of its ability to cope with drainage requirements. It has to

Plate 1 (A) Shing Mun dam forming Jubilee reservoir, Hong Kong, 1938: a composite rockfill dam (see Fig. 5.17). (B) Llyn Celyn earth dam in Wales for Liverpool water supply, 1964. (Engineers: Binnie & Partners.)

A

B

C

Plate 2 (A) Tai Lam Chung masonry faced concrete gravity dam, Hong Kong, 1958. (B) and (C) Kalatuwawa masonry faced concrete dam for Colombo water supply, Sri Lanka, 1960. (Engineers: Binnie & Partners.)

Plate 3 (A) Oros earth dam, Brazil, overtopped by flood during construction on 25 March 1960. (Photograph: *Water and Water Engineering.*) (B) and (C) Gleno buttress dam, Italy, before and after failure on 1 December 1923.

Plate 4 (A) Tarbela dam, Pakistan: part of overflow tunnel lining failure in 1974, believed due to cavitation. (Photograph: A. Garrod.) (B) Upper Neuadd masonry and earth-embanked dam completed 1902, showing failure of masonry joints in 1967. (C) and (D) Failure of Baldwin Hills earth-embanked reservoir, Los Angeles, on 14 December 1963. Excess flow in underdrains was noted about noon; silty water emerged from the toe at 1.00 p.m. and a hole in the upstream face was revealed at 2.30 p.m. The first photograph shows the upstream breach at 3.30 p.m.; the second shows the collapse of the roadway at 3.38 p.m. This was followed by complete breaching at 3.40 p.m.

cope with seepage, rainfall percolation, and changing reservoir levels. The movement of these waters must not dislodge, or take away, the materials of the dam, otherwise its condition and stability will deteriorate at a progressively increasing rate. It must perform this function despite long-term settlement of itself and its foundations. Hence unremitting care must be exercised in the construction of every part of an earth dam, using the right materials in their right context (see Section 5.16).

Outlet and overflow works

5.7 Provision for flood waters

Every dam must be provided with adequate overflow arrangements. An earth or rockfill dam must not be overtopped by water or it would fail completely in a few hours, nor must overtopping of a concrete dam occur except when the design specifically provides for this. Various overflow devices are used and these can be classified into two distinct kinds: (1) overflows which permit water to flow over the dam and (2) overflows which permit water to pass around the end of the dam, either in a tunnel driven through the abutment or through an open channel constructed on the natural ground.

For the UK the procedures for sizing the spillway and freeboard of a dam are set out in two 1978 publications: *Floods and Reservoir Safety: An Engineering Guide* (published by the ICE) and the *Guide to the Flood Studies Report* (published by the Institute of Hydrology). These are based upon the earlier five-volume *Flood Studies Report* financed by the Natural Environmental Research Council and published in 1975.

The ICE engineering guide recommends that in the first place dams should be classified according to the risk category they pose, as described in the first column of Table 3.9 (p. 79). For dams in the highest category of risk the spillway must pass the Probable Maximum Flood (PMF) safely, taking into account the effect of reservoir lag as water is temporarily stored above overflow level (Section 3.21). The PMF is not defined in relation to past (historic) experience of floods, nor is it precisely defined in any other way. It is the worst flood estimated to be possible from the Probable Maximum Precipitation (PMP) that could be realistically supposed to fall on the catchment area under some extreme, but possible, meteorological event, and also assuming that catchment to be in the worst possible flood-producing condition. The appropriate hydrograph to be used when estimating the flood flow is stipulated. It will be seen from Table 3.9 that whereas a dam whose failure would endanger the lives of persons living in a community below the dam must be designed for the PMF, in all other cases the spillway need be designed for only half or less than half of the PMF according to the risks posed to life and property downstream.

The guides also advise on the amount of freeboard that should be provided in any particular case, i.e. the additional crest height or other protection required, such as a wave wall etc., above maximum water level during a flood to protect the dam against wave and spray action.

For an earth dam the usual practice in the UK results in the top of the embankment being 2 m higher than the cill of the overflow weir, in addition to which a substantial wave wall capable of protecting the bank against spray from wave action would also be provided. For dams in other parts of the world—where reservoirs and floods can be of much greater magnitude than in the UK—substantially higher freeboards may be necessary, as for instance in the case of the Mangla Dam (Fig. 5.2) where 10 m is provided.

5.8 Direct discharges over a dam

Direct discharges are permissible on masonry or concrete dams. They are cheap because very little alteration to the profile of such a dam is necessary to accommodate the overflow section. Typical examples are shown in Fig. 5.3 and Plate 2. The water usually cascades down a smooth or stepped face of the dam. At the foot of the dam the water must be turned into a stilling basin to disperse some of its energy, because the main danger to avoid with this overflow device is the scouring away of the downstream toe of the dam during a high flood. Sometimes the water is ejected off the face of the dam by a ski-jump, which throws the water some distance away from the toe of the dam, thereby lessening the danger to the toe. It is essential, in all such work, that the construction is very massive and is soundly based upon solid rock. The design is not appropriate to a valley where good hard rock does not appear in the river bed at the dam and for some distance downstream.

Sometimes an additional feature of concrete or masonry dams is the inclusion of automatic crest gates to the overflow weir. The gates permit water to be stored above normal top water level of the overflow cill but, after a certain level has been reached, the gates are lowered automatically and permit an increasing discharge over the weir. Thus the greater the inflow, the greater is the discharge, and therefore the reservoir level is maintained sensibly constant at a higher top water level than would otherwise occur. It should be noted, however, that the ICE engineering guide (Section 5.7) suggests that gates should only be permitted under certain conditions. In fact gates have seldom been used on dams in the UK except in the case of concrete dams in remote locations.

Another method which is substantially more reliable because it is less dependent upon the mechanical operation of gates is the use of a siphon over the weir crest. Upon a predetermined level being reached in the reservoir the siphon comes into action, and its discharge per unit length of siphon crest is more than would occur for the same water head over a fixed weir. However, some doubts may be felt in this country as to their suitability in locations where ice formation is possible in winter.

5.9 Spillway channels

Flood waters are frequently discharged via spillway channels provided around the end, or both ends, of a dam, as shown in Plate 6D. They should be constructed on natural ground and not on the dam itself. The construction has frequently to be massive, of concrete or masonry, because of the need to prevent dislodgement of any part under flood conditions when the energy of the falling water may be very great.

At the upstream end of the channel there will be an overflow weir, usually with a rounded crest that will not be damaged by floating debris[9]. The collecting basin immediately downstream of the weir must be designed to take the water away without any backing up of water on the weir. At the further end of this section of channel across the abutment, critical depth of flow will develop (see Sections 10.8–10.13) acting as a control point for flow through the channel. The channel slope then increases as it falls down the abutment. Sometimes attempts have been made to destroy some of the energy of the falling water down this steep slope by constructing piers or steps in the channel, but they are not usually very effective. It is at the base of the spillway channel that, if necessary, a 'stilling basin' is constructed to dissipate some of the energy of the water. The most usual method is to induce the water to form a hydraulic jump within the stilling basin (see Section 10.14).

5.10 Bellmouth overflow

A bellmouth overflow (see Fig. 5.4 and Plate 4C) may be adopted where the expense of cutting a spillway channel is great or where a large tunnel has to be constructed in any case for the diversion of the river during the construction of the dam. The bellmouth, which is funnel shaped, must be constructed on firm ground, either clear or almost clear of the toe of the dam. If built within the body of the dam, its settlement might cause disruption of the fill of the dam or the different levels of fill against it might cause it to tilt. The discharge over the bellmouth lip may be computed as if the lip were a straight weir of equivalent length, provided that the size of the vertical shaft and the dimensions of the tunnel are sufficient to take all the water away up to the designed maximum capacity. The shaft from the bellmouth may be vertical or sloping and it may join to the tunnel with a smooth or sharp bend. The vertical profile of the bellmouth is usually designed so that no negative pressures, i.e. pressures below atmospheric, occur and may be calculated from Williamson's formulae[10]. The junction of the base of the shaft with the tunnel requires careful design: the falling water brings with it appreciable quantities of air and the design should be such that this air does not accumulate at the soffit of the tunnel reducing its discharge capacity.

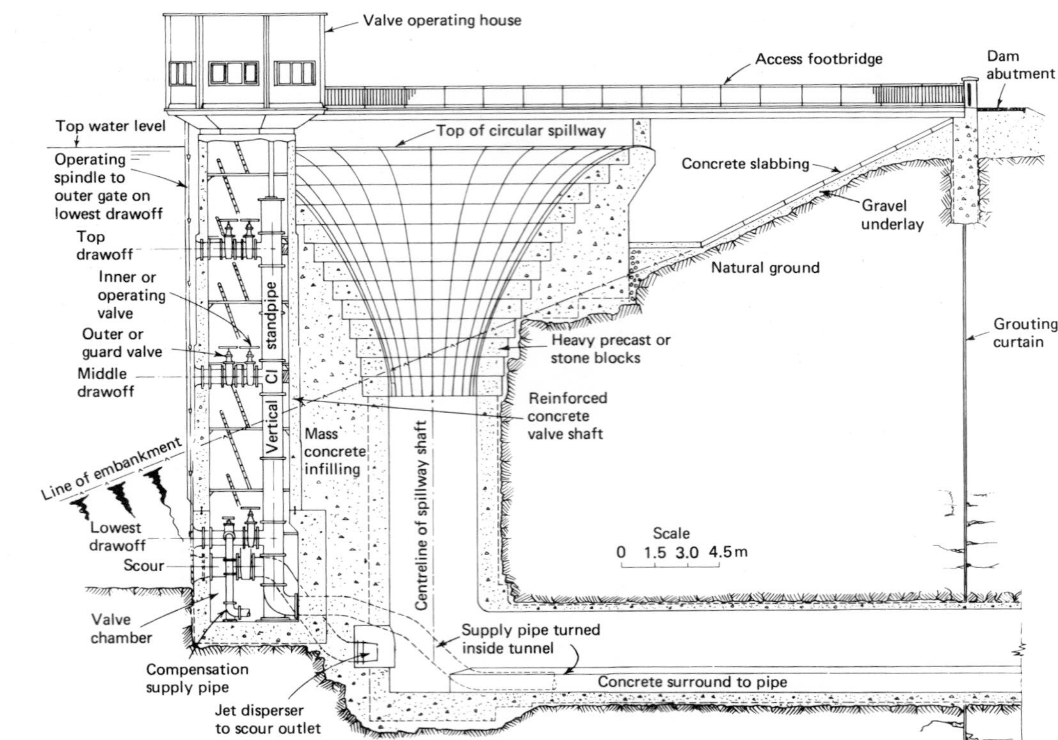

Fig. 5.4 The bellmouth spillway and drawoff tower of the Tittesworth Dam for the Staffordshire Potteries Water Board. Engineers: Binnie & Partners.

Also dangers arising from cavitation (see Section 8.26) must be avoided. At the Tarbela Dam in 1974 fast flowing discharge water eroded over 3 m thickness of tunnel lining within about 24 hours (see Plate 4A), the cause being attributed to cavitation[11].

An advantage of the bellmouth overflow and tunnel, apart from the use of the tunnel as a river diversion during the construction of the dam, is that the supply pipe and compensation water pipe may be led through the tunnel. The scour valve may discharge direct to the base of the bellmouth. By these means the pipes can be regularly inspected and maintained during periods when the reservoir is not overflowing.

The bellmouth overflow has a strictly limited capacity: when it is gorged with water it will take no more flow. This aspect needs to be taken into account when considering what the design flood should be. There is also doubt as to whether a bellmouth overflow is suitable for dams in climates where substantial ice may form on the reservoir during winter. For small reservoirs the size of the bellmouth shaft should not be reduced below a size large enough to allow the passage of any debris that might be brought down by the river. The bellmouth also has a disadvantage in that it represents a danger to sailing and fishing boats on the reservoir, and it then becomes necessary to put guards around the bellmouth which may, in turn, tend to collect floating debris which would restrict the discharge capacity.

5.11 River diversion and outlets

Where a concrete or masonry dam is to be built, the stream may be taken through an opening left at the base of the dam in the valley bottom, this opening being finally sealed up when the dam is completed. Even if floods occur during the course of construction and the partially built concrete dam is overtopped, no serious harm need come from the inundation, provided that the valley bottom is composed of rock.

It is not possible to follow this procedure with an earth dam. Firstly, the partially built dam must not be overtopped at any stage because it might be destroyed. Secondly, except in the case where an earth dam is on a sound rock foundation (which is quite often not the case), it is undesirable to have a central duct or tunnel left through the centre of an earth dam even if the tunnel is filled solid afterwards. An earth dam depends for its watertightness, and therefore for its security, upon a continuous watertight membrane existing through it from abutment to abutment, and from its crest to its foundations. To puncture this membrane at right-angles by a tunnel can be dangerous: for even if the tunnel is filled with concrete there is a danger that, unless it is on hard rock, the tunnel will settle more at the centre under the weight of the full height of the dam than it does at its ends where it emerges from the dam. So it may fracture somewhere about the middle of its length and cause disruptions and discontinuities in the dam at the centre-most deepest part, presenting passages for the water to find a way through and erode.

The near failure in 1970 of the Lluest Wen Dam[12], built in 1896, illustrates the danger of taking a tunnel through the centre of an earth dam, even when the foundation is sound rock. Due to mining of coal seams in the vicinity of the dam (a 'pillar' of unmined coal being left below the dam), there was differential settlement. The tunnel bulkhead (see Fig. 5.5) was fractured just in the zone of the puddle clay core. A small 150 mm drain pipe through this bulkhead was fractured also and, into this breakage, seepage must have occurred bringing with it clay from the core. Ultimately, after an unknown length of time, a 2 m deep hole suddenly appeared in the crest of the dam some 20 m higher up. Clay shortly afterwards began emerging from the fractured pipe, leading to the decision that the dam was potentially in danger of imminent failure so that emergency measures had to be undertaken immediately.

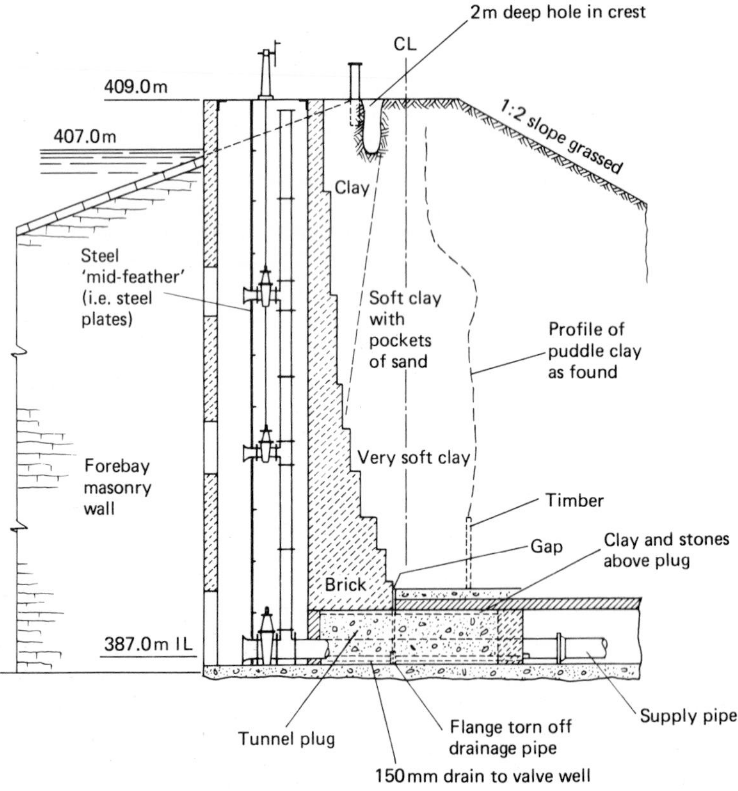

Fig. 5.5 Cause of the Lluest Wen Dam failure.

A tunnel through an earth dam on any sort of ground is therefore best avoided. Instead, it should be taken through one of the abutments. A diversion dam upstream may be needed to divert floods during construction into the tunnel, giving the benefit of some flood storage and surcharge head on the tunnel. Problems of this sort have to be considered with the greatest care, because if any misjudgement takes place the partially built dam may be overtopped and destroyed. This happened to the Oros Dam in Brazil[13]. A diversion tunnel had been built capable of taking 450 m^3/s, but 600 mm of rainfall fell on the catchment in a week, the reservoir level rose 3 m per day, and, despite all efforts to raise the half-built dam to beat the rising waters, it was overtopped on 25 March 1960 and virtually destroyed, having to be rebuilt.

5.12 Drawoff arrangements

Except when a reservoir is shallow, drawoff pipes are usually designed for withdrawing water from the reservoir at several different levels. At certain periods of the year the best water may be obtained from the middle depth of a reservoir; at other times the upper portion of the reservoir water may be better in quality. A usual provision is three levels of drawoff: upper, middle, and lower. The upper drawoff would be perhaps 4 m below top water level (it would be of little use only just below top water level). The lowest drawoff is usually about 1 m above the bottom of the reservoir to make some allowance for silting of the reservoir. A typical arrangement for a drawoff tower is shown in Fig. 5.4 (p. 158). The vertical standpipe has tee branches in it which communicate to the reservoir. On each branch are two valves, or else a valve and a sluice gate. The inner valve is normally used for frequent operation, the outer valve being closed if the inner valve has to be repaired. It is most important to keep all valves in good operating trim by practising a routine of opening and closing them regularly.

A drawoff tower is usually used with an earth or rockfill dam and, for the reasons given in Section 5.11, it should not be sited in the body of the dam but in or near an abutment (Fig. 5.4). With a concrete dam there is no objection to making the drawoff structure part of the dam. A scour pipe is also required for the purpose of keeping the channel approach to the lowest drawoff free of silt, and for the purpose of controlled fast emptying of the reservoir if this should ever be necessary. In the event of any weakness becoming apparent in the dam the ability to draw down the reservoir level at a reasonably fast rate is a great advantage. Wherever possible the scour capacity should be such that the water level in the reservoir can be dropped at least one-quarter of the full ponding height from overflow, in a period of 14 to 21 days, during a rainfall equal to the average daily rainfall for the wettest month. By this means both the volume of water stored and the stresses on the dam will be substantially reduced. Such a rate of lowering may not be feasible for very large reservoirs, in which case the aim should be to size the scour so that the water level can be dropped at a rate of 0.3 to 0.6 m per day during rainfall equal to the average daily rate of rainfall during the wettest month, the maximum drawoff to supply being taken into account.

Earth dams

5.13 Types of design

To avoid an elementary error it is as well to state that an 'earth' dam is not made of the tillable soil of gardens and fields. All such 'soil' must be excluded from earth dams because it contains vegetable matter which might eventually rot away leaving passages for water percolation, and the vegetable content reduces the strength of the earth. The first operation when undertaking any earth-dam construction is therefore to strip off all surface material containing vegetable matter. The dam itself may then be constructed, according to the decision of the designers, of any material or combination of materials such as clay, silt, sand, gravel, cobbles, and rock.

Earth dams of what might be termed 'early' design consisted of a central core of impermeable 'puddle clay', supported on either side by one or two zones of less watertight but stronger material. Two examples are shown in Fig. 5.6. The central puddle clay core is impermeable but relatively weak. This clay core must be continued as a cutoff of clay or concrete taken as deep as is necessary in a trench into the ground so that contact is made with impermeable rock. While the very earliest types of dam had clay-filled cutoff trenches, use was soon made of concrete for the trench. The shoulders of the dam usually had an inner and an outer zone. The inner zone, containing both clay and stones, would be impermeable to some extent, but the stony material would add to its strength. The outer zone would contain less clay again and might contain large boulders and much gravel. Boulder clay (a glacier-deposited mixture of clay, silt, and stones with boulders) was a favourite material for this outer zone in the UK because extensive deposits exist in those areas where impounding schemes were developed.

Many such dams have been and continue to be eminently successful. Their slopes were decided in the past by experience and were often 3:1 upstream and 2:1 or 2½:1 downstream with flatter slopes for the higher dams. The quality of the materials used, and the means by which they were laid down (compacted), were decided in the light of experience. A heavy responsibility for successful construction lay upon the supervising engineer. He had to locate the right materials necessary to make a sound job and see that compaction was properly achieved. The 'right' degree of compaction was judged by eye, by behaviour of the fill as it

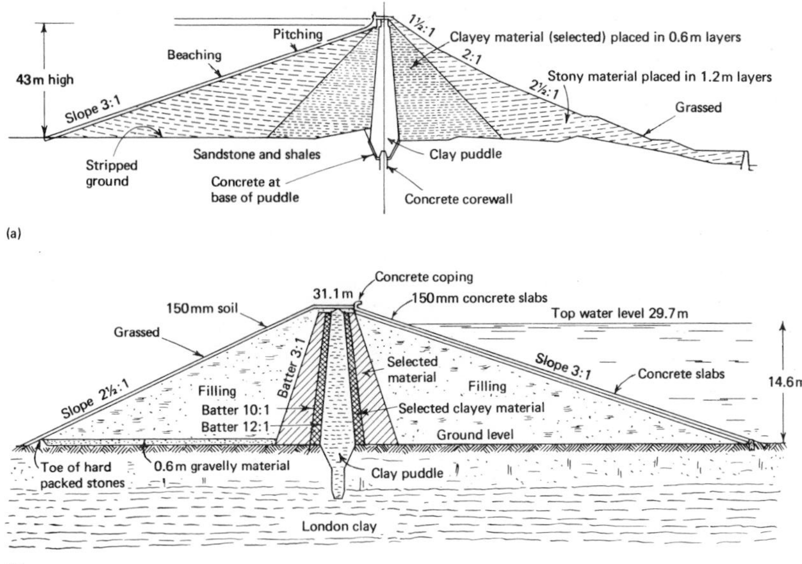

Fig. 5.6 Dams with puddle clay cores: (a) the Ladybower Dam (Hill, *JIWE*, 1949) and (b) the embankment of the King George VI Reservoir (*JIWE*, 1948).

was placed, and by a few empirical tests. Although many new kinds of tests are available nowadays, care in supervision of construction and skill in judging the quality of different kinds of earth and the amount of compaction achieved are still key factors in making a successful dam. The work of the designer is rendered useless if it is not supported by first-class site control.

Today the 'rolled clay core' made of a mixture of clay and coarser materials, 'rolled down' by compacting machinery, and laid as a much broader core, replaces the earlier puddle clay core. It is stronger, does not have to be placed at a high moisture content, is not so liable to erosion by seepage, and, being spread over a wider area at the base, is not so susceptible to leakage at foundation level. A dam of this sort is shown in Fig. 5.2 (p. 150). Sometimes a dam may consist wholly of rolled clay with only a thin outer protective zone of coarser material on the waterface for protection against wave action.

As an alternative, the articulated concrete corewall can be used (see Fig. 5.7). This tends to be more expensive than the rolled clay core, but it gives the advantage of being able to use a wide variety of materials for the shoulders and results in a dam of maximum strength according to the type of materials available for the shoulders. The core needs to be constructed in panels so that the joints between panels can be sealed after the concrete has contracted upon cooling, and also so that small

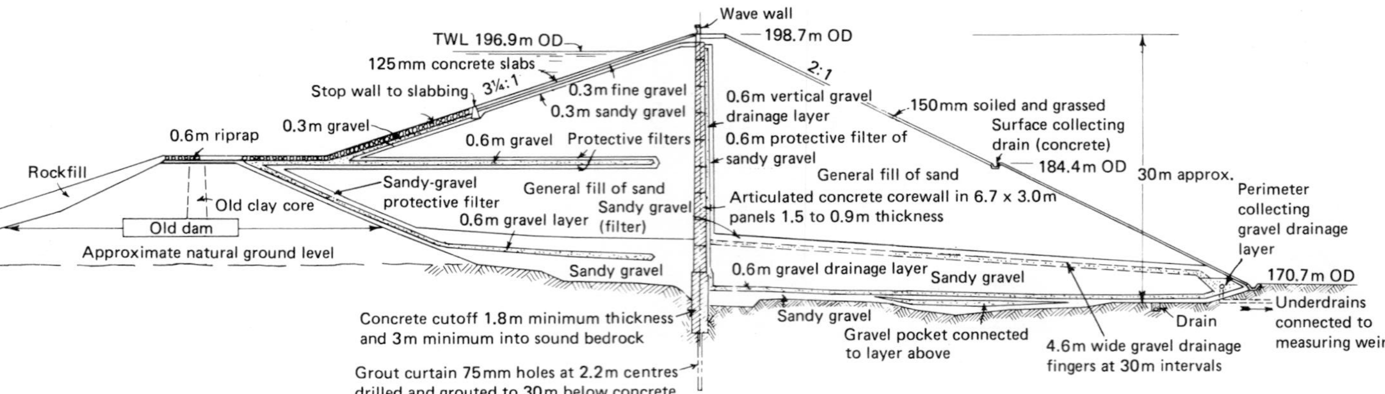

Fig. 5.7 The Tittesworth Dam for the Staffordshire Potteries Water Board showing the filter and drainage layers incorporated. Engineers: Binnie & Partners.

vertical movements due to compression of the foundation and small lateral movements caused by the water level rising and falling in the reservoir can be taken up without fracture of the core. A concrete corewall is not suitable where the foundations are soft.

With modern, computer-aided methods of analysis, quite complicated designs for earth dams can be adopted. The principal aim is always to use as much local material as possible because this contributes to cheapness of construction. The problem is how to incorporate the available materials in the dam in a manner which uses their different characteristics efficiently and safely.

5.14 Pore pressure in earth dams

The most essential matter in constructing an earth dam is to achieve compaction of the soils of which it is made. Various heavy compacting machines are used to achieve this and the soil has to be placed at its 'optimum moisture content'. For instance, when clay is dry it is hard and difficult to compact, but if it is too wet it is a slurry which cannot be compacted and which will shrink on drying. As layer upon layer of soil is placed upon the dam the earlier material will receive the load of later material upon it and will slowly consolidate further. The voids, or *pores*, in the soil will close as the soil particles move together and any water which previously occupied those voids must escape. If the water cannot escape, or if (which is the same thing) it cannot escape fast enough, the water pressure in the voids will build up, preventing the soil particles from consolidating. As a result the strength of the soil is less than it would be if the water from the pores could get out and the soil particles could consolidate.

Prior to World War II earth-dam construction was relatively slow so that the pore water drainage had time to take place as the dam was slowly raised. After the war the larger earth-moving machines which came into use speeded up construction, and it was found that an excessive pore water pressure could endanger the stability of an earthen embankment if it was raised too fast to give time for pore water dissipation. Hence a major technique in all subsequent earth-dam construction has been to measure the pore water pressure, and to adopt means of dissipating it or, if that is not possible, to slow down the rate of construction.

To measure pore water pressure the simplest apparatus comprises an unglazed ceramic or other porous pot buried in the fill to which a small pipe attached extends to the surface (see Fig. 5.8(a)). However, the actual amount of water in the soil is very small and, although ultimately the pot and tube will fill with water*, this takes considerable time. To

* Theoretically if there is '100% pore water pressure' the water will rise to a level *above* the formation equal to the depth of burial of the ceramic pot when the weight of the fill per unit volume is twice that of the water. Such a condition can be demonstrated.

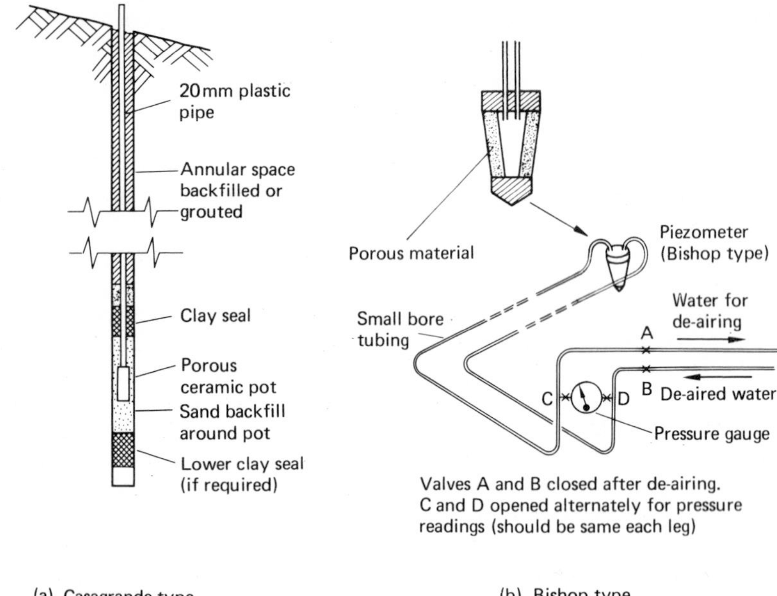

Fig. 5.8 Pore water pressure sensing devices.

overcome this difficulty, instead of a ceramic pot, a pressure sensing device can be inserted into the fill. One such device is shown in Fig. 5.8(b) and, in essence, it consists of a ready-filled circuit of de-aired water connected at one end to the 'piezometer' buried in the fill and, at the other end, to a pressure gauge in a gauge house. Being already filled with water, the pressure within the circuit stabilizes relatively rapidly with the pore water pressure in the fill via the ceramic membrane in the piezometer, hence the pressure can be read at the gauge.

During construction of the dam these piezometers can be positioned in the fill and therefore the pore water pressure within the fill can be measured. If higher values are recorded than the design allows for, these indicate that the strength of the material within the dam is less than the design strength assumed. Further embanking may have to be stopped temporarily to give time for the excess pore water pressure to dissipate or some additional drainage layer may be incorporated within the dam to shorten the distance water has to travel to a free-draining material, thereby speeding up pore water pressure dissipation and consequent consolidation of the fill.

The piezometers can also indicate where leakage of water is passing through the fill. Thus some piezometers are nearly always incorporated near the base of the dam, just downstream of the corewall. If these show a rise in water pressure when the reservoir is filled, they almost certainly indicate some degree of leakage, either through the dam or underneath it. By carefully positioning such piezometers in the downstream shoulder of the dam and in the foundation below, a good picture can be obtained of how and where any leakage is taking place. The pressures registered also permit the designer to estimate the stability of the dam.

Similarly piezometers can be placed in the upstream shoulder of the dam and there they will not only register the rise and fall of the water pressure within the embankment as the water level rises and falls in the reservoir, but they will also register the time lag occurring between the two. This time lag is important because if the reservoir water level is suddenly drawn down, but the piezometers show that the upstream shoulder does not drain out as fast, then a slip can develop in the upstream shoulder due to the extra weight of the saturated fill remaining undrained above the water level in the reservoir. This 'rapid drawdown' condition can be a dangerous one for the stability of the dam, especially during the early years of life of the dam when full consolidation of the material may not have been achieved. As a consequence it may not necessarily be the right policy to try to empty a reservoir at the fastest rate possible, in case in trying to avoid some possible trouble on the downstream slope, a worse trouble is induced of a slip on the upstream slope through taking the water level down too rapidly.

5.15 Slip-circle analysis of earth dams

An increasing knowledge of soils—their qualities and their behaviour under stress—has made possible radical changes in design from the early methods of earth-dam construction. Soils can now be tested, and can be classified and assigned certain mathematical values describing their composition, strength, and permeability. Using these values a mathematical approximation of the stability of an earth dam can be computed. To obtain such an estimate of stability it is necessary to assume that failure takes place in a certain way. One such method is known as the 'slip-circle failure' method and is illustrated in Fig. 5.9[14]. The unbalanced moment of the portion of dam within the slip circle is calculated. For stability this moment must be resisted by the strength of the dam material through and along the circle of failure.

Referring to Fig. 5.9, if the average pore water pressure along the base δl of a unit vertical slice of the fill above a possible slip plane is u (see Section 5.14), the shear strength along δl is $\delta l(c' + (N-u)\tan\phi')$ where $(N-u)$ is the effective intergranular stress normal to δl, c' is the apparent cohesion, and ϕ' is the angle of shearing resistance for this effective stress.

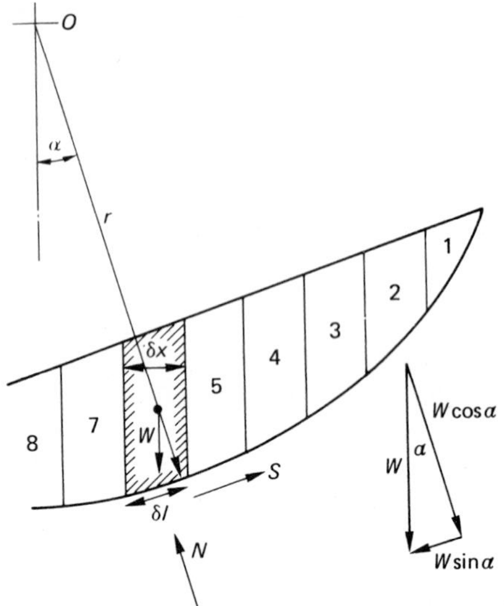

Fig. 5.9 Slip-circle stability analysis.

Hence, taking moments about O for the given slice, the factor of safety against the slice sliding along δl is given by

$$F = \frac{r\,\delta l(c' + (N-u)\tan\phi')}{r\,W\sin\alpha}$$

(neglecting forces between adjacent slices)

$$= \frac{\delta l\,c' + \delta l\left(\dfrac{W\cos\alpha}{\delta l} - u\right)\tan\phi'}{W\sin\alpha} \qquad \left(\text{since } N = \frac{W\cos\alpha}{\delta l}\right)$$

$$= \frac{\delta l\,c' + (W\cos\alpha - \delta l\,u)\tan\phi'}{W\sin\alpha}$$

If u is expressed as a percentage, p, of the weight of the fill per unit area, i.e. $u = pW/\delta x$, we can further reduce the expression to

$$F = \frac{\delta l\,c' + (W\cos\alpha - pW/\cos\alpha)\tan\phi'}{W\sin\alpha} \qquad \left(\text{since } \frac{\delta l}{\delta x} = \frac{1}{\cos\alpha}\right)$$

For safety against sliding of the slice as a whole (neglecting forces between the slices)

$$F_{\text{whole}} = \frac{\Sigma\,\delta l\,c' + \Sigma(W\cos\alpha - pW/\cos\alpha)\tan\phi'}{\Sigma\,W\sin\alpha}$$

and if c' and ϕ' can be taken as constant for all the slices we have

$$F_{\text{whole}} = \frac{c'\,L + \tan\phi'\,\Sigma(W\cos\alpha - pW/\cos\alpha)}{\Sigma W\sin\alpha}$$

where L is the whole length of the sliding path.

We have to compute F by some tabulation method to obtain $\sum_{i=1}^{n}(W_i\cos\alpha_i - p_iW_i/\cos\alpha_i)$ and $\sum_{i=1}^{n}W_i\sin\alpha_i$ for n slices.

Where partly submerged slices occur, it is necessary to substitute $(W_a + W_b)$ for W where W_a is the full weight of the portion above the water level and W_b is the submerged weight of the portion below. In addition $u\,\delta x/\cos\alpha$ is substituted for $pW/\cos\alpha$ where u is the *excess* of the pore water pressure above the water level in the fill in each particular case.

A large number of such slip circles are analysed so that the one giving the lowest factor of safety for a given design may be found (Fig. 5.10).

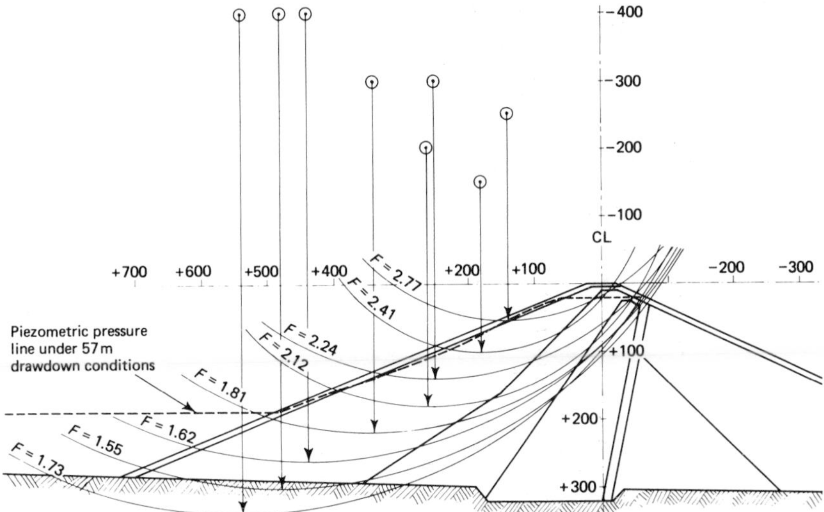

Fig. 5.10 Typical slip-circle stability analysis results for the upstream slope of a proposed design for the Mangla Dam. The results show a minimum factor of safety of 1.55 for the assumptions made. (This is 'Computer run no.111 (150 circles)' and is for static conditions with 57 m sudden drawdown.) Engineers: Binnie & Partners.

Other methods of failure can occur, such as the wedge type of failure illustrated by an actual case of failure[15] shown in Fig. 5.11. The factors of safety found, by whatever methods are used, must never be regarded as the actual factor of safety of the structure. Every analysis of this type is a simplified conception of what may happen in practice, and the data

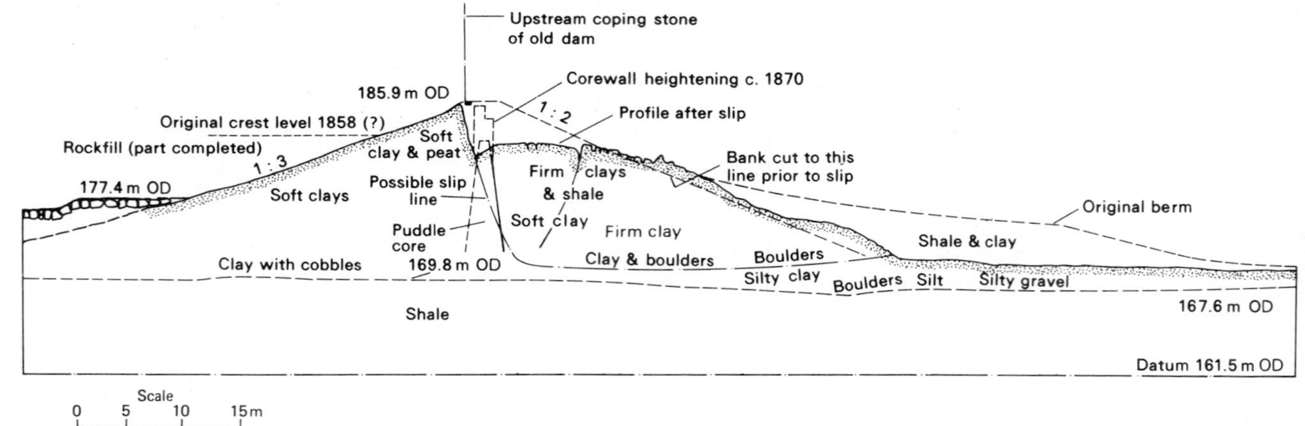

Fig. 5.11 Wedge failure of the old Tittesworth Dam.

fed into the calculations may never exactly represent the true characteristics of the materials as placed in the dam. Therefore a margin for error and inaccuracy must be allowed and it is seldom the case that any factor of safety less than 1.4 on a theoretical basis is adopted.

5.16 Durability of an earth dam

The description of pore water pressure in the previous section should have led the reader to appreciate that an earth dam is not a structure which is permanently saturated upstream of its core and permanently dry downstream. The water levels in the upstream shoulder rise and fall, following rises and falls of the reservoir level. The downstream shoulder has a natural moisture content which rises and falls with rainfall, and which may vary accordingly as there are leakages from the corewall, from the foundations, and from flows from the abutments either side. Far from being a static structure, an earth dam is continuously taking in, channelling away, and discharging to its underdrains varying amounts of water coming to it from many sources.

Drainage and filters Drains are necessary within a dam for accommodating the water movements mentioned above. However, a drain of coarse material, such as gravel, cannot be directly placed against a fine particle material, such as clay, or movement of clay particles into the voids of the gravel would occur with the movement of water. One of three serious consequences might follow. Removal of fines could cause voids in the fine material and ultimately these would enlarge to provide leakage paths. Deposition of the fines in the coarse material could clog the coarse material so that the water would be stopped from flowing, its pressure would rise, and it would tend to break out other paths for itself. If the fines were not held by the coarse material but passed out with the water, nothing would prevent larger and larger voids being created by the passage of water through the dam.

Every drain of coarse material, such as gravel, laid next to a fine material, such as clay, has therefore to be protected by a filter layer of intermediate size. Sometimes, if the difference in size between coarse and fine materials is large, two such filters are necessary. An important practical point of observation is that every drain leading from a dam must discharge to a trap and a measuring weir. The latter is for regularly recording the amount of flow; the former is for observing that the water does not carry material from the dam with it. Any flow emerging from a dam which is of cloudy water, thus denoting that the water carries suspended matter in it, is to be regarded with suspicion. All discharges should be of clear water and, if any are not, a most careful search must be made to determine the reason.

Protection of outer faces The upstream face of an earth dam must be safeguarded from erosion by water and wave action from the reservoir. Protection takes the form of covering the surface with large rock or with slabs of concrete (see Plate 5D). The rock must be large if the wave action expected is strong, and the rock should preferably be hand placed and locked together with small stones between the large stones. Beneath the rock must lie layers of material which act as filters between the large rock and the finer material of the body of the dam to prevent the finer material from being drawn through the rock surfacing. When concrete slabs are used as protection, they must not butt tight together, but must have a space between the joints. When the water level in the reservoir falls, water will be left in the dam at higher levels and this must be able to seep out of the dam without pushing up the slabs. Beneath the slabs are filter layers similar to those placed beneath the rock: usually there would be a layer of coarse gravel immediately below the concrete slabbing, then one or two layers of finer material protecting the inner material of the dam.

On the downstream face of the dam erosion by rain must be prevented. It is usual to achieve this by turfing the surface of the dam or by soiling and seeding it with a short-bladed strongly rooting grass. If the dam is more than about 15 m high then, according to the type of rainfall climate, berms should be constructed on the downstream slope, each berm having a collector drain along it so that surface runoff is collected from the area above and is not discharged in large amounts to cause erosion of the slope below. Cobbles or gravels are sometimes used to surface the downstream slope.

All drainage layers in the downstream shoulder should be so arranged that they discharge their flow over measuring weirs, surface drainage being kept separately measurable from underdrains. Continuous records should be kept of the underdrain discharges so that any signs of increased leakage through the corewall or cutoff may be detected.

5.17 Trees on embankments

A word may be said about tree growth on earth dams. Large trees or an extensive wooded area should not be permitted for the following reasons. Some tree roots can penetrate deeply in their search for water and such conditions should not be allowed: filter layers within the dam could be disrupted by extensive root growths and, if the tree should die, voids would be left in the dam through which drainage paths could form, capable of setting off cumulative erosion of the bank. An uprooted tree may cause a hole in the bank, into which surface runoff may penetrate in undesirable quantities. If the surface of the dam is obscured by a matt of roots and a carpet of rotting leaves, it is not then possible to observe any damp patch appearing on the surface of the dam which may be a sign of faulty interior drainage, nor can any bumps and hollows which may develop in the bank from uneven settlement be observed. These are serious objections because visual observation of the exterior surface of an earth dam plays an important part in keeping watch on the condition of the dam. No serious objection, however, could be taken to a few groups of flowering shrubs if it is thought that these will improve the appearance.

Concrete and masonry dams

5.18 Gravity-dam design

It is primarily the weight of a gravity dam which prevents the dam from being overturned when subjected to the thrust of the impounded water. For the prevention of sliding the dam must be sufficiently wide at the base and it must adhere strongly to foundations which must be of a strong sound rock. In carrying out the necessary calculations to determine the stability of any proposed section 'uplift' must be taken into account. Uplift is the vertical force exerted by water under hydrostatic pressure which finds its way into cracks in the body of the dam, or beneath it, as shown in Fig. 5.12.

It is prudent to take uplift pressure below the dam as of trapezoidal shape: the value at the upstream toe being taken equal to the maximum reservoir head (under flood conditions) and the uplift at the downstream toe equal to the tailwater level. If no tailwater level occurs then the ground level at the downstream toe should be taken. This amount of uplift may be more than actually occurs, especially when a deep cutoff and underdrainage is inserted, but much controversy surrounds the subject and it is not wise to run risks of underestimating the uplift force. At sections through the dam above the foundations, uplift is also assumed to exist within the concrete, and this again is taken as the maximum reservoir head at the upstream face, decreasing linearly to zero at the downstream face of the section.

Taking these three forces into account—uplift, water thrust, and weight of dam—the generally accepted rule is that the resultant of these should pass within the middle third of the section being analysed. The theoretical result of this computation, which assumes that the portion of the dam above the section is homogeneous and watertight, is that no tension forces are developed in the upstream face of the dam and that no cracks will appear on that account because there is no tension.

The reader will observe that this argument appears to be self-contradictory since the design, assuming uplift, assumes a crack to exist which it is the purpose of the design to prevent. However, firstly, perfect construction everywhere is not likely to be achieved and cracks might progressively develop; secondly, if a crack were to develop fully through part of a dam there should still be a factor of safety against the section above overturning; thirdly, even where no cracking exists there will be an uplift pressure of the water in the pores of a concrete. Figure 5.13 shows pore pressures measured in the Altnaheglish concrete dam, illustrating that concrete is a porous material. The pore pressure,

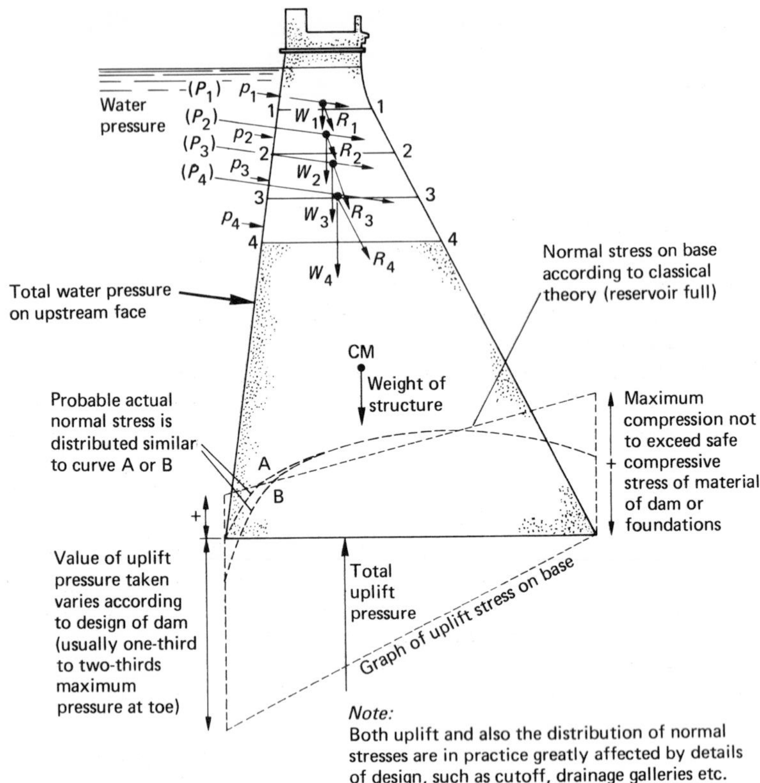

Fig. 5.12 Simple stability analysis of a gravity dam showing the effect of uplift. As well as taking into account uplift on the base of the dam, it may be necessary to take into account uplift acting below any portion of the dam above sections 1–1, 2–2 etc. (this is not shown on the diagram). If the resultants R_1, R_2 etc. of the forces P and W, together with uplift as deemed advisable, come within the middle third of sections 1–1, 2–2 etc. then theoretically no tension develops on the upstream face.

however, may be said to act only on that proportion of the concrete which consists of voids so that the total uplift force is some percentage of the concrete area multiplied by the pore pressure. What this percentage is, in effective terms, is a matter of considerable dispute and it is usual to be conservative and design on the assumption that uplift applies to the whole area.

In order to reduce the amount of uplift experienced below the dam arising from leakage through the foundations a concrete-filled cutoff trench may be sunk at the upstream toe of the dam, as shown in Fig. 5.13. Even if this cutoff does not make the foundation rock wholly

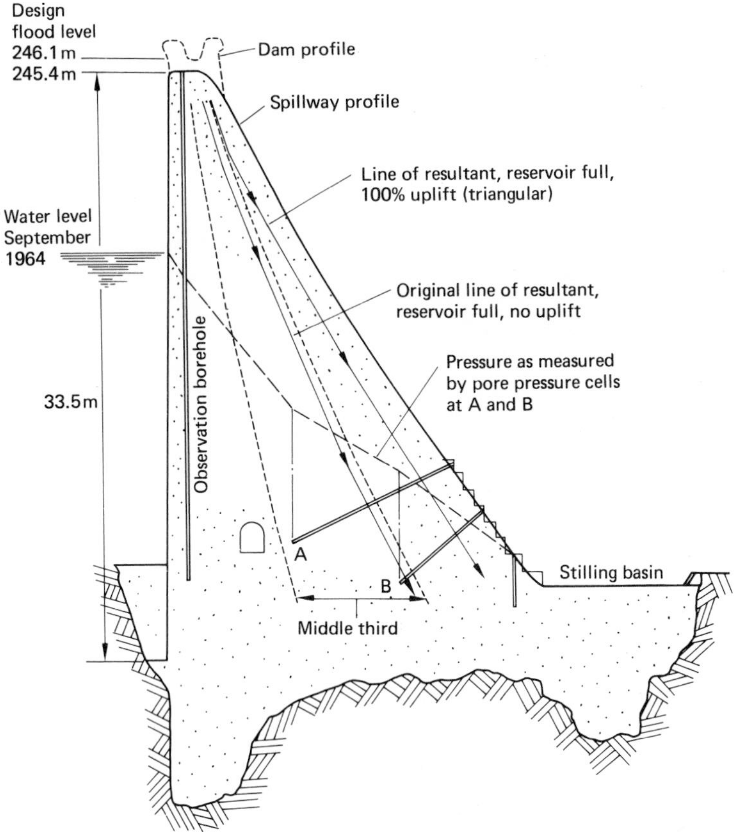

Fig. 5.13 The Altnaheglish Dam.

watertight, nevertheless it reduces uplift below the base of the dam by lengthening the path through which the water must travel from the reservoir to the downstream toe. Relief wells or drainage layers can be inserted below the base of the dam, commencing just behind the cutoff trench, their purpose being to release water which might otherwise create a hydrostatic pressure against the base of the dam. Some engineers take the view that relief wells are equally as effective as a cutoff; most engineers would prefer to adopt both methods. When a substantial cutoff at the upstream toe has been provided, with or without relief wells, then the amount of uplift taken into account may be reduced to some percentage of the full uplift, although it would not be usual to take the uplift value at the upstream toe as less than two-thirds of the difference in level between the maximum reservoir water level at the upstream toe and the downstream toe water level.

Ice thrust may have to be taken into account in the design of gravity dams in cold climates. Estimates of the thrust force vary according to anticipated ice thickness and range from 1.5×10^5 to 7.5×10^5 newtons per metre run of crest[16]. *Seismic* forces from earthquakes may also have to be taken into account, although this is not usually the case for dam design in the UK.

The design of gravity dams appears deceptively simple if only the overall principles are considered: in fact controversy raged for more than half a century over the subject of how stresses are distributed within a gravity dam. Much of the earlier controversy about the stress analysis methods to be used has died down[17] now that new methods of structural analysis (using relaxation methods) have been shown to apply, but there still remains considerable uncertainty concerning the amount of uplift that does occur, how it acts, and how best it may be prevented. In this field of civil engineering, as has been mentioned in connection with earth dams, it is always important for the engineer ultimately responsible for the final choice of design to pay particular regard to what has been done before, and he should not depart widely from what past experience has shown to be necessary unless the new evidence presented to him is unquestionably satisfactory.

5.19 Gravity-dam construction

Concrete The construction of concrete gravity dams is relatively simple. Most of the remarks concerning the preparation of foundations and cutoff trenches for earth dams apply also to gravity dams. The key problem with mass concrete dams is to reduce the amount of shrinkage which occurs when the large masses of placed concrete cool off. Heat is generated within the concrete as the cement sets and as this heat dissipates, which may take many months, the concrete cools and shrinks. In order to reduce this shrinkage the concrete is placed in isolated blocks and left to cool as long as possible before adjacent blocks are concreted, 'low heat' producing cement is frequently used, and in the larger installations water cooling pipes may be laid within the concrete to draw off the excess heat produced. Concrete dams are best constructed in valleys where rock abounds which is suitable for the making of concrete aggregates. A gravity dam, however, does not use concrete to its best advantage: it requires only the weight of concrete to achieve stability and therefore a large, and expensive, mass of concrete must be used. Some advantage accrues, however, from the fact that river diversion during construction is easily achieved and the spillway and outlet works can form part of the structure itself. There is no objection (as there is with an earth dam) to taking the drawoff and scour pipes through the body of the dam and this is the most usual practice.

Masonry The construction of masonry dams is expensive because of the large amount of hand-labour required. In Britain true masonry dams are no longer built, but some recent dams have had masonry facing and concrete hearting: primarily to improve the appearance of the dam, but secondarily to avoid the need for shuttering these faces with temporary shuttering. Overseas, where labour is less costly, masonry gravity dams continue to be built. Very great care must be taken with masonry dams to fill up all joints and beds with a watertight mortar mix and to pay special attention to the quality of the work on the upstream face.

5.20 Arch-dam design

The principle of design of an arch dam is greatly different from that of a gravity dam. The majority of the strength required to resist water thrust is obtained by arching the dam upstream and taking the load of the water upon the abutments. The abutments must therefore be completely sound, and where this criterion does not apply failure will occur. The theory of design is complex: the dam resists the water load partly by cantilever restraint at the base and partly by arching action from abutment to abutment, hence it can be much thinner than a gravity dam (see Fig. 5.14). Early designs were based upon the 'trial load' procedure. In this design the dam is assumed to consist of cantilevers from the base interpenetrating arches at every level across to the abutments. The water load is then divided so that deflections either way at any point of the dam are equal. The modern method is to adopt a three-dimensional stress analysis of the dam using mathematical relaxation methods of computation[18].

Models are often used to measure the distortion likely to apply and these are useful, partly as a check to the mathematical calculations and partly as an aid to these calculations by giving a first approximation to the distribution of likely stresses. There are many variations from the simple uniform arch shape, the most economic section being curved both vertically and horizontally, i.e. the horizontal arches vary in radii. A dam of this kind which curves 'both ways' is called a cupola dam and, although most economical in the use of concrete, it presents some difficulty in the incorporation of drawoff and scour valves within it because the spindles which operate these valves are usually arranged to give a direct vertical lift. Arch concrete dams are among the highest in the world and they are inherently stable when the foundations and abutments are satisfactory. River and flood diversions are usually taken in tunnels around the ends of the dam, the abutments being suitable for tunnel driving since they must consist of firm sound rock.

5.21 Buttress-dam or multiple-arch-dam design

Where a valley in rock is too wide for a single arch dam, the dam may consist of a multiple series of arches between buttresses, as in Fig. 5.15,

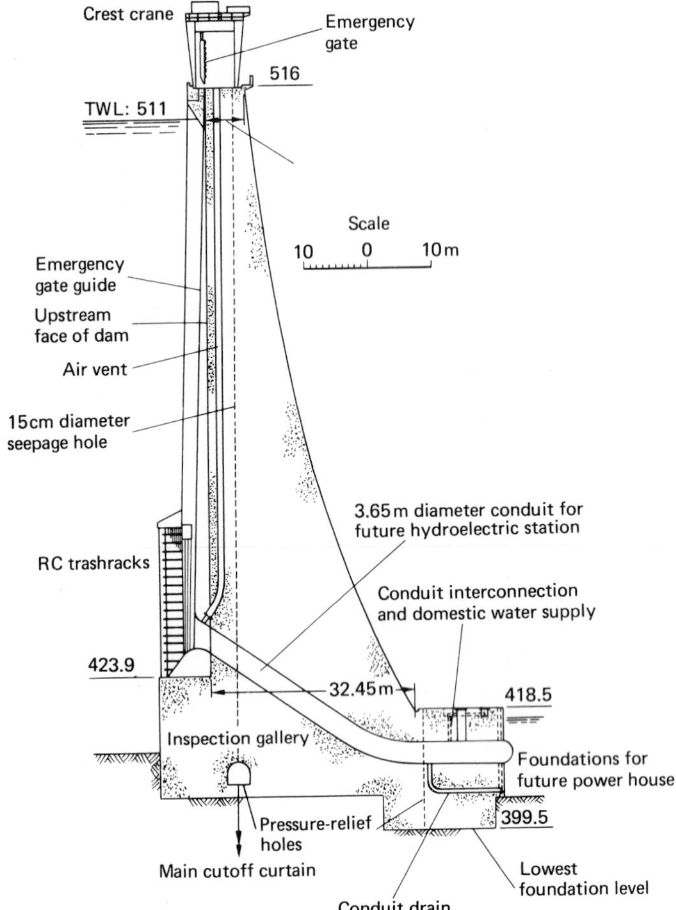

Fig. 5.14 The Dokan Dam for the Government of Iraq. Engineers: Binnie & Partners. (Binnie *et al.*, *Proc. ICE*, 1959.)

each section (consisting of a single arch and its buttresses) achieving stability in the same manner as a gravity dam. These buttress or multiple arch dams have the merit of relative simplicity of design, good stability, and good appearance. There is a considerable saving of concrete as against a gravity dam for the same height, but the expense thus saved is largely taken up by the cost of the extra shuttering required. Economic conditions, together with the depth of sound rock below ground surface, largely determine therefore when a buttress dam or multiple arch dam is used instead of a gravity dam. It should be noted that the buttress dam has a design advantage over the solid gravity dam in that the amount of

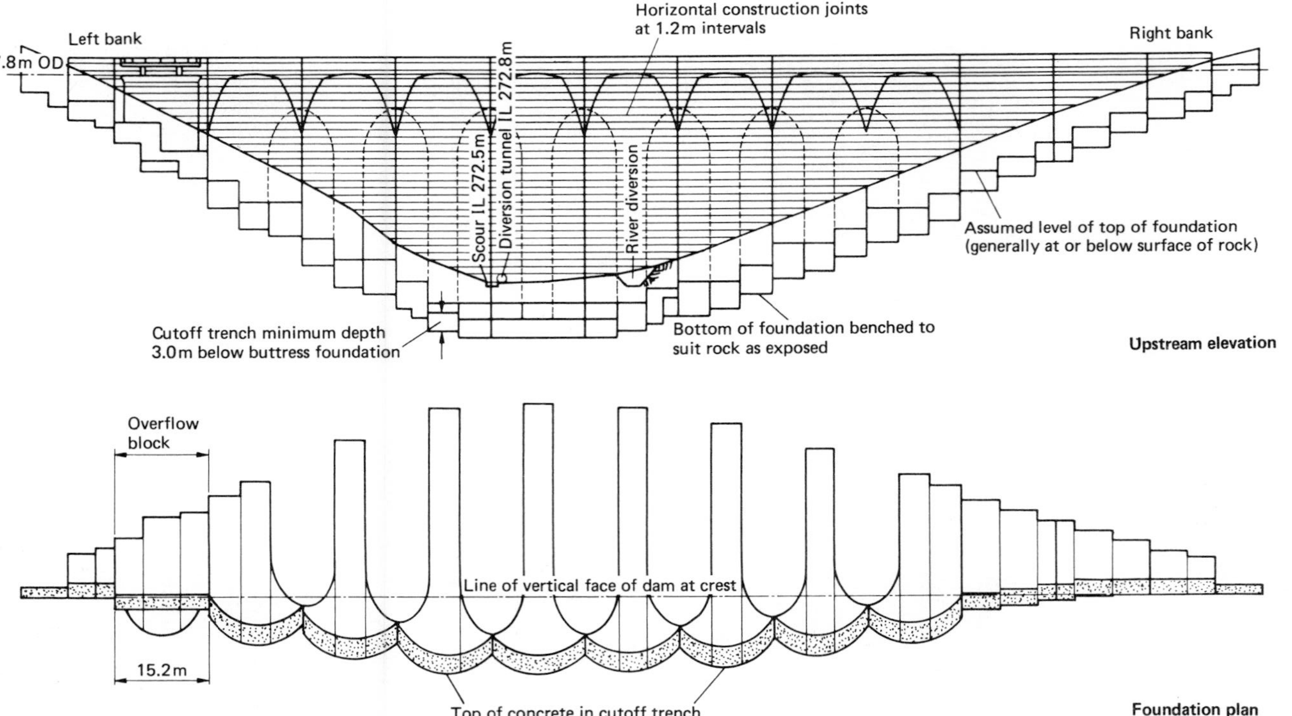

Fig. 5.15 Elevation and plan of the Lamaload Dam. Engineers: Herbert Lapworth & Partners.

foundation uplift to be taken into account is negligible because of the free space left between abutments through which uplift pressures may dissipate.

Rockfill dams

5.22 Design

A rockfill dam must have a watertight membrane incorporated with it. The main difficulty in construction is that this watertight membrane must be flexible because a rockfill dam is liable to continue to settle over a long period. The rockfill initially settles down under its own weight but, when the thrust of the impounded water occurs upon the dam, the rockfill will move, rock points will crush, and the fill will settle further. Upon the drawing down of the water the load will be released, and upon its reapplication further settlement will occur. This process may continue for many years, but the tendency can be reduced by ensuring that the fill has plenty of broken rock and fines in it to pack the voids as much as possible.

A corewall of rolled clay fill is frequently used in a rockfill dam. The protective zones of transitional material between the clay and the rock must not only perform the function of filters, but must also be capable of taking up progressive settlement of the rockfill. Usually two or more layers of transitional material are necessary, grading through from clayey-sand against the corewall to a crushed rock with fines against the rock, the rockfill zone itself being graded with its coarsest material on the outside.

Asphaltic membranes are sometimes used to form an impermeable facing or corewall to a rockfill dam. If the material is used on the face of the dam (the waterface) an advantage is that this membrane can be inspected whenever the reservoir is low and can therefore be maintained and repaired if necessary. A disadvantage is that the cutoff to such a dam must be constructed in a curve at the upstream toe of the dam so that the upstream face membrane can be connected into it. This makes the cutoff longer than along the centreline of the dam. An example of an internal asphaltic core to a rockfill dam is shown in Fig. 5.16, which is one of two dams at High Island, Hong Kong[19], 107 m high, which are, unusually, founded below the sea bed across a sea inlet. The core consists of 1.2 m thick asphaltic concrete, made of 19 mm aggregate mixed with cement and water, preheated and mixed with about 6% of hot bitumen. Below a certain level the core is duplicated, and above a certain level it is 0.8 m thick.

An unusual early design of a composite rockfill dam is the Shing Mun Dam, which is 83 m high, also built in Hong Kong about 1937[20], and shown in Fig. 5.17 and Plate 1A. The upstream watertight membrane consists of heavy concrete blocks securely jointed together. This is

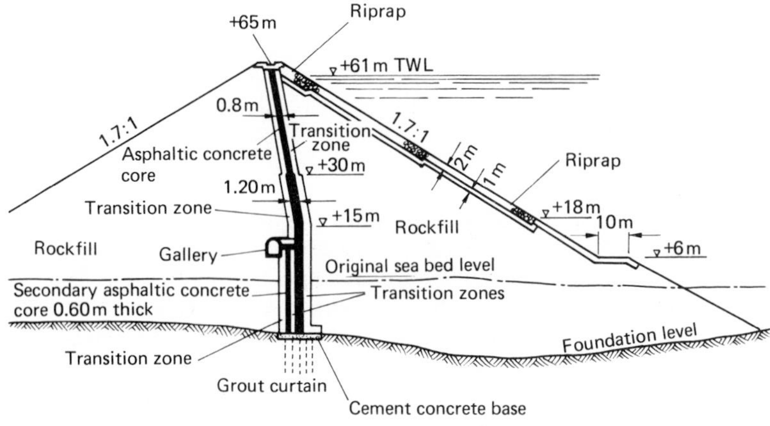

Fig. 5.16 High Island West Dam, Hong Kong, with asphaltic core membrane. Engineers: Binnie & Partners.

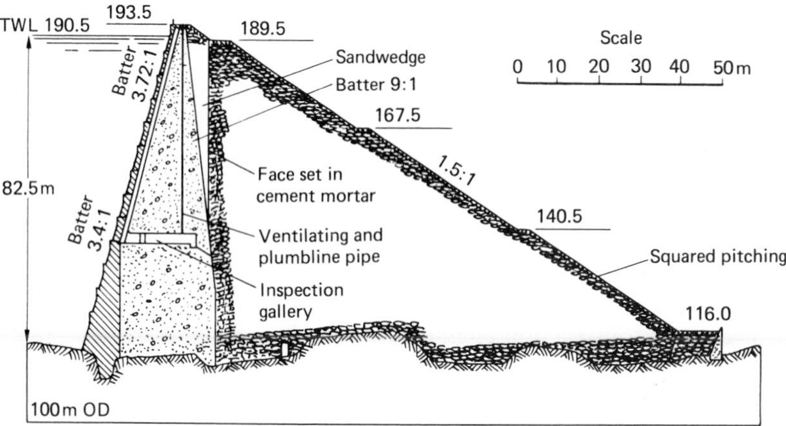

Fig. 5.17 The Shing Mun Dam forming the Jubilee Reservoir for Hong Kong. Engineers: Binnie & Partners.

supported by a concrete thrust block, which in turn is supported by the main rockfill. Between the thrust block and the rockfill a sand wedge is interposed which is designed to take up any movement of the rockfill. This bold and imaginative design proved successful.

Safety controls

5.23 Statutory control

Regulations to ensure that dams are regularly inspected, and are constructed or altered only under the charge of properly qualified engineers, are adopted in many countries. In Britain the Reservoirs Act 1975 was passed, but has not yet been brought into action*, principally because of the costs it would impose. Consequently the earlier Reservoirs (Safety Provisions) Act 1930 still currently applies. The 1930 Act requires reservoirs holding more than 22 700 m^3 (five million gallons) above the natural level of the ground to be inspected at not less than ten-year intervals, and continued impounding (or impounding for the first time in the case of a new or altered reservoir) is not permitted until a certificate has been issued by a qualified engineer stating that it is safe to do so. These qualified engineers are 'panel' engineers, as listed on panels of names approved by the Department of the Environment. The Act is remarkable in that the responsibility of the panel engineer is personal when he issues a certificate, i.e. he must issue it under his own name and must take personal responsibility for his decision, no matter for whom he is working. Other countries have adopted somewhat similar rules. The model rules of the USA Committee of ICOLD in 1970 (based largely upon Californian state law) require inspections at five-year intervals, the inspection being carried out by a government or state agency which can also require plans and designs to be submitted for approval before a new dam is built or an old one altered.

The 1975 British Act continued the provisions of the 1930 Act, but endeavoured to cure some weaknesses in it. The principal additions were as follows.

(1) The county and metropolitan local authorities are made responsible for keeping a register of all reservoirs to which the Act applies in their area, and must also act as the enforcement authority to see that the owners of such reservoirs comply with the Act.
(2) In the case of new or altered reservoirs, they must come under the control of a 'construction engineer' during design and construction and afterwards, until this engineer issues a Final Certificate. He must issue a Preliminary Certificate permitting first filling, and his Final Certificate must be issued not earlier than three years nor later than five years after his issue of the Preliminary Certificate.
(3) After the Final Certificate mentioned in (2) above, an independent 'inspecting engineer' must inspect the dam within two years and thereafter the same or another 'inspecting engineer' must inspect the reservoir at not less than ten-year intervals, or at such lesser interval as is stipulated in any previous certificate issued. The inspecting engineer cannot be the same person who acted as construction engineer, nor may he be an employee of the owner of the reservoir.
(4) Approved panel engineers who are permitted to act as above and

* An Act may contain provisions for it to come into force only after some 'appointed day' has been fixed, the Secretary of State (or some other Minister) being empowered to fix the day by regulation when the Government so decides.

issue certificates are appointed by the Secretary of State, and the appointments are made for five years only, although an existing panel engineer can reapply.

The problem with the implementation of this Act has been that it threatens to impose a large financial burden upon the local authorities, and complications have been foreseen with respect to private owners of small lakes. Notwithstanding the lack of implementation of the Act, all public water supply authorities voluntarily comply with its main provisions.

Although catastrophic failures of dams are relatively rare, failures as such are not. The history of dam failures shows that it is either want of care in construction or lack of attention to, or understanding of, observable defects that leads to premature failure. Few failures can be said to be due to the fact that no one could have known beforehand that there was anything wrong with a dam.

Many dams in Britain are approaching their 100-year age, and modern methods of examination frequently reveal weaknesses. Troubles often beset the earlier dam constructors and standards of building varied from builder to builder. Cases are known where dams have failed during construction and the remedial measures adopted did not get to the root of the trouble, leaving an inherent weakness behind.

In one known case the puddle clay core of a dam was inserted *after* the dam was fully built, in addition to which the dam was overtopped by a flood whilst the trench for the corewall was open. In another case a dam was inspected and reported as unsafe, but the remedial measures proposed were never adopted because, owing to changes of ownership, the inspecting engineer's report got filed away and forgotten. Many old dams have been repeatedly raised in the past by additions to their crest without records being kept, with the result that the original crest level remains unknown and the upper part of the corewall, if found, may not extend far enough into the abutments. Sometimes the true base level of the body of the dam may be higher than seems apparent from the level of the downstream toe. When such dams, in addition, show signs of leakage R C S Walters has rightly, if slightly ungrammatically, labelled these as 'Suspicious Old Dams'[21], and being suspicious of anything and everything about them is a most prudent attitude.

5.24 Earth-dam deterioration signs

Routine observations of an earth dam, which should be conducted annually, cover items too numerous to list here. Instead the items given below are those which are amongst the most important indicators of potential signs of deterioration.

(1) *Of the instrument recordings*
(a) higher than usual pore water pressures,

(b) readings remaining consistently high and not rising and falling with rainfall, especially those in the lower part of the downstream shoulder close to the formation and the corewall, and particularly if they appear to be influenced by reservoir water level,
(c) settlement and tilt gauge readings, if installed, which do not show declining range of movement with time.

(2) *Of the underdrain flows*
(a) increased flows compared with the pattern previously established,
(b) flows not decreasing in dry weather,
(c) outflows cloudy in appearance (test for clay and silt content),
(d) clay or silt in the weir boxes, or in any tunnel taking drainage from the dam.

(3) *Of the crest*
(a) any irregular settlement,
(b) the absence of any slight bow upwards along the crest of the dam towards its centre (denoting that if any settlement allowance had been given to the dam, as would normally have been the case, all this allowance has been taken up),
(c) tilting of any wave wall or of the roadway along the crest.

(4) *Of the shoulders*
(a) any damp patches on the downstream slope or at its toe, growth of rushes on any part or at the toe,
(b) irregular settlement, hollows, declevities, slight bulges, unusually luxuriant grass or weeds in any location,
(c) slabs or facings on the waterface irregularly settling,
(d) a water line against the dam which is not a consistent smooth curve.

(5) *Of the other works*
(a) tilting of the spillway crest or bellmouth as shown by uneven overspill,
(b) new leakage into any inspection gallery, shaft, or tunnel within the dam.

5.25 Fish passes

Where trout or salmon exist in a river it is necessary to provide a fish pass around a dam or weir constructed across the river. The design of fish passes is a matter on which specialist advice should be taken. In 1942 the Institution of Civil Engineers published a report of the Committee on Fish Passes which gives some suggested designs. The essential features of a fish pass consist of a series of shallow pools arranged in steps, down which a continuous flow of water is maintained. It is essential to keep this flow running with an adequate amount of water because when trout and salmon are in migration upstream they will only be attracted to the pass by a good flow of water which should be turbulent. Examples of fish passes are shown in Fig. 5.18. An alternative to the fish pass is the automatic fish lift. This consists of an upper and lower chamber connected by a sloping conduit through which a continuous flow of water is maintained. Fish migrating upstream enter the lower chamber. In due course the major discharge from this lower chamber is automatically closed and the water level then floods up through the conduit to the upper chamber. The fish rise with the water and then proceed out into the reservoir at higher level. The cycle of operations is repeated, being float- and time-controlled automatically.

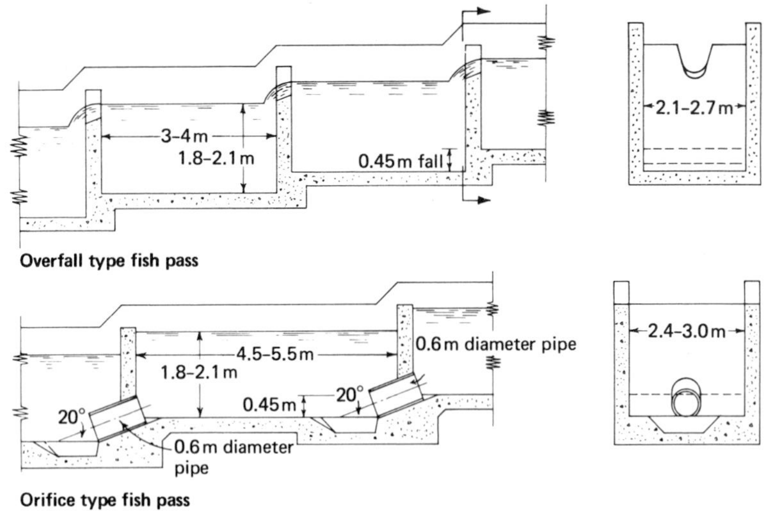

Fig. 5.18 Two types of fish pass used in Scotland. (Fulton, *Proc. ICE*, 1952.)

Part II
River intake works

5.26 Side intakes

It is surprisingly difficult to design a satisfactory river intake. Figure 5.19 shows the conventional solution of a weir across the river to raise its level, with a gated side intake just upstream. This solution is feasible only for rivers of moderate size which do not carry cobbles and boulders during a flood. On wide rivers the cost of the weir becomes prohibitive because of the massive works necessary to prevent the weir from being undermined and destroyed during a flood. On a rock foundation the weir will be less costly, but a frequent problem then is how to cope with the gravel, cobbles, and boulders brought down by the river. In some

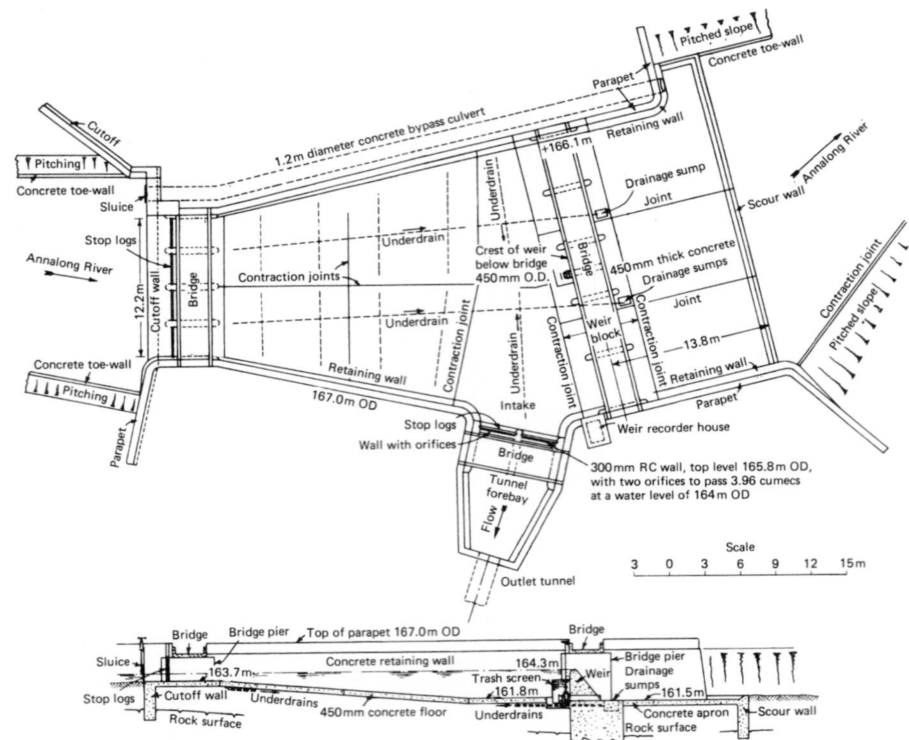

Fig. 5.19 The Annalong River intake for Belfast City and District Water Commissioners. Engineers: Binnie & Partners.

rivers this bed load may be sufficient to fill the weir basin completely every time there is a flood. In Cyprus where these conditions occur the 'Cyprus groyne intake' shown in Fig. 5.20 has been developed. This groyne weir, extending only partly in the river, is sufficiently low to be overswept in a flood. Many other devices have been tried for intakes in fast flowing streams which carry much bed load in an effort to beat the deposition problem. A rounded hump-backed weir with slots in its crest (or just upstream) feeding an offtake conduit is sometimes used. With all such designs it can be appreciated that the designer's principal problem is how to obtain a self-cleansing offtake which is effective from low flows to floods.

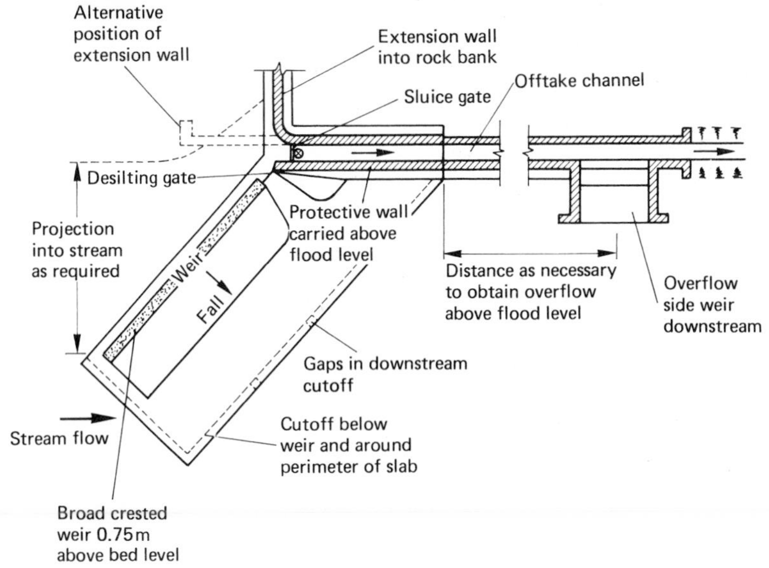

Fig. 5.20 Groyne intake for small abstraction on a flashy river (as used in Cyprus).

If the designer is fortunate and he can place his offtake at a point where the river is consistently deep and constrained between stable banks, an intake of the type shown in Fig. 5.21 is possible. However, in this case he then has the problems of siltation and fish entry to deal with. Usually the design is for about 1.0 m/s inflow rate, but this is a compromise: it will permit fish to swim out if they get in, it should keep siltation down to a reasonable amount, and it should, with luck, avoid dragging everything into the intake that the river is carrying. In practice, none of these things may be prevented entirely and, if screens of rather fine mesh (say 13 mm aperture) are installed to keep fish and debris out, the situation may be made rather worse, requiring the screens to be

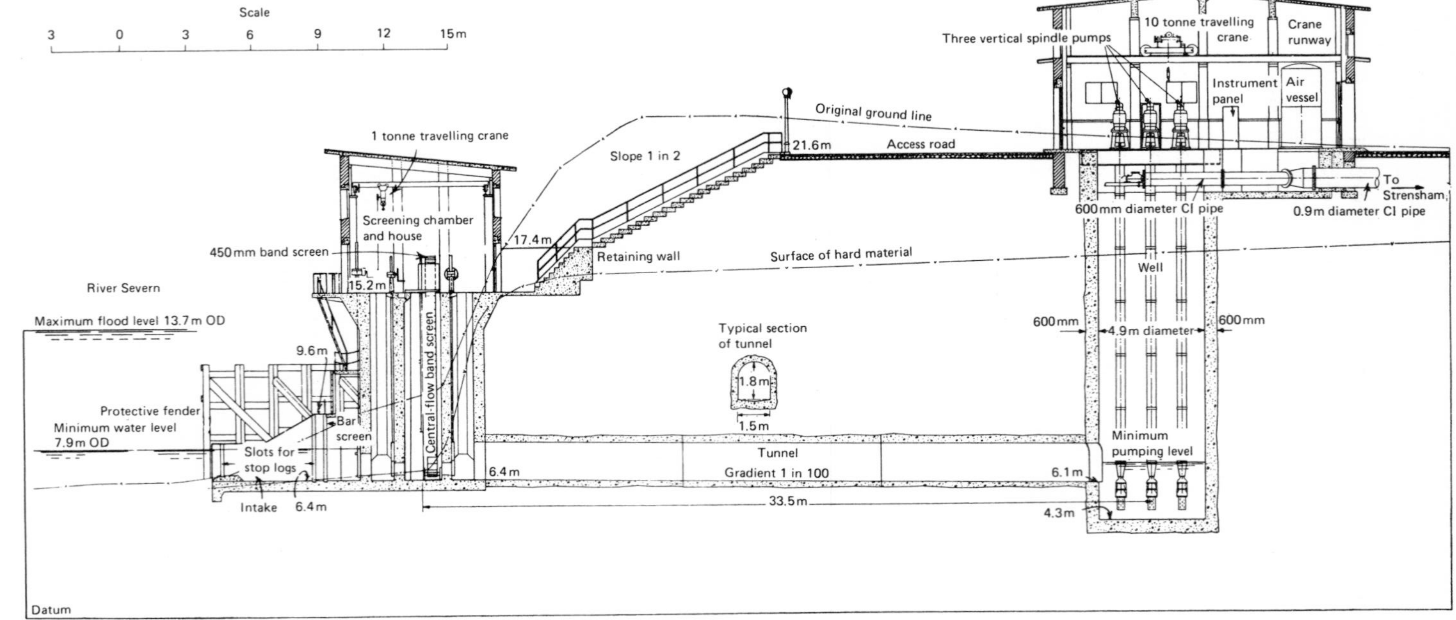

Fig. 5.21 The River Severn intake for Coventry Corporation. Engineers: Binnie & Partners. (Hetherington and Roseveare, *Proc. ICE*,1954.)

cleaned every day. The only answer in this case is the installation of heavy self-cleansing screens of the type shown in Fig. 5.22. These are expensive, but they can be expected to work without the need for constant attention. The plates forming the band screen are of 3 mm thick mild steel perforated with 4.75 mm square holes, supported on frames carried by the band links. Sturdy trash racks are fitted at right-angles to the perforated plates in order to pull out heavy floating debris. The band screen only moves when the differential head across it due to the build-up of debris exceeds a preset value, thus minimizing power consumption. High pressure jets wash debris off the screens.

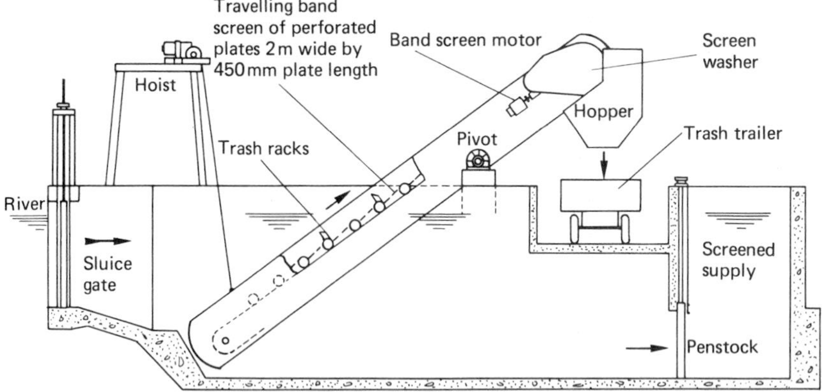

Fig. 5.22 Heavy duty band screen on river intake.

5.27 Other intakes

For a lake, and sometimes for a river, a piled crib is suitable. The pumps are suspended in the lake or river from the crib and the access bridge to it can support the delivery pipes. Submersible pumps can be used provided that the water is deep enough, the advantage of such pumps being that only the electricity supply cable has to be run out to them. Otherwise suction pumps must be used and care must then be taken to ensure that their strainers are of large size so that the pump can keep going even with a partially blocked strainer. Backflushing facilities are desirable to clean strainers. The crib intake is relatively cheap, but if there is any boat traffic on the river the cost rises because of the need to make the crib of sturdy construction in order to prevent damage by boats.

Floating pontoon intakes have been used, but usually only for relatively small abstractions. The principal problem with them is their secure anchoring to hold them in flood conditions.

River bed intakes have been used, the end of the pipe being buried in a gravel-filled box or similar. The problem with such intakes is that of preventing gradual loss of capacity through siltation.

A special type of intake[22] for unusual circumstances is shown in Fig. 5.23. In this case it was necessary to lead water from spring intakes down shafts to a collecting tunnel 100 m or so below. It was found that by inducing spiral flow down a vertical shaft the falling water could be made to cling to the walls of the shaft, thus increasing the discharge and reducing air entrainment. Nevertheless it will be noted that special

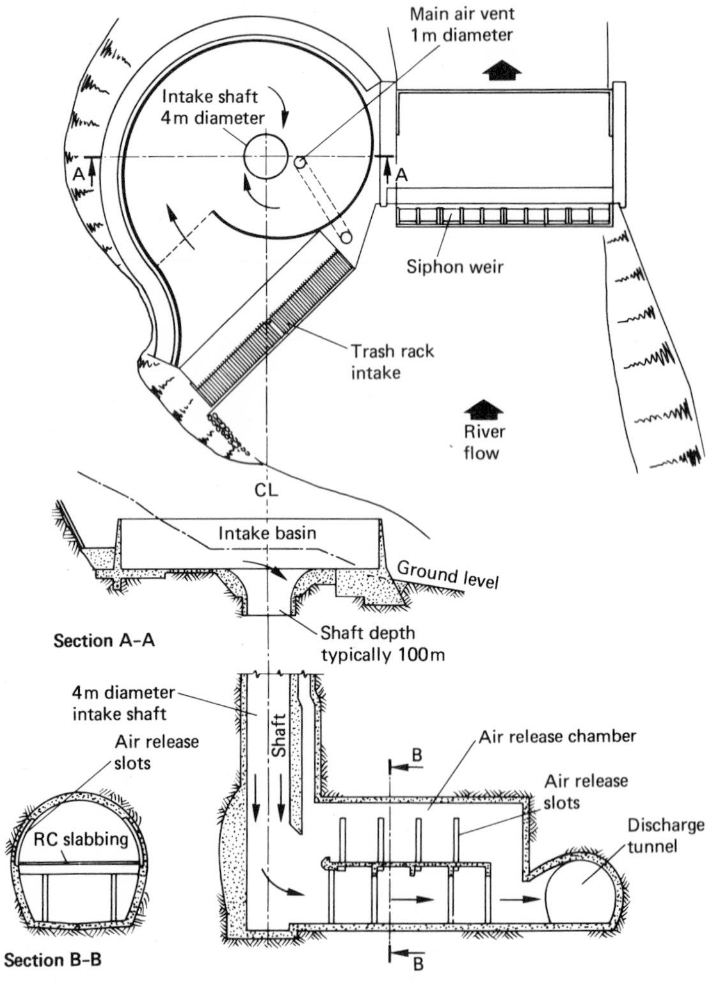

Fig. 5.23 Vortex intake drop shafts of the Plover Cove scheme, Hong Kong.

measures for releasing entrained air at the base of the shaft are incorporated.

Part III
Wells and boreholes

5.28 Choice of underground works

In Section 4.14 it was mentioned that the drilling of a small diameter pilot borehole at a proposed major development site is often not worthwhile. For most public supply purposes a hole capable of giving upwards of 5 Ml/d is usually required and, for this, a boring of at least 300 mm diameter is required to accommodate the pump. In practice a more usual size of hole would be 450 or 600 mm since allowance must be made for the lack of verticality in the hole, the need for good flow characteristics in the boring and through the rising main, and a possible need to reduce the boring diameter with depth.

Verticality (see Section 5.35) itself is not of so much importance as are 'kinks' in the boring down to the level of the pump. If any of these exist in the boring above the pump they may throw the siting of the pump to one side of the boring, making poor entry conditions to the pump suction and causing difficulty in lowering or removing the pump.

It is almost always necessary to line the upper part of a boring with solid tubes which are then grouted into the surrounding ground so that there is a seal preventing surface water penetrating into the borehole. This is both to prevent surface contamination of the borehole water occurring and to seal off water in the upper part of the strata that may not be of such good quality as that obtained from the main aquifer to be penetrated.

Where weakly cemented or uncemented sands, gravels, or clays are penetrated, further lining may be necessary: slotted where adjacent to water-yielding aquifers, solid when passing through clays. If such additional lining is required below the top sealing tube then it is usual to 'step in' the diameter of the lower lining so that it may be placed as a separate operation. Below the additional lining the borehole diameter may need to step in again in order to facilitate passage of the boring tools to drill the unlined boring below.

Where gravel packing is necessary (see Section 5.33) a further allowance for this must be made in the boring diameter, the thickness of the gravel pack usually being at least 75 mm so that the required diameter of the boring must be 150 mm greater than the nominal required lining size. In such circumstances the boring may need to start off at 675, 750, or even 900 mm diameter depending upon the need for gravel packs, step-ins of diameter for additional lining, the depth at which the pump will have to be sited, and the output which is anticipated.

The need to maintain good flow conditions to the pump suction is sometimes overlooked. With submersible pumps the suction strainer is above the pump motor so that the water has to pass through the annular space between the outside of the motor casing and the boring. It is important therefore to site such pumps centrally in the boring so that flow conditions to the pump are as even as possible. With vertical spindle pumps the strainer is below the pump and conditions of entry are slightly more favourable, but this is offset by the fact that the pump and rising main are a rigid assembly in which no deviation at the joints is permissible. When pumps have to be sited at considerable depth (irrespective of the water level) the size of the rising main should be carefully investigated. It should be such that the friction loss through it for something greater than the anticipated yield of the boring should still be a reasonable figure. For example, for a boring that might give up to 10 Ml/d a rising main of 250 mm diameter would cause about 1 m headloss in friction for a 30 m length of rising main: any less diameter than this would cause a large increase of friction loss.

5.29 Standby pumping arrangements

A single boring will only permit one pump to be installed and this may not adequately safeguard the supply. Even if no pump breakdown occurs, sooner or later some maintenance of the pump or its motor will be necessary. Therefore either two borings are necessary or a large diameter well must be sunk capable of taking two pumps. The latter is the most frequently adopted solution where the pumping water level will not be excessively far below ground level. In this case the well may be sunk over the boring, although it is easier to construct the well first and then to sink the borehole from the bottom of it. Alternatively, a well may be sunk adjacent to a boring, as shown in Fig. 5.24, and a connection may then be dug through from the well to the boring at some suitable depth or there can be a blown through connection from one boring to another using controlled explosives.

In aquifers which give consistent yields and it can be expected that any boring or well will tap the water-bearing formation, it is possible to plan the works ahead. In intermittently fissured rocks, such as chalk, limestone, and sandstone, one boring may intercept a large fissure yielding much water, whereas another quite close by may fail to tap the same fissure and give a much poorer yield. In such formations planning of the works has to follow upon results of the first boring and, if enough water has been found in one boring, an additional well or boring should be sited close by so that they can be interconnected.

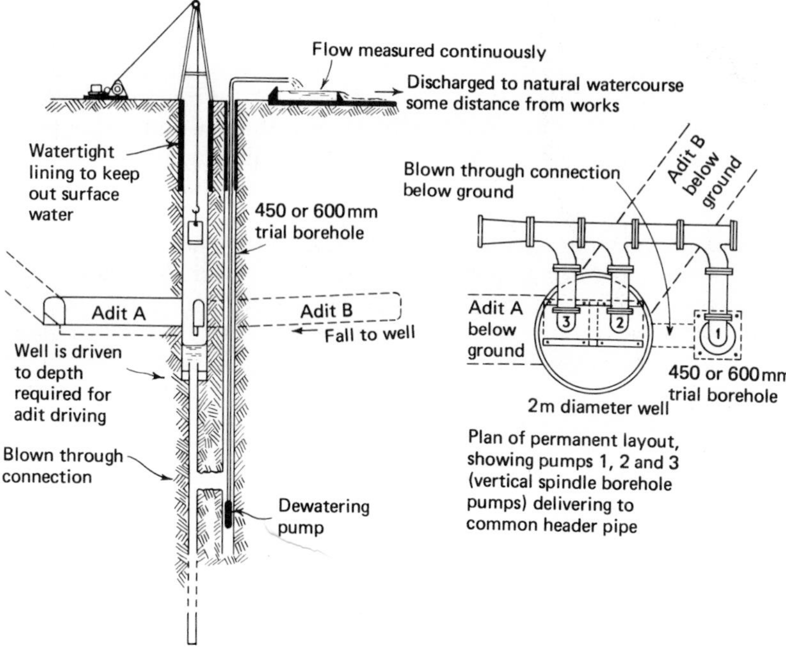

Fig. 5.24 Layout using trial borehole, deep well, and adits.

5.30 Well and adits

A well is expensive and would not be economic if the pumping water level is very low, say more than 50 m below ground. It is also important that the base of the well is taken substantially below the anticipated pumping water level: firstly, because the pump suction will need a substantial amount of submergence to prevent air entrainment and, secondly, because it is always possible that water levels in the well may at some time in the future be lower than when the first test pumping is conducted.

A well is also necessary if adits are to be driven. The depth and direction to which adits are driven depend upon the nature of the strata penetrated and, in this matter, hydrogeological advice should be sought. The purpose of an adit, which is simply a tunnel about 2 m high by 1.2 m wide (see Fig. 5.24 and Plate 8), is to intercept further water-bearing fissures in such formations as chalk, limestone, and sandstone. The adits usually rise slightly from the well so that any water entering the adit can be drained away to the well shaft by means of a grip, as shown in Plate 8A. To construct the adit the well must be kept dewatered, either by pumping from a borehole sunk from the bottom of it or by pumping from an adjacent borehole which has a connection to the well, as shown in Fig. 5.24. A very reliable pumping system must be adopted to keep the well dewatered when men are driving an adit, and it is quite common practice in the early stages of the work for the men to come up with every skipful of muck so that no man is left in an adit without a skip being ready at the adit entrance to lift him out of danger. On occasion, if a large fissure should suddenly be met in an adit, the men may simply have to run for the skip and be taken out of the well. However, the complexity of numerous long adits (often interconnected) that radiate from some old borings gives testimony to the fact that they are not always productive of much additional water, and some (at great expense) have proved virtually barren. They are not so frequently constructed nowadays because of their high cost and also because of a better understanding of underground water flow characteristics and the means of intercepting this water by boreholes.

5.31 Construction of wells and borings

Borings for water supply are usually 300 mm diameter upwards. They can be sunk by percussion or rotary drilling. In the percussive method a heavy chisel (see Plate 8D) is cable suspended and is reciprocated on the bottom of the hole by the action of the winch. The lay of the weighted wire cable as it tends to unwind gives a rotary action to the chisel after each blow; after a while a friction grip device in the rope socket attachment comes apart, the cable reverts to its natural lay, and the process of rotating the chisel continues. Every so often the chisel must be withdrawn and a bailer is inserted to clear the hole of slurried rock chippings. The bailer is a tube with a flap valve at the bottom permitting entry of the slurry; after being used as necessary to clear the hole, the chisel is again lowered and drilling continues. Another percussive method is to use a 'down-the-hole' hammer drill operated by compressed air. This is very much faster in hard rock formations than is a cable-operated chisel but is unsuitable for the softer formations and cannot be used in large depths of water.

For rotary drilling of large holes a roller rock bit is used, equipped with three or four toothed cutters of hard steel which rotate and break up the formation. Water is fed to the cutters down the drill rods and, rising upwards through the boring, brings the rock chippings with it. In the 'reverse circulation' method the water flow passes down through the boring and up through the drilling rods. This achieves a higher upward water velocity making it easier to bring chippings to the surface. Air can be used instead of water, but usually only for the smaller diameter holes. Another method of rotary drilling is to rotate a heavy core barrel at the base of the hole. This barrel has a thickened bottom edge. Chilled shot are fed with some water into the barrel and, becoming trapped under its bottom edge, they are rolled round thus cutting the rock. A

Plate 5 (A) Drawoff tower and bellmouth overflow of the Taf Fechan earth dam, South Wales, 1919. (B) 'Heeling in' puddle clay. (C) Pug-mill producing puddle clay. (D) New Tittesworth dam for Staffordshire Potteries Water Board, 1963: a sandfill dam with articulated concrete corewall. (Engineers: Binnie & Partners.)

Plate 6 (A) Dokan concrete arch dam 118 m high, Iraq, 1959. (B) Mudhiq concrete arch dam 73 m high, Saudi Arabia, 1980. (C) Mangla dam main spillway for 25 000 m^3/s flow, Pakistan, 1968. (D) Llyn Brianne rockfill dam with boulder clay core 91 m high, West Wales, 1971. (Engineers: Binnie & Partners.)

Plate 7 Repairing Lluest Wen earth dam for Taf Fechan Water Board, 1972, by insertion of a plastic concrete corewall through the old puddle clay corewall. Inset: the Kelly grab used for excavating the trench. (Engineers: Binnie & Partners.)

Plate 8 (A) An adit in the chalk at Falmer pumping station, Brighton Corporation, 1933. (B) Bailing out a borehole. (C) 'Down-the-hole' hammer drill. (D) A 450 mm cable-operated percussion chisel. (E) Roller rock bit with reverse circulation. (Engineers and photographs: George Stow & Co. Ltd.)

core of rock enters the barrel and this can be broken off and withdrawn by dropping some sharp pebbles into the boring which causes the core to jam inside the barrel. In small diameter holes chilled shot are not used: instead the thickened edge of the barrel has diamond cutters embedded. Rotary drilling methods are not suitable for soft formations, but the core-producing rotary methods are much favoured for drilling water holes in rock as they give a much better indication of the strata penetrated than do the chippings washed or blown out by percussion drilling or use of the roller rock bit.

The duties of the well or borehole driller are (1) not to lose his tools down the hole, (2) to note and log down every single change of operating circumstances—whether it be increased or decreased speed of boring, change of tools necessary, quick or sluggish recovery of water level on baleing out etc.—and (3) to note and log down the nature and depth of every piece of core extracted and every baleing out of chippings. The engineer needs to see that these logs of progress are meticulously kept and that every sample is labelled and put aside for inspection. Tools lost down a boring, such as drill bits which come off the rods, may cause days of delay in attempts to recover them, and if this 'fishing' is unsuccessful it may be necessary to reamer out the hole to a larger diameter down to the tool in order to recover it or, possibly, the hole may have to be abandoned.

5.32 Lining boreholes and wells

The upper part of any boring or well must be lined so as to prevent surface pollution entering. The boring is made slightly larger than the steel lining to be inserted and the outer annular space is sealed with grout or concrete. Depths of 'surface' linings vary from 6 to 30 m. Care must be taken not to carry linings so deep that good water-bearing strata are sealed off. When the boring is being sunk a careful watch should be kept for signs of large fissures: these signs are sudden drops of the chisel, no drop of water level when baleing out, and increased flow when dewatering. The sides of a boring may be too soft to stand up without support and in that case slotted linings must be inserted, which support the formation but permit water to enter the boring.

5.33 Gravel packing

Gravel packing may be necessary where pumping of water from a borehole may bring sand out of the formation. The borehole is drilled larger than the required finished size, a slotted tube is hung in the hole, and a specially graded gravel–sand mix is poured into the annular space. The grading of this gravel–sand mix is usually in the range of a coarse sand to 5 mm gravel (pea gravel). The grading of the gravel pack must be designed to match the grading of the formation. A good rule is to make the grading of the 50% size of the gravel pack four or five times the 50% size of the formation material. If the ratio is above ten the formation material will be drawn into the bore; if below four the yield of the bore may not be fully developed. The thickness of the gravel pack used must be such that the area exposed to the formation permits sufficient water to be drawn into the boring to give the yield required. Gravel packs or filters are especially necessary in the softer Bunter sandstones and the Greensands. No boring should be pumped at such a rate that sand or suspended chalk is carried away by the water for more than a very short length of time because collapse of the sides of the boring may follow from continued removal of these materials.

5.34 Surging

Surging is the name given to one of many different devices intended to increase the yield of a boring. Water is pumped from the boring at maximum rate and then suddenly stopped. The water in the rising main falls back and a reverse flow takes place from the borehole to the formation. In some cases water is pumped back into the formation. The intention of this process is to wash out any clay existing in fissures by causing the water to wash in, and out of, such fissures rapidly. A similar device is to explode a charge underground which forces the water to act in the same manner.

These 'development' methods can be dangerous for the boring or well and should never be attempted without first getting expert geological advice. In Britain many wells and borings would collapse if subjected to any violent surging.

5.35 Verticality

The importance of verticality is that it must be possible to get a pump down the hole. A guide tube, about 6 m long, must be used to start the boring off truly vertical. Thereafter the boring contractor must use his skill to keep the hole as vertical as possible. If the hole goes seriously off the vertical it may be necessary to reamer the hole out to larger size and so get the remainder of the boring vertical. It is especially difficult to keep a boring vertical when hard strata are encountered which are steeply inclined. A usual specification for verticality is that the hole should not be more than 100 mm off vertical in 30 m of depth. It is not necessary to demand that this specification be rigidly adhered to for depths which are below the lowest possible level that a pump will be set. Verticality is tested by lowering a wood or metal cylindrical cage down the boring, the diameter of the cage being just less than the diameter of the hole. To the centre of the cage a string is attached and passed over a pulley fixed about 3 m above, and central to, the top of the boring. The deviation of the string from its initial position at ground level as the cage moves down the boring indicates, by proportional triangles, the deviation of the cage from the vertical.

5.36 Choice of pumping plant

From the test pumping of the completed works the yield–drawdown characteristic curve of the source will be obtained. A typical characteristic curve is curve A in Fig. 5.25. This shows the *maximum* yield of the works, and some provision must be made for future deterioration of the yield (if this is counted to be a possibility) so that some modified yield–drawdown curve B is drawn. Curve B shows a reduction of the yield for any given pumping level or (what amounts to nearly the same thing) an increase of drawdown for a given yield. The judgement of the engineer determines the proportion of curve A represented by curve B. After this adjustment a further correction must be added to denote the range of fluctuation of the standing water level in the well to be expected seasonally. This is represented by the curves bb' and cc' and upon these we must also mark the range of pumping output required (shown shaded in Fig. 5.25).

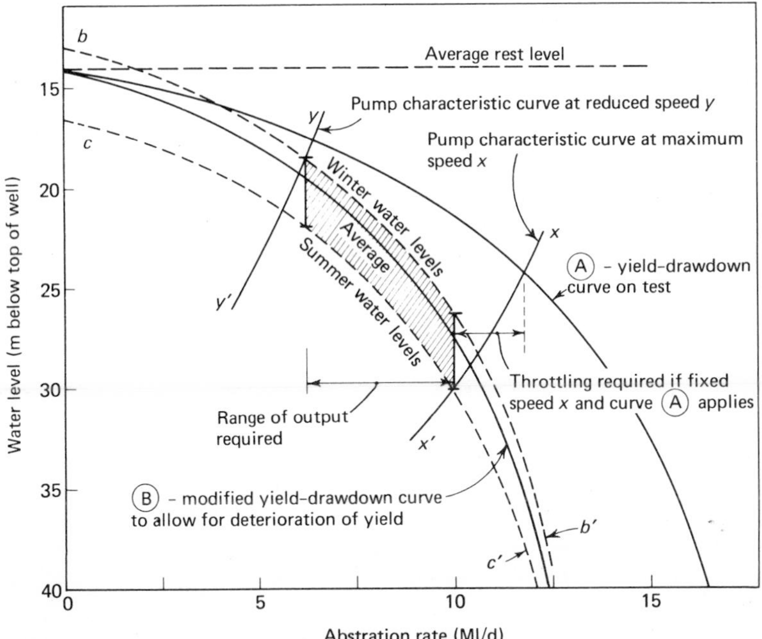

Fig. 5.25 Yield–drawdown curves for a well with pump characteristic curves superimposed.

When the characteristic curve of any pump is then superimposed upon any such yield–drawdown chart a curve similar to xx' results for a fixed speed pump. This indicates that if the pumping rate is to be restricted to not more than 10 Ml/d when the water table is high (as instanced by the curve *bb'*) then the pump must be partially throttled so as to increase the head on it and restrict the flow to the desired figure. If the pump is of variable speed, however, we can reduce the speed of rotation so that the pump curve shifts. The curve *yy'* represents the output at the lower speed of rotation and we can cover any point between the curves *yy'* and *xx'* by altering the speed of the pump. The decision as to whether a fixed speed or variable speed pump should be installed depends upon a costing of the alternatives available. A fixed speed pump may cause electrical energy to be wasted by having to pump against a closed valve when the standing water level in the well is high. A variable speed pump is greater in capital cost than a fixed speed pump and its average efficiency over its range may not be so high as that of a fixed speed pump running at design duty. The calculations need to take into account the fact that the yield characteristics of the well will appertain more closely to curve A for the first years of life of the well, and this can possibly be met by introducing a dummy stage into a fixed speed pump, the dummy stage being later replaced by an additional impeller if the drawdown increases. When calling for quotations for suitable pumps from manufacturers it is necessary not only to give the yield–drawdown characteristics of the well, but also to give the manufacturer a lead as to what will be the average duty to be met.

5.37 Pumps for wells and boreholes

These are in most cases either vertical spindle centrifugal pumps or submersible pumps. Vertical spindle pumps are robust with a long life and a high efficiency, they can cope with water which has suspended matter in it, and they can be variable speed. They are, however, expensive in initial outlay, need housing, and take two or three days to remove from a well, with as many days taken to install. In contrast, submersible pumps are relatively cheap, extremely easy to install, and require no housing. A submersible pump may be withdrawn from a well in a matter of hours and another replaced in the same time. This is an advantage for a waterworks where reliability of a supply is always a matter of importance. Provided that a spare pump 'head' is available it is a simple matter to withdraw a defective pump and substitute the spare. However, submersible pumps cannot cope with such a wide range of duties as can vertical spindle pumps, and their efficiency tends to be less due to the restricted size of electric motor that is necessary. Pumps are discussed in Chapter 12.

References

(1) Jaeger C, The Vaiont Rock Slide, *Water Power*, **17,** March 1965, pp. 110–111, and **18,** April 1965, pp. 142–144.

(2) *Engineering News Record*, 7 January 1926 and 25 November 1926.
(3) McIldowie G, The Construction of the Silent Valley Reservoir Belfast Water Supply, *Trans. ICE*, **239,** 1934–35, p. 465.
(4) Ischy E and Glossop R, An Introduction to Alluvial Grouting, *Proc. ICE*, **21,** March 1962, p. 449. See also Geddes W E N, Rocke S and Scrimgeoar J, The Backwater Dam, *Proc. ICE*, **51,** March 1972, p. 433.
(5) The Failure of the Bouzey Dam (Abstract of Commission Report), *Proc. ICE*, **CXXV,** 1896, p. 461.
(6) *Engineering News Record*, 29 March 1928, p. 517, and 12 April 1928, p. 596.
(7) Jaeger C, The Malpasset Report, *Water Power*, **15,** February 1963, pp. 55–61.
(8) *Engineering News Record*, 7 August 1924, pp. 213–215.
(9) For a typical design see *Design of Small Dams*, Chapter VIII, Section C, US Bureau of Reclamation, US Government Printing Office, Washington, 1977.
(10) Williamson J, Round-crest, Siphon and Trumpet Spillways, *Proc. 4th Congress on Large Dams*, **II,** 1951.
(11) Kenn M J and Garrod A D, Cavitation Damage and the Tarbela Tunnel Collapse of 1974, *Proc. ICE*, February 1981, p. 65.
(12) Twort A C, The Repair of Lluest Wen Dam, *JIWES*, July 1977, p. 269.
(13) The Breaching of the Oros Dam, North East Brazil, *WWE*, August 1960, p. 351.
(14) Bishop A W, The Use of the Slip Circle in the Stability Analysis of Slopes, *Geotechnique*, ICE, March 1955, p. 7.
(15) Twort A C, The New Tittesworth Dam, *JIWES*, March 1964, p. 125.
(16) Davis C V, *Handbook of Applied Hydraulics*, McGraw-Hill, 1969, and Ice Pressure Against Dams, *Proc. ASCE Symposium*, Paper 2656, Volume 119, 1954.
(17) See, for example, the 1949 Unwin Memorial Lecture by Pippard A J S, The Functions of Engineering Research in the University, *JICE*, **33,** February 1950, p. 265, and Zienkiewicz O C, The Stress Distribution in Gravity Dams, *JICE*, **27,** January 1947, p. 244.
(18) Allen D N, Chitty L, Pippard A J S and Severn R T, The Experimental and Mathematical Analysis of Arch Dams, *Proc. ICE*, **15,** May 1956, p. 198.
(19) Vail A J, The High Island Water Scheme in Hong Kong, *Water Power & Dam Construction*, January 1975, p. 15.
(20) Binnie W J E and Gourley H J F, The Gorge Dam, *JICE*, March 1939, p. 174.
(21) Walters R C S, *Dam Geology*, Butterworth, 1971.
(22) Townsend M I, Some Design Considerations of the High Island Water Scheme, *J Eng. Soc. Hong Kong*, Paper 2, 1973.

6

Chemistry and microbiology of water

6.1 Introduction

The more usual chemical substances and properties which concern the water supply engineer and scientist are listed in Part I, alphabetically. The chemical examination of a water can indicate the possible past history and whether it has been, or is now, polluted. Certain criteria are required for waters intended for human consumption so that the final product going into supply is pure and wholesome: substances which are potentially harmful to the consumer have to be limited and substances and properties which affect the general acceptability of the water have to be controlled. The WHO *International Standards for Drinking Water* of 1971[1] and their later *Guidelines for Drinking Water Quality*[2] of 1983, together with the European Commission (EC) Directive of 1980[3] on the quality of water intended for human consumption, are widely used as criteria for drinking water quality. A general discussion of these standards is given in Section 6.47, with a table comparing the main recommendations. A consideration of raw water classification, sampling frequency, and quality control in conditions of limited resources concludes Part II. Finally, bacteriological aspects of water supply are discussed in Part III of this chapter.

Part I
Significance of various substances in water

6.2 Acidity

An acid water is one which has a pH value of less than 7.0 (see Section 6.31). Acidity in an unpolluted water is usually from dissolved carbon dioxide which produces weak carbonic acid. Humic, fulvic, and other organic acids produced by decomposing vegetation can also give rise to acidity in waters, as in the case of waters from peaty moorland areas. When the acidity is from natural origins such as these, the pH value of the water is usually above 3.7. Waters polluted by industrial effluents can contain free mineral acidity from strong acids and their salts, with pH values of below 3.7. There is no defined limit for acidity in water, the main requirement being that the water is non-corrosive (see Section 8.28). Certain treatment processes, notably coagulation with aluminium sulphate, reduce the pH value of the water making it more acidic (see Section 7.20). The final water leaving a treatment works should be non-corrosive, with the pH value increased to correct for acidity when necessary. Where pH correction is not applied and an acidic water is allowed into the distribution system, corrosion problems can arise. This can result in attack of concrete pipes and the dissolution of heavy metals such as copper, zinc, and lead.

6.3 Alkalinity

Alkalinity is almost entirely due to the bicarbonate, carbonate, and hydroxide ions in the water, usually in association with calcium, magnesium, sodium, and potassium. Analyses often quote alkalinity in terms of $CaCO_3$ instead of carbonate and bicarbonate content. This is a convenient form of expression whereby the sum of the constituent salts is expressed in equivalent terms of calcium carbonate. Alkalinity can exist in a water below the neutral point of pH 7.0 because of the relationship between alkalinity, carbon dioxide, and pH value. In the simplest form of the relationship bicarbonate alkalinity is in equilibrium with carbon dioxide in the water between pH values 4.6 and 8.3. Above pH 8.3 free carbon dioxide ceases to exist and combines to give both carbonate and bicarbonate alkalinity. Between pH values 9.4 and 10.0 the alkalinity is all due to caustic or hydroxide alkalinity.

Alkalinity is the same as carbonate, or temporary, hardness when the value is less than the total hardness, the difference representing the non-carbonate, or permanent, hardness. In some waters, however, the total hardness is less than the alkalinity. These waters contain sodium bicarbonate alkalinity which does not affect the total hardness (see Section 6.20).

No limits are set for alkalinity levels in water, although high concentrations of sodium bicarbonate can give rise to taste problems. The level of alkalinity is also important in chemical coagulation because of the buffering capacity it imparts to the water.

6.4 Aluminium

Aluminium occurs in many natural waters in detectable amounts. The most usual sources of aluminium in drinking water, however, come from corrosion of aluminium utensils, tanks, or pipes (see Section 8.27) or from the incorrect dosing of aluminium sulphate as a coagulant at the treatment works: this should not occur with good operational control. There should also be no tendency for deposition of aluminium hydroxide after filtration if the treatment process is working efficiently. Ideally a water going into supply should contain less than 0.05 mg/l aluminium as Al, with a maximum acceptable concentration of 0.2 mg/l. The concentration of aluminium in drinking water is of importance for sufferers of kidney diseases who have to use dialysis[4].

6.5 Ammoniacal compounds

Ammonia is one of the forms of nitrogen found in water (see also Section 6.27) and is usually expressed in terms of mg/l N. Free ammonia, which is the same as free and saline ammonia or ammoniacal nitrogen, is the form most usually determined in water. It is so called because it exists either in the free state or as saline ammonium ion depending upon pH value. Albuminoid ammonia is the additional fraction liberated from organic material in the water by strong chemical oxidation. Ammoniacal compounds are found in most natural waters. They originate from various sources, some of which are completely harmless, for example decomposing vegetation. Deep well waters which are of good organic quality can contain high levels of free ammonia caused by the reduction of nitrates, either by bacteria or by the surrounding geological strata. However, ammonia can also indicate recent pollution of a waterbody, by either sewage or industrial effluent, the ammonia level in a typical UK sewage effluent being of the order of 50 mg/l N. The amount of ammonia in a raw water is of importance in determining chlorine doses for disinfection. Depending upon the type of treatment process being used, ammonia levels of up to around 3 mg/l N can be economically tolerated. For levels as high as this it is usual to investigate the possible sources of the ammoniacal compounds so as to minimize the risk of excessive bacterial pollution. Ammonia levels of up to 0.4 mg/l N are acceptable in treated water going into supply. This is because ammonia is sometimes used with chlorine in the final disinfection process to form chloramines. These give a more persistent residual than does chlorine alone (see Section 9.7).

6.6 Arsenic

Arsenic is acutely toxic to man and if detected in water should give cause for concern and investigation. The maximum permissible concentration in drinking water is set at 0.05 mg/l and is based upon current toxicological data. Arsenic can be found in surface waters from areas where there are certain types of metalliferous ore. More usually it is the result of pollution from weedkillers and pesticides containing arsenical compounds or from runoff from mining waste tips. A note in the WHO 1971 International Standards states: 'Figures higher than that quoted are found in a number of Latin American countries and levels up to 0.2 mg/l are not known to have caused difficulties.'.

6.7 Biochemical oxygen demand (BOD)

This test gives an indication of the oxygen required to degrade biochemically any organic material in a water as well as the oxygen needed to oxidize inorganic materials such as sulphides. It is purely an empirical test to compare the relative oxygen requirements of surface waters, wastewaters, and effluents and attempts to simulate the natural process of purification in a river or stream. The breaking down of organic matter by biological organisms uses up oxygen from the water so that if a sewage effluent of high BOD enters a stream, it will deplete the oxygen content of the water and destroy fish and plant life. The UK Royal Commission on Sewage Disposal in 1915 observed that unpolluted rivers rarely had BOD values of less than 2 mg/l and could accept added pollution up to a total BOD value of 4 mg/l with no apparent detriment. This gave rise to the maximum permissible BOD value of 20 mg/l in a sewage effluent entering a watercourse, with an 8:1 dilution of freshwater.

6.8 Bromide

Seawater contains 50 to 60 mg/l bromide so the presence of bromides in well supplies in coastal areas may be evidence of seawater intrusion. No limits have been suggested for bromides, as the quantities experienced in water supply tend to be negligible.

6.9 Calcium

Calcium is found in most natural waters, the level depending upon the type of rock through which the water has passed. It is usually present as the carbonate or bicarbonate and sulphate, although in waters of high salinity, calcium chloride and nitrate can also be found. Calcium contributes to the hardness of a water with the bicarbonate forming temporary, or carbonate, hardness and the sulphates, chlorides, and nitrates forming permanent, or non-carbonate, hardness (see Chapter 8). Calcium is an essential part of the human diet. However, the nutritional value from water is likely to be minimal compared to that from other food sources. There is no health objection to a high calcium content in a water, the main limitations being made on the grounds of excessive scale formation. The WHO recommends a maximum level of 200 mg/l as Ca (500 mg/l expressed as $CaCO_3$), above which deposition in water systems can cause major problems. A descriptive table relating to hardness (expressed as $CaCO_3$) is given in Section 6.20.

6.10 Carbon dioxide

Carbon dioxide is one of the components of the carbonate equilibrium in water. The free carbon dioxide content of a water depends upon the alkalinity and the pH value and can contribute significantly to the corrosive properties of the water. This is discussed more fully in Sections 6.31 and 8.28. Surface waters usually contain less than 10 mg/l free CO_2, although some groundwaters from deep boreholes may contain more than 100 mg/l. A simple means of reducing the free carbon dioxide content of such waters is cascade aeration (see Sections 8.15 to 8.19).

6.11 Chloride

Chlorides, i.e. compounds of chlorine with another element or radical, are present in nearly all natural waters and the range of concentrations can be very wide, but most combinations are with sodium (NaCl, 'common salt') and, to a lesser extent, with calcium and magnesium. They are one of the most stable components in water, with concentrations being unaffected by most natural physiochemical or biological processes. Chlorides are derived from natural mineral deposits, from seawater either by intrusion or by airborne spray, from agricultural or irrigation discharges, or from sewage and industrial effluents. Most rivers and lakes have chloride concentrations of less than 50 mg/l Cl and any marked increase in the chloride concentration of a waterbody is indicative of possible pollution. The chloride content of a sewage effluent under dry weather conditions is likely to be as much as 70 mg/l more than the original water supply.

The main problems caused by excessive chlorides in water concern the acceptability of the supply. Whilst chlorides are not normally detrimental to health, the salt intake, i.e. the common salt or NaCl intake, for people suffering from heart or kidney disease has to be restricted. Although the salt intake from drinking water is very small compared to that from foodstuffs, it is still desirable to limit the chloride concentration in water. From Table 6.1 (p. 222) it will be seen that the WHO 1971 International Standards set a desirable level of 200 mg/l Cl for chloride, with a maximum permissible level of 600 mg/l. The EC Directive and the WHO 1983 Guidelines recommend a much lower guide level of 25 mg/l Cl with the comment that undesirable effects might occur above 200 mg/l. The former levels of 200 and 600 mg/l were set in relation to taste and corrosion in hot water systems and in this respect they are still relevant. A sensitive palate can detect chlorides at as low a level as 150 mg/l and concentrations above 250 mg/l can impart a distinctly salty taste to a water. However, waters containing more than 600 mg/l are drunk in some arid or semi-arid places where there is no alternative supply.

6.12 Chlorine

Chlorine gas is used as a biocide and disinfectant in water (see Sections 9.1 to 9.15). The maximum amount of residual chlorine which may be permitted in a water going into supply depends upon the taste, odour, and corrosion that may result. One of the disadvantages of dosing chlorine is that chlorinous and chlorophenol tastes are occasionally produced. The presence of biological material can also give rise to earthy or other tastes. Trihalomethanes (see Section 6.44) can be produced if chlorine is used as a pretreatment measure on highly coloured waters or waters containing organic material. Typical chlorine residuals for chlorinated waters going into supply are of the order of

0.2 mg/l. In an emergency where a suspect water has to be distributed to consumers and emergency sterilization by chlorine is adopted, 2 mg/l or more of free chlorine may be permitted to remain in a water.

6.13 Colour
Colour in an unpolluted water can be caused by humic and peaty material and by naturally occurring metallic salts, usually of iron and manganese. Waters subject to industrial pollution can also contain a wide variety of coloured material. The colour of a water is usually expressed in Hazen units which are the same as the platinum–cobalt scale. Chromatic variations of an individual colour-forming substance can also be determined. A water often appears coloured because of material in suspension so true colour can only be determined after acceptable pretreatment, such as filtration. The method of pretreatment should always be reported. The level at which colour becomes unacceptable depends largely upon that to which the consumer is accustomed. The WHO 1971 International Standards suggest a desirable level for colour of 5 units with a maximum permissible level of 50 units. The EC Directive recommends a much lower level for water going into supply, with 1 mg/l on the platinum–cobalt scale as a guide level, but with many waters it is not possible or necessary to achieve this standard.

6.14 Conductivity
Conductivity is the measurement of the ability of a solution to carry electric current. As this ability is dependent upon the presence of ions in solution, a conductivity measurement is an excellent indicator of the total dissolved solids in a water. The unit of conductivity is μS/cm (microsiemen per cm). For most waters a factor in the range 0.55–0.70 multiplied by the conductivity gives a close approximation to the dissolved solids in mg/l. The factor may be lower than 0.55 for waters containing a lot of free acid or higher than 0.70 for highly saline waters. Conductivity is temperature-dependent and a reference temperature (usually 20 or 25 °C) is used when expressing the result. One of the advantages of conductivity determination is that it can be very easily measured in the field or used for continuous monitoring.

6.15 Copper
Copper is rarely found in unpolluted water, although trace amounts are sometimes found in very soft, acidic moorland waters. Copper salts are used to control algal growths in reservoirs and are the usual source of copper in unpolluted surface waters. An initial dose of 0.5 mg/l followed by doses of 1 to 2 mg/l at intervals of a day or so, as necessary to effect reduction, is recommended[5]. More usually the copper found in domestic water supplies is from the corrosion of copper and copper-containing alloys used as pipes and fittings within the system. Copper is

an essential element in the human diet and has to be present in excessive concentrations (20 mg/l) to create health problems. Water containing about 1 mg/l or less of copper can cause green stains on sanitary fittings and, according to Campbell[6], much lower concentrations can cause accelerated corrosion of other metals in the same system. The rate of corrosion of galvanized steel can be considerably increased by the presence of 1 mg/l of copper and as little as 0.02 mg/l can cause pitting of aluminium in some waters. The WHO 1971 'highest desirable level' for copper in drinking water is 0.05 mg/l, with a maximum permissible level of 1.5 mg/l. The EC Directive gives a guide level of 0.01 mg/l in water leaving a treatment works, but up to 3 mg/l of copper is permissible in water that has been standing in copper piping for up to 12 hours. Above this level astringent taste problems, corrosion, and 'green staining' may occur.

6.16 Corrosive quality

There is no definition of the corrosive quality of a water because many factors concerning the water itself and the materials with which the water comes into contact will determine whether or not corrosion occurs: each case must be examined individually. Nevertheless, among a great variety of causes of corrosion, three special characteristics of a raw water can lead one to expect consequential corrosion of metals in common use in a water scheme. These characteristics are:

(1) a low pH value, i.e. acidity,
(2) a high free CO_2 content, i.e. a lot of carbon dioxide,
(3) an absence of temporary hardness, or alkalinity.

Waters which are frequently corrosive are soft moorland waters, shallow well waters of low pH with little temporary hardness but a lot of permanent hardness, water from iron-bearing formations, chalk and limestone waters having a high CO_2 content, waters from greensands, waters from coal measures, and waters having a high chloride or residual free chlorine content. To test the aggressive nature of a water it can be placed in contact with powdered marble or chalk for a defined period. The initial and final tests for pH alkalinity will then show whether the water has readily taken up calcium or whether calcium has been deposited. If calcium is taken up, the water is likely to be corrosive to iron and steel as well as cement. An important matter which must always be checked when examining a new water for its corrosive tendencies is to investigate whether it is plumbosolvent, i.e. whether it will take lead into solution. As lead is a cumulative poison (see Section 6.23) problems can arise with plumbosolvent waters in areas where lead service pipes have been used in the distribution system. The subject of corrosion is discussed more fully in Chapter 8.

6.17 Cyanide

Cyanide and cyanide complexes will only be found in waters polluted by industrial or mining effluents involving processes using cyanide. At pH 8 and below the cyanide exists mainly as undissociated hydrogen cyanide (HCN), which is generally more toxic to aquatic life than is the undissociated cyanide ion. Most cyanides are biodegradable and can be removed by chemical treatment before an effluent is discharged into a river.

6.18 Detergents

There are several substances which can cause foaming in water, the largest group being detergents or surfactants. In recent years the increased use of synthetic and biodegradable detergents, which can be removed by normal sewage purification processes, has led to a reduction in the residual detergent discharged in sewage effluent. The main limitation on detergents is to prevent foaming in drinking water, although some of the components of anionic surfactants are toxic to aquatic life. Methylene blue active substances (MBAS) relate to the more common anionic surfactants found in detergents, which react with methylene blue. Many surface waters downstream of urban areas contain detergent and the EC Directive recommends a maximum limit of 0.2 mg/l in treated water.

6.19 Fluoride

Fluoride may occur naturally in a water or it may be added in controlled amounts during the treatment process. It is now generally accepted that fluoridation of water supplies to a level of 1 mg/l F is both safe and effective in substantially reducing dental caries. The practice has been adopted by many water authorities in the UK[7], the USA, and elsewhere. The greatest reduction of dental decay occurs if fluoridated water is drunk in childhood during the period of tooth formation. Fluoride levels have to be closely controlled as excessive amounts can lead to fluorosis with resultant mottling of the teeth or even skeletal damage in both children and adults. Thus the optimum fluoride concentration has to be related to climatic conditions and the amount of water likely to be consumed. Specialized treatment can be used to remove excess fluorides from water, but it is costly to operate and it is often more economical to abandon a source if the fluoride levels become unacceptable. There are some waters which naturally contain several mg/l of fluoride in use in the UK. They are mainly deep well waters from chalk or limestone under clay formations shown on the 'solid' geological map[8], particularly the London Clay and Oxford Clay.

The addition of fluoride to water supply is a new principle which many persons do not accept.

6.20 Hardness

A full discussion of hardness is given in Sections 8.1 to 8.11. We may note here that total hardness consists of carbonate, or temporary, hardness and non-carbonate, or permanent, hardness. Temporary hardness is precipitated by boiling and forms the scale found inside kettles. Permanent hardness is due to calcium and magnesium sulphates and chlorides which are not precipitated on heating. Hardness is usually expressed in mg/l as $CaCO_3$ and a comparison of hardness levels may be given as follows.

Range (mg/l)	*Hardness level*
0– 50	Soft
50–100	Moderately soft
100–150	Slightly hard
150–200	Moderately hard
Over 200	Hard
Over 300	Very hard

The problems caused by excessive hardness are mainly economical in terms of scale formation in boilers and hot water systems. Conversely, waters softer than 30–50 mg/l tend to be corrosive and should always be examined for plumbosolvency (liability to take lead into solution). A statistical connection between the hardness of water and the incidence of cardiovascular diseases has been found. No causal relationship appears to have been established—the softer the water the higher the incidence of the diseases (see also Section 8.1)—and the Department of Health and Social Security in the UK drew attention to this[9]. It is not usually economic to soften a water when the hardness is less than 150 mg/l, except for certain industrial requirements. The WHO 1971 'highest desirable level' of 100 mg/l seems to be unduly low and is not consistent with the corresponding desirable level of calcium which has a hardness equivalent of 187.5 mg/l $CaCO_3$. Where softening or demineralization is used for domestic supply the EC Directive recommends a minimum hardness concentration of 60 mg/l Ca (equivalent to 150 mg/l $CaCO_3$) in softened water.

6.21 Iodide

Most natural waters contain only trace amounts of iodide, usually in μg/l quantities. Higher concentrations may be found in brines or in waters treated with iodine as a disinfectant. The levels of iodide found in most natural waters are unlikely to contribute significantly to dietary requirements.

6.22 Iron

Iron is found in most raw waters and trace amounts can usually be found in distribution systems where the water has been in contact with iron pipes. Iron can be present in water in numerous forms: in true solution, as a colloid, in suspension, or as a complex with other mineral or organic substances. The element is not harmful but undesirable on aesthetic grounds: it can impart a bitter taste when present in large amounts, making the water unpalatable. When water containing a lot of iron takes up oxygen on exposure to air, the iron is likely to be precipitated, causing brown stains on laundry and plumbing fixtures. Even small amounts of iron can lead to the accumulation of large deposits in a distribution system. As well as being unacceptable to the consumer, such deposits can give rise to iron bacteria which in turn cause further deterioration in the quality of the water by producing slimes or objectionable odours. The WHO 1971 International Standards recommend 0.1 mg/l Fe as the highest desirable level for total iron, with 1.0 mg/l as the maximum permissible level. The EC Directive is more stringent with a guide level of 0.05 mg/l and a maximum concentration of 0.2 mg/l. It should be noted, however, that derogations from the EC Directive can be granted to take into account the nature of the ground from which the supply emanates (see Section 6.47). This takes into account the good quality groundwaters which are acceptable in all other respects but can contain more than 0.2 mg/l of iron.

6.23 Lead

Lead is a cumulative body poison and the hazards of exposure to lead in the environment have been well documented over a number of years. It is rarely detectable above 0.02 mg/l in most natural waters except in areas where soft, acidic waters come into contact with galena or other lead ores. As with many other cumulative poisons, water is rarely the major source of intake. A lead level of 0.05 mg/l in drinking water would contribute approximately 25% of the total daily intake based upon a consumption of two litres per day. The main problem of lead in domestic supplies comes from the dissolution of old lead plumbing. On this basis the EC Directive differentiates between plumbic and other types of pipework. The maximum admissible concentration in running water is given as 0.05 mg/l. Where lead pipes are present this should not be exceeded after adequate flushing. The EC Directive then goes on to say: 'If the sample is taken either directly or after flushing and the lead content either frequently or to an appreciable extent exceeds 0.1 mg/l, suitable measures must be taken to reduce the exposure to lead on the part of the consumer.'. The WHO 1971 International Standards increased the tentative level for lead from 0.05 to 0.10 mg/l on the grounds that the higher level had been accepted in many countries without apparent ill effects for a number of years. It was also felt that the lower level was difficult to achieve in areas using lead pipes.

Although plumbosolvency tends to be associated with very soft, acidic waters from upland and moorland catchments, it can also occur in hard water areas, especially if the hardness is mainly non-carbonate. Thus the lead content of drinking water should always be monitored in areas where lead pipes are known to be in use. Unacceptable levels of lead can also occur with certain types of plastic pipes if the lead compounds used as stabilizers are leached out.

6.24 Magnesium

Magnesium is one of the earth's most common elements and forms highly soluble salts. Magnesium contributes to both carbonate and non-carbonate hardness in water, usually at a concentration considerably lower than that of the calcium component. Excessive concentrations of magnesium are undesirable in domestic water because of problems of scale formation and also because magnesium has a cathartic and diuretic effect, especially when associated with high levels of sulphate. The two parameters are so interrelated that the WHO 1971 International Standards recommend a magnesium level of 30 mg/l for sulphates of more than 250 mg/l, with a maximum permissible magnesium level of 150 mg/l providing the sulphate concentration is less. Magnesium hydroxide is only sparingly soluble at high pH values (pH 10.6 and above), a factor utilized in lime–soda softening (see Sections 8.3 to 8.5).

6.25 Manganese

Manganese can be a troublesome element in water, even when present in small quantities. It can deposit out from water in the presence of oxygen or after chlorine has been added, coating the interior of the distribution systems with a black slime. These slimes occasionally slough off, giving rise to justifiable consumer complaints. Large quantities of manganese are toxic, but a water requires treatment on the grounds of taste and aesthetic quality long before such levels are reached. The concentration of manganese in solution rarely exceeds 1.0 mg/l in a well aerated surface water. However, much higher concentrations can occur in groundwaters subject to reducing conditions. In impounding reservoirs the bottom water may take up manganese from the reservoir floor soil or bottom deposits, especially if the water becomes deoxygenated. This can lead to an increase in the manganese content of the supply when reservoir 'overturn' occurs in spring and autumn (see Section 7.1).

In general a lesser amount of manganese can be tolerated in a supply system than iron because, although the deposition of manganese is slow, it is continuous. The onset of serious troubles may not become apparent for some 10 to 15 years after putting the supply into service. The EC Directive gives a guide level of 0.02 mg/l and a maximum admissible concentration of 0.05 mg/l.

6.26 Mineral constituents

The dissolved solids make up the mineral constituents of a water. These include all of the soluble anions and cations and any dissolved silica present. The results are usually expressed in mg/l, although the term milligram equivalents per litre (meq/l) is often used, especially in anion–cation balance calculations to check whether or not an analysis is correct. (The milliequivalent weight is the molecular weight divided by the valency.) The use of milliequivalents also simplifies the working out of the hypothetical or probable composition of compounds in a water, whereby anions and cations are combined to give the salts most probably taken up by and dissolved in the water. The usual order of combination is calcium, magnesium, sodium, potassium for cations, and carbonate or bicarbonate, sulphate, chloride, nitrate for anions, which also represents the order in which the dissolved salts come out of solution on evaporation. In particular the figure for calcium carbonate or bicarbonate indicates the amount of that compound which will be deposited as scale on boiling. The sum of the combined salts and dissolved silica should approximate to the total dissolved salts measured quantitatively by evaporation.

6.27 Nitrite and nitrate

Nitrite (NO_2) and nitrate (NO_3) are usually expressed by water analysts in terms of nitrogen, i.e. mg/l N: the total oxidized nitrogen is the sum of nitrite and nitrate nitrogen. Nitrite is an intermediate oxidation state of nitrogen in the biochemical oxidation of ammonia to nitrate and in the reduction of nitrates under conditions where there is a deficit of oxygen. Surface waters, unless badly polluted with sewage effluent, seldom contain more than 0.1 mg/l of nitrite nitrogen. Thus the presence of nitrites in conjunction with high ammonia levels in surface waters generally indicates pollution from sewage or sewage effluent. Whilst the presence of nitrites in a groundwater may be a sign of sewage pollution, it may have no hygenic significance. Nitrates in groundwater can be reduced to nitrite, especially in areas of ferruginous sands, and new brickwork in the wells is also known to produce a similar effect.

Nitrate is the final stage of oxidation of ammonia and the mineralization of nitrogen from organic matter. Most of this oxidation in soil and water is achieved by nitrifying bacteria and can only occur in a well oxygenated environment. The same bacteria are also active in percolating filters at sewage treatment works, resulting in large amounts of nitrate being discharged in sewage effluents from such works. The extended cultivation of land and the increased use of nitrogenous fertilizers over the last few decades have given rise to increases in nitrate concentrations in both surface waters and underground waters. This is causing concern in parts of the UK, especially in chalk and limestone areas dependent upon borehole water[10,11]. Nitrate levels in surface waters often show marked seasonal fluctuations, with higher concentrations being found during the winter months compared to the summer months. With the onset of winter rains, catchment runoff of nitrates will increase at a time of reduced biological activity in the river leading to an overall increase in nitrate concentration. During summer the nitrate levels are likely to be reduced by algal assimilation and biochemical mechanisms. Algal assimilation also helps to reduce nitrate levels in a stored water. In addition bacterial denitrification and anaerobic reduction to nitrogen at the mud interface can substantially reduce the nitrate levels in a stored water.

Waters containing high nitrate concentrations are potentially harmful to infants and young children. Bacteria in the digestive tract can reduce nitrates in food and water to nitrites which are then absorbed into the blood stream and convert the oxygen-carrying haemoglobin into methaemoglobin. Whilst methaemoglobin itself is not toxic, the effects of reduced oxygen-carrying capacity in the blood can obviously be very serious, especially for infants who have a high fluid intake relative to body weight. Concern is currently being expressed on the possible formation of nitrosamines, which are potentially carcinogenic, within the digestive tract. Ingested nitrites, some of which may be formed by bacterial reduction of nitrates, can react with secondary and tertiary amines found in certain foods to give nitrosamines. The EC Directive suggests a guide level of 25 mg/l for nitrate as NO_3 (5.6 mg/l as N), with a maximum admissible concentration of 50 mg/l NO_3 (11.3 mg/l as N) and a maximum admissible level of 0.1 mg/l for nitrite as NO_2 (0.03 mg/l as N) in treated water. There are no simple methods for treating a water to reduce the nitrate–nitrite concentration. Two processes currently in use are demineralization and biological removal under controlled conditions using denitrifying bacteria. Neither method is easily suited for large-scale treatment.

6.28 Odour

Odour and taste are closely related so that the subject is considered in Section 6.41 under taste and odour and also in Sections 8.23 and 8.24. Odour is largely subjective and it is usual to specify that the odour of a water should be unobjectionable.

6.29 Organic matter

The organic matter found in water can come from a variety of sources such as plant and animal life, partially treated domestic waste, and industrial effluents. The total organics present in a water can be estimated from the chemical oxygen demand, or from the total organic carbon content, or from the carbon–chloroform extractable material. The oxygen demand tests—the oxygen absorbed permanganate value, the biochemical oxygen demand (BOD), and the chemical oxygen demand

(COD)—all determine the relative oxygen requirements of a waterbody and do not directly measure the organic content. Total organic carbon and the carbon–chloroform extract directly determine the concentration of organic contaminants. Certain organic compounds such as pesticides and trihalomethanes are determined by specific tests (see Sections 6.30 and 6.44).

Many organic compounds found in water and effluents undergo biochemical degradation and are relatively harmless. However, the list of controlled organics in water is growing as more toxicological data become available.

6.30 Pesticides

Pesticides cover a wide variety of compounds used as insecticides, herbicides, fungicides, and algicides. They include inorganic compounds such as copper sulphate; chlorinated hydrocarbons such as DDT, dieldrin, aldrin, and lindane; organophosphorus compounds such as parathion and malathion; and many others. Pesticides find their way into natural waters from direct application for aquatic plant and insect control, from percolation and runoff from agricultural land, from aerial drift in land application, and from industrial discharges. The organic compounds are potentially toxic even in trace amounts and some, particularly the chlorinated hydrocarbons, are very resistant to chemical and biochemical degradation. Thus the effects of pesticides are often cumulative and can give rise to long-term problems. Where pesticides and algicides are used for aquatic control they can cause deoxygenation as a result of the decomposition of the treated vegetation, which in turn can cause other problems such as the dissolution of iron and manganese and the production of taste and odour.

In the UK the use of pesticides is controlled, but traces can be found in sewage effluents and in some rivers used for public supply purposes. However, the amounts which may be consumed from drinking water are only a small fraction of those likely to be consumed from foodstuffs (see Croll, 1969[12], and Lowden *et al.*, 1969[13]). Accidental discharges of pesticides in bulk to watercourses can be more serious: they occasionally occur, causing fish death and making it necessary for a temporary shutdown of waterworks intakes.

Many of the chlorinated hydrocarbon pesticides are rapidly adsorbed on to sediment or suspended material. This property can be utilized for removing pesticides in treatment processes involving coagulation followed by sedimentation. A better removal may be achieved by oxidative treatment using ozone, chlorine, chlorine dioxide, or potassium permanganate. However, in some cases the use of chlorine may convert some of the pesticides into products more toxic, or chlorine may react with some solvents to give odorous compounds. For more effective removal adsorption on to activated carbon should be adopted.

Powdered activated carbon may be dosed to the water ahead of clarifiers or filters for later removal in subsequent processing units, or granular activated carbon may be used in final filtration units. It is preferable to restrict the levels of pesticides in a water prior to treatment, rather than to remove them in the treatment process.

6.31 pH value

The pH value or hydrogen ion concentration is a measurement of the acidity or alkalinity (basicity) of a water. It is one of the most important determinations in water chemistry since many of the processes involved in water treatment are pH-dependent. Pure water is very slightly ionized into positive hydrogen (H^+) ions and negative hydroxyl (OH^-) ions. In very generalized terms a solution is said to be neutral when the numbers of hydrogen ions and hydroxyl ions are equal, each corresponding to an approximate concentration of 10^{-7} moles/l. This neutral point is temperature-dependent and occurs at pH 7.0 at 25 °C. When the concentration of hydrogen ions exceeds that of the hydroxyl ions, i.e. at pH values of less than 7.0, the solution or water has acid characteristics. Conversely when there is an excess of hydroxyl ions, i.e. the pH value is greater than 7.0, the solution or water has basic characteristics and is sometimes described as being on the alkaline side of neutrality (but see Section 6.3).

The pH value of an unpolluted water is mainly determined by the interrelationship between free carbon dioxide and the amounts of carbonate and bicarbonate present (see Sections 6.3 and 6.10). Thus the pH values of most natural waters are in the range 4 to 9, with soft, acidic waters from moorland areas having low pH values and hard waters which have percolated through limestone having high pH values. Waters of low pH tend to be more corrosive (see Section 6.16) and if the pH value is very low a water can have a sour or acidic taste.

6.32 Phenols

The phenolic compounds found in surface waters are usually a result of pollution from trade wastes such as petrochemicals, washings from tarmac roads, gas liquors, and creosoted surfaces. Natural phenols can be released in water from decaying algae or higher vegetation. Phenols can also be found in groundwaters, especially in areas with oil-bearing strata. Most phenols, even in minute concentrations, produce chlorophenols on chlorination, which are objectionable in both taste and odour. The WHO 1971 International Standards suggest a phenol concentration of 0.001 mg/l as the highest desirable level before problems arise. However, as the composition of the phenols likely to be in water is very complex, the actual presence of chlorophenols often provides a more reliable criterium. In water treatment phenols can be removed by superchlorination, whereby excess chlorine chemically decomposes the phenol, or by oxidation with ozone, or by adsorption on to activated carbon.

6.33 Phosphates

Phosphates in surface waters mainly originate from sewage effluents which contain phosphate-based synthetic detergents, or from agricultural effluents including runoff from inorganic fertilizers, or from industrial effluents[14]. Groundwaters usually contain insignificant concentrations of phosphates, unless they have become polluted. Phosphorus is one of the essential nutrients for algal growth and can contribute significantly to eutrophication of lakes and reservoirs[15,16].

6.34 Polycyclic aromatic hydrocarbons (PAHs)

These organic compounds are produced from tar and tar products and several are known to be carcinogenic. Trace amounts of PAHs have been found in industrial and domestic effluents. Their solubility in water is very low but can be enhanced by detergents and by other organic solvents which may be present. Whilst PAHs are not very biodegradable, they tend to be taken out of solution by adsorption on to particulate matter. To date there is no evidence of the effects of oral ingestion of PAHs[17] and the intake from food and cigarette smoke is likely to be far greater than that from drinking water. PAHs are usually removed by adsorption during conventional coagulation treatment followed by treatment with activated carbon. The WHO 1971 International Standards suggest that the total concentration of six (named) representative PAH compounds should not exceed 0.0002 mg/l (0.2 μg/l) in water after treatment.

6.35 Potassium

Although potassium is one of the abundant elements, the concentration found in most natural waters rarely exceeds 20 mg/l. The EC Directive recommends a maximum concentration of 12 mg/l with a guide level of 10 mg/l.

6.36 Radioactive substances

Many waters contain trace amounts of radioactivity caused by naturally occurring radioactive isotopes of elements such as potassium (^{40}K). In granitic areas waters can contain trace amounts of radium (a radioactive daughter product of uranium), although no evidence of an effect on the health of consumers has been found (see Turner, 1962[18], and Kenny *et al.*, 1966[19]). Usually the levels of natural radioactivity found in drinking water are too low to constitute a health risk, although it has been known for many years that some 'spa' waters contain significant levels of radioactivity.

Artificial radioactivity caused by fallout from atmospheric nuclear explosions has contaminated surface waters throughout the world. This has been monitored in the UK on a national basis under government auspices by various bodies such as the Metropolitan Water Board[20], now part of the Thames Water Authority. The subject was reviewed by Windle Taylor and other authors in a publication by the Royal Society of Health, 1971[21]. In addition to radioactive fallout, increased radioactivity can be created in waters by effluents from nuclear industries such as uranium mines and nuclear fuel reprocessing plants, and by radioisotope users. In most cases there are stringent controls on the disposal of such effluents to minimize the amount of pollution.

The organization which gives detailed recommendations on the control of radioactive material and the permissible concentrations of radiation is the International Commission on Radiological Protection[22]. These recommendations have been utilized by the WHO in their 1971 International Standards to derive the maximum permissible concentrations for radioactivity in drinking water. The values are based upon a consumption of 2.2 litres of water per day and include naturally occurring radioactivity in addition to radioactivity from artificial sources. The recommendations apply to α- and β-radiation as they are likely to become concentrated in specific parts of the body, for example strontium 90 in bones, causing localized damage. The maximum level for gross α-activity is recommended as 3 pCi/l and 30 pCi/l for gross β-activity. If either activity level is exceeded, further radioanalysis is recommended into the types of radioisotope present. The 1983 WHO Guideline recommendations give similar values expressed in becquerel (Bq). (1 pCi $= 3.7 \times 10^{-2}$ Bq.)

6.37 Silica

Silica can be found in water in several forms caused by the degradation of silica-containing rocks such as quartz and sandstone. Natural waters can contain between 1 mg/l of silica in the case of soft moorland waters and around 40 mg/l (as SiO_2) in hard waters. Volcanic and geothermally-heated waters often contain very high levels. Whilst there is no evidence of silica constituting a health hazard in drinking water, it is a troublesome material in a number of industrial processes because it forms a very hard scale which is difficult to remove.

6.38 Sodium

In addition to being very abundant, sodium compounds are very soluble so that the element is present in most natural waters. Levels can range from less than 1 mg/l to several thousand mg/l in brines. The threshold taste for sodium in drinking water depends upon several factors such as the predominant anion present and the water temperature. The threshold taste concentration for sodium chloride is around 350 mg/l (138 mg/l as Na)[23], whereas the threshold taste concentration for sodium sulphate can be as high as 1000 mg/l (348 mg/l as Na). The use of base-exchange or lime–soda processes to soften hard waters can lead to a significant increase in the sodium concentration. Apart from the possibility of imparting taste, such increases could be detrimental to consumers on sodium-restricted diets. The EC Directive recommends a sodium concentration guide level of 20 mg/l and a maximum concentration of 175 mg/l reducing to 150 mg/l from January 1984. The EC Directive accepts that a softened water sodium content could exceed this maximum concentration and recommends that: 'An effort must be made to keep the sodium content at as low a level as possible and the essential requirements for the protection of public health may not be disregarded.'.

6.39 Sulphates

The concentration of sulphate in natural waters can vary over a wide range from a few mg/l to several thousand mg/l. Sulphates can come from several sources such as the dissolution of gypsum and other mineral deposits containing sulphates, from seawater intrusion, from the oxidation of sulphides, sulphites, and thiosulphates in well aerated surface waters, and from industrial effluents where sulphates or sulphuric acid have been used in processes such as tanning and pulp paper manufacturing. Sulphurous flue gases discharged to atmosphere in industrial areas often result in acidic rain water containing appreciable levels of sulphates. Sulphates in domestic water contribute the major part of the non-carbonate, or permanent, hardness. High levels can impart taste and when combined with magnesium or sodium can have a laxative effect. Naturally occurring sodium sulphate (Glauber's Salt) and magnesium sulphate (Epsom Salt) are both well known laxatives. Consumers can become acclimatized to high sulphate waters and in some parts of the world waters with very high sulphate contents have to be used as there is no alternative supply. Bacterial reduction of sulphates under anaerobic conditions can produce hydrogen sulphide, which is an objectionable gas smelling of bad eggs. This can occur in deep well waters and the odour rapidly disappears with efficient aeration.

6.40 Suspended solids (*see also turbidity*)

The suspended solids content or filter residue of a water quantifies the amount of particulate material in a water sample. This includes both organic and inorganic matter such as plankton and clay and silt. The suspended solids content of a surface water can vary widely depending upon flow and season, with some rivers under flood conditions having several thousand mg/l in suspension. The measurement of suspended solids is usually on a weight–volume basis and gives no indication as to the type of material in suspension, the particle size distribution, or the settling characteristics.

6.41 Taste and odour

Taste, like odour, is a subjective test which relies upon description rather than quantitative results. Tastes tend to be closely related to odours, although there are certain non-volatile substances, such as sodium chloride, which give rise to tastes without causing odours. There are many potential causes of tastes and odours in water, the principal ones being algae, decaying vegetable matter, products resulting from chlorination, such as chlorophenols, and stagnant water in dead ends of the distribution system. Definitions for taste tend to be more simple than those for odour, there being only four true taste sensations, and the main requirement for a treated water is that the taste is unobjectionable.

6.42 Total solids in solution

Total dissolved solids is a quantitative measurement of the dissolved salts in a water. For a given water the dissolved solids concentration can be directly related to the conductivity (see Sections 6.14 and 6.26).

6.43 Toxic substances

Several of the toxic substances which might be present in waters have been discussed under individual headings, such as arsenic (Section 6.6), cyanide (Section 6.17), and pesticides (Section 6.30). The EC Directive contains a list of thirteen toxic substances, eleven with maximum admissible concentrations. This will no doubt be subject to revision as more toxicological data become available. The toxicity of chemical substances in water depends upon several factors other than the actual concentration. Some substances, which are highly toxic, are unstable in water and break down into innocuous byproducts. Some, such as pesticides, become adsorbed on to particulate matter and can be removed in the sedimentation stage of conventional water treatment processes. The degree of toxicity also has to be assessed on the daily intake from sources other than water, for example lead pollution in air. Most of the toxic substances listed in the EC Directive enter natural waters as a result of pollution from inefficiently treated or poorly controlled industrial effluents. As conventional water treatment processes generally have little effect in reducing the concentrations of such substances, the maximum limits usually apply to both treated and untreated waters. Full details of concentrations are given in Table 6.1 (p. 222).

6.44 Trihalomethanes (THMs)

When chlorine is used to disinfect waters containing organic substances a range of chloro-organic compounds is formed. One group of these halo-organics which has attracted considerable attention recently is the trihalomethanes or haloforms (THMs). They are believed to be formed from the interaction of chlorine with certain organic compounds, such as humic and fulvic acids, which are usually present in coloured surface waters. The principal trihalomethanes are chloroform, bromodichloromethane, dibromochloromethane, and bromoform. There is evidence that these compounds are potentially carcinogenic, but the subject is complex and research is continuing. The EC Directive gives a guide level of 0.001 mg/l for organochlorine compounds which do not come within the definition of pesticides (see Table 6.1) and also states that 'haloform concentrations must be as low as possible'. However, notwithstanding much research work, no conclusive evidence on the risks associated with trace concentrations in water has yet been produced and exposure from other sources, such as chloroform in certain types of toothpaste, is likely to be considerably higher.

The most effective ways of reducing THMs in drinking water are to reduce organic levels before disinfection with chlorine and to restrict the use of chlorine for prechlorination. Granular activated carbon can also be used to remove THMs after formation or to remove their precursors.

6.45 Turbidity

Turbidity is an indication of the clarity of a water and is defined as the optical property that causes light to be scattered and absorbed rather than transmitted in straight lines through a sample of water. Although turbidity is caused by material in suspension, it is difficult to correlate it with the quantitative measurement of suspended solids (Section 6.40) as the shape, size, and refractive indices of the particles in suspension all affect their light-scattering properties. Early turbidity measurements compared the strength of a transmitted beam of light through the water sample with the strength of the scattered or reflected light. This forms the basis of the standard method of determining turbidity by the Jackson Candle Turbidimeter which consists of a special candle and a flat bottomed glass tube which has been graduated in Jackson Turbidity Units (JTUs). The Jackson Candle method is of historical importance but is limited to turbidity values of more than 25 units. Since treated water turbidities are usually required to have less than one unit other methods had to be developed for measuring low turbidities. Several commercial instruments are available using either nephthelometry or absorptiometry. Nephthelometers measure the intensity of light scattered in one particular direction, predominantly at right-angles to the incident light. They are highly sensitive for measuring turbidities as low as 0.05 unit and are usually unaffected by dissolved colour. Absorptiometric methods assess the amount of light absorbed by the particles, the incident light passing through the sample. Absorptiometry is not as sensitive at very low turbidities as nepthelometry and dissolved colour can interfere. As a result of the differences in the optical systems the two types of instrument often give dissimilar turbidity readings even after calibration with the same turbidity standard. Therefore the type of instrument used to measure turbidity should be specified, although this is rarely done. A primary standard called 'Formazin' has been developed for turbidity measurements, which can be used to correlate Jackson Candle units with the methods for low turbidity measurement. When formazin is used to calibrate nephthelometric instruments the terms formazin turbidity unit (FTU) and nephthelometric turbidity unit (NTU) become interchangeable. Turbidity measurements are often used to monitor the performance of treatment works processes. Turbidimeters are frequently installed on-line to check the amount of flocculant material being carried over from sedimentation tanks and to check individual filter performance.

6.46 Zinc

Zinc tends to be found in only trace amounts in unpolluted surface waters and groundwaters. However, it is often found in domestic supplies as a result of corrosion of galvanized iron piping and tanks and dezincification of brass fittings. The concentrations usually found in drinking water are unlikely to be detrimental to health. Zinc has a threshold taste at approximately 5 mg/l and can also cause opalescence above this value.

Part II
Water quality standards

6.47 Drinking water standards (physical and chemical)

The most widely used standards before 1983 were the WHO *International Standards for Drinking Water* first published in 1958 and revised in 1963, 1968, and 1971. These have been further revised and were reissued in a new form in 1983, now retitled *Guidelines for Drinking Water Quality*. These WHO 1983 Guidelines are summarized in Table 6.1 following, but the WHO 1971 International Standards are also reproduced since they have formed the basis for many national standards. The WHO also published European Standards, the latest edition of which was issued in 1970, but these no longer apply as they are merged into the WHO 1983 Guidelines.

Within the European Community an EC Directive issued in 1980 on the quality of water intended for human consumption applies to the member states. The provisions of this EC Directive are also summarized in Table 6.1 so far as they relate to physical and chemical characteristics. The EC Directive is essentially a legal document, setting out end results to be achieved by the member states, leaving each state to decide on the necessary measures. In the UK the Department of the Environment (DoE) has issued a circular[24] explaining the provisions of the EC Directive and how they are to be implemented. No new legislation is required in the UK to meet the provisions of the EC Directive since water undertakers already have a statutory obligation under the Water Acts 1945 and 1973 to supply 'wholesome water' and the EC Directive can be regarded as a contribution to the definition of 'wholesomeness'. In this respect, other requirements, such as those set out in *The Bacteriological Examination of Drinking Water Supplies* published by the DoE in 1982 (see Section 6.64), have also to be taken into account in the UK. The provisions of the EC Directive are complex and, except in the case of toxic substances, a member state may grant 'derogations from the provisions of the Directive' to individual water undertakers when various circumstances apply. This in effect means that, with respect to non-toxic substances and characteristics, the Maximum Admissible Concentrations (MAC) quoted may be exceeded in the UK with the permission of the DoE in 'exceptional meteorological conditions', or in 'situations arising from the nature and structure of the ground' from which the supply emanates, or in emergencies. Where such derogations relate to supplies of less than 5000 m^3/d or serving less than 5000 population they need not be reported to the EC.

Other standards of importance are those used in the USA, formerly issued by the US Public Health Service, now superseded by the Drinking Water Regulations published by the US Environmental Protection Agency (EPA) in 1977. The old US Public Health Service Standards were first issued in 1913 and the 1962 updated version of them was reflected in many of the provisions of the second issue of the WHO International Standards in 1963.

Many other countries have their own national standards, most of which are based upon the WHO International Standards of 1958, 1963, 1968, or 1971 with only slight modifications to allow for local in-country conditions.

Comment on the use of standards The widespread use made of the WHO International Standards shows their considerable usefulness. However, neither those standards nor any other should be applied as the sole criterium for determining whether or not a water should be used or, if used, how it should be treated. It is sometimes better to use a water that is bacteriologically safe despite the fact that it contains some excess of non-toxic material, such as iron, chlorides, or hardness, than to use a water which has a higher risk of conveying water-borne disease, even though it otherwise complies with the standard. Account must be taken of all the circumstances applying, including an assessment of the

Table 6.1 Drinking water standards.

Substance or characteristic		Unit	WHO 1983 Guidelines: Guideline value[a]	WHO 1971 International Standards: Upper limit of concentration (tentative)	EC Directive 1980 relating to the quality of water intended for human consumption: Guide level (GL)	EC Directive 1980: Maximum admissible concentration (MAC)
Inorganic constituents of health significance						
*Antimony	Sb	mg/l				0.01
*Arsenic	As	mg/l	0.05	0.05		0.05
*Cadmium	Cd	mg/l	0.005	0.01		0.005
*Chromium	Cr	mg/l	0.05			0.05
*Cyanide	CN	mg/l	0.10	0.05		0.05
Fluoride	F	mg/l	1.5	0.9–1.7[b]		1.5[b]
				0.6–0.8[c]		0.7[c]
*Lead	Pb	mg/l	0.05	0.10		0.05[d]
*Mercury	Hg	mg/l	0.001	0.001		0.001
*Nickel	Ni	mg/l				0.05
Nitrates		mg/l	10 (as N)	45 (as NO_3)	25 (as NO_3)	50 (as NO_3)
*Selenium	Se	mg/l		0.01		0.01
Organic constituents of health significance						
*Pesticides and related products[e]		mg/l	Volume I of the Guidelines lists 18 compounds and their guideline values			
individually		mg/l				0.0001
in total		mg/l				0.0005
*PAH—six reference substances		mg/l				0.0002
Other organochlorine compounds additional to pesticides etc.		mg/l			0.001	Haloform concentration must be as low as possible

				Highest desirable level	Maximum permissible level		
Other characteristics or substances							
Colour		°Hazen	15	5	50	1	20
Odour			Inoffensive	Unobjectionable			2 or 3 TON[f]
Taste			Inoffensive	Unobjectionable			2 or 3 TON[f]
Suspended solids						None	
Turbidity		JTU	5	5	25	0.4	4
pH			6.5–8.5	7.0–8.5	6.5–9.2	6.5–8.5	9.5 (maximum value)
Temperature		°C				12	25
Aluminium	Al	mg/l	0.20			0.05	0.20
Ammonium	NH_4	mg/l				0.05	0.50
Barium	Ba	mg/l				0.10	
Boron	B	mg/l				1.0	
Calcium	Ca	mg/l		75	200	100	
Chloride	Cl	mg/l	250	200	600	25	
Copper	Cu	mg/l		0.05		0.10[g]	
Hydrogen sulphide	H_2S	mg/l	Not detectable				Not detectable
Iron	Fe	mg/l	0.30	0.10	1.0	0.05	0.20
Magnesium	Mg	mg/l		30[h]	150	30	50
Manganese	Mn	mg/l	0.10	0.05	0.50	0.02	0.05
Nitrite	NO_2	mg/l					0.10
Phosphorus pentoxide	P_2O_5	mg/l				0.40	5.0
Potassium	K	mg/l				10	12
Silver	Ag	mg/l					0.01
Sodium	Na	mg/l	200			20	175[i]
Sulphate	SO_4	mg/l	400	200	400	25	250
Zinc	Zn	mg/l		5.0	15	0.10[g]	
Anionic detergents		mg/l		0.2	1.0		0.20[j]
Mineral oil		mg/l		0.01	0.30		0.01[k]
Phenolic compounds		mg/l		0.001	0.002		0.0005[l]
Total dissolved solids		mg/l	1000	500	1500		1500
Conductivity		μS/cm				400	
Total hardness as $CaCO_3$		mg/l	500	100	500		—[m]

Notes to Table 6.1 on p. 224.

Notes to Table 6.1:

(a) The guideline value must be interpreted according to comments made in Volume I of the Guidelines. It is not to be interpreted as a maximum permissible value.
(b) For water temperature 10–12 °C (WHO); 8–12 °C (EC).
(c) For water temperature 26.3–32.6 °C (WHO); 25–30 °C (EC).
(d) Applies to running water or after flushing lead pipes.
(e) Includes insecticides, herbicides, fungicides, PCBs, and PCTs.
(f) TON = threshold odour number: dilutions 2 at 12 °C; 3 at 25 °C.
(g) Measured at outlet of works.
(h) May be up to 150 mg/l as sulphate reduces below 250 mg/l.
(i) To be reduced to 150 mg/l by year 1987.
(j) Value quoted for 'surfactants (reacting with methyl blue)'.
(k) Value quoted for 'dissolved or emulsified hydrocarbons (after extraction by petroleum ether): mineral oils'.
(l) Excludes natural phenols not reacting with chlorine.
(m) Minimum hardness after softening must be 60 mg/l Ca (equivalent to 150 mg/l as $CaCO_3$).
*Classified as 'Toxic Substances' in the EC Directive.

difficulty of maintaining consistent treatment and disinfection in the light of the level of maintenance and quality of skilled labour likely to be available. Sometimes physical and financial constraints will make it impossible to procure a water which complies in all respects with the WHO or other standard. In all such cases the risks attaching to each possible course of action must be identified and duly weighed so that the safest water from a practical point of view is adopted.

Two other factors must be borne in mind. Firstly, it must be expected that continuous development of standards will occur in the future as more becomes known of the effect of trace compounds in water upon health. The subject is complex and the implications may be very far from clear for a long time. In these circumstances the water engineer is forced to be realistic and practical and, if he is wise, he will use the simple guideline that his predecessors have used—namely that the less a source is liable to pollution from both domestic and industrial wastes, the more it is to be preferred, and he will adopt systems of treatment and disinfection that experience shows are most likely to be safe as well as being economic. Secondly, the importance of palatability must be emphasized. This applies especially to public supplies in poor or under-developed areas where the public can still gain access to inferior sources liable to carry disease. To encourage such people to use the public supply, even when it involves nil or minimal payment on their part, the public supply has to be in all senses of the word 'palatable': it should be taste and odour free, clear and attractive in appearance. It is especially important for the water to be carefully chlorinated and, if necessary, afterwards dechlorinated so that it remains free of the chlorinous tastes and odours which so many consumers find objectionable. The public supply must gain a reputation for being good, both for the palate and for the health of consumers, so that it is wanted and people are willing to pay for it, thus helping the water undertaking to become financially self-sufficient.

6.48 Standards for raw water classification

The classification of raw waters can sometimes be useful in indicating under what conditions a water source could be used or whether it is inadvisable to use it at all for public supply purposes. In any such classification the bacteriological content of a water plays a dominant part. Bacteriological aspects of water supply are dealt with in Part III of this chapter; it will suffice for the present to mention that the two principal analyses used to denote bacterial quality are:

(1) the *total coliform count* which estimates the numbers of bacteria of the 'coli-aerogenes' group in a sample, these being of both faecal and non-faecal origin,
(2) the *faecal coliform count* which estimates the numbers of *E. coli*

bacteria in a sample, these being a particular strain of bacteria within the coli-aerogenes group which are definitely of faecal origin.

Hence the total coliform count denotes the likelihood of sewage pollution, and the faecal coliform count confirms any pollution as being of human or animal origin. Also the numbers of such bacteria present per unit volume of water (usually 100 ml) indicate the degree of pollution.

In an early edition of the WHO European Standards (see Section 6.47), now withdrawn, an attempt was made to classify raw waters according to their degree of bacterial contamination, as shown in Table 6.2. Four classes of water were identified, labelled I–IV, of which the most heavily polluted, IV, having over 50 000 coliform bacteria per 100 ml, was defined as a 'source to be used only when unavoidable'.

Table 6.2 Classification of raw waters according to bacterial numbers (as proposed at one time in the WHO European Standards).

Classification		Total coliform bacteria per 100 ml	Faecal coliform bacteria per 100 ml
I	Bacterial quality applicable to disinfection treatment only	0–50	0–20
II	Bacterial quality requiring conventional methods of treatment (coagulation, filtration, disinfection)	50–5000	20–2000
III	Heavy pollution requiring extensive types of treatment	5000–50000	2000–20000
IV	Very heavy pollution, unacceptable unless special treatments designed for such water are used, source to be used only when unavoidable	Greater than 50000	Greater than 20000

An EC Directive[25] of 1975 continues a similar classification, but for surface water sources only. Three categories of surface water are designated, A1, A2, and A3, and the bacterial guide levels set for each are the same as for the WHO categories I, II, and III respectively. Mandatory chemical limits for toxic substances are also set and these correspond with the mandatory limits set for toxic substances in drinking water in those instances (the majority) where reduction by normal treatment processes is unlikely to occur. The EC Directive treatment provisions required for each category of water are similar to those prescribed by the WHO classifications, except that in the case of category A3 water, equivalent to WHO grade III water, the EC Directive mentions that adsorption with activated carbon must be included in the treatment process. The EC Directive states that surface water falling short of A3 category, i.e. WHO grade IV water, should not be used unless exceptional circumstances apply.

6.49 Sampling frequency and type

Four types of analyses can conveniently be distinguished:

(1) checks for residual chlorine,
(2) simple bacteriological testing,
(3) simple chemical analysis,
(4) full chemical and bacteriological analysis.

Checks for residual chlorine are normally applied several times a day on final water ex works, or else continuous monitoring by means of a chlorine residual recorder is adopted. It is highly desirable to make a daily check as well at some point in the distribution system.

Simple bacteriological tests are of paramount importance and consist of the total coliform count and the faecal coliform count (see Section 6.48). In general the frequency for these tests stipulated by the various standards depends upon the size of population supplied. Daily total coliform counts on disinfected water ex works and on a sample from the distribution system are recommended by the WHO 1971 International Standards for systems supplying over 100 000 population, reducing to monthly tests of the same kind for systems supplying under 20 000 population. The 1980 EC Directive requires daily total and faecal coliform counts on disinfected water 'at the point where it is made available to the user' (presumably from the distribution system) for systems supplying 150 000 population, reducing to fortnightly tests of the same kind for systems supplying 10 000 population. The WHO emphasizes the importance of taking simple tests frequently, rather than more sophisticated tests infrequently.

Simple chemical analyses are usually interpreted as covering relatively easily measurable but important parameters such as colour, odour, taste, turbidity, pH, conductivity, and residual chlorine, together with any other parameters that are regarded as being particularly important with respect to the specific source. Among the latter might be included chlorides to test for salt-water intrusion, nitrates and ammonia to indicate pollution, iron or lead for monitoring purposes in special cases, and residual aluminium and hardness for checking treatment performance. The WHO 1971 International Standards suggest a minimum of monthly simple chemical tests for supplies serving more than 50 000 population and twice-yearly tests for smaller supplies. The EC Directive suggests six simple chemical analyses per annum per 50 000 population supplied up to one million served, with three samples per annum for the small supply serving 10 000 population.

Full chemical analyses, including testing for toxic substances, are needed for new supplies or whenever circumstances require this. On existing supplies the WHO recommends annual examination of all supplies, whilst the EC Directive suggests a frequency of one analysis per annum per 50 000 population supplied, up to a population of one million.

In practice considerable latitude is exercised by water undertakings in deciding the frequency of testing since raw waters have very wide-ranging characteristics, some needing a far more frequent testing than others. There is much to be said for this approach, provided that the aim is to comply with the intentions of the various standards to safeguard the wholesomeness of the water supplied to the consumer.

6.50 Water quality testing in conditions of limited resources

The provisions of the various standards with regard to the type and frequency of testing tend to be difficult to interpret because of the wide range of needs to be covered. Also the published standards do not allow for the limitations of equipment, money, and skilled labour or remoteness of access that many large and small water undertakings suffer from throughout the world. In such cases there is a want of guidance as to what an undertaking should do when it simply cannot match the published standards. Where it is difficult or impossible to mobilize the full resources that are implied by the WHO and similar standards, it is suggested that some policy along the following lines should be adopted.

Principal aims of testing

(1) To ensure that all water produced from sources has adequate clarity and is adequately dosed with chlorine.

(2) To ensure that the bacteriological quality of the raw and treated waters is checked by bacteriological examination at regular intervals.

(3) To ensure that sources of pollution or potential pollution are known.

Chemical testing

(4) At all sources the chlorine dosage rate and the residual chlorine in the final water should be checked a minimum of twice daily, a.m. and p.m.

(5) At all sources at least two full chemical analyses of the raw water should be made at different seasons of the year, e.g. at the high monsoon period and at the dry period, to give an indication of the possible range of characteristics of the water.

(6) (a) Simple physical examination and chemical testing of the raw water should take place at each source to cover the few most important parameters (as ascertained from (5) above) which are likely to indicate potential change of raw water quality necessitating change of treatment. Typical tests would be for pH, conductivity, turbidity, hardness, or chlorides, together with reporting appearance, colour, and odour. These tests should be conducted as the plant operator thinks necessary and not less than monthly on a regular reporting basis.

(b) Where filter plant is used the clarity and pH of the water ex clarifiers

and ex filters should be tested daily, the clarity if necessary by simple visual reporting but preferably by means of some tubidity-measuring device.
(7) From the distribution system at least weekly samples should be taken and examined for residual chlorine, visual appearance, colour, and odour.

Bacteriological testing
(8) At least two full bacteriological analyses of the raw water should accompany the full chemical analyses made under (5) above.
(9) Where possible it is desirable that total and faecal coliform counts should be conducted on samples taken at least weekly from the treated water (dechlorinated) ex works and monthly from the raw water and from water drawn from the distribution system.
(10) Where the frequencies under (9) cannot be met, resources should be directed towards making total coliform counts on treated water ex works, taking into account the need to give priority to systems which supply large populations and which use raw water sources liable to experience serious pollution.
(11) Where tests under (9) and (10) cannot be conducted in a laboratory set up for the purpose, consideration should be given to conducting them by means of the membrane filter technique, using labour properly trained in field testing, visiting works as required on a routine basis.

Catchment supervision
(12) Sources of pollution liable to affect the water quality should be identified and recorded. Regular visits should be paid to examine these sources of pollution and sufficient watch should be kept to observe new potential sources of pollution.

Special actions required
(13) Where toxic substances are revealed in any raw water, special monitoring and other precautions must be adopted additionally.
(14) The use of waters having a very high total coliform count or dangerously sited with respect to any effluent or waste discharge should be avoided.

In general a given water has a set range of characteristics. Once these are known and effective treatment and disinfection processes have been set up, the primary requirements of routine testing are to check that the treatment processes are operating correctly and that there is no change of raw water quality likely to give rise to the need for changes in treatment and disinfection.

6.51 Methods of chemical analysis

It is important that methods of analysis should be standardized in order to achieve comparability of results. In the UK chemical and physical methods were set out in the *Analysis of Raw, Potable, and Waste Waters* 1972 published by the Department of the Environment. Subsequently a Standing Committee of Analysts representing a wide range of interests was set up by the DoE in 1974, and this committee has been producing more detailed and up-dated guidance in a series of publications issued under the general title of *Methods for the Examination of Waters and Associated Materials*[26]. Each publication takes a single method or a linked group of methods and a substantial number has already been published by the DoE. Another useful publication is *Field Testing of Water in Developing Countries* by Hutton[27]. In the USA a comprehensive book setting out American methods of analysis is *Standard Methods for the Physical and Chemical Examination of Water and Wastewaters*[28]. This book is a valuable reference and these methods have had a wide influence on standards adopted in other countries. A useful paper on waterworks laboratory design and equipment based upon the experience of a number of UK laboratories was given by Wallwork to the Society of Water Treatment and Examination in 1975[29].

Part III
Water microbiology

Diseases in man caused by water-borne bacteria and other organisms

6.52 Introduction

Diseases in man can be caused by the presence of certain bacteria called pathogenic bacteria, and also by other organisms which are not bacteria, such as viruses, protozoa, and worms. For convenience we consider here the diseases other than those caused by chemical characteristics of a water supply.

The following sections give the classical intestinal bacterial diseases which are commonly, although not invariably, water-borne. Also included are amoebiasis and some virus diseases which may, although rarely, be water-borne, and schistosomiasis, a disease widespread and serious in warm climates.

Many other parasitic diseases of the tropics are related to the sources and storage of water, e.g. streams, canals, dams, lakes, and reservoirs. These waters may either promote the growth of intermediate hosts of parasites (snails, fish, water plants) or encourage the breeding of vectors of disease (mosquitoes and other biting insects). These diseases may impinge upon the work of water undertakings, and engineers are therefore urged to consult local health authorities before beginning work on water supplies in such areas.

Details of the life histories of parasites are given by Manson-Bahr and Apted[30].

6.53 Bacterial diseases

Cholera The cause is the bacterium *Vibrio cholerae* and its variant the *El tor* vibrio. Infection is usually contracted by ingestion of water contaminated by infected human faecal material, but contaminated food and personal contact may also spread infection. In recent years cholera has moved from the Far East to the Near East, Africa, and southern Europe, and could enter Britain or other European countries with carriers or persons in the incubation period of the disease. It is not likely to spread in communities with controlled water supplies and effective sewerage, but shellfish inhabiting polluted seawater, and eaten uncooked, can carry the vibrio (and also the virus of hepatitis) and lead to serious outbreaks.

Typhoid fever The cause is the bacterium *Salmonella typhi.* Infection is usually contracted by ingestion of material contaminated by human faeces or urine, including water and food, e.g. milk, shellfish, or canned meat improperly processed. *Salm. typhi* occasionally continues to proliferate in the gall bladders of a few patients who have recovered from the primary infection, and these carriers continue to excrete the organisms in their faeces or, occasionally, urine for long periods, even for life.

The most recent major water-borne outbreak of typhoid fever in Britain, investigated by Suckling[31], killed 43 people in Croydon in 1937. It was caused by a 'concatenation' of circumstances which included a person who was a carrier of *Salm. typhi* working down a well which was pumping into supply when the filtration and chlorination plants were bypassed. For the last forty-seven years the memory of this has continued to serve as a salutary warning which has helped to avoid any further major water-borne outbreaks in this country. The outbreak of typhoid in Aberdeen in 1964[32] was traced to unchlorinated river water used for cooling cans of corned beef at the manufacturer's plant in South America; thanks to modern antibiotics, nearly all patients, who numbered more than 400, recovered without complications. An outbreak involving 210 cases occurred in 1973 in Dade County, Florida, USA, and another occurred in 1974 at Poitiers in France involving 60 cases[33]. Both are believed to have been due to water contamination coinciding with inefficient disinfection.

Paratyphoid fevers These are caused by *Salmonella paratyphi A, B*, or *C.* Infection may exceptionally be via contaminated water, but is more commonly due to ingestion of contaminated food, especially milk, dried or frozen eggs, and other dairy products. *Salm. paratyphi B* has occasionally been isolated from animals, but *Salm. paratyphi A* is confined mainly to man: it is largely found in the East. *Salm. paratyphi*

C is common in South America. Carriers of these infections are not rare.

Bacillary dysentery This is caused by bacteria of the genus *Shigella—Sh. dysenteriae 1, Sh. flexneri, Sh. boydii* and *Sh. sonnei*—there are several subspecies. Infection can occasionally be contracted via water contaminated by human faeces, but more commonly it is due to ingestion of foods contaminated by flies or by unhygienic food handlers who are carriers. The most virulent is *Sh. dysentariae 1* (formerly known as *Sh. shigae*) which produces an exotoxin and has often proved fatal. The commonest in Britain is *Sh. sonnei*, which is constantly responsible for outbreaks in children, but is relatively mild. It is not usually water-borne, but epidemics due to infected water have occasionally been reported.

Travellers' diarrhoea (Turista) The cause is not definitely known, but may be some forms of pathogenic *Escherichia coli* or, rarely, *shigellae.* It is probably transmitted in the same way as bacillary dysentery and water may sometimes be the vehicle. Infant diarrhoea is probably related to this.

Leptospirosis Diseases of this group, which vary from mild fever to severe jaundice, are caused by very numerous serogroups of the motile, spiral organisms known as *Leptospira.* They commonly infect rats, dogs, pigs, and other vertebrates, and are shed in the urine of these animals. They are very often present in ponds and slow flowing streams haunted by animals, and people who bathe in, fish in, or sail on these waters are at risk, becoming infected via the mouth, nasal passages, conjunctiva, or abraded skin through which the organisms can enter. Workers in rat-infested sewers and abattoirs are particularly at risk. Normal water treatment eliminates these organisms.

6.54 Protozoal diseases

Amoebiasis and amoebic dysentery The cause is the protozoon *Entamoeba histolytica*, which usually lives in the human large intestine, producing cysts which are the only infective forms and which are passed in the faeces. Infection takes place by ingestion of these cysts, usually from carriers who may be healthy, but whose fingers convey the cysts to other people directly or to food; flies also carry them. It is rarely water-borne except when untreated water is used from grossly contaminated sources. It flourishes where sanitation is poor. Liver abscess is a serious *sequela.*

6.55 Virus diseases

Viruses differ from bacteria—they are very much smaller and they multiply only within suitable host cells in which they produce changes which give rise to a range of diseases. More than one hundred different types of virus have been identified in faeces. Bonde[34] lists eleven groups and their disease effect and states that in polluted water there may be found enteroviruses (polio, coxsackie, and ECHO*), adenoviruses, and reoviruses. The agent causing hepatitis is believed to be an enterovirus, but knowledge concerning it is limited because of difficulties in finding a culture medium for it. There is clear evidence that hepatitis can be transmitted via sewage polluted drinking water, but the evidence with respect to the transmission of poliomyelitis and other virus diseases via drinking water is not absolutely clear despite the fact that all these viruses are frequently found in raw waters which, after treatment, are used for drinking.

Poliomyelitis This virus persists in the intestines of infected, not necessarily paralysed, persons for a short time after infection, and is shed in the faeces; it can often be found in untreated sewage and even in the effluent from sewage disposal units. Like other viruses, it does not multiply in the absence of living cells. Infection probably takes place from contaminated fingers directly or on food; there have been a few reports of water-borne infection, but little confirmation. It is common where sanitation and food hygiene are poor, and in such communities children are widely infected, but paralysis is rare. The virus has been found in oysters.

Infectious hepatitis The virus inhabits the intestine and is discharged in the faeces. Carriers may be infective for long periods. Transmission is probably as for poliomyelitis. Several water-borne epidemics have been reported, especially where water treatment has broken down, where the distribution system has been disturbed, or where badly constructed wells have been contaminated from cesspits or as a result of heavy rainfall. Shellfish are almost certainly involved in some cases. The virus of *serum* hepatitis is usually transmitted differently.

6.56 Helminthic (worm) diseases

Schistosomiasis (bilharzia) The cause is a group of trematode worms (flukes) of the genus *Schistosoma—S. haematobium, S. mansoni*, and *S. japonicum*—which inhabit the veins of the bladder or large intestine and discharge eggs into the urine or faeces. If these reach warm, slow flowing streams or ponds containing certain genera of freshwater snails, the embryos from the hatched eggs enter the snails, undergo development, and are discharged as second-stage embryos which can penetrate the human skin or mucous membranes. From there they mature and finally reach the veins. The disease can be serious and it is widespread in Africa, the Near and Far East, and Central and South America. In the Far Eastern form many animals, including rats, convey the infection. Irrigation canals encourage the prolific growth of the requisite snails and control in the field is difficult. In public water supplies filtration plus

* Enteric Cytopathogenic Human Orphan viruses, contracted to ECHO.

chlorination is effective if carefully controlled. These infections often cause widespread damage to the liver and other vital organs.

Swimmers' itch (schistosome dermatitis) The cause is a group of schistosomes of birds, whose embryos, emitted from certain snails, can penetrate the human skin, but cannot develop further in the human body. The eggs are present in the faeces of infected birds and reach water in their droppings. There have been many small outbreaks in Britain and North America.

6.57 Comments

An abundant supply of safe water promotes personal cleanliness and has been found to reduce diseases spread by contagion, including trachoma, yaws, and ringworm, as well as the usual intestinal water-borne diseases and schistosomiasis.

The details in Sections 6.53 to 6.55 show that the great majority of water-borne diseases are caused by sewage (faecal or urine) contamination of water from man or animals. Water need not be the immediate cause of an outbreak of disease, but a contaminated supply may be the first of a series of events leading to disease originating from meat, milk, watercress, or vegetables and fruits washed in contaminated water.

Examination and testing of water for pathogenic organisms

6.58 Difficulty of detecting pathogenic bacteria

Pathogenic bacteria and other organisms are usually difficult to detect in a water supply because of their small numbers. Even in a sewage effluent or polluted river water they may be present only infrequently or at irregular intervals depending upon one or more carriers who excrete the bacteria.

Searching for pathogenic bacteria directly is not therefore a practicable safeguard for a water supply. Instead, evidence of any pollution by the excreta of man or animals is sought, and if the evidence is positive it is assumed that the water can also contain pathogenic bacteria and must therefore be regarded as unsuitable for public supply purposes.

6.59 Prevalence of *E. coli*

The number of bacteria in the faeces of man and animals runs into thousands of millions per gramme. Of these the majority are *Escherichia coli*, a natural inhabitant of the intestines. Crude sewage contains some millions of *E. coli* per 100 ml, but the number of pathogenic bacteria present will normally be relatively very small. To obtain a water absolutely free from *E. coli* is to ensure that faecal pollution is non-existent or has been eliminated and the resultant risk of the water containing pathogenic bacteria is therefore remote. Although *E. coli* has been considered harmless, certain strains have now been implicated in infantile diarrhoea and travellers' diarrhoea.

Hence the testing of a water supply for the presence of *E. coli* is pre-eminently the test for safeguarding the bacterial quality of a water. It is a test that is universally adopted as the first standard of purity and its regular repetition is of the utmost importance, but see Section 9.4 regarding chlorination.

6.60 The test for *E. coli*

Escherichia coli (formerly known as *Bact. coli Type 1*) is a bacterium belonging to the 'coli-aerogenes' group of bacteria. These terms are used because *E. coli* is the predominant bacterium existing in faecal matter, whereas the other members of the coli-aerogenes group are found predominantly in soil.

A high percentage of all bacteria found in faecal matter are *E. coli.* The presence of this bacterium in a water supply therefore indicates pollution by faecal contamination. Other coliform bacteria will usually be present as well but if *E. coli* are absent then the inference is that pollution primarily arises from the soil or vegetation. The presence of any coliform bacterium must, however, be regarded as an advance warning that more serious pollution may follow, especially after rain.

6.61 Routine tests for bacterial contamination of water

The routine tests to distinguish the kinds of bacteria in water and to estimate the probable number of *E. coli* present are as follows.

Test	*General inference from a positive reaction*
Agar plate count at 20 to 22 °C (agar is the medium employed)	An indication of the numbers of bacteria, mostly non-pathogenic, which occur. These bacteria are from soil, dust, vegetation etc.
Agar plate count at 37 °C	An indication of the numbers of bacteria which thrive at body temperature and which therefore include those of faecal origin.
The presumptive or 'total' coliform count at 37 °C	An indication of the numbers of bacteria present of the coli-aerogenes group.
The differential or 'faecal' coliform test at 44 °C	An estimation of the numbers of *E. coli* present and therefore a positive indication of the degree of faecal pollution.

The plate counts are made by counting the numbers of colonies of bacteria which develop and are visible under a magnifying glass. The presumptive and differential counts are made by inoculating a series of tubes of media with different quantities of the water to be tested and thereby deducing (by statistical methods which have been worked out) the Most Probable Number (MPN) of bacteria which exist per 100 ml of the original sample. Alternatively the result may be expressed as 'present in 10 ml' or 'present in 1 ml' etc. of water, quoting the smallest quantity of water which has given a positive reaction.

The *membrane filter technique* is an alternative method. In this a sample is drawn by vacuum through a membrane which retains all bacteria in the sample on it. The membrane is then saturated with, or placed on, a suitable culture medium and placed in an incubator so that each bacterium multiplying develops into a visible colony. The number of colonies is counted (as in the agar plate count, but using a magnifying glass), each separate colony being assumed to be originated by one bacterium; total coliform and faecal coliform can be distinguished. The method permits bacteriological testing 'in the field', provided that there is also available a portable incubator, a supply of sterilized glassware, distilled water, and (if any extensive testing is necessary) an autoclave for resterilizing glassware. The method is unsuitable for turbid waters having low coliform counts or for waters having many non-coliform bacteria which also grow on the same medium. If many coliform bacteria are present the sample must be diluted with distilled water.

For frequent examination of a treated water the presumptive coliform count is always undertaken because it will indicate the numbers of coliform bacteria present. If a positive result is indicated the differential coliform tests should then be undertaken. The agar plate counts are not always required for routine simple examination, but for full examinations, especially those undertaken to show the efficacy of a treatment plant for removing bacteria, agar counts are taken both before and after such treatment.

6.62 The test for *Cl. perfringens* (*Cl. welchii*)

The particular significance of *Cl. perfringens* (*Clostridium perfringens*) in water testing is that it is a spore-forming bacterium of faecal origin which may remain in water for a much longer period than *E. coli.* Its presence in a water almost certainly indicates faecal pollution at some time or place. The test for it, which is conducted in a manner somewhat similar to the presumptive coliform count, is particularly useful in well or borehole supplies since the water drawn from underground may have travelled far or spent much time travelling from an original source of faecal contamination. When present it then acts as an indication that the supply is connected to some source of pollution which may at any time increase in severity.

6.63 The test for faecal streptococci

Faecal streptococci are bacteria which principally occur in the faeces of man and animals. They are not as numerous as *E. coli* in all normal cases, hence the test for faecal streptococci offers no advantage over the *E. coli* test, except in cases of doubt.

6.64 Standards of bacterial purity

The DoE publication *The Bacteriological Examination of Drinking Water Supplies* 1982[35], which is a revised edition of Ministry of Housing Report No. 71 1969, not only sets out detailed procedures to be adopted for testing the bacterial purity of a water, but also makes recommendations concerning the standards of bacterial purity that should be adopted by a public water supply authority. This publication should, of course, be in the hands of every water engineer, chemist, and bacteriologist, but the standards applied are of such importance that they are summarized below.

Chlorinated water, as it leaves the works, should always be free from coliform bacteria in 100 ml. Piped supplies of unchlorinated water are not recommended, but if small supplies are in use they should be free from *E. coli* in 100 ml.

The following is quoted from the publication.

> 'Ideally all samples taken from the distribution system, including those from consumers' premises, should be free from coliform organisms. However . . . this is not always attainable in practice, and tolerance may be allowed up to the following limits for routine samples
>
> —*E. coli* should not be detectable in any sample of 100 ml,
> —no sample of 100 ml should contain more than three coliform organisms,
> —coliform organisms should not be detectable in any two consecutive samples of 100 ml from the same or a closely related sampling point,
> —for any given distribution system, coliform organisms should not occur in more than 5% of routine samples, provided that at least 50 samples have been examined at regular intervals throughout the year.
>
> When any coliform organisms are found, the minimum action necessary pending confirmation is to check that the disinfection process is operating satisfactorily and to re-sample immediately from the same point.'

Similar recommendations are given by the WHO in the 1971 International Standards. In practice most water undertakings would initiate investigatory action immediately any single sample of treated water showed the presence of even one coliform organism per 100 ml, but an isolated low count of this nature in a series of regular samples that otherwise show coliforms absent is frequently due to some poor technique of sampling.

The *frequency of sampling* is discussed in Sections 6.49 and 6.50 above.

6.65 Technique of sampling

Care must be taken to prevent contamination of a water sample during the process of sampling. The best method is to sample from a sampling tap used only for this purpose. The length of service pipe to the tap from the main should be as short as possible. Any tap must be 'flamed', i.e. flames from a blow-lamp or burning methylated spirit must be applied to it for about one minute, so that all bacteria on it are destroyed. Water should then be run off for a further two or three minutes while the tap cools off before the sample is taken. Only laboratory prepared sterilized sample bottles should be used, and the stopper should be lifted out by its handle for as short a time as possible. No sample should be taken when dust is blowing about. The time between sampling and examination should be as short as possible: the WHO 1971 International Standards suggest that examination should preferably be started within one hour of collection and that the interval of time should never exceed 24 hours.

The importance of chlorination should not be overlooked when sampling. If a well or borehole is chlorinated direct, no true raw water sample can be taken, even if the chlorination is stopped for a while (which it should not be). All sample bottles should contain sodium thiosulphate so that samples containing chlorine are immediately dechlorinated when filled into the bottles, and it has also been found[36,37] that thiosulphate itself tends to have a stabilizing effect on the numbers of coliform bacteria and *E. coli*, providing that its concentration in the bottle is at least 100 mg/l, as recommended by the American Public Health Association[28].

Sampling from consumers' taps is liable to give erroneous results because of contamination of the tap which may not easily be sterilized. Leakage from tap glands must not enter the bottle. If a new tap is inserted care must be taken to see that both tap and jointing material have been sterilized. Some authorities prefer to install special sampling taps at certain points in the distribution system: these are used only for sampling and are kept locked. The metals copper and zinc have a bactericidal action and so they should be avoided for the service pipes connecting these special taps.

6.66 Biological examination

The most usual form of biological examination consists of identifying and enumerating algae in samples of reservoir waters. Sometimes the necessity arises of identifying low forms of life in samples from consumers' taps such as Nais worms, *Asellus aquaticus*, and chironomid larvae which can be alarming although harmless. A more comprehensive examination can be made by using a net of nylon or other fabric to strain a large volume of water flushed from a hydrant.

6.67 Virological examination and associated problems

It is much more difficult to achieve virus isolation from water than it is to make bacteriological examination. One problem is that large sample volumes must be used because viruses are much less numerous than are bacteria in water. Hence special techniques have to be adopted to concentrate the viruses, after which complicated identification procedures can follow. These are time-consuming, require the use of extensive equipment, and can only be reliably accomplished by a qualified specialist. Hence virological examination cannot be applied as a routine laboratory test, at least at present. The WHO 1971 International Standards advise that there should be at least one laboratory in each country or region capable of carrying out virus examinations and also of pursuing further research in this subject. In the UK several of the water authorities are able to do this work. The WHO 1971 International Standards do not stipulate any virological standard for treated water; the EC Directive does not mention viruses.

Despite this lack of progress the importance of understanding more about viruses in water cannot be ignored. In Section 6.55 it was mentioned that the 'agent' causing hepatitis is believed to be an enterovirus, but a culture medium for it had not been found (by 1982), yet it can undoubtedly be transmitted via water and, as is well known, hepatitis outbreaks have increased over the past few years. There are also problems associated with ensuring that a water is virus-free.

The survival time for some viruses in chlorinated water appears to be longer than for *E. coli*[38], also Slade[39] examining a number of raw waters used to supply London in 1975 found a lack of correlation between *E. coli* counts and the incidence of viruses. Later he found that viruses had passed through slow sand filters in the winter months, being found in the filtrate in concentrations as high as 1 PFU (plaque-forming unit) per 6 litres[40]. In this case, however, he found only one case of filtered water in which a virus was detected in the filtered water when the *E. coli* count was zero (see the comment under Section 9.5). The degree of virus kill achieved by chlorination is dependent upon a number of factors. Amongst other things Dyachkov[41] mentions that the antiviral effect of free chlorine is about 50–100 times greater than the effect of combined forms, and he reports that: 'It has been proved that the resistance of coliforms to chlorine is far lower than the resistance of viruses.'. Whilst there is some doubt as to the precise meaning of these findings (since a great number of variables affect the killing power of chlorine, there are many types of virus, and laboratory tests do not always simulate field conditions), the general evidence is that viruses are likely to be at least as long living in water and more resistant to at least some forms of chlorine than are *E. coli.* This conclusion is perhaps confirmed by an outbreak of gastroenteritis and hepatitis which occurred in Georgetown, Texas, in 1980 when human enteric viruses were isolated from the water supply, probably due to faecal contamination of groundwater, 'in spite of a total chlorine residual of 0.8 mg/l . . . [and the fact that] bacteriological samples of tap water taken by local and state officials were consistently free of coliform bacteria'[42].

The absence of *E. coli* in the standard sample size of 100 ml gives a very high likelihood that viruses will be absent also in the usual sample sizes of about 15 litres, but this is not always certain. The reason may be the large difference in sample size or the differing resistance of *E. coli* and viruses to the water conditions and the type of chlorine present.

References

(1) WHO, *International Standards for Drinking Water*, 1971, obtainable from HMSO, London, or WHO, Geneva.

(2) WHO, *Guidelines for Drinking Water Quality 1983:* Volume I—*Recommendations*, Volume II—*Health Criteria*, Volume III—*Sanitary Survey and Bacteriological Analysis for Rural Water Supplies*, HMSO, London, or WHO, Geneva, Volume I printing 1984.

(3) Council of European Communities (EC), Directive of 15 July 1980 relating to the quality of water intended for human consumption, *EC Official Journal*, L229/11, 1980.

(4) Moore B, Medical Aspects of the EC Directive on Water Quality, *Proc. IWES Symposium on EC Directives*, March 1980.

(5) *Manual of British Water Engineering Practice*, 4th Ed., Volume III, IWE, 1969, p. 168.

(6) Campbell H S, Corrosion, Water Composition and Water Treatment, *JSWTE*, **20,** 1971, p. 11.

(7) *Fluoridation Studies in UK and Results Achieved After Eleven Years*, Report of a Research Committee of the MoH, No. 22, HMSO, 1969.

(8) Hoather R C, *J Roy. San. Inst.*, **73,** 1953, p. 202.

(9) DHSS, Circ. 71/159, 17 August 1971.

(10) Tomlinson T E, Trends in Nitrate Concentrations in English Rivers, *JSWTE*, **19,** 1970, p. 277.

(11) Green L A and Walker P, Nitrate Pollution of Chalk Waters, *JSWTE*, **19,** 1970, p. 169.

(12) Croll B T, Organo-chlorine Insecticides in Water—Part I, *JSWTE*, **18,** 1969, p. 255.

(13) Lowden G F, Saunders C L and Edwards R W, Organo-chlorine Insecticides in Water—Part II, *JSWTE*, **18,** 1969, p. 275.

(14) Cooke G W and Williams R J B, Losses of Nitrogen and Phosphorus from Agricultural Land, *JSWTE*, **19,** 1970, p. 253, and Owens M, Nutrient Balances in Rivers, *JSWTE*, **19,** 1970, p. 239.

(15) See articles by Lund J W, Ridley J E and Bellinger E G, *JSWTE*, **19,** 1970, pp. 332, 374, and 400.

(16) Vollenweider R A, *Scientific Fundamentals of the Eutrophication of Lakes and Flowing Waters with Particular Reference to Nitrogen and Phosphorus*, OECD, Paris, 1968.

(17) Borneff J, Polycyclic Aromatic Hydrocarbons in Water, Special Subject No. 1(a), *Proc. IWSA Congress*, 1976.

(18) Turner R C, *Brit. J Cancer*, **16,** 1962, p. 27.
(19) Kenny A W, Crooks R N and Kerr J R W, Radium, Radon and Daughter Products in Certain Drinking Waters in Great Britain, *JIWE*, **20,** 1966, p. 123.
(20) Windle Taylor E, 36th to 44th Annual Reports of the London Metropolitan Water Board, 1953 to 1971.
(21) *Radiation and Health 1971*, Royal Society of Health, London.
(22) *Recommendations of the International Commission on Radiological Protection*, ICRP Publication No. 9, Pergamon Press, 1966.
(23) Windle Taylor E, *The Examination of Water and Water Supplies*, 7th Ed., Churchill, 1958, Chapters IV and VIII, pp. 28 and 100.
(24) Joint Circular from the DoE and Welsh Office of 19 August 1982.
(25) Council of European Communities (EC), Directive of 16 June 1975 concerning the quality required of surface water intended for the abstraction of drinking water in the member states, *EC Official Journal*, L194/26, July 1975.
(26) DoE, *Methods for the Examination of Waters and Associated Materials*, HMSO.
(27) Hutton L G, *Field Testing of Water in Developing Countries*, WRC, Medmenham, 1983.
(28) *Standard Methods for the Physical and Chemical Examination of Water and Wastewaters*, American Public Health Association, New York, 1971, and subsequent editions.
(29) Wallwork F, Waterworks Laboratory Design and Facilities, *JSWTE*, **23,** Part 4, 1974, p. 316.
(30) Manson-Bahr P E C and Apted F I C, *Manson's Tropical Diseases*, 18th Ed., Baillière Tindall, 1982.
(31) Suckling E V, *The Examination of Water and Water Supplies*, 5th Ed., Churchill, 1943.
(32) *The Aberdeen Typhoid Outbreak*, HMSO, 1964.
(33) Windle Taylor E, The Relationship Between Water Quality and Human Health: Medical Aspects, *J Roy. Soc. Health*, June 1978.
(34) Bonde G J, Water Quality and Health, General Report No. 1, *Proc. IWSA Congress*, 1980.
(35) DoE, *The Bacteriological Examination of Drinking Water Supplies*, HMSO, 1982.
(36) Hoather R C, *J Appl. Bact.*, **20,** 1957, p. 180.
(37) Noble R E, *JAWWA*, **55,** 1963, p. 115.
(38) Poynter S F B, Slade J S and Jones H H, The Disinfection of Water with Special Reference to Viruses, *JSWTE*, **22,** 1972, p. 194.
(39) Slade J S, Enteroviruses in Partially Purified Water, *JIWES*, May 1977, p. 219.
(40) Slade J S, Enteroviruses in Slow Sand Filtered Water, *JIWES*, November 1978, p. 530.
(41) Dyachkov A V, Recent Advances in Disinfection, Special Subject No. 5, *Proc. IWSA Congress*, 1976.
(42) Hejkal T W *et al.*, Viruses in a Community Water Supply Associated with an Outbreak of Gastroenteritis and Infectious Hepatitis, *JAWWA*, June 1982, p. 318.

7

Storage, sedimentation, coagulation and filtration

7.1 Raw water storage

This may often be regarded as a first stage in treatment as it involves a complex combination of physical, chemical, and biological changes. Traditionally, raw water storage has been regarded as a major or almost essential 'first line of defence' against the transmission of water-borne diseases; this aspect is still of major importance if the unstored water is liable to excessive bacterial pollution from sewage, even though such pollution may only occur occasionally, e.g. if storm-water sewage overflows discharge into a river. When a water is stored in a reservoir for a period of one month to several months, there is usually a very great decrease in the numbers of bacteria of intestinal origin and the specific organisms of typhoid and other water-borne diseases also disappear. To some extent this effect may be due to sedimentation, but the bactericidal action of ultraviolet radiation and of visible light is of major importance near the surface and there are numerous biotic agencies which probably play a major role in reducing the population of enteric micro-organisms[1]. Even a few days storage of a polluted water will improve its physical and microbiological characteristics. Furthermore, a fairly short time of storage may allow of a river intake being shut down in order to avoid or investigate any unusual pollution which might, for example, be indicated by the death of fish or by other information[2].

There are, however, some disadvantages in prolonged storage of raw waters and the most obvious of these is the growth in large numbers of various forms of algae which may increase the difficulties of treatment. Waters which contain sufficient nutrient materials to support prolific growths of plankton are usually described as eutrophic; such prolific growths are usually seasonal.

In most reservoirs in temperate climates, assuming they are more than about 10 m deep in parts, thermal stratification occurs on a seasonal basis. As the water warms up in the spring, any winter stratification is first overcome and then the water being warmed in the upper part of the reservoir tends to remain there because of its lower density. Much colder water remains in the bottom part. The upper and lower layers are known respectively as the epilimnion and hypolimnion; inbetween there is a zone known as the thermocline in which there is a relatively steep change with depth from the higher temperature of the epilimnion to the lower temperature of the hypolimnion. This thermal stratification is often of major importance with reference to water quality and, in any large reservoir, there are usually arrangements for withdrawing the water for treatment at several different levels which can be chosen as circumstances indicate.

In some large reservoirs the water in the hypolimnion is of a high standard of purity, as well as being cool, e.g. in some places in Scandinavian countries, in the Lake District of England[3], and in Lake Constance (Bondensee). However, in eutrophic reservoirs organic impurities derived from the entering water and from the botton mud or soil tend to accumulate in the bottom water and, as a result of bacterial activity, the concentration of dissolved oxygen falls and may even become zero. If it approaches zero major chemical changes begin to take place involving the dissolving of many materials from the bottom of the reservoir. The most prominent of these is often iron and a major reason for this is that the solubility product of ferrous hydroxide is much greater than that of ferric hydroxide. If manganese is present it also is converted to lower states of oxidation and dissolves in the water, often in combination with organic colouring matter. The concentration of phosphates may also increase in such bottom water; taste and odours may develop. In such circumstances abstraction for treatment of water in the hypolimnion is avoided if possible. However, when the surface water cools down in the autumn and wind becomes effective, the reservoir water mixes again. This 'turnover' sometimes results in sudden deterioration in the quality of the water near the surface of the reservoir in respect of its colour and its content of iron and manganese. A more gradual effect, however, is that nutrients carried up from the bottom increase algal growths in the water nearer the surface of the reservoir.

Since in eutrophic waters it is usually the water above the thermocline that is most suitable for use, measures have been taken in recent years to control the stratification of a reservoir. Thus the Metropolitan Water Board[4,5] (now part of the Thames Water Authority) has installed arrangements for carefully controlled raising of water from below the thermocline in some of its reservoirs to increase the volume of satisfactory water above it. Other large reservoirs have recently been constructed with the inlet arrangements specially designed to promote circulation as the water enters. Large-scale air lift pumping arrangements, known as 'bubble guns', have been used in some reservoirs to circulate the water. A fairly recent development, simple but cheap, is to use perforated air lines laid along the bottom of the reservoir and an alternative air injection system is the use of a porous medium, such as a collection of ceramic domes, to produce fine air bubbles. Mechanical pumps and air lift pumps have been used fairly widely, particularly in the USA, with varying degrees of success, but the systems are usually expensive to install. These systems are intended to have a dual purpose of oxygenating the bottom layers of a reservoir as well as controlling stratification.

In some circumstances it may be a great asset to have an emergency bypass so that water can be taken directly from a river instead of from a reservoir: this could be used in the case of exceptional algal growths or in the case of some exceptional pollution having occurred or being suspected in water in the reservoir.

Screening

7.2 Bar screens

Practically all intakes are screened, even though the screens may be of the simplest type of bar grille. The bars must be quite substantial in size (of about 25 mm diameter) and are normally spaced at 75 to 100 mm centres. If the bars are inclined it is easier to clean them with a rake. Should a smaller mesh be necessary, it is best to group bars into frames so that each frame can be lifted out of the water, cleaned, and lowered back into position. To prevent unscreened water from passing through the intake when the screens are lifted, they should be provided in duplicate or else provision should be made for stop log insertion upstream for temporary shutting off of water. At river intakes a great deal of trash may collect. Rags, plastic bottles, plastic sheets, paper bags, and branches of foliage are all particularly difficult to remove from bar screens, even with automatically operated rakes. It is best in such cases to install a robust band screen equipped with perforated steel plates, as shown in Fig. 5.22 (p. 189).

7.3 Band and drum screens

If fine screening is adopted, some means must be found of continuously cleaning the screens or they would soon clog up. For this reason fine screens are usually arranged as endless bands or rotating drums of material perforated with holes of about 6 mm diameter. Plate 9B shows drum screens and Plate 9C shows an endless band screen. The fabric is in continuous motion and passes over water jets which wash off the screened material into a trough. A pressure supply of clean water is needed for the washwater jets, and this may have to be pumped from the strained water. The total amount of water required for washing may be of the order of 1%.

7.4 Microstrainers

These are revolving drums of stainless steel wire fabric or other material having a very fine mesh. Two of the grades widely used have limiting apertures of 23 microns* and 35 microns respectively. Such a strainer is

* 1 micron = 0.001 mm.

shown in Plate 9D. Water jets for cleaning the fabric use about 1% of the total quantity of water strained, and this washwater should be filtered and chlorinated. The loss of head through a microstrainer varies from 150 to 450 mm and single units may deal with 45 000 l/h to a maximum of 1.35 Ml/h for a 3 m diameter × 3 m wide drum. Microstrainers are useful for screening stored waters which do not contain a large amount of suspended matter, but which contain plankton, algae, and other microscopic-sized particles. When used in this case they may lighten the loading upon any sand filters to follow, such that the length of run of these filters between backwashings is considerably extended. In the case of algae removal it is important to undertake pilot plant trials or, if this is not possible, laboratory tests to ensure that the microstrainers are effective: this is because of the wide variation in size of different species of algae and of the difficulty in selecting the correct mesh size in advance. A microstraining screen can easily be damaged if too great a loading is placed upon it; hence alarms should be incorporated to operate when the head across the screen approaches the maximum permissible.

Microstraining cannot remove colour or finely divided matter such as clay, nor is it effective in removing chemical floc from a treated water, but it is excellent for preparing a reasonably pure water for efficient sterilization by chlorine, thus avoiding the need for filtration in some cases. Sometimes it is followed by ozonization for the removal of colour.

The ideal water for a microstrainer is a lake supply or a large storage reservoir; the ideal position for a microstrainer is in advance of rapid gravity or slow sand filters whose output may thereby be increased by as much as 50%. The installation of microstrainers may represent a considerable financial advantage when the installation or extension of a sand filter plant is contemplated, and may also be of particular advantage where an industrial supply only is desired.

Sedimentation and settling tanks

7.5 General design considerations

Sedimentation tanks are designed to reduce the velocity of flow of water so as to permit suspended solids to settle out of the water by gravity. There are many different designs of tanks and most are empirical—'seasoned with a moderate amount of theoretical computations', as E Sherman Chase has put it. No hard and fast rules can be laid down and many contradictory results have been reported. A sedimentation tank which may be very successful on one kind of water may make a poor showing when dealing with a different kind of water. The success of a tank may be judged on the basis of its ability to maintain the claimed throughput and the agreed effluent water quality under adverse conditions of raw water quality. A specified effluent quality for suspended solids and turbidity of less than 5 mg/l and 5 FTU respectively would be acceptable to most designers. The amount of suspended solids in a water, the nature of these solids, their shape and relative density, the extent of clarification required, the temperature of the water, the rate of flow that must be handled—all these matters influence the performance of a tank, so that after doing laboratory settling tests on samples of raw water, the best guide to be had in choosing a tank for a particular water is to find out what type of tank has been successful before under similar conditions, and then to add the 'seasoning of theory' that Sherman Chase mentions. This could lead to the conclusion that there is only one solution of tank design for the treatment of a particular water, but this is far from the truth. The designer's task is to find the most economical and efficient solution for all conditions of water quality, although it would be foolish not to take due note of the type of tank which has been successful before in similar circumstances.

7.6 Plain settling

In plain settling (or sedimentation) suspended solids in a water are permitted to settle out by gravity alone: no chemicals are used. For this purpose the water can be left to stand in a tank, but for a continuous supply two such tanks have to be used alternately. Such fill-and-draw tanks are seldom used in modern plants, except for filter washwater recovery. Instead plain sedimentation tanks are designed for continuous throughput, the velocity of flow through the tank being sufficiently low to permit gravitational settlement of some portion of the suspended solids to occur. In practice the application of plain sedimentation in waterworks is very restricted because impurities, such as algae, vegetable debris, and finely divided mineral matter, do not settle at a rate sufficient to be utilized in a tank of reasonable size. Plain sedimentation is most frequently used as a preliminary treatment for fast flowing river water liable to carry a high content of mineral matter in suspension, such as in the case of intake works on water transfer schemes where it is desirable to minimize the amount of suspended material passing into the system, otherwise chemically assisted sedimentation is adopted, as described later, which is a more complex process.

The velocity with which a particle in water will fall under the action of gravity depends upon the horizontal flow velocity of the water, the size of the particle, the relative density of the particle, the shape of the particle, and the temperature of the water. The theoretical velocity of falling spherical particles in slowly moving water is given[6] by

$$V\,(\text{cm/s}) = \frac{g}{18}(r-1)\frac{d^2}{\gamma}$$

where $g = 981\,\text{cm/s}^2$, r is the relative density of the particles, d is the diameter of the particles in mm, and γ is the kinematic viscosity of water in centistokes*, which varies with the temperature of the water as below.

Temperature (°C)	0	5	10	15	20	25
(°F)	32	41	50	59	68	77
Value γ (centistokes)	1.79	1.52	1.31	1.15	1.01	0.90

This applies only for Reynolds numbers less than 0.5.

A number of different (mainly empirical) formulae have been given for the settlement of sand and soil particles in *still* water. Some guides are as follows.

For sand of relative density 2.65 in water at 10 °C

Table 7.1

Diameter of particle (mm)	Falling speed (cm/s)*	Falling speed (cm/s)†
1.0	10.0	14.0
0.6	6.3	—
0.5	—	7.0
0.4	4.2	—
0.2	2.1	2.2
0.1	0.8	0.67
0.06	0.38	—
0.05	—	0.17
0.04	0.21	—
0.02	0.062	—
0.01	0.015	0.008
0.005	—	0.0016
0.004	0.0025	—

* From *Water Treatment Plant Design*, published by the AWWA, 1969.
† From *Disposal of Sewage* by K Imhoff, published by Butterworth, 1971.

For soil particles (0.012 to 0.075 mm size)
Parker[7] quoted Wiley as suggesting:

$$V = \frac{1}{10}\sqrt{\left(\frac{d}{0.0314}\right)}$$

where d is the diameter of the particles in mm and V is the velocity of fall in cm/s.

* 1 stoke = 1 cm^2/s. 1 centistoke = 0.01 cm^2/s. The coefficient of kinematic viscosity = coefficient of absolute viscosity (poises)/density. 1 poise = 1 g/(cm s).

For particles of relative density values 1.50 and 1.20

Table 7.2

	Settling speed (cm/s) in still water at 10 °C	
Diameter of particle (mm)	Coal (RD = 1.5)	Domestic sewage solids (RD = 1.2)
1.0	4.0	3.0
0.5	2.0	1.7
0.2	0.7	0.5
0.1	0.2	0.13
0.05	0.04	0.03
0.01	0.002	0.0008
0.005	0.0004	0.0002

Note: From *Disposal of Sewage* by K Imhoff, published by Butterworth, 1971. (After Fair.)

Clays generally have a grain diameter of under 0.01 m to less than 0.001 mm (1 micron) so that it is impracticable to remove them from a water by simple sedimentation, or even by filtration, without prior chemical coagulation treatment as described in Sections 7.12 to 7.23. The smaller sized materials may be regarded as colloidal suspensions.

7.7 Maximum velocity to prevent bed uplift or scour

Apart from the falling rate in still water it is, of course, essential that once a particle has reached the base of the tank it shall not be picked up again by the velocity of flow of water over the bed. Camp[8] gives the channel velocity V_c required to start motion of particles of size d as

$$V_c = \sqrt{\left(\frac{8\beta g}{f}(r-1)d\right)}$$

where r is the relative density, d is the diameter of the particle, f is the friction factor in $(4flv^2/2gd)$, and β is in the range 0.04 to 0.10 for sticky flocculent materials, and 0.10 to 0.25 for sand. The formula is in SI units.

7.8 Maximum horizontal velocity of flow

A third flow measure which must be taken into account is that the horizontal velocity of flow must not be so great as to prevent, by turbulence, the settling of particles under gravity. There is general agreement that this velocity should not be more than 0.3 m per second to allow sand grains to fall out. This is, of course, too high a velocity for the settling of particles of light relative density (1.20 and less), but this is the figure normally used for sewerage grit chambers where the heavier material is to be deposited and the lighter material left to carry over. At 0.2 m per second faecal matter, i.e. organic matter, will begin to settle.

7.9 Theory of design of tanks

Suppose we have a tank of length L, water depth d, breadth b, and let the inflow rate be Q (= outflow), Fig. 7.1.

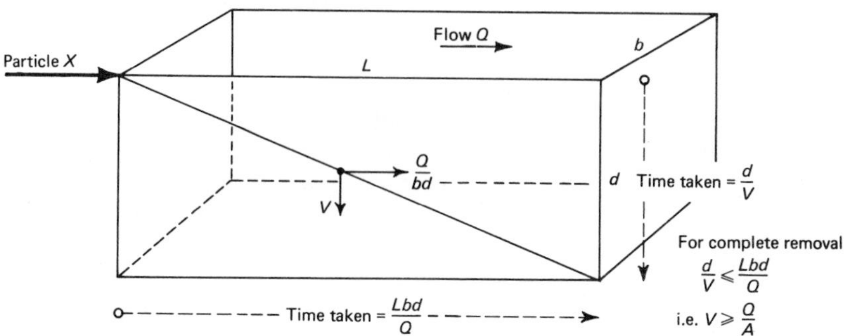

Fig. 7.1 Theoretical flow through a rectangular sedimentation tank.

Let us take a particle of silt entering the tank to have a vertical falling speed of V. Then

$$\text{speed of horizontal flow} = \frac{Q}{bd}$$

and $$\text{time of horizontal flow} = \frac{L}{Q/bd} = \frac{Lbd}{Q}$$

Now time for falling distance d is d/V, and for the particle to reach the bottom before the water leaves the tank the time of fall must equal the time of horizontal flow, i.e.

$$\frac{d}{V} = \frac{Lbd}{Q}$$

from which

$$V = \frac{Q}{Lb} = \frac{Q}{A}$$

where A is the surface area of the tank.

This is the limiting speed of fall to enable the particle to reach the bottom of the tank. All particles with speed greater than Q/A will reach the bottom before the outlet end of the tank. Particles with a speed less than Q/A will be removed in the same proportion as their speed bears to Q/A (see Fig. 7.1), e.g. if the speed V is only half Q/A then only half the particles falling at this speed reach the bottom. Hence Q/A for any tank is a measure of the effective removal of the particles in that tank. Suppose we have a tank of 300 m^2 surface area and the rate of inflow is 1.2 m^3/s, then $Q/A = 0.40$ cm/s. Thus, theoretically, all particles having a falling speed of 0.40 cm/s or more will be removed, 50% of those having a falling speed of 0.20 cm/s, 25% of those with a falling speed of 0.10 cm/s, and so on.

The foregoing theory, however, assumes that the falling particles do not hinder each other, but McLaughlin[9] has shown in laboratory experiments with clay and alum that the faster particles settling through the slower ones gather some of the latter up and drag them out of suspension. This causes an initial increase of the mean settling velocity for all particles which declines with time. Hence the performance of a tank in which this phenomenon occurs is related not so much to its overflow rate as to the time of detention. These findings relate to still water. Thus, dependent upon the nature and size of the settling particles, the ranges of sizes, the degree of concentration of the suspension, and the amount of turbulence, the performance of a sedimentation tank may relate to its overflow rate or to its detention time, or partly to both.

In practice a tank does not work as efficiently as its theoretical capacity would indicate. Normally it will only work at about 33% efficiency although, in the case of chemically assisted sedimentation, actual achievement is generally much closer to results obtained by laboratory jar tests. Hence in general the size of a plain sedimentation tank must be three times the theoretical size to get the same results. In rare cases a tank of 1.5 times theoretical capacity will work, but this would be of very special design. The reason for this is the difficulty of achieving uniform flow through a body of water. If the water is simply delivered by a pipe to one end of the tank, and taken out at the other, it will shoot straight across from inlet to outlet in one-tenth to one-twentieth of the time it would do if evenly spread. To prevent short-circuiting, a number of inlet penstocks can be provided instead of only one, orifices can be used to direct the incoming water downwards, and baffles may be put in the tank. Whilst these measures may help, they may have little effect when the temperature of the incoming water differs from that in the tank: even a difference of ± 0.2 °C can have a noticeable effect. Furthermore, even if baffles do succeed in spreading out the inflow, the change of motion of the water up and over, or down and under, such baffles creates turbulence which must subside before the particles can resume their fall in quiescent water.

7.10 Grit tanks

Where tanks are required for the removal of grit and sand only, the size of tank can be small. We see from Table 7.1 that fine sand of 0.20 mm

C

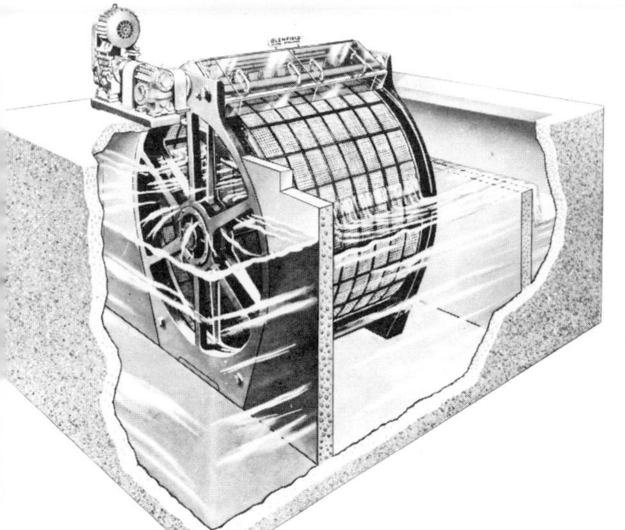

D

Plate 9 (A) Faughan river intake weir for Londonderry RDC. (Engineers: Binnie & Partners.) (B) Large diameter cup-screen. (J. Blakeborough & Sons Ltd.) (C) Bandscreen. (J. Blakeborough & Sons Ltd.) (D) Microstrainer. (Glenfield & Kennedy Ltd.)

Plate 10 (A) Filter bed under backwash and air scour, Bretton treatment works, Central Flintshire Water Board. (B) Horizontal pressure filters. (C) Filter interior at Bretton treatment works showing filter nozzles. (Contractors and photographs: Paterson Candy International Ltd.)

Plate 11 (A) Filter gallery at Iver treatment works of Three Valleys Water Company. (B) Filtration plant control desk at River Itchen works of Portsmouth Water Company. (C) Sulphuric acid plant and (D) liquid alum dosing plant at Purton treatment works, Bristol Waterworks Company. (Contractors and photographs: Paterson Candy International Ltd.)

A

B

C

Plate 12 (A) Inlet aerator at Grafham Water for Great Ouse Water Authority. (B) Lime weigh batcher and (C) lime silo at Purton treatment works, Bristol Waterworks Company. (Engineers: Binnie & Partners. Contractors (B) and (C): Paterson Candy International Ltd.)

diameter will fall in still water at about 2.0 cm/s. Suitable grit chambers will be 1.2 to 2.0 m deep, having a horizontal flow velocity through them of about 0.2 to 0.3 m/s. Grit tanks or detritus tanks are always included in sewage works, but less often in waterworks.

Chemically assisted sedimentation or clarification

7.11 Chemically assisted sedimentation

This comprises several separate processes of treatment which go to make up the complete system known as 'clarification'. This system is designed to clear a water as much as is desirable, or possible, of suspended materials before it passes to rapid gravity filters. It is a delicate and chemically complex phenomenon having three stages: (1) the addition of measured quantities of chemicals to water and their thorough mixing, (2) the formation of a precipitate which coalesces or coagulates and forms a floc, and (3) sedimentation.

Chemical mixing and flocculation are considered in the next two sections, and sedimentation in Sections 7.14 to 7.16.

7.12 Chemical mixing

The principal objective in chemical mixing as applied to water treatment is to obtain uniform dispersion of the chemical in the main flow of water, so as to avoid local over- or under-dosing, which in turn can prevent the chemical reaction process going to completion. For example, inadequate mixing of aluminium sulphate can impair the formation of a good floc in the clarifier and would result in poor plant performance or wastage of chemicals, or both. This can be of particular importance in sludge blanket and high rate clarifiers using, for example, dissolved air flotation and lamella flow designs, and in extreme cases inadequate mixing can render the process inoperable. Despite the importance of ensuring efficient chemical mixing in water treatment, particularly for 'difficult' waters, it seldom receives adequate attention. The addition and mixing of chemicals to the main flow of water is a continuous process and is frequently described as either rapid or flash mixing. The design of mixers is often based upon the concept of velocity gradient and its value is used to express the degree of mixing at any point in the liquid system. The velocity gradient G is defined in terms of power input by the following relationship developed by Camp and Stein[10] for flocculation:

$$G = \left(\frac{P}{\mu V}\right)^{1/2}$$

where G is the velocity gradient (s^{-1}), P is the useful power input (watts), V is the volume (m^3), and μ is the dynamic viscosity ($N\,s/m^2$).

Mixing efficiency is directly related to the local flow turbulence created and should give a high degree of chemical-in-water homogeneity within a short time, with low absorption of power. Mixing designs should aim to achieve the optimum combination of maximum turbulence and low power input. The methods used for mixing can be hydraulic or mechanical.

Hydraulic mixing makes use of the turbulence created due to the loss of head across an obstruction to flow such as from an orifice plate, taper, bend, tee, valve, or by a sudden drop in level such as when water flows over a weir or flume. The best results are usually obtained by introducing a chemical immediately upstream of the point of turbulence. The usual power input for mixing is based upon a typical mixing time of 2 to 3 seconds, with a G value of about 800 to $1000\,s^{-1}$. The power is related to headloss by the equation:

$$P = Q\rho gh$$

where P is the useful power input (watts), Q is the flow (m^3/s), h is the headloss (m), ρ is the density of water (kg/m^3), and g is the acceleration due to gravity (m/s^2).

In practice it is generally found that adequate mixing is obtainable with a headloss of between 0.25 and 0.40 m; in any case it should not be less than 0.20 m.

Hydraulic mixers are usually simple and particularly suitable where some headloss can be tolerated. They have the advantages of having no moving parts or direct power consumption, and maintenance is therefore negligible. A disadvantage is that the efficiency of mixing drops if works throughput is lowered.

Mechanical mixing is achieved in purpose built chambers equipped with mechanical rotary mixers such as turbines or propellers. Typical residence times are of the order of 20 to 30 seconds and the velocity gradient value G varies between about 500 and $600\,s^{-1}$. This gives a range equivalent to about 4 to 10 kW per m^3 per second of water flow and the value selected depends upon the particular chemical itself, the strength of solution at the dosing point, and the efficiency of dispersion. Mechanical mixers have the advantages that they produce thorough mixing with minimal headloss and are not affected by flow variations. To maintain a uniform velocity gradient at varying works throughputs a variable speed mixer is needed, but this tends to be expensive.

7.13 Chemical coagulation and flocculation

The terms coagulation and flocculation are sometimes used with the same meaning, but it is probably more correct to use the former to mean the first stage in the formation of a precipitate and the latter to consist of the building up of the particles of floc to a larger size which can be removed either by sedimentation or by filtration, or by the two processes in series. The aluminium sulphate, or other chemical or combination of chemicals as will be discussed later, is mixed thoroughly with the water and a precipitate consisting basically of aluminium hydroxide is formed. Flocculation is usually achieved by a continued but much slower process of mixing the floc with the water in one of many possible types of plant which may precede or include the sludge blanket types described in Section 7.15. In some low turbidity waters, where relatively low doses of aluminium sulphate are required, the sedimentation stage is omitted and the filters, usually pressure filters (see Section 7.30) or sometimes open gravity filters, are then required to remove all the floc with associated matter in suspension. The more widespread practice, however, is to use coagulation and sedimentation, followed by rapid filtration through open filters. One of the main reasons for this is the greater versatility in dealing effectively with the various types of impurities encountered, although this is not always without difficulty.

In the theory of flocculation the rate at which it takes place is directly proportional to the velocity gradient and the same equation as used in Section 7.12 for mixing is used for determining the velocity gradient G for flocculation. In fact the equation was first developed by Camp and Stein for flocculation applications and it has since been used for chemical mixing. If the retention time in a flocculation chamber is t seconds then the extent of flocculation which takes place, or the number of particle collisions which occur, is a function of the dimensionless expression Gt which is given by the equation:

$$Gt = \frac{1}{Q}\left(\frac{PV}{\mu}\right)^{1/2}$$

where the symbols have the same meaning as in Section 7.12.

For the common coagulants of aluminium and iron salts the value of G for flocculation is usually in the range 20 to 75 s^{-1} with retention times in flocculation chambers varying from about 10 to 60 minutes. The value of Gt would then be in the range 12 000 to 270 000.

The agitation required for flocculation is usually provided by either hydraulic or mechanical means. The most common hydraulic flocculator is the baffled basin equipped with either horizontal or vertical baffles. After determining the power input from an assumed Gt value the required headloss can be calculated from $P = Q\rho gh$. The advantage of the baffled basin is its simplicity because there is no mechanical or moving equipment. A disadvantage is that most headloss occurs at the 180° bends and therefore the value may be too high at bends and inadequate in straight channels for efficient flocculation. A second disadvantage is that the G value will vary as the flow Q varies. When designing baffled or 'sinuous' channels some allowance must be made for water quality and suggested velocities to minimize settlement are:

for highly turbid waters	0.40 m/s
for slightly turbid waters	0.30 m/s
for waters of low turbidity	0.25 m/s

There are several types of mechanical devices for flocculation, the most common being the paddle type which is mounted either horizontally or vertically in the flocculating chamber. The power input is given by the equation:

$$P = F_D V = \tfrac{1}{2} C_D \rho A V^3$$

where P is the power input (W), F_D is the drag force (m kg/s^2), C_D is the drag coefficient, A is the submerged area of the paddles (m^2), and V is the relative velocity of the paddles (m/s). V may be approximated to 0.75 times the peripheral velocity of the paddle or equals $1.5\pi rn$ where r is the effective radius of the paddle (m) and n is the number of revolutions (s^{-1}). ρ is the density of water (kg/m^3).

The speed of rotation of the paddles varies from 2 to 15 revolutions per minute and the peripheral velocity of the paddles ranges from 0.2 to 0.8 m/s. Some clarifiers, particularly horizontal flow tanks, employ rotating paddles in series to give a decreasing rate of flocculation which is sometimes referred to as tapering flocculation.

Impurities which require coagulation and flocculation before they can be removed in a waterworks by sedimentation and filtration can be of many types and origins, but broadly they can be classified as either inorganic or organic material. Inorganic material is usually easier to remove and the efficiency of removal can be measured without much difficulty using conventional laboratory analytical techniques and equipment. For example, inorganic matter in suspension can be measured before and after chemical treatment using the parameters of turbidity and suspended solids, and metals such as iron, aluminium, zinc, and lead can all be measured quickly using well established laboratory techniques. The more usual type of inorganic material encountered in water treatment settles easily, especially when it is of particulate size; when plain settling is used one or two hours retention will usually remove at least 40% to 50% of particulate matter and treatment with a coagulant such as aluminium sulphate or an iron salt will remove at least 99% of the remainder. On its own a primary coagulant, sometimes assisted by a coagulant aid, will usually remove 98% or more of the suspended particulate matter for lightly loaded waters with suspended solids not exceeding about 300 mg/l, but such a removal rate may not be adequate for those river waters typical of the Middle East, India, the Far East, and parts of Africa where suspended solids are regularly measured in excess of 1000 mg/l. In such places it may be necessary to introduce a plain settlement stage before chemical sedimentation or, alternatively, to have two separate stages of chemical treatment and settlement in series because the quantity of solids in suspension, and incidentally the sludge formed by the addition of chemicals, is too great to remove in one stage of treatment. Not all inorganic matter settles out quickly in plain settlement and in some waters, such as are found in Central Africa, it may be weeks or even months before any significant settlement of colloidal material takes place. With such a water there is no option but to use chemical coagulation, at both the presettlement and the main clarification stages. Finely divided mineral matter, including various forms of clay, comes in this category. A suspension of clay in water is a hydrophobic (water hating) colloidal suspension, or 'sol', in which the surfaces of the particles are considered to have a negative charge which contributes to the stability of the sol by helping to prevent the particles coalescing into larger particles which would have a relatively rapid rate of settling. It has often been thought that the excellent coagulating effect of aluminium sulphate is due to the triple positive charge on the trivalent aluminium ion neutralizing the negative charges on the clay particles. The literature on the coagulation of clay and similar materials has been reviewed by Packham[11], who has also described his own investigations which have developed the theories in a somewhat different direction. His experiments have confirmed that it is not mainly the aluminium ions in solution which react with the clay particles, but it is the mass of rapidly precipitating and flocculating aluminium hydroxide which enmeshes them. This mechanism does not exclude an electrochemical effect because it has been shown that the precipitating aluminium hydroxide has a weak positive charge which will enhance the initial deposition of the hydroxide on to the surfaces of the clay particles.

Generally, therefore, the conditions required for the rapid coagulation and flocculation of a suspension of clay are those for the complete precipitation and flocculation of aluminium hydroxide: these conditions include a sufficient concentration of aluminium sulphate and the correct pH range of approximately 6.0 to 7.0 in which aluminium hydroxide is insoluble. It should be mentioned that modern theories do not envisage simple molecules of aluminium hydroxide in a floc, but complex hydrated molecules possibly also influenced by the anions including sulphate.

The variety of impurities to be removed in treatment explains the dictum that all waters are different and that water treatment is an 'art' as well as a science. A plant attendant experienced with a particular water can be extremely capable in adjusting treatment to variations in water quality, but in planning new treatments or operating some plants a qualified and experienced chemist is needed who has developed a 'feel' for his subject. This is particularly the case for those waters whose chief natural impurity is organic in origin, as these are very often the most difficult and obstinate to treat. Miscellaneous fragments of animal and vegetable matter contribute towards the organic content, as does the organic colouring matter derived from peat and similar sources consisting largely of humic and fulvic acids and of more complex compounds, partly in true solution and partly in colloidal form. The optimum chemical treatment for such waters is sometimes very difficult to achieve, particularly when it is necessary to remove dissolved iron or manganese. Sometimes the best results are obtained with a high coagulation pH in excess of 9.0 using, say, lime and chlorinated ferrous sulphate treatment, whereas other waters respond better to an acid or slightly acid pH range of 4.5 to 6.0. The dose of coagulant is often closely dependent upon the initial concentration of colour, e.g. a colour of 50 Hazen units may require 20 to 25 mg/l of alum, and there are probably definite chemical reactions between the molecules, such as humic and fulvic acids, and the aluminium (or ferric iron) ions.

Plankton, mainly phytoplankton, i.e. low forms of plant life, can give rise to tremendous problems of treatment. Their removal by flocculation, sedimentation, and filtration is basically similar to, but more complicated than, that of other forms of organic impurities because of the widely diverse types, densities, and shapes of plankton. On the whole plankton do not settle very readily even with an alum floc and some types are liable to pass through the rapid filters; others are liable to cause a rapidly increasing loss of head in the filters if they are present in large numbers. Improvements to the settleability of plankton can sometimes be brought about by pretreatment with chlorine or an algicide, such as copper sulphate, or by the use of a polyelectrolyte as a coagulant aid. Since blue-green algae contain minute air vacuoles within their cells, they tend to be buoyant and live in the top layers of the water where light penetration aids their reproduction by photosynthesis. The dissolved air flotation process is therefore particularly suited to blue-green removal because of the natural tendency for the algae to rise to the surface. The process can also be effective with other less buoyant algae, such as diatoms.

Of the other impurities which are encountered in raw waters, it is much less easy to be precise as to removal efficiency by chemical coagulation (and filtration). In many lowland waters there are to be found significant quantities of substances derived from sewage and trade effluents, particularly the latter. Many of the organic substances are in a complex form and are usually classified as volatile or non-volatile with many more subdivisions, as for example whether they are biologically degradable. Special laboratory equipment is necessary to measure their presence so that detection in the raw water is usually beyond the capability of the normal waterworks laboratory. Bacteria and viruses also come in this category and, whilst it is known that chemical coagulation and rapid filtration are at least partially effective in their removal, there is an absence of clear scientific evidence from existing treatment works. Much attention is currently being devoted to research and the collection of operating data to evaluate removal efficiency of artificial organic impurities and viruses by different unit processes. Some useful information has been collected together and published[12] as a manual of treatment techniques for the removal of some of the better known inorganic, organic, and radioactive contaminants from drinking water.

7.14 Clarifiers generally

Some simple types of settlement tank are in use for the clarification of flocculated waters. For large volumes of water containing a relatively heavy load of suspended solids a relatively dense floc is formed which settles easily, and in warm climates where the viscosity of the water is lower, so permitting more rapid settlement of floc, the large horizontal (or 'cross-flow') sedimentation tank can be an economical solution for clarification. Although a large tank is necessary in order to keep velocities low enough to permit settlement of floc to the base of the tank, its construction is simple and relatively cheap because it need not be very deep and few internal walls are required. In the simplest design of horizontal tank, floc is allowed to accumulate on the floor of the tank until such time as the increasing velocity of water above the accumulated sludge stirs up some of the floc, thereby affecting the clarity of the effluent. When this occurs the tank should, at least in theory, be cleaned out. Moving scrapers operated continuously or intermittently add to the overall efficiency by pushing the settled floc to outlets in the base of the tank where it may be drawn off as sludge.

Accepting their initial low rating and correspondingly high civil construction cost, horizontal tanks are nevertheless versatile clarifiers and modifications to an original simple design to meet changing conditions of raw water can be fairly easily accommodated if provision has been made at the design stage. For example, rotating flocculators and sludge scrapers can be added, and some tanks have had their rate of flow increased by the addition of tube settlers (see Section 7.16). Some horizontal tank designs use the principle of shallow depth sedimentation with a sloping tray so that the direction of flow of water reverses up and over the tray and exits near the inlet. Yet another variation is where the depth of the rectangular tank is divided by as many as four inclined trays and flow takes place in parallel streams between the trays. In this way the capacity of a single multiple tray tank can be made to equal the flow capacity of a more sophisticated design of sludge blanket or sludge recirculation tank. Flow in a horizontal direction is frequently maintained in circular configurations, instead of rectangular, so that after chemical flocculation, which usually takes place in a central compartment, flow is radial and upwards to peripheral collecting launders or to a combination of peripheral and radial launders. This type of circular 'once-through' sedimentation tank is nearly always fitted with rotating sludge scrapers, either with the drive mechanism mounted on a central platform or bridge or, in the case of larger tanks, with the drive unit mounted on the outside wall and the scraper bridge pivoted on a central support. There are circular radial flow tanks in operation in many parts of the world, particularly for the treatment of heavily silted waters, and their use should always be considered as an alternative to rectangular horizontal flow tanks for the treatment of such waters. One such design by Walton and Key for the settlement of Nile water before construction of the Aswan Dam is shown in Fig. 7.2. This particular design has the water entering tangentially at the base and overflowing at the top over a one-third peripheral overflow trough. A simpler but less efficient version of this type of design, and also used quite frequently for the settlement of Nile water prior to construction of the Aswan Dam, omitted the scraping mechanism but included a system of internal walls forcing the water to follow a longer spiral flow path. Since the Aswan Dam now intercepts most of the suspended debris in the upper reaches of the river, Nile water is now lightly laden in terms of suspended material. Instead, algae occur in large numbers, particularly in the lower reaches, and it becomes necessary to extend the options for treatment in addition to horizontal and radial flow clarifiers to include also sludge blanket and solids contact clarifiers which are described in the next section.

7.15 Sludge blanket tanks or solids contact clarifiers

Probably the earliest type of sludge blanket tank is that often referred to as the 'hopper bottomed tank'. It is believed that the first of these tanks were built in India during the 1930s, but there have been hundreds of them built since, mostly in places where there has been a British influence. The sludge blanket effect is obtained simply by allowing the chemically treated water to flow upwards in an inverted pyramid type of tank with the angle of slope usually 60° to the vertical. They are usually square in plan (occasionally circular), although rectangular multiple hopper tank variations have been constructed for large capacity works. As the water rises in the tank of increasing area, its velocity progressively decreases until, at a given level, the force of the upward flow on the particles is so reduced that it only just counterbalances the downward weight of the particles, which therefore hang suspended in the water. Their presence forms a kind of blanket in the water in which chemical and physiochemical reactions can be completed and in which a straining action to remove some of the finer particles may also take place. Whilst the top of the blanket is very often well defined there is no distinguishable bottom to the blanket because of the varying density of the particles forming the blanket. The concentration of the sludge

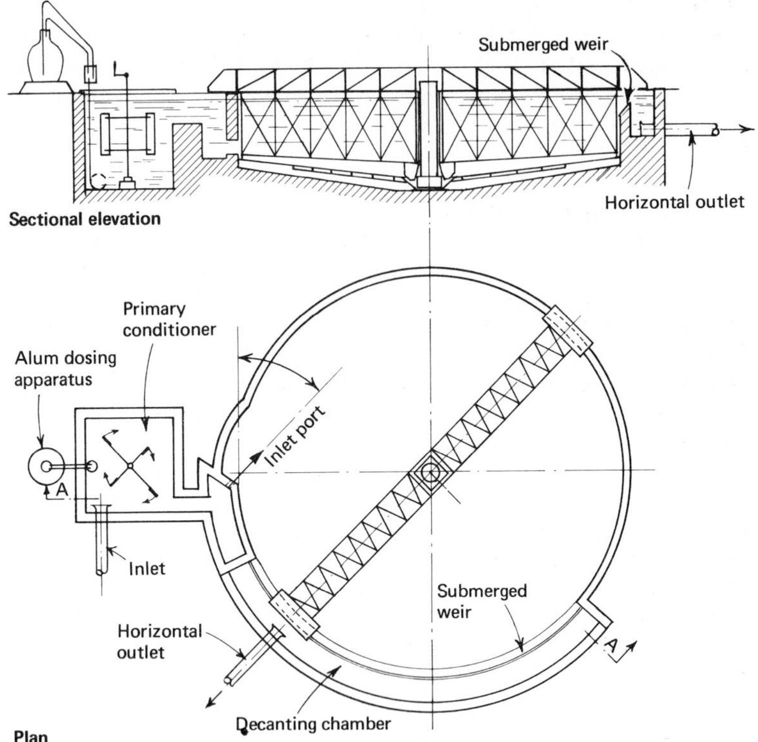

Fig. 7.2 A circular clarifier as designed by Walton and Key. (Walton and Key, *Proc. ICE*, 1939.)

blanket is controlled at a suitable amount, e.g. 5% by volume, by allowing it to bleed off from suitably placed 'concentrators'. The hopper bottomed tank is not cheap to construct as the total water depth is governed by geometry and is usually about the same dimension as the side of the square in plan. During construction care must also be taken to ensure that the inverted pyramid is reasonably accurate and the inlet pipe discharges at the geometric centre; if it is not then streaming of the water to one side can occur and this can upset the stability of the sludge blanket. Some sludge blanket tanks have been criticized for the tendency of the blanket to 'boil' and this can happen particularly in hot climates or where the tank is in an exposed position and subject to high winds. However, the effect on water quality leaving a tank due to temperature boil is usually less severe than that due to wind effect. The flow rating of hopper bottomed tanks is in the range 1.3 to 4.3 m per hour, the lower velocity mentioned being sometimes needed for an alum floc formed in the removal of colour from a soft reservoir water[13].

To reduce construction costs there has been a continuous striving towards simplifying the shape of the tank, and the most recent addition to the sludge blanket family is a flat bottomed variety, usually rectangular in plan and illustrated in Fig. 7.3. In this type of design chemically treated water is directed downwards on to the base of the tank through

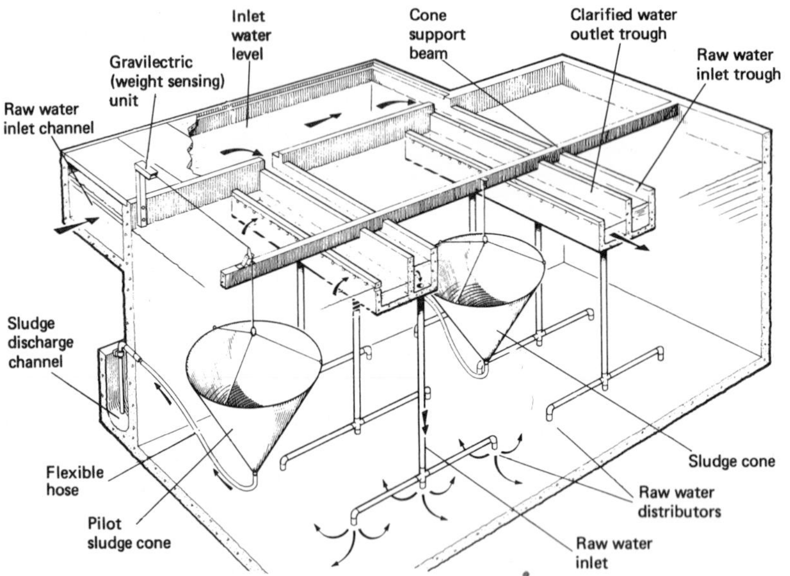

Fig. 7.3 A flat bottomed upward flow sludge blanket clarifier. (Paterson Candy International Ltd.)

suspended pipe 'chandeliers' before passing upwards through the blanket to the surface collecting launders. The reversal of flow which takes place at the base of the tank assists chemical flocculation in the same way as in a hopper bottomed tank. The way in which sludge is removed from the clarifier is referred to in the next section.

Another special type of upward flow sludge blanket tank is the 'Pulsator' through which an intermittent flow is achieved by means of a vacuum chamber built on to the main inlet channel (see Fig. 7.4). The pulse is adjustable, but generally it gives a rapid flow for 5 to 10 seconds at intervals of about 30 seconds. The sludge blanket is thus alternatively raised up and allowed to subside. This type of plant has been used in many parts of the world and has gained considerable popularity in Britain during the last decade.

Sludge blanket clarifiers are not advisable for heavily silted water unless preceded by simple rectangular or circular basins as described in

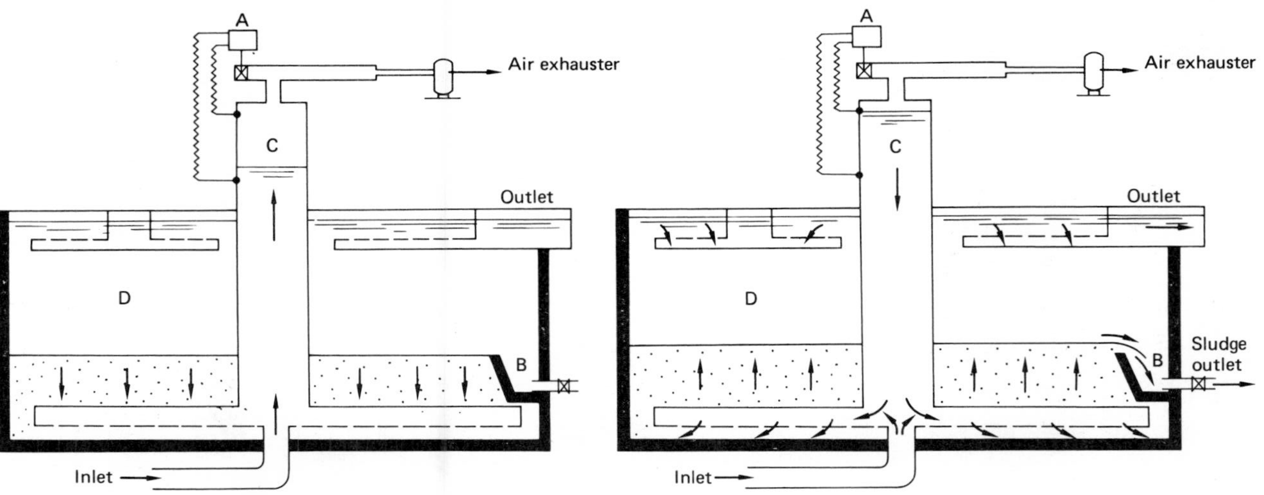

First half-cycle: Air valve A closed and water rises in vacuum chamber C. Water in clarifier D at rest and sludge settles.

Second half-cycle: Water in C rises to upper contact and air valve A opens. Water in C falls and enters D raising sludge which enters concentrator B. When water falls to lower contact, air valve A closes.

Fig. 7.4 The 'Pulsator' clarifier. (Degremont Laing Ltd.)

Section 7.9. The operation of tanks can be 'tricky' and more so in hot climates overseas where there can be a wide difference in day and night temperatures, thus affecting viscosity and the settleability of light suspended particles. They are not particularly suited to 'stop–start' conditions of operation and it will usually take twelve hours or more to obtain a stable blanket. However, so long as their sensitivity is appreciated and they are operated intelligently they will produce a good quality effluent and are very tolerant to changing conditions of raw water quality which would be detrimental to the operation of many other types of clarifiers.

Apart from those sludge blanket tanks already mentioned there are many other types of designs which endeavour to achieve the same high level of performance by providing extra mixing energy for flocculation or by the recirculation of sludge, or by a combination of both these methods. Frequently the resultant design is something halfway between a wholly vertical flow sludge blanket tank and a solids recirculation radial flow tank. Nearly all variations have circular configurations and are equipped with bottom sludge scrapers.

Mixing energy for flocculation may be located internally or externally and sludge recirculation to 'seed' the incoming raw water may be either an internal or an external impeller or pump. The recirculation flow of preformed floc can be as high as 20% of tank throughput, the power rating of flocculators can be up to 4 kW, and the drive motors of sludge scraping gear can also be as high as 2.5 or 3 kW. Clearly, therefore, a tank which has, at first sight, all the apparent signs of a simple and cheap design can in fact be an expensive unit to operate. In evaluating different tank designs it is very necessary for the engineer to take all costs into consideration before coming to a conclusion, and this should include the cost of civil construction, the costs of mechanical plant, and the costs of chemical mixing energy, sludge scraping, and sludge recirculation where used.

7.16 High rate clarifiers

Commercial competition between treatment plant manufacturers has provided the main incentive for the development of new techniques for obtaining higher flow ratings in sedimentation tanks. The flow rating of a clarifier is usually expressed in terms of the drawoff rate from the surface, sometimes referred to as the overflow rate in $m^3/(m^2 h)$ or more often as the rising velocity in m/h. Whereas two decades ago the maximum rate for a sludge blanket tank was about 2 m/h, the upper limit is now 4 to 5 m/h. This increase is partly attributable to the wider use of polyelectrolytes, partly to closer attention to the mixing of chemicals and flocculation, and partly to better control of sludge blanket formation and sludge removal. In all proposals for the adoption of high rate clarification it is important to check the probable performance at the lowest water temperatures likely to apply when the viscosity of the water will be highest. This can be the limiting criterium for performance.

Some success has been achieved to increase flow rates through existing clarifiers by the use of tube settlers, which make use of the principle established long ago by Hazen[14] that a settling tank should be as shallow as possible in order to shorten the distance of falling. The concept of shallow depth settlement by multiple trays is still used in horizontal tank designs in Europe (Section 7.14) and the use of all plastic tube settler packs is a logical development which helps lessen the sludge removal problem of wide shallow trays. The 'tubes' are square, about 50 mm × 50 mm, made of plastic baffles inclined at 60° to the horizontal, and located between sheets of plastic (Fig. 7.5(a)). The

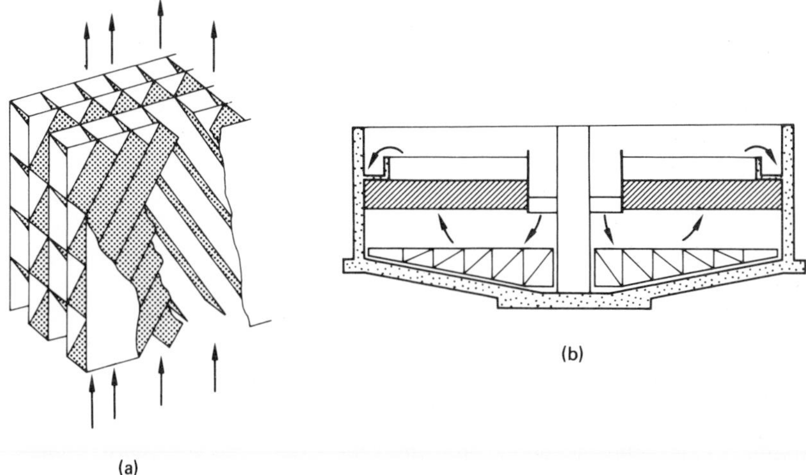

Fig. 7.5 The tube settler: (a) tube module construction and (b) conversion of clarifier with tube module. ((b) Clarke Chapman–John Thompson Ltd.)

tubes have a large wetted perimeter relative to the wetted area and thereby provide lamella flow conditions which theoretically offer optimum conditions for sedimentation. The solids settle to the inclined wall of the tube and leave the tube by sliding downwards along its lower side, the clarified water flowing in the opposite or counter-current direction. Most tube settlers have been used in circumstances where it was desired to increase the flow through an existing clarifier, particularly horizontal and radial flow clarifiers, so that the increased flow rates achieved were still in the range 2 to 5 m/h. Additionally some improvement in water quality was usually obtained. There are few examples of purpose built tube settler clarifiers, and Fig. 7.5(b) shows the way in which a radial flow clarifier can be converted.

More recently the principle of shallow depth sedimentation has been extended to the design of parallel plate systems, sometimes referred to as lamella flow sedimentation. Much of the basic research and development has taken place in Scandinavia, in particular in Sweden, but with practical application in several European countries. Clarifiers using the plate system are usually purpose built to take advantage of the high settlement rates which can be obtained and the greater density of sludge provided. To ensure good performance it is usually necessary to include additional stages of chemical mixing and flocculation and this can be critical for successful operation. There are claims of rise rates well over 20 m/h for normal operation and over 40 m/h in exceptional conditions, which can give a much reduced surface area compared with more conventional designs of clarifier. This also means that the retention time within the clarifier is low, sometimes 20 minutes or less, so that control of chemical treatment becomes more exacting.

There are two basic types of lamella plate designs. The first operates on the principle of co-current sedimentation in which the flocculated suspension passes downwards through the parallel plates, the sludge settling on to the plates and sliding down to the sludge collector. The angle of inclination of the plates is about 30° to 40° and the plates are about 35 mm apart. Clarified water is withdrawn at the bottom of the plates and is then made to pass upwards to outlet collecting launders on the surface. The second type of design operates on the principle of counter-current sedimentation with the flocculated suspension entering the sides of the plates and moving upwards, the sludge settling on to the plates and sliding down by gravity to the sludge collector. The inclination of the plates is steeper at about 55° as the sludge has to flow against the liquid flow.

An interesting application of the co-current plate system of sedimentation is a clarifier which combines the lamella flow principles with a conventional sludge thickening stage in the same tank. Sludge from the plates falls into the tank bottom which is equipped with gate and thicket stirrers and bottom sludge scrapers. In this way sludge can be thickened within the clarifier itself up to 8% dry solids and it has been claimed that solids concentrations of up to 15% can be achieved. The design of the clarifier is illustrated in Fig. 7.6.

An example of a sludge blanket clarifier which has made practical use of the plate system of sedimentation is the 'Super Pulsator' (see Fig. 7.7). The clarifier has many features in common with the pulsator from which it has been developed and retains the principle features of the raw water feed and distribution and formation of a sludge blanket. Parallel plates are located within the blanket at an angle of 60° and coagulated water passes upwards. The plates are fitted with deflectors to create internal sludge recirculation and to thicken the sludge; by this arrangement of counter-current flow and plates and deflectors it is possible to

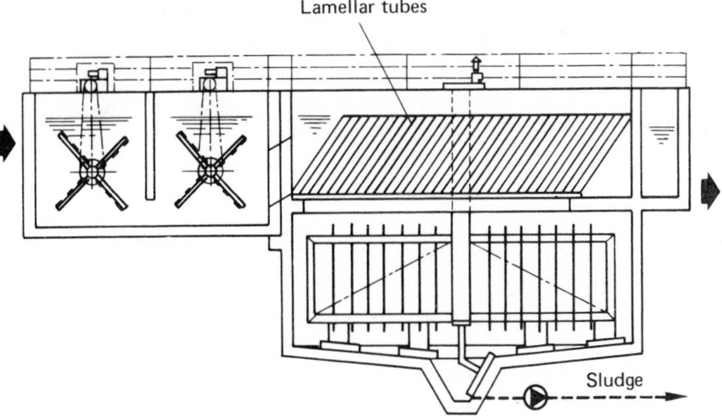

Fig. 7.6 Lamella separator and thickener. (Passavant Ltd.)

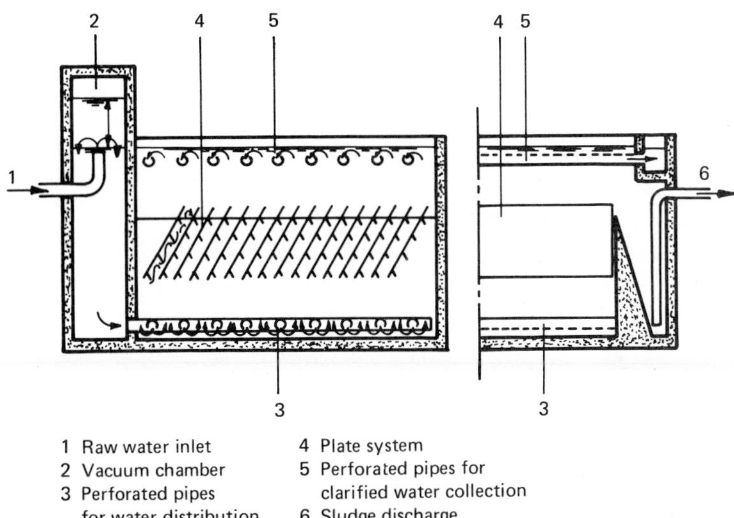

Fig. 7.7 The 'Super Pulsator' clarifier. (Degremont Laing Ltd.)

obtain, with raw waters which respond to this type of treatment, clarification rates in the range 5 to 20 m/h with enhanced sludge concentration. It is also claimed that this type of clarifier is capable of start-up in a matter of hours from a completely shutdown condition.

In recent years there has been increasing interest in the application of dissolved air flotation as a clarification stage of treatment. The process has been used for many years in industry for liquid–solids separation and its potential for application in potable water was first recognized in Sweden. There are now about twenty large plants operating in Scandinavia and other parts of Europe including the UK. The process is particularly suited to algae laden waters, but treatment of some lowland river waters and upland coloured waters has also been successful[15]. Before using dissolved air flotation a pilot plant study is advisable, as chemical mixing and flocculation requirements upstream of the flotation unit can be more exacting than with, for instance, a sludge blanket clarifier. Also the authors know of a case of an eutrophic water with very high counts of algae where dissolved air flotation was not successful, so that caution is necessary in choosing the process. The retention period in the mixing and flocculation compartments is about 20 minutes, followed by about another 20 minutes in the flotation cell itself—this 40 minutes total is not too long a time if things start to go wrong with chemical treatment and corrections have to be made. Sludge is removed from the surface of the flotation unit by overflow, or by surface scraping, and a big advantage of the process is that sludge is more concentrated than in most types of clarifier, concentrations of 2% to 4% solids being common. For the process to work it is necessary to recycle clarified water, which has been pressurized and air saturated, back to an inlet distribution system at a rate equivalent to between 8% and 15% of throughput, and this extra power consumption partly offsets the advantages of higher clarification rates. Surface ratings are in the range 8 to 14 m/h and the clarifiers can usually be started up from cold and can achieve their rated output in a few hours.

Another high rate process which was successful for a time (a number of installations are still in operation), but which did not seem to achieve widespread popularity, is one which involves the dosing of polyelectrolyte to a suspension of fine sand, injected into the raw water entering a clarifier after initial conditioning with primary coagulant. The sand increases the solids contact within the clarifier tank and the sludge containing some sand falls to the bottom of the tank, is removed by scrapers, and is then passed through hydrocyclones to separate sand from sludge. The separated sand is then recycled to 'seed' incoming water again. Rise rates of 6 m/h and above have been achieved with this type of clarifier.

7.17 Sludge removal from clarifiers

Effective removal of sludge is very important for the efficient operation of clarifiers, but it is a subject frequently overlooked by designers. With a raw water having suspended solids not greater than about 250 mg/l—most waters used in the UK fall in this category—then the sludge volume to be removed from the tank should be a maximum of about 3% of throughput. For raw waters having solids up to about 1000 mg/l the sludge may be as high as 10% by volume of throughput; for very silty waters the sludge to be removed continuously in order to keep the tanks functioning, even at a reduced output, and maintain an acceptable water efficient quality can be as much as 20% to 25% of throughput. Special measures for sludge removal must obviously be provided for heavily silted waters and for those having above 1000 mg/l solids it is necessary to provide scraping equipment for all designs if throughput and quality are to be maintained. Scrapers in horizontal tanks are either bridge mounted with side drives or the scrapers are drawn backwards and forwards by cable. Circular tanks have the scraper and bridge supported and driven from the centre and, in larger tanks, support is also provided by a travelling wheel on the outside wall. In some square tanks corners are curved on the bottom and the scraper is then arranged to fold back on itself when traversing the four sides. In another square tank design still working in Egypt and treating Nile water the whole bridge structure, which is centrally mounted, is made to 'telescope' when scraping the four sides. Some designs include a pump mounted on the bridge structure with the suction drawing sludge from points just in front of the scraper blades (called squeegees); others are aided by submersible pumps or down pipes acting as ejectors. In both cases the sludge pumps are usually in addition to the central or radial collecting pockets where sludge is directed by the scrapers. For those designs not equipped with scrapers it is most important to try to provide sensible and practical sludge evacuation arrangements. Many old designs of horizontal flow tanks, and some circular ones as well, were provided with 'ridge and furrow' bottoms over their whole length in the expectancy that the sludge would distribute itself evenly along the bottom of the tank. This unfortunately does not happen and the error was compounded by providing 'plug' or 'carrot' valves to be operated from above the water surface, forgetting that it is almost impossible to move a valve of this type against a head of water of 5 m or more. Valves for sludge removal are nearly always better placed outside the walls of the tank. Both valves and pipework should be adequately sized to pass the maximum sludge withdrawal rate, which can be 400% or 500% greater than the average rate.

Sludge removal from blanket tanks is generally easier than with other designs, although the same rules for valve and pipe sizing must be used. Certainly the positioning of sludge pockets can be less critical than with other designs as a sludge blanket is in continual movement and will migrate towards the space left by evacuated sludge. In the original hopper tank design sludge collectors were usually located near the bottom of the vertical wall section of the hopper, the sludge being removed every few hours by the hydrostatic head of water above the sludge concentrator. Almost the full depth of head is used for sludge removal in the pulsator where the collectors or concentrators are located along a central axis on the bottom of the tank. Many attempts have been made to improve the efficiency of sludge removal and concentration. Some sludge removal systems operate continuously, but this is usually wasteful except in the case of high solids waters. Nowadays, even in what may be essentially a manually operated works, sludge removal is usually operated by an automatic system having adjustable timers for varying the duration of opening of sludge discharge valves at preselected but adjustable times. The equipment is relatively simple and straightforward.

Some success has been obtained using photoelectric cells for the detection of sludge build-up, but this method is unlikely to work for heavily silted waters. However, a method which has met with considerable success over the last few years is one patented and developed by PCI Ltd, and relies upon sensing the differential weight of concentrated sludge in water. The equipment is shown in Fig. 7.3 and consists of one or more flexible sludge cones suspended in water and connected by a cable to a load cell. The load cell is sufficiently sensitive that when the weight of the sludge reaches a preset value (usually when the cone is about two-thirds full of sludge) then the load cell initiates the opening of the desludging valve on the cone outlet. In theory this particular design, which uses the principle of weighing concentrated sludge, should completely satisfy the requirement for a stable suspended blanket if the rate of sludge removal is the same as the rate of sludge blanket formation.

7.18 Detention times

It may be observed from Section 7.9 that the depth, *d*, of a sedimentation tank does not enter into theoretical calculations, and this fact is made use of in tube settler and lamella plate design. In a so-called conventional tank the depth, *d*, is usually decided by experience and long and slim horizontal tanks for grit and sand removal usually have overflow rates in the range 1000 to 3000 m^3/day per m^2. In chambers for grit and sand removal the horizontal speed of flow should not exceed 0.3 m/s and the detention time should be about three times that required to allow the smallest sand grains required to settle. Horizontal chemical sedimentation tanks for the removal of organic matter and light flocculated particles, with length to breadth ratios between 2 : 1 and 3 : 1, have overflow rates between 20 and 40 m^3/day per m^2. Detention times for these particles of low relative density, i.e. the theoretical time of travel of water in the tank, varies from a minimum of 1½ hours to an average of 4 hours.

In the case of large circular cross-flow clarifiers the detention time can be as high as 4 hours, but is more usually about 2½ hours, and these clarifiers generally operate with overflow ratios up to about 1.5 m/h (36 m^3/day per m^2). It should be noted that this rate is near the top end of the overflow rate for horizontal tanks. For sludge blanket tanks or solids contact clarifiers the detention time can be as low as 1 hour, but normally averages between 1½ and 2 hours. Overflow rates for circular solids contact clarifiers usually have an upper ceiling of about 3 m/h, whilst the upper limit for sludge blanket tanks is usually about 5 m/h. Before deciding on a high flow rate for a clarifier it is as well to remember that when the detention time for a particular design is shortened there is less time for the operator to deal with any change in raw water conditions which requires a change of chemical rate of application, before the effluent from the clarifier passes to the filters.

Overflow rates in clarifiers are often obtained by dividing the daily or hourly flow by the gross surface area of the tank. This can be misleading as it may not take into account the area occupied by the mixing and flocculation compartments, where these are provided, and the effluent launders which usually account for at least 12% of the gross area. A better guide to the true overflow rate is to take the horizontal clarification area available at a depth of about 1 m below the surface.

7.19 Chemical dosing plant

As described in Section 7.12 the addition of chemicals to the water and their thorough mixing is the first stage in the clarification process. Chemical dosing must be accurate and therefore the total dosage must vary according to the rate of flow. The most usual practice nowadays of achieving this is the use of positive displacement pumps for the metering of chemicals. The pumps can be provided with diaphragm, piston-diaphragm, or plunger heads depending upon the chemical to be handled and the accuracy required.

On base load treatment works, where the works throughput is usually steady for long periods, manual dosage adjustment of the metering pump output is adequate. Dosage adjustment is carried out by manual adjustment of the pump stroke, the stroking speed being fixed. Where plant throughputs are variable due to demand requirements, the pump stroke may be adjusted manually to give, say, 1 mg/l of chemical in the water; the stroking speed is, however, automatically controlled through a variable speed drive or similar system in proportion to the flow through the treatment works by a control device such as a weir or Venturi tube. Thus, whatever the flow, 1 mg/l is consistently maintained. Other methods are possible, varying from simple mechanical devices to complicated electronic control systems.

Chemicals are usually applied as solutions or suspensions in water. In the former case they are diluted to a standard density which the chemical injection pumps meter by volume. In the latter case the dry chemical may be metered out by weight before a suspension is made. Once the water has entered the clarification chamber, and in its subsequent passage to the filters, its flow should be kept as smooth as possible to prevent turbulence causing carry-over of floc to the filters or break-up of a fragile floc.

Coagulants in common use

7.20 Alum

In discussing coagulation only aluminium sulphate has been mentioned so far. In waterworks practice this material, commonly referred to as 'alum', is the hydrated material of composition $Al_2(SO_4)_3xH_2O$ where the number x may range from about 14 to 20. As a result of this variation it is preferable to refer to the alum dose in terms of mg/l of aluminium (expressed as Al) or as alumina (expressed as Al_2O_3). In some plants situated fairly near to chemical works aluminium sulphate is now supplied economically in liquid form as a strong solution in water. When dissolved in water, aluminium sulphate tends to hydrolyze into aluminium hydroxide and sulphuric acid, hence strong solutions of alum are acid and corrosive. For the precipitation of a floc of aluminium hydroxide it is necessary also to have an alkaline material present: this very often consists of the natural calcium bicarbonate alkalinity in the water. The reaction may then be expressed by the equation:

$$Al_2(SO_4)_3 + 3Ca(HCO_3)_2 = 2Al(OH)_3 + 3CaSO_4 + 6CO_2$$

To avoid the need for excess alum to bring the pH down to about 7.0 it is sometimes economical to use a small dose of sulphuric acid with it. With so-called thin waters which are lacking in natural alkalinity, a solution of sodium carbonate ('soda ash') or lime in the form of a slurry is added in the required amount. After filtration a further dose of lime or soda ash can be added to combine with the free carbon dioxide which has been liberated by the alum. Calcium or sodium bicarbonates are thus added to the final water and the pH is raised, minimizing its possible action on metals. This aspect will be mentioned in the next chapter in relation to corrosion and plumbosolvent action. When soda ash or lime is used with alum in the coagulation of a soft water, which may itself have a slightly acid reaction, the effect of the chemical on pH is continuously checked in a modern plant by a pH recorder. As lime is the stronger alkali, it has the disadvantage that the amount used must be very exactly controlled, particularly for the treatment of a soft water low in natural alkalinity. Although not yet in common use, polyaluminium chloride (PAC) is a chemical which can, with some waters, combine the settlement and clarification properties of alum with the floc strengthening properties of a polyelectrolyte.

7.21 Sodium aluminate

Sodium aluminate, $NaAlO_2$, which is alkaline in reaction, is used very much less often than alum. It is sometimes used after alum in the second stage of 'double coagulation' of highly coloured waters and is occasionally used in lime–soda softening, particularly at the relatively high pH required for precipitation of magnesium hydroxide. It is more often used, particularly in the Far East, as an alternative to lime for pH conditioning prior to alum coagulation.

7.22 Ferrous sulphate

Hydrated ferrous sulphate, $FeSO_4\,7H_2O$, has been traditionally referred to as 'copperas'. It can be used as a coagulant under alkaline conditions in water softening. However, ferrous sulphate has too high a solubility to act as a satisfactory coagulant at the usual pH ranges and is generally used in 'chlorinated copperas' treatment in which it is first oxidized to ferric sulphate and ferric chloride by mixing its solution with the feed from a chlorinator before administering it to the bulk of water. Since ferric hydroxide has a very low solubility, ferric sulphate (or 'chlorinated copperas') can be used for coagulation over a wide range of pH values from 4.0 to 11.0. The lower range is useful for dealing with highly coloured moorland waters, but it is sometimes preferable to operate at the higher and less corrosive end of the range. Exactly the same equations for the precipitation of the hydroxide from ferric sulphate can be given as were given for alum, but writing Fe instead of Al. Treatment using iron salts needs close control including analyses for iron in the filtered water. The presence of iron over 0.3 mg/l will quickly cause complaints from consumers, whereas deposits from a similar amount of aluminium are less noticeable. It should be remembered that nowadays the presence of aluminium in filtered water can be particularly harmful to users of renal dialysis units and, for this reason, iron coagulants may be preferred.

7.23 Coagulant aids or polyelectrolytes

Coagulant aids have been used for many years to assist the action of alum or ferrous sulphate. In recent years the term polyelectrolyte has been coined and materials of natural origin have for the most part been supplemented by numerous synthetic products. The coagulant aid most used for a number of years was activated silica; other coagulant aids were the sodium alginates and some soluble starch products. These substances had the advantage of being well known materials already used in connection with foodstuffs and were thus recognized as harmless in the treatment of waters. The assessment of synthetic polyelectrolytes is more complicated, but in the UK an appropriate expert committee in the Department of the Environment considers them and issues detailed lists of those considered safe. With polyacrylamides it is important that the unpolymerized material should be absent.

The amount of a polyelectrolyte used is very small in relation to the amount of the primary coagulant. For activated silica it is commonly in the range of 2 to 5 mg/l, but for other polyelectrolytes the concentration is normally less than 0.1 mg/l and in some circumstances may even be as low as 0.01 mg/l. Although polyelectrolytes are occasionally added just before, or with, the primary coagulant, the timing of the application can be critical and they are often added after the main coagulant and sometimes only just prior to filtration. They should preferably be introduced in a solution of low concentration and rapid mixing with the bulk of water is important. The theory of the action of polyelectrolytes has been reviewed by Packham[16]. They are essentially water soluble polymerized molecules with ionizable groups. From their action it would be more correct to refer to them as flocculation aids (or sometimes filtration aids), rather than as coagulant aids. It appears that in general the main mechanism of their action consists of bridging together adjacent particles. Some polyelectrolytes have been developed with the intention of omitting the primary coagulant, although the dose level then applied has usually to be increased.

In waterworks treatment activated silica is prepared immediately before use by the partial neutralization of sodium silicate to form a silica sol, i.e. a colloidal solution. Sulphuric acid was one of the neutralizing agents first used and the necessity for careful control may be illustrated by the fact that the authors know of early instances in which a batch was inadvertently allowed to set into a gel. Chlorine and aluminium sulphate solutions have been used with the idea of achieving economy by preparing the activated silica with chemicals already required in the treatment. In recent years the partial neutralizing agent probably used most often for activating the silica has been sodium bicarbonate. Most other coagulant aids, including the large number of synthetic products now available, need only to be dissolved using a high speed mixer (to avoid forming sticky lumps) before being added to, and rapidly mixed with, the water. Most of the fairly new synthetic polyelectrolytes are based upon polyacrylamide or polyacrylic acid; a few are based upon polyamides and a few upon quaternary ammonium compounds.

The practical effect of introducing polyelectrolytes in many existing waterworks has been to allow substantially greater output through the sedimentation tanks and filters[17], and an additional advantage has been improvement of sludge treatment.

Basic concepts in rapid filtration

7.24 Mechanism of rapid filtration

Sedimentation with or without chemical coagulation is usually followed in a waterworks by rapid filtration, and the basic principles determining the removal of particles by such filtration will now be discussed. Usually rapid filtration is preceded by chemical treatment of the water; without chemical treatment rapid filtration is effective for relatively few waters, one of its main uses being 'primary' filtration before *slow* sand filtration. The latter is discussed in Sections 7.35 to 7.39.

In rapid filtration the removal of particles is largely by physical action, although physicochemical considerations may enter. The size of grain of the filter media, usually sand, are normally within the range of 0.4 to 1.5 mm, whereas particles which may be removed by simple filtration, e.g. mineral particles or diatoms, may be at least twenty times smaller. In fact a proportion of particles even several hundred times smaller than the size of the sand grains may be removed. In achieving effective removal of the smaller particles the addition of a coagulant to form a floc, containing aluminium or iron hydroxides, is usually necessary, but even the particles of floc may be very small compared with the size of grain of the filter media. It is therefore evident in the first place that filtration is quite different from a simple straining action such as that of microstrainers. In many instances there may be some straining action due to a coating on the surface of a filter, but in general filtration is a process in which some depth of the filter media is utilized. In filtration of water the flow within the filter bed is laminar or streamline; the loss of head through the media is proportional to the velocity of flow of water.

It is evident that it is necessary to consider by what mechanisms relatively small particles are removed by simple filtration, and a considerable amount of research has been carried out on this matter in recent years. Some of this work has been carried out and the concepts have been conveniently summarized by Ives *et al.*[18,19], with a list of references. The general conclusion has been reached that the principal mechanisms of simple filtration are physical and they may be considered under the headings of gravity (or sedimentation), interception, hydrodynamic effects, and diffusion.

Research on filtration has also considered mechanisms of possible attraction (or repulsion) between particles and filter grains; in the absence of any attraction some particles would tend to become detached. Such considerations are, however, complex and any clear conclusions are at present of very limited application. Van der Waals forces are well known as attractive forces between molecules and theoretically they apply to nearly all materials in water, but their range is usually limited to minute distances of less than 0.05 micron.

Finally it should be mentioned that in some water treatment processes physicochemical or chemical reactions occur in contact with filter grains. One example is the deposition of calcium carbonate from waters having a positive calcium carbonate saturation index. Another is the oxidation and deposition of compounds of iron and manganese. After such reactions have commenced the coated grains may provide active surfaces for their continuance. In a few instances the oxidation of ammonia to nitrate or other biochemical action may be an incidental effect of passing a water through a rapid filter, due to the development of the necessary bacterial flora in organic impurities on the filter grains.

With some waters the oxidation of ammonia to nitrate can occur when the water passes through rapid gravity sand filters. The principle of biological reaction is now used in potable water treatment for reducing the concentration of nitrates to an acceptable level by the development of a biomass on the surface of filter sand. The reaction takes place within a fluidized sand bed after the incoming raw water has been treated with a source of carbon, usually methanol, to convert the nitrates to nitrogen and carbon dioxide. The methanol dose required for conversion is about 2.5 mg per mg of nitrate nitrogen.

7.25 Design and construction of rapid gravity filters

The working part of a rapid gravity sand filter or the part that removes the solids from the incoming water is the sand. As already discussed in Section 7.24, there are many different mechanisms (sedimentation, interception, hydrodynamic, diffusion, attraction and repulsion etc.) which contribute towards the removal efficiency of a filter. Some have a greater role to play than others according to the nature of the water and the chemical treatment which has been used in previous treatment stages. Rapid gravity filters usually constitute the last physical stage (as opposed to chemical stage) in a treatment cycle for drinking water, and the objectives with all designs of filters are to reduce the solids content to a negligible amount, the turbidity to less than 0.4 FTU (in the USA 0.1 FTU), and the suspended solids to less than 1 mg/l—these being the usual targets.

To achieve this high standard of purity there is some limited but nevertheless important scope for varying the basic design of a rapid gravity sand filter: for instance the sand size can either be (nominally) constant, i.e. monograde, or it can vary from fine to coarse; the depth of the sand can be shallow or deep; the direction of flow of water can be downflow, upflow, or can even be brought into the middle of the sand bed to flow both upwards and downwards. A further variation in design is to use three or more layers of sand and pebbles or, alternatively, to use other materials, such as anthracite or garnet, having differing grain size and relative density. All these variations of filter design have been incorporated from time to time in different commercial (sometimes patented) filters. When a fine sand is used, such as 0.4 mm size, then the collection of solids during filtration, and hence the build-up of headloss, tends to be within the top layers of sand. In contrast, with coarser sands, such as 0.9 mm size, the solids penetrate to a greater depth and the lower layers of the sand bed are then called upon to do some of the work of solids removal. So long as there is an adequate factor of safety in bed depth against complete dirt penetration it makes good sense to utilize at least some of the bed depth for solids capture, but the proviso must be that the backwashing system can be relied upon to remove accumulated solids and achieve thorough cleansing of the sand before its next working cycle. In theory the ideal sand grading arrangement is to have a decreasing sand size in the direction of flow as this will bring about the greatest degree of solids capture. Attempts to do this are discussed in Section 7.31 dealing with the use of anthracite media.

The design of filter most frequently used in the past in this country employs a sand bed between 0.45 and 0.75 m thick supported on several layers of fine to coarse gravel so that the total bed thickness is about 1.0 to 1.3 m in a concrete, or steel, filter shell of 3.0 to 3.5 m depth (see Fig. 7.8 and Plate 10). The sand grains are as uniform as can be practically arranged, of a size between 0.5 and 1.5 mm. A 'uniformity coefficient' is sometimes specified for the sand*, usually lying between 1.3 and 1.8. During the past decade there has been a gradual acceptance of the type of design in common use on the Continent which has a bed about 0.9 m thick of a single grain size of 0.85 mm, overlying a shallow band 2.5 cm deep of coarser media.

The collector system of perforated pipes, ducts, or plates in the lowest layer of gravel is usually a proprietary system and may take any of a number of forms. The aims of the collector system are to collect water from the underside of the bed in an even manner and to spread air and water scour evenly through the bed when backwashing. The overall size of filter varies, the limiting factor for size being the difficulty of maintaining an even filtration flow and even backwash and air scour. Single sand beds have been built up to 40 m × 20 m, but larger filter units of over 150 m^2 total sand area are made possible by having dual beds with a common washout channel between the beds.

In the downflow filter design with upflow washing it is usual for the filter to operate with about 2 m head or more of water over the bed; however, there are some proprietary designs which operate with much less depth of water and even with negative head conditions, but the latter is not regarded as good practice as difficulties can occur with cracking and mud balling in the filter bed, especially with high filtration rates. When a cleaned filter is put back into supply the loss of head through the bed and underdrain system should be less than 0.25 m. If an outlet controller is fitted it throttles the discharge so as to maintain a flow which does not exceed a permitted maximum. As the bed becomes dirtier and the headloss through it increases, the outlet controller opens to keep the output constant. There are also mechanical, electrical, or pneumatic linkages from this controller back to the filter, so arranged as to maintain a substantially constant water level during the course of a filter run. The principal reason for an outlet controller is to maintain a predetermined constant output from each filter under varying headloss conditions, the amount being such that the total inflow to the filtration plant is divided equally among all operating filters.

$$\text{*Uniformity coefficient} = \frac{\text{size of aperture through which 60\% sand passes}}{\text{size of aperture through which 10\% sand passes}} \quad \text{(by weight)}$$

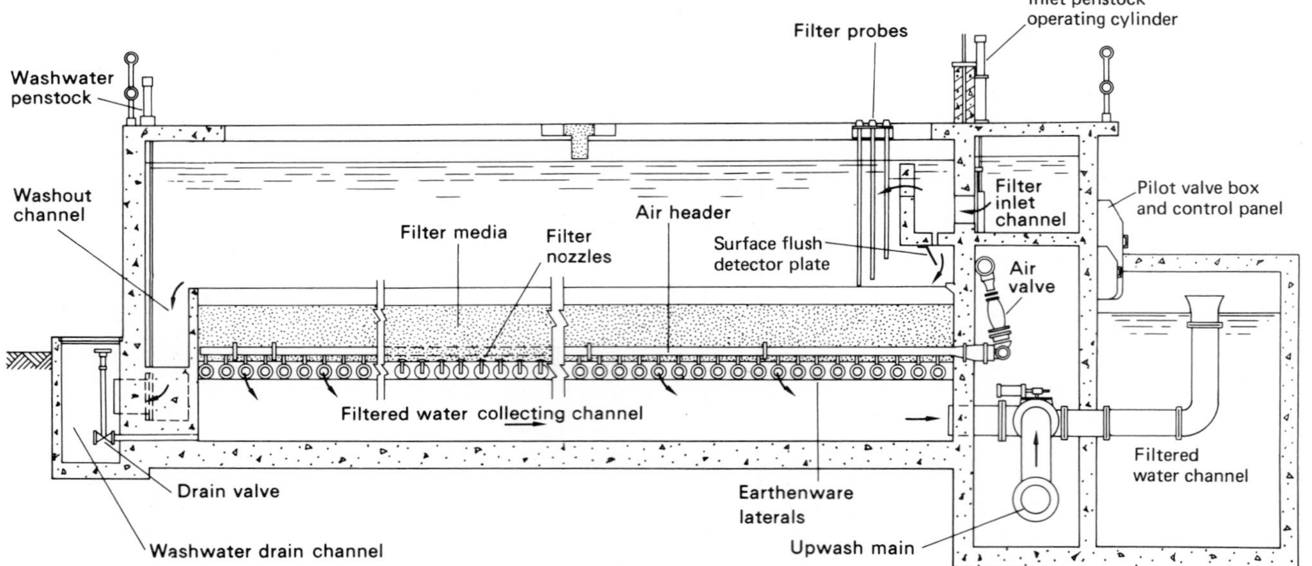

Fig. 7.8 Rapid gravity filter. (Paterson Candy International Ltd.)

The same result of equal flow can, however, be achieved by using flow splitting weirs proportioning the flow equally to all filters or it may be achieved by sizing the outlet pipework and valves to each filter to limit the maximum flow hydraulically. In such cases, after a filter is backwashed, the level of water in the filter box will rise to such a level above the outlet head on the filter as will be sufficient to overcome the headloss through a clean filter and its underdrainage system. As the sand becomes progressively dirtier during a filter run, increasing the head lost through the filter, so the water level in the filter box rises to the maximum possible. The simplicity of such a design is attractive, but it suffers from the absence of any flow control on the outlet valve at start-up after a filter wash. At this time it is desirable to have a reduced rate of flow through the bed for a while because the full output through the newly cleansed sand can result in a temporary breakthrough of turbidity.

Declining flow filters form another type of simple design for a group of filters. The system is best suited for a group of six or more filters so that the additional flow to be shared when one filter is taken out of service for washing is not excessive. There are no hard and fast rules for design and the installation must suit particular circumstances. It is reasonable to design on the basis of a maximum flow range through each filter of ± 35% of the average filtration rate, the average being taken as the total output divided by the total sand bed area provided. The inlet valve or penstock to each filter is usually submerged and, to restrict the filtration rate to the maximum, it is necessary to install some form of restricting orifice or valve on the outlet. Filters are washed in a fixed sequence and individual filter instrumentation for loss of head or quality of filtrate, i.e. turbidity, is used only to detect a 'rogue' filter, i.e. a filter whose behaviour is out of line with the rest for some reason. In terms of hardware the system is very simple and this is its chief merit; claims of longer filter runs, improved quality, and a less costly installation compared with other systems need to be critically examined as there is doubt as to whether they are obtained in practice.

7.26 Backwashing

Rapid gravity filters of the design illustrated in Fig. 7.8 are washed by first sending air and afterwards water upwards through the bed by reverse flow through the collector system. The first operation is to allow the filter to drain down until the water lies some few centimetres above the top of the bed. Air is then blown back through the collector system at a rate of about 1.0 to 1.5 m^3 free air per minute per square metre of bed area for about 2 to 3 minutes. The surface of the sand should show an even spreading of bursting air bubbles coming through the sand. The water over the bed quickly becomes very dirty as the air-agitated sand breaks up surface scum and dirt is loosened from the surface of the sand grains. Following this an upward flow of water is sent through the bed at a carefully designed velocity, sufficient to expand the bed and cause the sand grains to be agitated together so that deposits are washed off them, but not so high that the sand grains are carried bodily away in the rising upflush of water.

Upward washwater rates are usually of the order of 5.0 to 6.5 mm/s, but higher rates are sometimes used, particularly in warm climates where the temperature of the water is higher than in the UK and the viscosity of the water lower. A bed expansion accompanies the application of the backwash water and this is usually of the order of 20% to 30% of the depth of the bed at backwash rates of 6.5 mm/s.

Most authorities now recognize that properly controlled air scour makes a major contribution to the cleansing of a bed and in the USA air scour is now beginning to be used, whereas formerly reliance was placed almost entirely upon high velocity wash together with a system for water jetting the surface of the sand. Continental practice, particularly in France and Germany, relies upon a backwash sequence which includes the joint application of an air and water wash together. The combination of air in water makes an explosive mixture, hence whilst the air rate is increased to about 16 mm/s the wash rate has to be reduced to 4 mm/s: if this were not done the sand media would be blown out of the bed into the washout channel during backwash, and the pressure generated can be so great as to rupture a filter floor.

Some older rapid gravity filters shaped circular in plan, and still made in Egypt, use a system of rakes hanging vertically from two radial arms which are rotated during backwashing, thus physically raking the top layers of the bed. The idea is simple and may have merits, but the shape of the beds is not the most economic when a bank of filters has to be constructed.

An unusual washwater system for rapid gravity filters is found in South Africa and Lesotho. The filters are designed to receive a conventional wash, using separate air followed by water. After air agitation through the underdrain system the same air blower is then used to direct air to the top of the filtered washwater tank which is located directly underneath the filter. The air displaces the stored washwater up a pipe connected to the underdrain system and water washing then proceeds for a normal 5 or 6 minutes. A disadvantage of the design is that because washwater is stored below the filter the head of water over the sand media has, for practical reasons, to be reduced to a metre or less to limit the depth of the filter bed.

It is sometimes very difficult to drain off the whole of the dirty backwash water from a bed before the filter is filled up again and put to use: to aid in this some of the washwater may be caused to flow across the top of the bed (called 'surface flushing') during the last stages of backwashing in order to flush the dirty water to waste from above those parts of the bed which are remote from the overflow discharge channel. Even so, the water remaining in the bed itself after backwashing will be dirty and, before a filter is put back into use, either the first part of the filtrate must be discharged to waste or time must be allowed for the impurities within the bed to settle out by starting the operation of the filter at a slow rate. The latter can be made to operate automatically so that the full discharge rate is not reached until some 15 to 20 minutes have elapsed.

Water used for backwashing should be filtered water, i.e. not the raw supply, and the total amount used has an important bearing upon the economy of a treatment works, especially in relation to the net yield of a source. The total washwater used should normally not exceed 2% of the treated water output and should preferably be less.

7.27 Filtration rates

These were 3.4 to 3.9 m/h in the past allowing for filters out of service for backwashing, but there is now a tendency to use higher filtration rates and some recent large installations are designed for 12 m/h. The higher rates of filtration are more commonly used in warm climates where the higher temperature of the water and its lowered viscosity permit prior sedimentation treatment of the water to be more effective so that a water containing less impurities passes to the filters. Higher rates still are possible with multilayer filters (see Section 7.31). It is not possible, however, to make a valid comparison of filtration rates without reference to the quality of the water being filtered, the type of bed used, and the size of sand grains adopted, all of which are interrelated together with other factors.

7.28 Operation of filters

If a filter has no outlet controller, or if a system of equal division of influent flow is not used, then the filter output will gradually decline: this is the principle of declining rate filtration and is referred to in Section 7.25. Figure 7.9 illustrates typical relationships which exist between output, time, turbidity of effluent, and headloss during the course of a filter run for both constant flow and declining flow operation. The breakthrough of turbidity on start-up of a filter after backwashing is quite usual (see Fig. 7.9(b)), even though precautions may be taken to effect a 'slow start'. The length of filter run between backwashings varies according to headloss build-up and Fig. 7.9(c) shows a typical headloss build-up for a sand–anthracite filter. The relatively high loss through the anthracite where the greater part of the dirt is retained and across the anthracite–sand interface should be noted.

It is usual to aim for a washing frequency not greater than once every 24 hours and this interval is often convenient for the operating staff who are on day shift. With some waters, particularly those containing

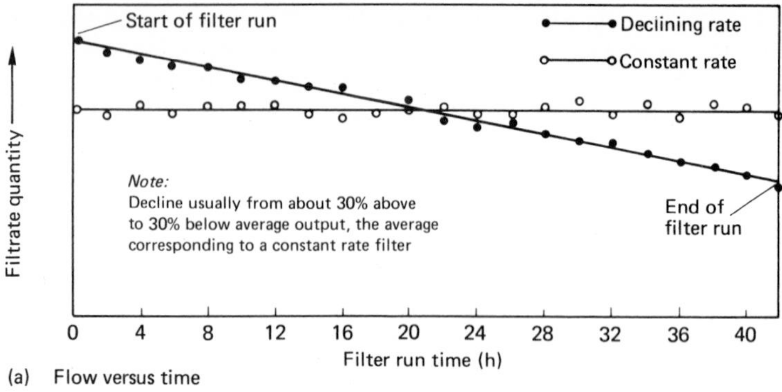

(a) Flow versus time

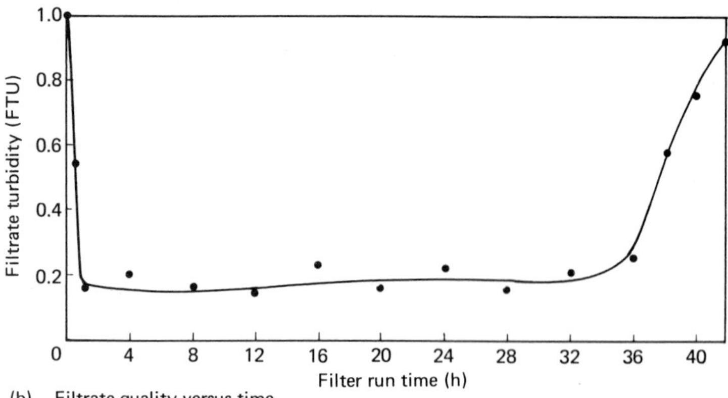

(b) Filtrate quality versus time

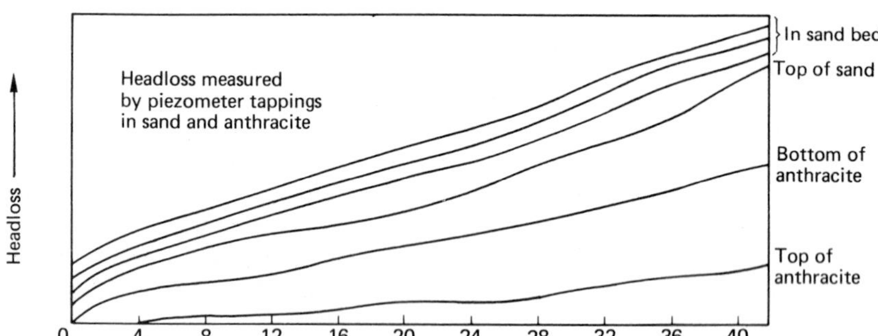

(c) Headloss versus time in anthracite-sand filter

Fig. 7.9 Typical output characteristics of a rapid gravity filter.

organic material, there is a strong case for not permitting the filter run to extend too long, certainly not more than 48 hours. At the other end of the time scale, it can prove difficult to maintain full output from a works if each filter has to be washed more than about once every 6 hours. The sequence of backwashing usually takes one hour, so that for a bank of filters requiring to be washed every 6 hours it could prove necessary to be draining one filter whilst another is being backwashed. In difficult circumstances, when filter runs are very short, it is often the practice to discharge the total contents of the filter to waste in order to speed up the wash cycle: this is sometimes referred to as 'dumping' the filter. The primary indications that a filter needs backwashing are increase in filter run time, loss of head, or reduction in filtrate quality measured by turbidity. In some modern installations all three parameters are monitored and can be used to initiate the washing cycle automatically and eventually return the filters to service. More often than not, only the first two parameters are used for automating the start of a filter washing cycle and the combined filtrate quality is monitored, either manually or by a sensing device, to initiate an 'alarm' condition. In automatic filter plants manual control of washing should always be possible as an alternative so that special cleansing measures can be taken when necessary.

7.29 Filter equipment

For air scouring, air blowers or compressors working in conjunction with air storage cylinders are necessary. For backwashing it is usually most economic to use gravity flow from a large elevated storage tank, since the rate of flow required is large and electrical demand charges are minimized by keeping such a tank topped up by a relatively small, continuously running pump drawing from the filtered water supply. Meters to measure the rate or amount of water filtered (in total and from each filter) are usual, and measurement of the washwater used is also advisable to keep a check on this. Flow rate meters on both the air scour and the backwash flow are important and these should have alarms preventing a maximum from being exceeded. On large filters the valves and penstocks are too big for frequent manual operation and power assisted cylinders to open and close a valve become necessary. Air is often used for operating the power cylinders, but hydraulically operated valves using oil or water are also successfully employed. In temperate and warm climates it is unnecessary to cover rapid gravity filters, but in cold climates covering is required to prevent freezing. The water from a rapid gravity filter is not completely bacteriologically pure and, before the water passes into supply, it must be disinfected by the addition of chlorine or other disinfecting agents. The clarity of the water from some filtration plants is now automatically monitored by turbidity recorders.

7.30 Construction and operation of pressure filters

Pressure filters are similar in bed construction to open rapid gravity filters, except that they are contained in a steel pressure vessel (see Fig. 7.10). Perforated pipes or a steel plate with nozzles are used for collecting the filtered water and for distribution of the washwater and air scour. The steel pressure vessel is cylindrical, arranged horizontally (see

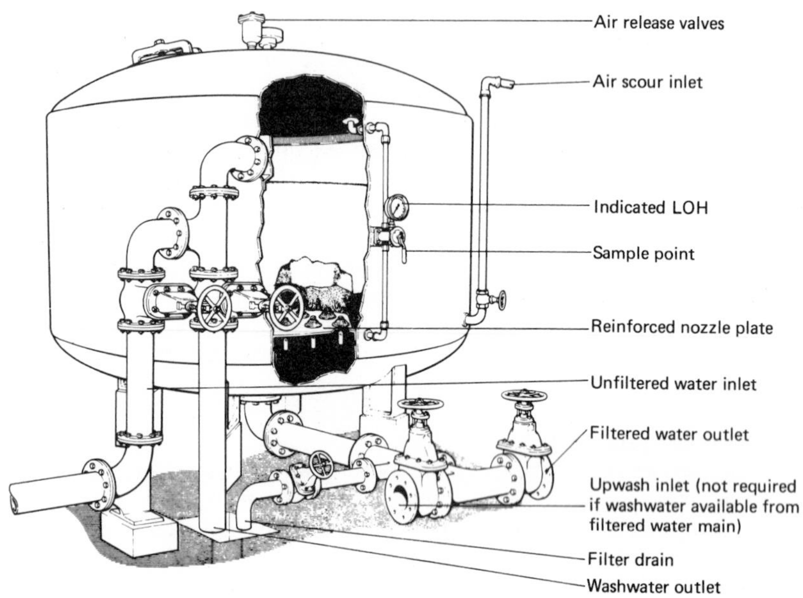

Fig. 7.10 Sectional view of a pressure filter. (Paterson Candy International Ltd.)

Plate 10B) or vertically. With a pipe underdrain system the bottom of the vessel is usually filled with concrete so as to obtain a flat base. In a horizontal vessel vertical plates are welded inside to give a rectangular shaped sand bed within the cylinder so that the bed may be washed evenly and there are no 'dead' areas beneath which air scour and water pipes cannot be placed. Ideally, the top of the sand in a horizontal vessel should coincide with the horizontal diameter to give the maximum sand surface area. The whole of the cylinder is kept filled with water under pressure and at the highest point an air release valve is inserted for the release of trapped air. For practical reasons the maximum diameter of vertical filters is limited to about 2.75 m; in the case of horizontal filters, to the same diameter and to a length of about 12 m.

The backwashing of such filters is very similar to that of an open rapid gravity filter. A bellmouth and pipe can be used for the removal of dirty washwater in a vertical filter; for most horizontal filters a single vertical plate located near to one of the dished ends will facilitate washwater removal, but for the larger filters a central washout channel formed by two vertical plates is necessary. There are still many vertical pressure filters in operation which have mechanical rakes which rotate in the sand bed whilst backwashing is in progress—although this was an effective method of cleaning, this type of filter is not made now because of the extra cost of equipment.

The advantage of pressure filters is that the pressure of water in the mains is not lost when the filter process takes place, as is the case with an open rapid gravity plant. About 3 m head may be lost in friction through the sand bed and the inlet and outlet controllers, but this is the total loss of head. Thus pressure filters may be interposed on a pumping line or a gravitational line without a large loss of pressure on the supply.

Pressure filters suffer from the disadvantage that the state of the bed under backwashing conditions and when the plant is working cannot be directly observed. It is of vital importance, therefore, that every pressure filter is fitted with an open box or dish in the front of it into which the washwater is turned so that any washing out of the filter sand may be observed and the backwash rate immediately reduced. A further disadvantage is that chemicals for the coagulation of the raw water and the production of a suitable floc prior to filtration must also be injected under pressure.

It may be found that some further time after chemical injection and mixing must be allowed for coagulation and flocculation, and this can only be achieved by passing the water through a large pressure vessel where time is given for these processes to take place. The time given must necessarily be short otherwise the size of the pressure vessel would have to be so large that the expense of the installation would be great, but some time for flocculation, even if only 5 to 10 minutes, is essential for producing a water in suitable condition for passing on to the filters.

In marked contrast to rapid gravity filters it is very rare to find a pressure filter installation with any form of flow control on the inlet or outlet to each filter. It is not often appreciated that without this flow control each filter in a battery of pressure filters operates as a declining rate filter for at least part of its filtration cycle. Thus the only restriction to output through a clean, newly washed filter is the hydraulic constraint imposed by the size of pipework at inlet and outlet. This is probably not a serious defect for a plant with, say, six filters or more, but break-through of turbidity can be troublesome with a plant of only two or three filters since the filters in service then take a large share of the output.

Equipment for air scouring pressure filters is similar to that used for rapid gravity filters, and the same applies to the type of equipment provided if power assisted opening and closing of filter valves is called for. For backwashing, a system is often devised to take advantage of the pressure water available from a battery of filters. For a filter washed by separate air and water the required rate of application is about four times as great as the conventional rate of filtration which is about 5 m/h. Thus in a large filter battery groups of five filters can be taken out of service at a time and the combined filtrate from four of the filters can be used to wash the fifth filter, and so on, until all filters in the group are washed. In large pressure filter plants arrangements are usually made to wash filters in groups at a specific time each day, and the washing and monitoring of individual filters for loss of head and turbidity is seldom done. In fact, because a large battery of filters is usually supplied by a common inlet bus main and the outlets from filters also connect to a common outlet bus main, it is only meaningful to measure the headloss across the battery of filters.

The cost of pressure vessels is similar to the cost of concrete work for open rapid gravity tanks so that the capital cost of construction of both types of filtration plant is, by and large, equal. The steel shells, however, require careful maintenance so as to prevent both internal and external corrosion. Condensation upon the outside of the tanks is a continual nuisance as it will cause corrosion of the steel shell and staining of the floor below. The pressure applied to steel pressure filters is not usually in excess of 80 m head of water. This should normally be adequate for most distribution systems. Above this pressure the thickness of plates used for the steel shell may be so great that cost rises rapidly.

Multilayer and other methods of filtration

7.31 Use of anthracite media

As referred to previously in Section 7.25, the most efficient form of media grading for a rapid gravity downflow filter in order to obtain maximum capture of solids would be to have the sand decreasing in size in the direction of flow. Clearly this is not possible because hydraulic regrading takes place during backwash, so that the finer sand collects at the surface of the bed. This can be countered by using separate layers of different filter materials having different density and grain size, the denser materials being at the bottom of the bed and the less dense at the top. One type of multilayer filter bed in wide use at present is the two-layer filter using anthracite over sand. The relative density of anthracite is lower than that of sand: beds of anthracite can be expanded or partially fluidized at about the same backwash rate as a sand bed when the size of the anthracite grains is about twice that of the sand grains. After backwashing, the two-layer bed settles down again with the anthracite on top (subject to a relatively small amount of mixing). With these two-layer filters following coagulation and sedimentation, it has been found that the filtrate quality can be as good as by conventional sand filtration, but filter runs can be 1.5 to 3 times as long. Alternatively higher filtration rates, even up to 17 m/h, can be achieved in some instances. A summary of the effects has been given by Miller[20]. The size of the anthracite used is usually 1.25 to 2.50 mm placed over 0.5 to 1.0 mm sand, the depth of the anthracite bed being of the order of 300 mm. Larger size anthracite has been tried, but variable results have been reported[21].

7.32 Use of anthracite to uprate filters

In recent years advantage has been taken of these higher filtration rates to uprate the output of existing filter plants by changing their media from sand to anthracite–sand, the previously conventional filtration rates of 4 to 6 m/h for sand filters being increased to rates of 6 to 12 m/h. Increases of plant output capacity in the range 33% to 50%, and in some cases 100%, have been achieved. However, these increases have frequently been associated also with the improvement of floc characteristics by the addition of a polyelectrolyte. It is important to pay attention to floc size when adopting anthracite filters: too fine a floc may pass through the anthracite layer and cause too large a load of impurities to be thrown on to the sand below; too large a floc in relation to anthracite size may throw too large a load on to the anthracite defeating the object of gaining filtration in depth by using anthracite and sand. A number of associated matters have also to receive attention. Whilst it is possible for the backwash rate to remain the same as before for a number of filter runs, it becomes necessary to apply a higher wash rate every so often to remove the larger load of impurities held in depth by the anthracite and to regrade the media. Some mixing of the anthracite and sand does occur during a filter run and the higher wash rate brings the anthracite to the top of the bed once again. It is usual to have the ability to increase the backwash rate by about 50% when anthracite is used: thus, for a filter design using separate water and air, a maximum washwater rate of 7.3 mm/s is desirable in a temperate climate and 12.2 mm/s for high temperature conditions. Since anthracite is an expensive material compared to sand it is important to ensure a filter is adequately designed hydraulically before anthracite is used for uprating. This means checking that the inlet, outlet, and flow control pipes and valves can accept the higher flow rates; that the distribution of water at the inlet to the filter does not cause excessive scouring of the anthracite–sand bed because of the higher flow rate; that the higher backwash rate required does not result in loss of anthracite over the washwater weir; that the washwater discharge channel is of sufficient size and gradient to accept the increased backwash without backing up. It is important to ensure that the anthracite is of good quality, correctly graded in relation to the underlying sand media, and sufficiently hard to resist abrasion so that it does not break down into finer material with use.

With a water which produces a fragile floc at the sedimentation stage it may be necessary to add a dose of polyelectrolyte to the water before it passes on to the anthracite–sand beds in order to prevent excessive penetration of floc into the bed. The same precaution may be necessary to prevent the penetration of algae, as reported at the Three Valleys treatment plant[22] where polyelectrolyte was used to arrest penetration through the anthracite and sand by small green algae, dominated by minute species *Nannochloris* and *Ankistrodesmus* with cell diameters of 4 to 8 microns.

7.33 Upward flow filtration

Upward flow filtration with upflow washing has been used for a few potable water treatment plants in the UK, but its use is more appropriate to industrial water applications or to tertiary sewage filtration where a high standard of filtrate quality is not so important. The principle used in upflow filters is to have progressively finer sand in the direction of flow, which allows the filter to carry a greater load of impurity before backwashing because the larger particles tend to be held in the lower, coarser part of the filter, leaving the upper layers to deal with the smaller particles. However, unless the finer grades of sand are restrained they would be washed away at higher rates of filtration, as well as during the backwashing stage, and consequently designers have introduced a number of techniques to stop this occurring. One method is to use a filtrate collector pipe system just buried in the top layer of fine sand, with strainers located on the side of the pipes so that filtrate water flow has to change from a vertical to a horizontal direction, thus preventing expansion of the sand. During backwashing the filtrate collector is not used and dirty washwater escapes from an elevated trough. A modification to this system has the filtrate collector located about halfway down the sand bed and the inlet water is divided into an upflow stream and a downflow stream. Such a design can be unnecessarily complicated as it means one-half of the filter operates with an increasing sand size in the direction of flow, whilst the other half does the very opposite—consequently each half has its own particular characteristic of headloss build-up. The design of upflow filter most commonly used in the UK is one which contains a grid, square in section, located about 10 cm below the surface of the sand. During filtration the sand arches between individual members of the grid and prevents expansion of the sand, whilst during backwashing the arches are intentionally broken by successive applications of air and backwash water.

Claims have been made that upflow filters can be operated at filtration rates greatly in excess of conventional downflow rates: in practice the rates are not greatly or always in excess of what can be achieved with multilayer filtration (see Sections 7.31 and 7.32). A sequence of upward flow filtration followed by downward flow filtration has also been tried.

7.34 Miscellaneous filters

Radial flow filters are believed to have a potential for development. They operate on the principle of decreasing velocity as water flows through the sand bed and this should, at least in theory, improve filtration efficiency. The best known type operates continuously with unfiltered water flowing radially from the central core of a double walled vertical cylinder through a sand bed to the outer section where it is collected. Sand is continuously removed by a hydraulic ejector from a cone section at the bottom of the cylinder to a separate washing section at the top of the cylinder where it is washed by part of the incoming unfiltered water. The sand then falls back on to the top of the cylinder, the dirty washwater being discharged to waste. Filtration rates for radial flow filters barely exceed 5 m/h, but the fact that they can be made very tall to fit in a small plan area may be attractive where space is limited. The filters are designed as single units or, alternatively, as a group of four units positioned centrally round an inner tank of unfiltered water.

A design of filter mainly to be found in the USA consists of a number of conventional, open type, rapid gravity filters, constructed in concrete, which surround a central control section, constructed in concrete or steel, which houses all the filter controls. The principal difference in operation is that siphons are used for inlet flow control and backwash instead of backwash pumps or an overhead tank. Water is maintained at a constant level by a weir in the effluent chamber and as one of the filter units becomes progressively dirty the level in the filter shell rises until the inlet flow is stopped, either manually or by automatic operation of a siphon break. Washing is usually performed by a separate air application, followed by a water wash using some of the filtered water from the remaining filters in a group and the head created by the outlet weir in the effluent chamber. This head is about 1.5 m above the top of the sand bed which is low for backwashing a sand filter, particularly if it is in a dirty condition. A compact arrangement can be obtained using this design of filter, but it would be prudent not to use it for the treatment of difficult waters because of the backwashing constraints.

Filtration of water and other fluids can be achieved with diatomaceous earth filters, or precoat filters, where the filtering medium is formed on septums or 'candles' of metal or other materials inside a type of pressure vessel. Excellent removal of suspended matter may be achieved. This type of filter is well known for its use in portable plant under wartime and other emergency conditions. In the UK it is sometimes used for swimming baths (see Chapter 8) and for small private supplies, but only occasionally for public water supplies. A good account of the process is given in *Water Quality and Treatment*[23].

Slow sand filtration

7.35 Introduction and history

The subject of slow sand filtration has not been discussed before in this chapter, principally for the reason that the considerable merits of the treatment can be better appreciated after a knowledge has been acquired of the complexities involved in treatment by coagulation and rapid gravity filtration. Slow sand filters were the first effective method devised for the purifying in bulk of surface waters contaminated by pathogenic bacteria. They remain equally effective today and there is a growing number of circumstances which suggest that their use in new works will tend to increase in the future.

'Slow' sand filters are so called because the rate of filtration through them may be only one-twentieth or less of the rate of filtration through rapid gravity or pressure filters. They were first constructed in the UK in the early years of the nineteenth century. Their capacity for purification is well illustrated by the history of the cholera outbreak in Hamburg in 1892. In that year some emigrants for America arrived at Hamburg docks from Russia, but they missed the sailing of their ship and therefore encamped on the banks of the River Elbe. Among these emigrants were some carriers of cholera. The river formed the source of water for Hamburg and it was also the supply to the town of Altona, a short distance downstream. Altona had already installed slow sand filters for the treatment of the river water, but at Hamburg, although filters were being built, they had not been brought into commission. Four days after the emigrants had set up their camp a cholera epidemic broke out at Hamburg during which over 16 000 cases occurred and 8600 people died. At Altona no outbreak of cholera occurred because of the protection given by the slow sand filters.

This history vividly illustrates the ability of slow sand filters to purify a water bacteriologically, as well as physically, and many slow sand filters are in use today. The whole of London's surface derived supplies are treated by slow sand filters, a 490 Ml/d plant being added in 1972[24], and they have also been adopted in works to treat water from Lough Neagh in Northern Ireland. Since the 1930s water treatment by coagulation and rapid gravity filtration tended to outpace slow sand filtration in new plants and, in many cases, slow sand filters were replaced by rapid gravity filters following coagulation.

7.36 Construction and cleaning of slow sand filters

The bed of sand in a slow sand filter is 0.6 to 0.9 m thick and it is laid over a supporting bed of fine gravel, beneath which a collector pipe system is constructed. The water passes downwards through the sand bed, through the gravel, and is collected by the pipes beneath, the whole arrangement being sited in a shallow watertight tank of large size. It is important to note that the bed is 'drowned'; it is not a trickling filter as in sewage treatment works where air is permitted to get to the bed. At the Ashford Common Works of the Thames Water Authority the sand bed is 675 mm thick (similarly at the 1972 Coppermills Works) and lies upon 75 mm of fine gravel which, in turn, rests upon a bed of porous concrete (see Fig. 7.11). Below the concrete, collector drains take the filtered water to the main effluent pipes. The sand grading is uniform: not more than 4% is finer than BS No. 410 mesh 0.3 mm and not more than 4% is coarser than BS No. 410 mesh 1.6 mm. The rate of filtration is 0.198 m/h and each filter constructed is about 90 m long by 34 m wide, capable of dealing with over 15.9 Ml/d.

The raw water is led gently on to the filter bed and percolates downwards. Directly after a bed has been cleaned a head of only 50 to 75 mm of water is required to maintain the design rate of flow through the bed. However, as suspended matter in the raw water is deposited on to the surface of the bed, so a mat (or 'schmutzdecke' as it is called) of organic and inorganic material builds up on the surface of the sand and increases the friction loss through the bed. To maintain the flow rate uniform as far as possible the head across the bed is gradually increased and when this reaches some predetermined value between 0.6 and 0.9 m the bed must be taken out of service and cleaned.

When a sand bed requires cleaning it is drained of water and the top 12 to 25 mm of the sand surface are carefully scraped off. The filter is then returned into service by gradually increasing the flow over 24 hours, sometimes longer. When the sand bed again requires cleaning a further 12 to 25 mm of sand are scraped from the surface, and this process is repeated until the bed is thinned to the minimum practical thickness for efficient filtering. When this stage is reached the bed is then topped up with clean new sand to its original level or the old sand may be replaced if it has been adequately washed and cleaned in a sand cleaning machine.

The interval between cleanings may vary from several months during the winter when prefiltration is installed to 10 days where no prefiltration occurs and algal growth is at a maximum. Originally slow sand filters were invariably scraped by manual labour, but the increased cost and decreased availability of this type of labour has led to the introduction of mechanical methods. Some of these are fairly simple, e.g. using small vehicles with a skimmer bucket discharging to trucks at the end of a filter. One of the difficulties of mechanizing the cleaning of the older slow sand filters is that their sizes and positioning were often not uniform since these factors were unimportant in the days of manual cleaning. However, with new installations, provided that the filters are constructed with the correct dimensions, it is possible to span them with a standard sand lifting bridge which runs on tracks along the sides of the

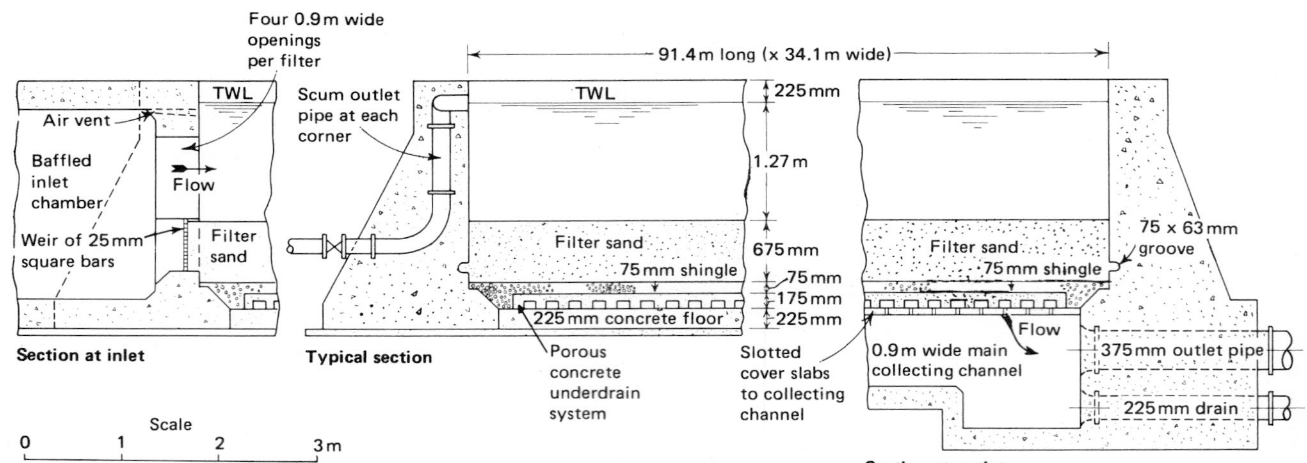

Fig. 7.11 Slow sand filters installed at the Ashford Common Works of the Thames Water Authority, 1958, and similarly at the Authority's Coppermills Works, 1972.

filters, thus greatly reducing the labour necessary for scraping. Grey[25] reports details of such an installation at Lough Neagh.

7.37 Mode of action of slow sand filters

The slow sand filter does not act by a simple straining process. It works by a combination of both straining and microbiological action of which the latter is the more important. The mode of operation is complex and is from time to time a subject of argument between experts. There is no doubt, however, that purification of the water takes place not only at the surface of the bed but also for some distance below. Van de Vloed[26] has given as clear an account as any of the details of the purification process. He distinguishes three zones of purification in the bed: (1) the surface coating, the 'schmutzdecke', (2) the 'autotrophic' zone existing a few millimetres below the schmutzdecke, and (3) the 'heterotrophic' zone which extends some 300 mm into the bed.

When a new filter is put into commission and raw water is passed through it, during the first two weeks the upper layers of sand grains become coated with a reddish brown sticky deposit of partly decomposed organic matter together with iron, manganese, aluminium, and silica. This coating tends to absorb organic matter existing in colloidal state. After two or three weeks there exists in the uppermost layer of the sand a film of algae, bacteria, and protozoa, to which are added the finely divided suspended material, plankton, and other organic matter deposited by the raw water. This skin is called the schmutzdecke and it acts as an extremely fine meshed straining mat.

A few millimetres below this schmutzdecke is the autotrophic zone, where the growing plant life breaks down organic matter, decomposes the plankton, and uses up available nitrogen, phosphates, and carbon dioxide, providing oxygen in their place. The filtrate thus becomes oxidized at this stage.

Below this again a still more important action takes place in the heterotrophic zone which extends some 300 mm into the bed. Here the bacteria multiply to very large numbers so that the breakdown of organic matter is completed, resulting in the presence of only simple inorganic substances and unobjectionable salts. The bacteria act not only to break down organic matter but also to destroy each other and so tend to maintain a balance of life native to the filter so that the resulting filtrate is uniform.

7.38 Use of prefilters with slow sand filters

Slow sand filters may operate as the sole form of filtration or they may be preceded by rapid gravity filters (without the use of coagulants) or by microstrainers. The object of prefiltration is to lighten the load on the slow sand filters and so permit a longer period between cleanings or a faster rate of filtration, or both. A summary of the effects of prefiltration is given by Ridley[27], who compared three London supply works—Hanworth Road with slow filtration only, Kempton Park with rapid gravity filters preceding slow sand filters, and Ashford Common with microstraining preceding slow sand filters. Figures for the period 1 April to 30 September (when algal growths are usually most prominent) for the five years 1959 to 1963 inclusive are shown in Table 7.3.

Table 7.3 Performance of slow sand filters for London water supply (summer seasons 1959–63).

Station and type of prefiltration adopted	Rate of filtration (m/h)	Quantity of water filtered per hectare of sand bed cleaned (Ml/ha)*
Hanworth Road (no prefiltration)	0.004–0.059	315–705
Kempton Park (rapid gravity prefiltration)	0.132–0.152	1260–1600
Ashford Common (micro-strainer prefiltration)	0.132–0.137	1000–1570

* 1 Ml = 10^3 m^3. 1 ha = 10^4 m^2. So 1 Ml/ha = 0.1 m^3/m^2.

During the same period Kempton Park rapid gravity filters used 1.15% to 1.62% washwater (as a percentage of the water filtered) while Ashford Common used 1.64% to 2.03%. The extra washwater used and the capital and running costs of prefiltration are usually taken as economically justified in view of the very considerable increase in water filtered through the slow sand filters between cleanings.

7.39 Limitations and advantages of slow sand filters

Slow sand filters were originally often the only line of defence against pollution and their function as purifiers of water from the bacteriological aspect was paramount. With the introduction of treatment of waters by coagulation and rapid gravity filtration, together with the practically universal use of effective disinfection (usually by chlorine), slow sand filters tended to be ignored when new plants were being considered because of the amount of land and labour they used and their relatively large capital cost. However, there is still a place in water treatment for slow sand filters, providing their advantages and limitations are carefully weighed in any particular case.

Considering the limitations first, slow sand filters do not materially reduce the 'true colour' of a water. (The term 'true colour' may be taken as the colour of the filtrate after removing colloidal clay, if necessary, by passing through a membrane filter of aperture approximately 0.5 μm.) They are thus only suitable for dealing with waters of relatively low colour. What the limiting colour should be depends upon the standards set for the final water, but generally one would say that raw waters of colour 10 to 15 Hazen units would, other things being equal, be quite suitable for slow sand filtration. With a colour of 20 to 25 Hazen units, slow sand filters might still be used, but in this case it would probably be advisable to bleach the colour in the filtrate by chlorination or ozonization. If the colour were higher than 25 to 30 Hazen units one would hesitate to recommend slow sand filters. Another factor is that slow sand filters cannot be expected to be effective in removing any high concentration of manganese in solution, and they are adjudged to have a marked reduction in efficiency for the removal of contaminants at low temperatures.

Slow sand filters are also not very suitable for dealing with any substantial amount of finely divided inorganic suspended matter. With prefiltration using primary filters, however, they may function successfully for years treating a water which is intermittently laden with fine silt. An example is the River Severn, where the city of Worcester was the original user of this river as a source of supply of drinking water. At this time slow sand filters were the only effective method of treating polluted river waters and so slow sand filters were built at Worcester. With the primary filters they are still producing a uniformly satisfactory final water, although all other users of River Severn water now employ chemical coagulation and rapid filtration. In another large plant with a direct river intake the runs on the primary filters were occasionally very short due to silt, but this difficulty could be minimized by the provision of a few days raw water storage.

An important disadvantage discovered at one treatment works was that the filtrate from one treatment stream of slow sand filters contained more nutrient in solution (orthophosphate) than did a parallel stream using chemical coagulation sedimentation and rapid gravity filters. Not all raw waters will necessarily behave in the same way, but nutrient 'slip' can encourage aftergrowths of organisms in the distribution system and is to be avoided if possible.

Apart from the above, the other stated limitations of slow sand filters often appear of doubtful validity on closer inspection and four examples are given to illustrate this. Firstly, it is pointed out that slow sand filters occupy a greater area of land than do coagulation and rapid gravity plants. This is true so long as one ignores the question of sludge disposal, but if *adequate* sludge disposal facilities are included the land required for slow sand filters may well be no greater than that for a coagulation and rapid gravity plant. Secondly, it is maintained that slow sand filters are expensive in capital costs. However, when the cost of chemicals is taken into account, slow sand filters may be cheaper in annual costs than coagulation plants[28]. Thirdly, whereas it was usual to design slow sand filters for the rates shown in Table 7.3, it has been found that, with adequate pretreatment, rates as high as 15 m/d can be used. Fourthly, slow sand filters are said to be labour intensive, but with modern properly designed slow sand filters permitting the latest methods of mechanical cleaning, they may be no more labour intensive than coagulation plants, taking into account the work involved in sludge disposal.

Turning to the advantages of slow sand filters, provided that the water they treat, either directly, following storage, or following rapid gravity filters, has relatively good physical and chemical characteristics then they will produce excellent quality water. In quite a few continental countries[29] they are used as a final polishing stage of treatment, being preceded by as many as five or six stages of treatment including chemical coagulation and sedimentation. To employ slow sand filters in this way is expensive, but it illustrates the value water authorities attach to this type of filter to give the best possible results. Recent studies on the removal of viruses from contaminated reservoir waters indicate their efficiency in this respect, adding to their well proven ability to remove bacteria. It has been reported[30] that tests on experimental slow sand filters gave highly effective removal of viruses at up to 2.5 times the normal filtration rate of 4.8 m per day and at temperatures as low as 5 °C. Some preliminary results of work carried out by the Thames Water Authority also indicate that slow sand filters aided by bankside storage and prefiltration through rapid gravity filters can bring about a very substantial reduction in the number of polynuclear aromatic hydrocarbons.

Finally, the disposal of washwater from slow sand filters (and from their prefilters) is much easier and cheaper than the disposal of chemical sludges from coagulation plants and generally presents little difficulty in relation to amenity considerations.

Disposal of sludge from treatment works

7.40 Types of sludges

Types of sludges may be classified according to the type of water treatment process adopted.

Non-chemical sludges These arise from microstrainers and presettlement clarifiers, and from rapid gravity and slow sand filters which are used for the treatment of low turbidity waters, usually stored waters. These sludges are for the most part fairly inoccuous and can be discharged to a watercourse without treatment; an exception occurs when the raw water has high populations of algae when treatment such as the use of drying beds may be required.

Coagulation by the addition of chemicals Sludges from treatment plants using coagulation, whether by means of aluminium or iron salts, are the most difficult to treat for disposal because of the relatively large volumes involved and the difficulties of dewatering. These are dealt with in detail below.

Softening sludges These arise from lime or lime–soda treatments and their treatment is dealt with in Section 8.7.

7.41 Quantities

Where coagulants are used, the water is usually filtered by rapid gravity or pressure filters. The coagulated water may pass directly to the filters, in which case there is only the sludge content of the filter backwash water to deal with, or there may be a clarification stage in sedimentation tanks before filtration, when there is the sludge drawn off from these tanks to be dealt with as well. Where there is no sedimentation before filtration, the backwash water can range from 1% to 10% of plant throughput, with a quoted average of 2%. With sedimentation, sludge from the sedimentation tanks may range from 0.25% to 1.50% of water treated and an extra 0.5% to 1.0% may be added from the filter backwash water. The actual values are dependent upon raw water conditions and the dosage of coagulants and are usually lower if a coagulant aid is used. Filter backwash water usually contains 0.01% to 0.10% *W/V** dry solids, while sludge withdrawn from sedimentation tanks may contain 0.1% to 2.0% *W/V* solids[31].

For presettlement tanks and clarifiers which have to treat waters with heavy silt loads of over 1000 mg/l, the sludge removed from a clarifier can be well over 10% of throughput, even as high as 25%, and concentration of sludge can be 5% or 6% *W/V* solids. These are exceptional cases where special removal methods are necessary, but it is as well to know that they can occur with some raw waters, especially in overseas countries.

7.42 Methods of disposal

In a well run sedimentation and filtration plant using a coagulant, some 80% to 95% of the suspended matter may be removed in the sedimentation tanks, leaving 20% to 5% to be removed by the filters. Where raw water is scarce, it is usual to recycle the whole of the filter backwash water so that only sludge from the sedimentation tanks needs to be dealt with. Even if raw water is plentiful, it is usually better to keep the filter backwash water separate from the sedimentation tank sludge. This is because there is little point in diluting the sedimentation sludge which has already been partially concentrated in the clarifier. The first treatment of sludge is very often settling, usually aided by slow stirring in a circular tank equipped with fence and picket stirrers. The addition of polyelectrolytes at this stage usually enables the sludge density to be increased. Experiments carried out by the Fylde Water Board showed that 0.03% to 0.04% *W/V* solids could be thickened to 2.5% to 3.0% *W/V* initially, and with the introduction of polyelectrolytes to 5.0% to

* *W/V* = weight per volume (grammes/100 ml).

6.0% *W/V*[32]. At Manchester's Arnfield works[33], with polyelectrolyte dosing and slow stirring, a sludge of 2% *W/V* solids was achieved. A comprehensive review of methods of sludge disposal was published by the Water Research Association in 1970[34].

Although at present some water authorities are discharging some form of sludge direct to watercourses, it is unlikely that with the general move to tighter control of effluents this practice will be permitted indefinitely. In some cases it is possible to discharge waterworks sludge to the public sewerage system where the sewage treatment processes are able to cope with it. Surface spreading on land is another method of disposal which can be very useful for softening sludges, but has limitations for alum and iron sludges. Only a thin layer is possible, otherwise vegetation will tend to be stifled and evaporation will not be effective.

Sludge disposal by lagooning or on shallow drying beds is still in wide use. Dewatering in lagoons is mainly by surface evaporation and some drainage, and over a lengthy period sludges may give up to 10% or 15% solids content, but there is a tendency for a surface crust to form under which the sludge may still be liquid. Final disposal is therefore often a problem. A common fault is to make lagoons too deep; preferably the depth of sludge should not exceed 1 m and the total lagoon area should be sufficient for the lagoon to hold at least six months sludge production. It has been reported[35] that in Florida, USA, artificial lagoons are about 2.5 m or more in depth, but such a depth would not be feasible in the temperate climate of the UK. Shallow drying beds, unlike lagoons, have a permeable base of sand or clinker and an underdrain system and should be filled with not more than about 0.3 m of sludge. Dewatering generally follows a pattern of initial drainage followed by cracking and shrinking, the various steps being very much influenced by weather and the condition of the sand bed. The ability of a sludge to dewater by gravity varies a great deal and is very much dependent upon the characteristics of the individual sludge. Alum sludges usually drain more slowly than do iron coagulant sludges, but it is suggested that the rate of draining for alum can be increased by 50% or more by using a conditioning agent[36].

Filter pressing of sludge is becoming increasingly used for clarification sludges, almost invariably combined with a pretreatment stage for thickening. In the process the sludge is forced by pressure against a battery of filter cloths through which liquor is expressed as filtrate, while the sludge remaining on the filter cloth forms a cake. With clarification sludges it is usual to aim at a solids content of 20% to 25% in the cake, as cake of this solids content can then be disposed of on the land. Tests in Scotland[37] indicate that the choice of filter cloth and the operating pressure used in the press are important to achieve satisfactory results. After exhaustive trials in the USA, filter pressing was the preferred method chosen by both the Eerie County Water Authority for treating sludge from one treatment plant and by the Monroe County Water Authority for treating sludge at a single site from three treatment plants.

Centrifuging has always had a place in the UK for the dewatering of softening sludges and in recent years it has also been used successfully for alum sludges. At the Egham Works[38] of the North Surrey Water Company clarification sludge is first concentrated to about 3% *W/V* by storage in disused slow sand filter beds for about forty days and is then pumped to a centrifuge. Sludge cake of 22% *W/V* consistency is produced after dosing with 125 mg/l polyelectrolyte. Only minor operating problems have occurred with the equipment since 1979 and the capitalized cost of centrifuge treatment is some 0.14 pence per m^3 of treated water or about 0.38 pence per m^3 for the whole sludge operation, which is approximately 5% of the treated water costs. Freezing and thawing have also provided an effective method of dewatering clarification sludges, but the method is more appropriate to countries with severe winter conditions than to the UK. Vacuum filtration has not been used very often in this country except for dewatering lime softened sludges, although rotary drum vacuum filters have been used in the USA for dewatering alum sludges.

It seems certain there will be a continual striving towards finding new and more efficient methods of sludge dewatering, the impetus being the need in many countries to comply with more stringent standards of discharge to watercourses. A new method used in Japan for the treatment of alum sludge is the pellet flocculation process[39]. Here the sludge is subjected to successive stages of chemical treatment and thickening using sodium silicate and a polymer, followed by further thickening in a machine called a dehydrum with the possibility of adding oil-fired drying as a final dewatering stage. It is stated that a sludge of 25% to 30% solids can be produced without the need for further mechanical dewatering equipment. The high rate lamella sedimentation tanks referred to in Section 7.16 claim sludge concentrations of up to 15% *W/V* direct from the tanks themselves which is far better than can be achieved with conventional thickening tanks. Finally, a new process[40] known as Sirofloc and developed in Australia for the treatment of highly coloured organic waters goes a long way towards eliminating the production of any sludge. In the process a slurry of magnetite particles previously activated with caustic soda is used to absorb colloidal and coloured organic contaminants. After absorption the slurry is passed between magnets to separate the sludge from clear water. This magnetite sludge is then regenerated with caustic soda before reuse.

7.43 Alum recovery from sludge

The recovery of aluminium sulphate from sludge is considered under the 'disposal of sludge' heading as it not only accomplishes reuse of the coagulant chemical but it also substantially reduces the volume of sludge to be disposed of from a treatment works. Recovery using sulphuric acid was investigated by Issac and Validi[41]. A plant for recovery, for the purpose of simplifying the disposal of sludge, is in use at the Daer works of the Lanarkshire Water Board[42] and, although constructed as long ago as 1969, this plant is probably still the most complete treatment for a clarification sludge at present in operation in the UK. It involves freezing and consists of: (1) settlement of the sedimentation sludge and filter backwash water in sludge holding tanks; (2) dosing with sulphuric acid to reduce the pH to 3.5 followed by slow stirring, the acid alum liquor then gravitating to a storage tank; (3) further thickening in separator tanks with the alum liquor gravitating to the alum storage tank; (4) freezing and thawing of the acidified sludge with more alum liquor recovered, and final disposal of the sludge on to an adjacent hillside covered with trees. By this means, for an output of 140 Ml/d, some 0.45 Ml/d of sludge from the sedimentation tanks and filter backwash water containing not more than 0.5% *W/V* solids is reduced to 24 000 l/d (24 m^3/d) of approximately 4% to 5% *W/V* solids by processes (1), (2) and (3), whilst process (4) further reduces the quantity to a maximum of 5400 l/d in the wet state which will dry out to about 2.8 m^3 of granular solid material. The Daer plant has the extra stage of freezing, but most conventional alum recovery plants will have the three basic steps of thickening, including acidification with sulphuric acid at a pH between 2 and 4 and separation of liquid alum usually by some mechanical dewatering device.

Alum recovery has the added advantage of reduction to sludge volume so the economics of partial or complete recovery are now being examined much more closely, particularly in Europe and the USA. For small and medium sized treatment works up to about 90 Ml/d it is usually cheaper to purchase new alum, although the extra cost of recovery may well be offset by less difficult disposal problems. Above this capacity it is sensible to carry out a full economic appraisal, especially in some overseas countries where alum has to be imported, but then other factors such as additional costs of equipment and operation have to be assessed. One treatment plant manufacturer in the UK has designed a full scale pilot plant for concentrating and acidifying alum sludge with direct recycling to the clarifier inlet and results so far are promising. The system avoids the use of a lot of extra mechanical plant. In the USA research is proceeding on a method for alum recovery using a liquid ion-exchange solvent extraction process[43]. The quality of a raw water can also have a marked influence on whether alum recovery is feasible. Waters containing high concentrations of iron and manganese may not be suitable as the reuse of the alum recovered from this sludge will tend to build up to unacceptably high concentrations which will not be removed in the treatment processes. Low to moderate concentrations should be acceptable. Similar considerations apply to coloured waters; raw waters with wide variations in quality may also be unsuitable for alum recovery, particularly where the parameters of turbidity and colour vary substantially. The most suitable water is likely to be one whose quality remains fairly consistent throughout the year, with low concentrations of colour, iron, and manganese.

References

(1) Sykes G and Skinner F A (Eds), *Microbial Aspects of Pollution*, Academic Press, 1971, p. 160.
(2) Young E F, Wallingford F E and Smith A J E, Raw Water Storage, *JSWTE*, **21**, 1972, p. 127.
(3) Lund J W G, Mackereth F J H and Mortimer C H, *Phil. Trans. Roy. Soc.*, **B246**, 1963, p. 260.
(4) Windle Taylor E, Reports on the Bacteriological and Biological Examination of the London Waters, Metropolitan Water Board, 1961–68.
(5) Ridley J E, Cooley P and Steel J A P, Control of Thermal Stratification in Thames Valley Reservoirs, *JSWTE*, **15**, 1966, p. 225.
(6) Camp T R, Sedimentation and the Design of Settling Tanks, *Trans. ASCE*, **III**, 1946, p. 898.
(7) Parker P A M, *The Control of Water*, Routledge & Kegan Paul, 1949, p. 555.
(8) Reference 6, p. 913.
(9) McLaughlin R T, The Settling Properties of Suspensions, *JASCE*, HY12, Paper 2311, December 1959.
(10) Camp T R and Stein P C, Velocity Gradients and Internal Work in Fluid Motion, *J Boston Soc. Civil Eng.*, **30**, 1943, p. 219.
(11) Packham R F, *JSWTE*, **11**, 1962, pp. 50 and 106, and **12**, 1963, p. 15.
(12) US Environmental Research Agency, Office of Research and Development, *Manual of Treatment Techniques for Meeting the Interim Primary Drinking Water Regulations*, US Government Printing Office, Washington, 1978.
(13) Capacity and Loadings of Suspended Solids Contact Units, Committee Report, *JAWWA*, 1951, p. 263.
(14) Hazen A, On Sedimentation, *Trans. ASCE*, **53**, 1904.
(15) *Water Clarification by Flotation*, WRC Technical Report TR 114, April 1979.
(16) Packham R F, Polyelectrolytes in Water Clarification, *JSWTE*, **16**, 1967, p. 88.
(17) Atkinson J W, Hilson M A, Bell F and Hunter R W, Papers on Practical Experience in the Use of Polyelectrolytes, *JSWTE*, **20**, 1971, p. 165.
(18) Ives K J and Gregory J, Basic Concepts of Filtration, *JSWTE*, **16**, 1967, p. 147.
(19) Ives K J, Theory of Filtration, *Proc. IWSA Congress*, **1**, p. K.1, 1969, also Filtration: the Significance of Theory, *JIWE*, **25**, February 1971, p. 13.
(20) Miller D G, Rapid Filtration Following Coagulation Including the Use of Multi-layer Beds, *JSWTE*, **16**, 1967, p. 192.

(21) Miller D G, Filtration: Two Experimental Developments, *JIWE*, **25**, 1971, p. 21, also remarks on p. 49 by Carr W and p. 62 by Jeffery J.
(22) Crowley F W and Twort A C, Current Strategies in Water Treatment Developments, *Proc. ICE Conf. Water Resources—A Changing Strategy*, 1979.
(23) AWWA, *Water Quality and Treatment*, 3rd Ed., McGraw-Hill, 1971.
(24) The Coppermills Works of the Metropolitan Water Board, *WWE*, August 1972, p. 277.
(25) Grey I W, Lurgan and District Waterworks Joint Board's New Works at Castor Bay, *WWE*, July 1971, p. 297.
(26) Van de Vloed A, Report to the 3rd Congress of the IWSA, Subject No. 7, 1955.
(27) Ridley J E, Experiences in the Use of Slow Sand Filtration, Double Sand Filtration and Microstraining, *JSWTE*, **16,** 1967, p. 170.
(28) English E, Water Treatment Problems at Lough Neagh, *JIWE*, June 1972, p. 201.
(29) Schalekamp M, The Development of the Surface Water Treatment for Drinking Water in Switzerland, *Proc. IWES Symposium: The Water Treatment Scene—The Next Decade*, 1979.
(30) Poynter S F B and Slade J S, The Removal of Viruses by Slow Sand Filtration, *Prog. Wat. Tech.*, **9**, Pergamon Press, 1977, p. 75.
(31) Russellman H B, Characteristics of Water Treatment Plant Wastes, *Proc. 10th San. Eng. Conf.*, 1968, pp. 10–20, Urbana Illinois University College of Engineering, 1968.
(32) Hilson M A, *Sludge Conditioning by Polyelectrolytes*, Fylde Water Board, Blackpool, 1970.
(33) Sankey K A, The Problem of Sludge Disposal at the Arnfield Treatment Plant, *JIWE*, **21**, 1967, pp. 367–384.
(34) Chappell T W J and Burley M J, A Review of Sludge Treatment and Disposal Practice in the Water Industry, Technical Paper TP 81, WRA, Medmenham, 1971.
(35) Reh C W, Disposal and Handling of Water Treatment Plant Sludge, *JAWWA*, February 1980, p. 115.
(36) Novak J T and Langford M, The Use of Polymers for Improving Chemical Sludge Dewatering on Sand Beds, *JAWWA*, February 1977, p. 106.
(37) *Dewatering of Waterworks Sludge*, Engineering Division, Scottish Development Department, September, 1977.
(38) Personal Communication from North Surrey Water Co.
(39) Pigeon P E, Linstedt K D and Bennelt E R, Recovery and Release of Iron Coagulants in Water Treatment, *JAWWA*, July 1978, p. 397.
(40) Clayton R C, Sirofloc—A Novel Technique for Water Clarification, *Proc. ICE Symposium*, Series No. 59.
(41) Isaac P C G and Validi I, The Recovery of Alum Sludge, *Proc. SWTE*, **10**, 1961, pp. 91–117.
(42) Webster J A, Paper on the Recovery of Aluminium Sulphate from Sludge at Daer Waterworks, presented at the 60th AGM of the British Waterworks Association, June 1971, Lanarkshire Water Board, 1971.
(43) Westerhoff G P and Cornsell D A, A New Approach to Alum Recovery, *JAWWA*, December 1978, p. 709.

8

Softening and miscellaneous water treatment methods

Softening of water

8.1 Hardness compounds

Certain materials in water react with soap, causing a precipitation which appears as a scum or curd on the water surface. Until enough soap has been dissolved to react with all these materials, no lather can be formed. A water which behaves like this is said to be 'hard'. When synthetic detergents are used instead of soap, the inconvenience of a hard water is not so strongly felt because the detergents do not cause a precipitation of the hardness compounds. Nevertheless there are still many objections to a hard water, the principal one being that when the water is heated in boilers and heating systems, the hardness compounds precipitate to form a hard scale on the surface of the boiler and the interior of pipes. The portion of the total hardness which is deposited when water is boiled is called temporary hardness; the remainder is called permanent hardness or more exactly non-carbonate hardness. The hardness compounds are as given in Table 8.1—from which we see that calcium and

Table 8.1

Causing temporary hardness (carbonate hardness)	Causing permanent hardness (non-carbonate hardness)
Calcium bicarbonate $Ca(HCO_3)_2$	Calcium sulphate $CaSO_4$
Magnesium bicarbonate $Mg(HCO_3)_2$	Magnesium sulphate $MgSO_4$
	Calcium chloride $CaCl_2$
	Magnesium chloride $MgCl_2$

magnesium are the cause of hardness, the bicarbonates producing temporary hardness and the sulphates and chlorides (and nitrates) permanent hardness.

A large proportion of waters from underground sources are hard, particularly waters from the chalk and limestone which are of major importance. The carbonate hardness, e.g. about 250 mg/l expressed as $CaCO_3$, results from the solution of calcium and magnesium carbonates by carbon dioxide which is formed in soils by the oxidation of organic matter. A major source of non-carbonate hardness in surface waters is the calcium sulphate present in clays and other deposits. However, a large proportion of the surface runoff from older geological formations in the western and northern areas of the UK is soft or very soft, e.g. 15 to 50 mg/l, because the rocks are impermeable and insoluble.

In recent years a striking statistical relation has been found between soft waters and mortality from cardiovascular diseases[1,2]. Although no causal connection has been proved, this finding at present indicates that public water supplies should not be softened or should only be partially softened if they are very hard. In the light of existing knowledge it is difficult to know how far to soften a hard water: for the time being, until more medical evidence is available, most authorities consider a total hardness of 150 mg/l as $CaCO_3$ to be acceptable.

The hardness of a water does not detract from the palatability of a water—on the contrary many people consider that the sparkling clear hard waters obtained from the chalk are second to none for drinking. Hardness can be measured by the 'soap destroying power' of a water, but more precisely by the total of the hardness-producing constituents expressed as the chemically equivalent amount of calcium carbonate* ($CaCO_3$). The softness or hardness of a water, as experienced by the consumer in relation to the mg/l equivalent amount of $CaCO_3$, is set out in Section 6.20. Hardness has also been measured in the past as 'degrees Clarke' which in fact are grains per gallon or parts per 70 000. Thus 100 mg/l hardness is equivalent to 7 degrees Clarke. In the majority of waters the greater portion of the hardness is temporary. When a water is excessively hard there is usually a substantial proportion of permanent hardness as well.

8.2 Means for removal of hardness

There are two different methods available for changing a hard water into a soft water. In the first method chemicals (lime and soda) are added to the water which change the hardness compounds so that they become insoluble and precipitate, the water is then sedimented and filtered to remove the precipitate. In the second method the nature of the hardness compounds is changed by passing the water through a bed of ion-exchange resins so that the changed compounds do not react with soap and the water therefore appears soft. The difference between the two processes should be noted: in the former process of chemical addition the hardness compounds are *removed*; in the latter process they are *changed* by substituting non-hardness compounds. The result of this is that the total dissolved solids are reduced with the lime and soda process, but they are not reduced by the ion-exchange process. This difference is of importance to many industrial users of water who frequently wish to have a water with as low a total dissolved solids content as possible.

8.3 Lime–soda reactions in softening

In this process the aim is to make the calcium and magnesium contents

* Calcium *carbonate* is only slightly soluble in water and any excess above about 15 to 25 mg/l, depending upon the pH value of the water, will precipitate out.

of the hard water take their insoluble forms so that they precipitate out and the remainder can then be removed from the water by the use of filters. The insoluble forms of calcium and magnesium are:

calcium carbonate $CaCO_3$ (partially soluble)

magnesium hydroxide $Mg(OH)_2$

Both these compounds are insoluble in cold water. Magnesium carbonate ($MgCO_3$), unlike calcium carbonate, does not precipitate in cold water. The chemical additions for the removal of the various kinds of hardness are as follows (the compounds underlined are those precipitated).

To remove calcium temporary hardness add LIME.

calcium bicarbonate + lime = calcium carbonate + water

$$Ca(HCO_3)_2 + Ca(OH)_2 = \underline{2CaCO_3} + 2H_2O$$

To remove calcium permanent hardness add SODA ASH.

calcium sulphate + soda ash = calcium carbonate + sodium sulphate

$$CaSO_4 + Na_2CO_3 = \underline{CaCO_3} + Na_2SO_4$$

calcium chloride + soda ash = calcium carbonate + sodium chloride

$$CaCl_2 + Na_2CO_3 = \underline{CaCO_3} + 2NaCl$$

To remove magnesium temporary hardness add LIME + LIME.

Stage (1)

magnesium bicarbonate + lime = magnesium carbonate + calcium carbonate + water

The calcium carbonate precipitates, but the magnesium carbonate does not, so we add further lime for Stage (2).

Stage (2)

magnesium carbonate + lime = magnesium hydroxide + calcium carbonate

The magnesium hydroxide and the calcium carbonate precipitate. The chemical equations for these two reactions are:

(1) $Mg(HCO_3)_2 + Ca(OH)_2 = MgCO_3 + \underline{CaCO_3} + 2H_2O$

(2) $MgCO_3 + Ca(OH)_2 = Mg(OH)_2 + \underline{CaCO_3}$

To remove magnesium permanent hardness add LIME + SODA ASH.

$$\begin{Bmatrix}\text{magnesium chloride}\\ \text{magnesium sulphate}\end{Bmatrix} + \text{lime} = \begin{matrix}\text{magnesium}\\ \text{hydroxide}\end{matrix} + \begin{Bmatrix}\text{calcium chloride}\\ \text{calcium sulphate}\end{Bmatrix}$$

$$\begin{Bmatrix}MgCl_2\\ MgSO_4\end{Bmatrix} + Ca(OH)_2 = \underline{Mg(OH)_2} + \begin{Bmatrix}CaCl_2\\ CaSO_4\end{Bmatrix}$$

To complete the softening we must add soda ash to convert the calcium chloride and the calcium sulphate to calcium carbonates, as in the case of the removal of permanent calcium hardness mentioned above, e.g. $CaCl_2 + Na_2CO_3 = CaCO_3 + 2NaCl$.

In most waters hardness due to calcium is predominant. The addition of lime alone removes only the carbonate hardness portion of the calcium hardness. Soda ash is required to remove the non-carbonate hardness. Often the addition of lime alone will achieve sufficient reduction of hardness because in many waters the greater portion of the hardness is temporary.

8.4 Application of the lime–soda process

Although the lime–soda process was formerly used for the softening of public water supplies, few plants of this type have been installed in recent years. The process is more frequently used for the softening of waters for industrial use. After adding lime the water is coagulated, clarified, and filtered by rapid gravity filters using plant similar to that described in Sections 7.11–7.29 for the clarification of river waters. In order to precipitate the calcium and magnesium more lime must be added than is strictly required by the chemical equations for softening. This normally results in coagulation and precipitation several times faster than clarification without softening, but is a slower process when magnesium hardness must be removed also. It is sometimes found helpful to return some of the precipitant sludge to the incoming raw water as this assists in the quick formation of a heavier floc.

In the case of a water requiring clarification for the removal of colour and suspended solids etc. and also softening, as for example a hard river water, then both stages of treatment are sometimes completed in the same reaction tank. However, due to the complex nature of softening this is not always possible or desirable and two-stage treatment in series may be necessary or split treatment may be adopted in which only part of the water is softened.

8.5 Stabilization after lime–soda softening

The softening reactions and precipitation are not usually completed in the clarification tanks and therefore the water tends to form further deposits, mainly of calcium carbonate, in the rest of the treatment plant. If calcium carbonate deposits only slowly on the filter sand without cementing it together in lumps this can be an acceptable method of stabilizing the water, but the filter sand will need renewal every few years. The more acceptable method of stabilization is to inject carbon dioxide into the water. Carbon dioxide can be derived from the burning of coke or by using submerged combustion equipment permitting mixtures of gases such as propane and air to be burned under water, but direct injection of carbon dioxide via diffusers is simpler using cylinders of liquified carbon dioxide (see Section 8.35). Sulphuric acid is used in some waterworks, but it converts carbonate to sulphate and care must be taken to ensure that it does not reduce the final calcium carbonate in the water too low (or reduce the pH too low) as this might render the water corrosive. Another method of avoiding after-precipitation is to add a few mg/l of a suitable polyphosphate such as 'Calgon'.

8.6 Form of chemicals used

Lime is usually used in the form of hydrated lime which is a dry powder and can be fed by 'dry-feed' equipment or added as a made-up slurry. It is only two-thirds to three-quarters as effective per kilogramme weight as is quicklime, but the latter needs slaking before use and does not keep so well. Dosages of lime for softening are high, of the order of 100 to 200 mg/l. Soda ash (sodium carbonate) can also be added to the water as a dry feed or a solution.

8.7 Sludge disposal

Probably the greatest disadvantage of the lime–soda process of softening is the large amount of sludge that results and the difficulty of its disposal. The problem is acute for inland softening plants and an extremely difficult one to solve when the plant is in a built-up area.

The possible methods of disposal are as follows.

(1) Direct disposal of wet sludge to a watercourse, river, or the sea.
(2) Drying of the sludge in open lagoons and carting away dried material to tips, e.g. in quarries. This material can only be said to be relatively dry.
(3) Efficient drying of the sludge by mechanical means so that it may be sold as fertilizer or used in industry.
(4) Recalcining the sludge so that it may be reused for softening, the surplus being sold as fertilizer or for industry.

Direct disposal is virtually prohibited to all rivers and streams in the UK. Riparian owners downstream of the point of disposal would suffer deterioration in the quality of the water and the amenities of the river would be affected by unsightly deposits of lime along the banks. For lagooning a large area of land may be required by a large undertaking. At least two lagoons will be necessary: one being filled and the other drying out. Even so the dampness of the British climate may prevent the final material from improving beyond a sticky mess which is difficult to handle. To some extent the use of different coagulants in the water treatment controls the physical characteristics of the sludge, e.g. activated silica gives a more granular product. However, the method is probably the most economic provided that sufficient land can be found for the lagoons and there is a place for the final disposal of the dried sludge. In some works it may be possible to dry the sludge by vacuum filters or filter pressing; alternatively centrifuging may be employed. The resulting sludge may be useful for agricultural purposes, but for industrial purposes it must contain a high percentage of calcium carbonate, very little magnesium, and no alum or silt. Such pure material is obtained from a few plants softening chalk well waters and is dried and sold for high grade products such as cosmetics. If the softening plant is large it may not be possible to find a market for all the dried sludge produced. Recalcining the sludge is practicable for some large undertakings provided that the resulting material is sufficiently pure. There is still, however, a good proportion of the material to dispose of in any case.

8.8 The 'Spiractor' or pellet reactor process of softening

This is a clever adaptation of the lime or lime–soda process. The hard water is injected with lime and is then passed to the base of an inverted conical vessel through which the water flows spirally upwards at a decreasing velocity. Within the vessel are grains of sand which act as a catalyst. The grains are held in suspension by the upward flow of the water and the calcium carbonate causing temporary hardness is deposited on to them so that they gradually increase in size until they can no longer be held in suspension and fall to the base of the tank. The pellets are about 2 to 3 mm diameter and are easily disposed of. The supernatant water, if not perfectly clear, may be sand filtered; if removal of permanent hardness is required in addition, the water may be passed through a base-exchange softener (see Section 8.9). The process is quick and simple and is particularly useful for the removal of calcium temporary hardness[3,4].

8.9 Base-exchange softening

In the base-exchange process of softening the total dissolved solids in the water undergo little change. The calcium and magnesium are 'exchanged' for sodium. When this happens the water no longer contains those salts which react with soap to form a curd or scum and it therefore feels and acts as soft water. The exchange is achieved by passing the water through a granular bed of special material which has the ability to exchange ions. These materials were formerly loosely called 'zeolites', but are nowadays usually resins which, when used in the sodium form for softening, have the property of removing calcium and magnesium and substituting sodium. When the resin is exhausted it is then regenerated by passing through the bed a strong sodium concentration in the form of brine. The reverse action then takes place, the calcium and magnesium being released from the resin and the sodium being substituted. The brine wastewater is very hard with a high concentration of dissolved salts and its disposal may therefore present problems.

The ion-exchange materials formerly used and known as zeolites were mainly processed greensand or synthetic zeolites made from sulphonated coals and condensation polymers. These have almost entirely been superseded by ion-exchange resins which are cross-linked polystyrene spherical particles or 'beads', sulphonated in the case of cation exchangers and aminated for anion exchangers. There are many different types but it is usual to describe them either as strong or weak acid in the case of cation resins, or as strong or weak base in the case of anion resins. Many resin types now exist commercially, each type having been developed for a particular reason, as for example the macroreticular resin which has a high resistance to organic fouling. A recent development is the use of polyacrylamide resins which are claimed to be superior in performance to those made from polystyrene.

The 'exchange capacity' of a base-exchange material is most often stated in terms of the hardness removed by a specific volume of resin, for example 1 kg of $CaCO_3$ per m^3 of resin. A second measure of performance in base-exchange softening is the amount of salt that must be used in regeneration per unit of hardness removed. The theoretical figure for regeneration is 11 g of salt per 100 g of $CaCO_3$ (or its equivalent) removed, but an excess of salt is required in practice.

8.10 Plant for base-exchange softening

The ion-exchange softening media are usually contained in a cylindrical steel vessel, the size of which depends upon the rate of flow of water to be softened. For public waterworks the size and construction is almost identical with that of a pressure filter (see Fig. 7.10), being about 3 m diameter. The bed of the media is usually about 1.2 m thick, resting upon supporting layers of gravel in which are embedded the collecting effluent pipes. Flow during softening is usually downwards. In regeneration a diluted brine solution is passed through the bed, also downwards, followed by sufficient washwater to wash the brine out of the resin. When the salt content of the effluent diminishes to a permissible level the unit is put into service. The brine solution is made up from common salt which must, however, be completely free from any pollution.

The process is simple to operate and can easily be made automatic, as shown in Fig. 8.1. Also the process can be conducted under pressure, so avoiding repumping of the water after softening. The loss of head is usually between 4 and 5 m through the bed. However, the process is not suitable for a turbid water or for one which contains more than a certain amount of iron.

8.11 Demineralization of water by ion exchange

The base-exchange process of softening described above is only a particular example of ion-exchange treatment and is more specifically an example of cation exchange with the resin in the sodium form.

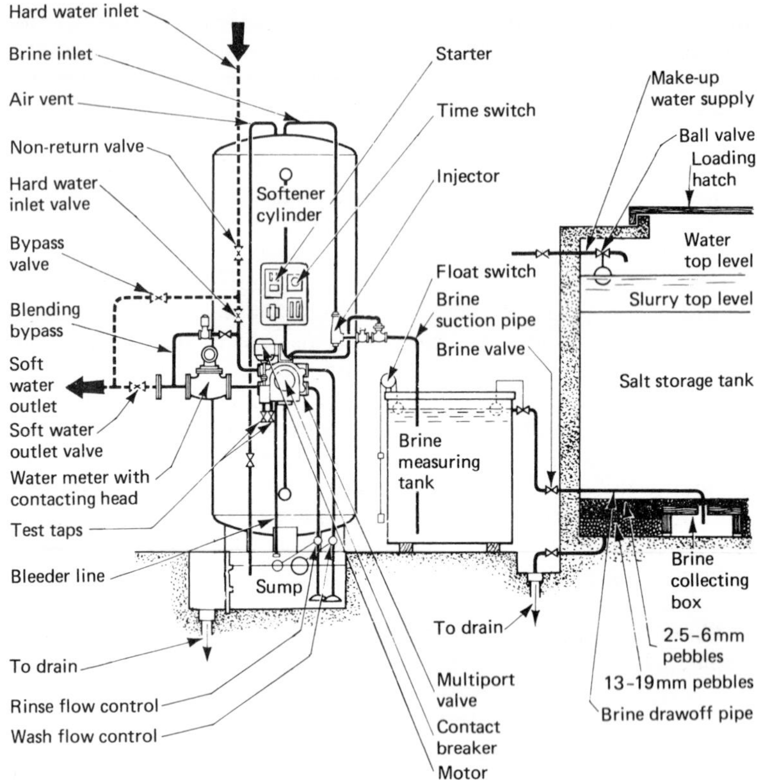

Fig. 8.1 Diagrammatic arrangement of an automatic water softener. (Permutit-Boby Ltd.)

An alternative form of cation exchange is a process in which the calcium, magnesium, sodium, and potassium are all replaced by hydrogen ions: the resin is then regenerated with hydrochloric or sulphuric acid. After this the same water can be treated by an anion-exchange process in which the chloride, sulphate, and nitrate are removed. This anion-exchange resin is regenerated with sodium carbonate or caustic soda solution. Carbon dioxide formed from the bicarbonate is often removed by 'degassing' or by aeration as an intermediate stage between the cation and anion vessels.

A 'mixed bed' base-exchanger containing both cation-exchange and anion-exchange resins as described above is widely used as a final stage to produce a water having total dissolved solids as low as, or lower than, ordinary distilled water. The technology of ion-exchange processes is now very complex with numerous variations in design and detail according to the precise quality of water required. The combination of procedures for removing all the mineral constituents is usually referred to as 'demineralization'.

It is common practice to examine demineralization plant proposals in great detail before making a final selection, taking into account both capital plant cost and running costs. Optimization of running costs, sometimes by computer analysis, can lead to a change in resin-type selection because of a better exchange capacity or a more economical use of regenerant. The cost of ion-exchange treatment increases in proportion to the concentration of constituents in solution. In some cases the cost is increased by the effect on some of the anion-exchange resins of organic materials in the water, including organic acids derived from peat or other natural sources. It follows that in some circumstances there are industries requiring to demineralize a public water supply for use in boilers or for other processes which are more critical of the quality of the public supply than are the domestic consumers. However, with the usual range of mineral and saline constituents in public water supplies, demineralization is not necessary for domestic purposes and would not be economic. In circumstances where freshwater supplies are not available dimineralization can in theory be applied to brackish waters but, because of the expense, the process only finds application for waters having less than about 1500 mg/l dissolved solids. For waters having dissolved solids greater than this the ion-exchange process would be too expensive.

Iron and manganese removal

8.12 General

Traces of iron and manganese are found in many waters. Occasionally quantities may range up to 20 mg/l of iron and up to 5 mg/l of manganese, but at these high quantities it is usual to find that most of the metals are in suspension so that they may be relatively easily removed. It is the dissolved iron and manganese which can be troublesome and the disadvantages arising from their presence above certain levels are described in Sections 6.22 and 6.25 above. Removal is therefore often necessary.

8.13 Removal of iron and manganese from underground waters

In Chapter 7, discussing the use of ferrous sulphate as a coagulant, it was mentioned that ferrous hydroxide is much more soluble than is the more highly oxidized compound ferric hydroxide. The very large difference is expressed quantitatively by the solubility products* which are 1.6×10^{-14} and 1.1×10^{-36} respectively. When iron occurs in

* The solubility product is the product of the molar concentrations of the ions concerned.

underground waters it is usually in solution in the ferrous form in a water which is devoid of oxygen. Such waters are fairly common in water-bearing formations which are underneath an impermeable stratum, e.g. chalk or greensand and other sand formations in situations where they are covered by a clay formation.

Many, but not all, waters from deep boreholes in sands thus contain ferrous iron in solution. When a sample is first drawn it may appear perfectly clear, but after it has been exposed to air for some time it gradually acquires a turbid appearance, and after a longer time a brown precipitate of ferric hydroxide is formed. These waters may also be slightly acid in reaction due to their content of free carbon dioxide, but the solubility of ferrous hydroxide is sufficient for the iron to remain in solution even if the water is neutral in reaction. (At a pH above 8 ferrous iron is likely to precipitate in the form of ferrous carbonate which is less soluble than is the hydroxide[5].) To remove iron from these waters it is therefore only necessary to add sufficient air to provide the amount of oxygen required to allow the ferrous iron to oxidize to ferric iron, and to provide a suitable filter or other situation in which the ferric hydroxide can be removed, leaving a clear water. As the reaction of oxygen with the ferrous iron in solution tends to be slow, it is necessary either to allow time for it to take place in a sedimentation tank or to encourage the reaction to proceed more rapidly on the surface of sand or other material which tends to act as a catalyst as well as retaining the precipitated iron.

When the amount of iron in a water of this type is fairly low, i.e. less than about 1 mg/l, the removal of iron is often achieved in a pressure filter containing the usual bed of sand, but sometimes incorporating a layer of proprietary material such as 'Polarite' or an iron ore which acts as a catalyst. As the amount of oxygen required is small, it is provided merely by introducing air into a space at the top of the filter shell. Much faster rates can be used for this type of filtration than are possible for chemical coagulation preceding sand filters: rates of over 50 m/h have been successfully used. As an alternative to sand filtration, it is sometimes possible to use semicalcined dolomitic limestone (trade name 'Akdolit') as the filter medium for iron removal. This material can be effective for the removal of low concentrations of soluble iron by precipitating ferric hydroxide at its alkaline surface and the material can also help to render the water less aggressive. For manganese removal a filter can be used containing manganese prepared filter material, such as 'Akdolit-mangan', with provision for dosing potassium permanganate as well as to improve oxygenation. An alternative to a pressure filter comprises trays of coke, pumice, or ceramic rings or other material over which the water is allowed to trickle. The effective aeration achieved in such a plant also assists in the removal of free carbon dioxide.

When the amount of iron in a well water is larger, up to the unusually high figure of 20 mg/l which was mentioned above, it is more usual to provide an aerator as the first stage, lime is sometimes added to raise the pH, and then a sedimentation tank is provided to allow a large proportion of the ferric hydroxide to settle out so that it may be removed, leaving only a relatively small amount to pass on to the filters which form the final stage in treatment. Figure 8.2 shows a plant for the removal of iron and manganese, commencing with aeration, followed by chlorination, passage over Polarite, and followed by sand filtration. Although ferrous iron can be oxidized very rapidly by chlorine, this reaction alone may give a colloidal solution and in practice chlorine is not widely used in iron removal. An exception to this case occurs where the iron is present as an organic complex, in which case it may be necessary to use chlorine to release the iron before aeration.

Manganese in any appreciable amount occurs in only a minority of those raw waters which contain iron, although there are exceptions to this. A recent survey of 34 wells in the Delta region of Egypt, where there were widespread complaints of iron and manganese discoloration and taste, showed that the iron concentration was never more than 1 mg/l, but manganese was present up to 1.5 mg/l and proved the more troublesome of the two impurities.

Occasionally manganese is found without iron. The basic chemical consideration in its removal is that the hydroxide of manganese in its lowest state of oxidation has an appreciable solubility similar to that of ferrous hydroxide. Oxidation to the insoluble manganese dioxide is therefore necessary, but this oxidation is very slow under most conditions at pH values below 9.5. Manganese is not removed as easily as iron, but is often removed from underground waters in iron removal plant when it is present in a small proportion with the iron.

In dealing with a well water with very high contents of iron (over 10 mg/l) and manganese (over 1 mg/l) Bolas[6] described a sludge blanket process in which the complete oxidation and removal of the manganese was achieved by the addition of potassium permanganate. This compound contains manganese in its highest state of oxidation and is particularly effective in oxidizing manganous compounds to the insoluble manganese dioxide, but the dose required in this plant was very much lower than the theoretical and indicated also a catalytic action in the sludge blanket.

8.14 Removal of iron and manganese from river and reservoir waters

As river and reservoir waters frequently receive treatment which includes rapid sand filtration, often preceded by coagulation and sedimentation, the removal of iron and manganese is normally included, when necessary, in the same plant. Most river waters used as sources for water supplies are well oxygenated, if not saturated with oxygen. Usually, therefore, iron and manganese in such river waters are present

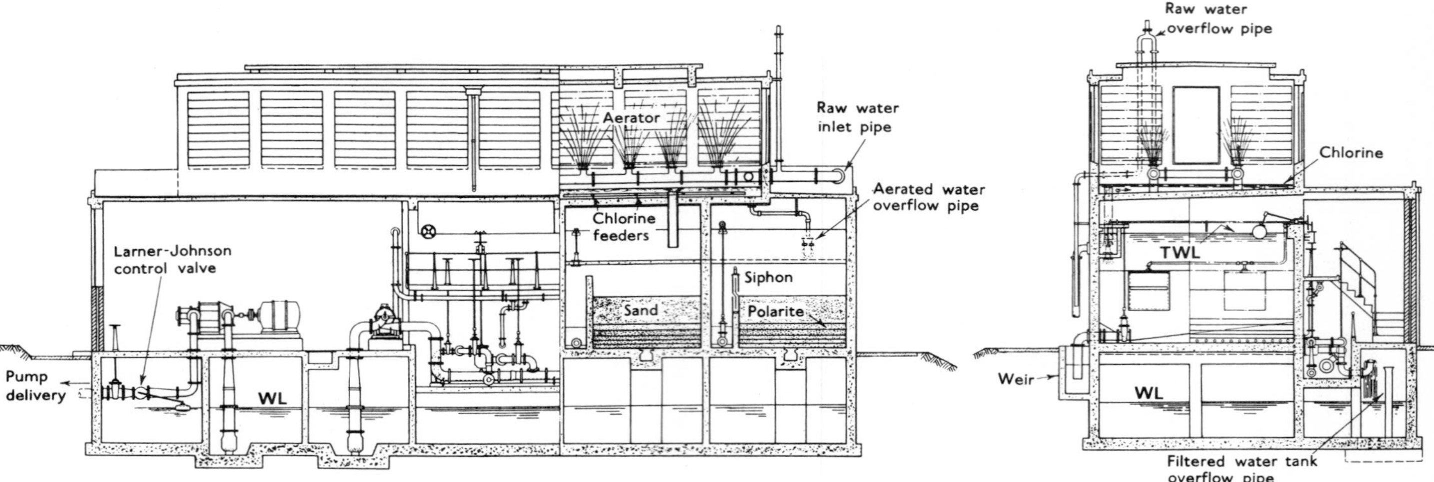

Fig. 8.2 Iron and manganese removal plant at the New Brancepeth Works for Sunderland and South Shields Water Company. (Carey and Mawson, *Proc. ICE*, 1951.)

in the suspended matter in insoluble form and are necessarily removed by the filtration treatment. However, this does not always apply.

The circumstances in which the removal of iron and manganese from a surface water require special consideration are generally when the water is drawn from a reservoir in which these elements have been dissolved in the bottom water under conditions of deoxygenation, as described in Section 7.1. Generally the iron is fairly easily oxidized, but sometimes the iron and, more often, the manganese are combined with organic matter in a very stable form. One method by which both the iron and the manganese may be removed is by applying the lime–soda softening process. Sometimes it has been found that with a soft reservoir water, which does not require full coagulation treatment, the manganese can be removed simply by the addition of lime before rapid sand filtration[7]. Caustic soda can sometimes be used more conveniently than lime. A more general method for soft waters is to coagulate the water with ferrous sulphate (copperas) or with chlorinated copperas at a pH of about 9.5 before sedimentation and filtration. More often, however, when manganese is found in a stable condition in a reservoir water the most convenient and satisfactory procedure is to coagulate it with alum plus a small dose of potassium permanganate at the usual pH of approximately 7 or just below 7. Prechlorination is often less effective, but may allow economy in permanganate. In some cases where a trace of manganese, such as 0.1 mg/l, has passed through a filtration plant it appears that the subsequent chlorination has assisted in precipitating it to form objectionable deposits in the mains.

Aeration

8.15 Purpose

Aeration has a large number of uses in water treatment. Listing the more usual, these are:

(1) to increase the dissolved oxygen content of the water,
(2) to reduce tastes and odours caused by dissolved gases in the water, such as hydrogen sulphide, which are then released and also to oxidize and remove organic matter,
(3) to decrease the carbon dioxide content of a water and thereby reduce its corrosiveness and raise its pH value,
(4) to convert iron and manganese from their soluble states to their insoluble states and thereby cause them to precipitate so that they may be removed by filtration.

Water at a temperature of 20 °C and a pressure of 1 atmosphere will, when exposed to air, tend to reach an equilibrium content of dissolved gases from the air, at which state it will contain 9.1 mg/l O_2 and approximately 0.5 mg/l CO_2. The purpose of aeration is to speed up this process and four main types of aerators are in general use: free fall aerators, spray aerators, injection aerators, and surface aerators. For them to be effective it is essential that they provide not only a large water–air interface but, at the same time, the water at the interface must be rapidly moving to facilitate transfer of oxygen across the 'liquid film' which restricts the rate of transfer.

8.16 Cascade aerators

Cascade aerators are the simplest type of free fall aerators. Figure 8.3(a) shows the design of a cascade aerator for treating about 9 Ml/d of water so as to reduce the CO_2 content. Such aerators are widely used: they will take large quantities of water at low head, they are simple to keep clean, and they can be made of robust and durable materials giving a long life. The plates can be made of cast iron, of reinforced concrete, or even of glass. This aerator should preferably be in the open air or, for protection against pollution, in a small house which has plenty of louvred air inlets. Reduction of CO_2 content is usually in the range of 50% to 60%. For increase of dissolved oxygen content the cascade aerator is one of the most efficient of all types of aerator, but it must be designed so as to create a real turbulence and break-up of the falling water.

Weirs and waterfalls of any kind are, of course, cascade aerators. Where a river passes over such artificial or naturally occurring obstacles a large contribution occurs to the self-purification of natural waters by the increase of the dissolved oxygen content which allows the process of breakdown of organic matter to proceed at a faster rate. From observations on a number of rivers Gameson[8] has given an approximate formula for the ratio, r, of the oxygen deficit just above a weir to that just below which is:

$$r = 1 + \frac{a \times b \times h}{2}$$

where a is 1.25 in slightly polluted water, 1.00 in moderately polluted water, and 0.85 in sewage effluents; b is 1.00 for a free fall weir and 1.30 for a stepped weir; h is the height of the fall in metres.

The 'oxygen deficit' is the difference between the actual oxygen content and the equilibrium oxygen content for the water at that temperature and pressure. Best aeration occurs when the water splashes off each step of the stepped weir. If the water clings to the steps the efficiency is reduced. It appears likely that in some rivers more oxygen enters the rivers at weirs than in the reaches of stream inbetween.

Cascading through beds of coke, limestone, or anthracite is believed to give more efficient CO_2 removal than do other methods. This may be due to the longer contact time between the water and the air. This type of aerator is expensive in first cost and entails some additional head on the water supply because the top of the bed may have to be as much as 6 m above ground. Water is discharged on the top of the bed and trickles

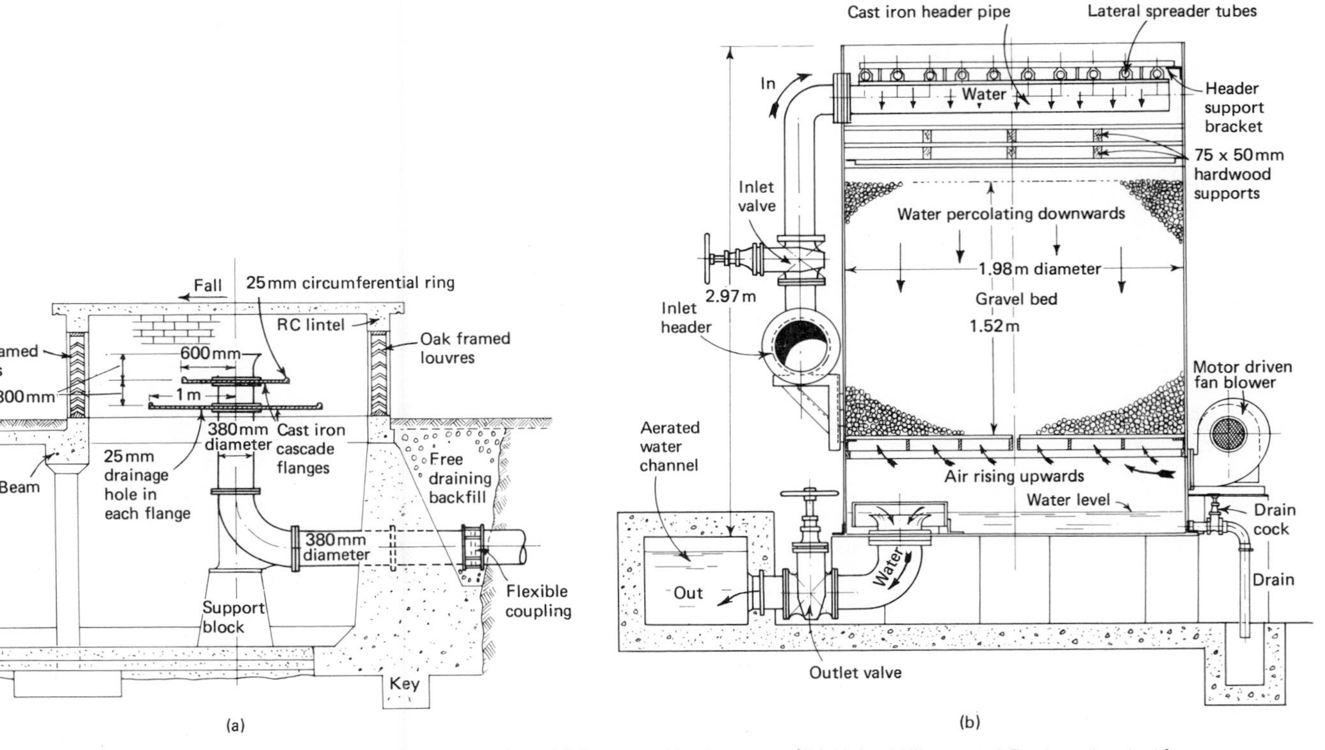

Fig. 8.3 (a) a simple cascade aerator at the inlet to a tank and (b) a gravel bed aerator. [(b) United Filters and Engineering Ltd.]

downwards while air is blown upwards, as shown in Fig. 8.3(b). After a number of years the bed material will need replacing, dependent upon the deposits which have accumulated on the surfaces of the material. Initial cost, maintenance costs, and power costs all tend to make the bed aerator unpopular, although it may be necessary where a high CO_2 content has to be dealt with. With careful design 90% removal of CO_2 can be achieved, but this performance is not always possible or necessary.

8.17 Spray aerators

Spray aerators work on the principle of dividing the water flow into fine streams and small droplets which come into intimate contact with the air in their trajectory. About 70% removal of CO_2 content can be obtained with the best type of spray nozzles. There are a number of different types of nozzle on the market and the correct type of nozzle to be used must be chosen with care so as to prevent troubles caused by clogging or excessive pressure required to force the water through them. Up to 10 m head of water may be required and a large collecting area is necessary because many hundreds of spray nozzles have to be provided. At one installation 1000 nozzles are provided for an output of 9 Ml/d. Extra efficiency is obtained in some types of plant in which the spray is broken up by impinging on a plate.

8.18 Injection aerators

Injection aerators avoid the need to break the pressure of the water if this is particularly inconvenient or wasteful of energy. The water may be sprayed into a compressed air space at the top of a closed vessel under pressure, such as into the top of a pressure filter. The air will, of course, have to be circulated by a compressor. Alternatively compressed air may be injected into the flowing water in a pipe or air at atmospheric pressure may be drawn into the pipe where a constriction, such as the throat of a Venturi tube, reduces the water pressure below atmospheric. In the latter case the Venturi tube is of special design, having a much narrower throat than usual and a much longer divergence cone downstream than in the case of a Venturi designed for flow measurement. A defect of aeration under pressure is that it will not remove CO_2; its effect is to increase the oxygen content of the water and to saturate it with nitrogen at the operating pressure. The latter can be a disadvantage because air can appear as bubbles when the pressure is released again at the consumer's tap.

8.19 Surface aerators

Where water is not under pressure but flowing through a tank a surface aerator may be used. This consists of a circular steel plate fixed to a vertical shaft rotated by a motor. The plate has vertical blades fitted to the underside, set at an angle to the radius of the plate; with high speed rotation a vortex is caused in the water and air is drawn through suction ports in the circular steel plate and is dispersed into the water in the form of minute bubbles. Such surface aerators are used predominantly in sewage treatment works, but also in water treatment works.

Miscellaneous water treatment methods

8.20 Addition of fluoride

Fluoride is added to water to reduce the incidence of dental caries (see Section 6.19). It is best added to the water by using a solution because in the powder form as sodium fluoride or sodium fluosilicate it is toxic and must be contained in dust-tight hoppers and containers. For safer handling hydrofluosilicic acid is better since, even if this is spilled on to the skin, it can be removed easily and quickly by immediate washing in cold water. Tanks for the storage of the acid must have corrosion-resistant linings, rubber being suitable. The point of injection of fluoride should preferably be after any treatment, such as softening, that might interfere with the fluoride content of the water. It can be added prior to the base-exchange process of softening without loss. In general it is best to apply fluoride after other treatments, but at a point where thorough mixing of the fluoride with the water can still take place before the water leaves the treatment works. Close control of the dose is required. In addition to carefully controlled dosing equipment, a daily check should be kept on the quantity of fluoride in store to show the amount used and frequent samples of the treated water are necessary to show the actual fluoride content of the finished water. The normal dose is that required to bring the fluorine content up to 1.0 mg/l in the finished water, but a range of limits recommended by the WHO for international use depends upon air temperatures (see Table 6.1).

8.21 Removal of detergent

In the UK it has been found that the removal of detergent by the slow sand filtration method has varied from 30% to 75% according to the amount originally present in the raw water, the higher removal being achieved with a higher initial detergent content. Rapid gravity filtration is not as effective in the removal of detergent as is slow sand filtration. Thirteen progress reports of a standing Technical Committee on Synthetic Detergents have been published for the Ministry of Housing and Local Government (now the Department of the Environment) between 1958 and 1972. During the period covered by these reports the use of synthetic detergents has greatly increased, but the introduction of biodegradable types has recently reduced the amounts passing through sewage works into rivers. Therefore no recommendations have been found necessary for treatment in waterworks for the removal of detergent. Where reduction or removal is necessary treatment with activated carbon or with ozone may be adopted.

8.22 Purification of water for swimming baths

The purification of water for swimming baths is best achieved by continuous forced circulation of the water through pressure filters followed by superchlorination. The rate of circulation should be such that the whole contents of the bath are changed in four to six hours. A dose of aluminium sulphate should be applied when necessary to the water before it passes on to the filters and, in order to combat the consequent reduction in pH value, soda ash or sodium bicarbonate should be added immediately after the filters. Simple straining of the water is necessary prior to filtration. The chlorination must be sufficient to create a water which is bacteriologically as pure as a drinking water. The usual aim is to maintain free residual chlorine by means of a sufficiently high dosage to destroy ammoniacal compounds in a 'break-point' reaction (see Chapter 9). A heavy dose of chlorine is necessary, of the order of 3 to 4 mg/l, or occasionally even higher when the bath is crowded. The total residual chlorine maintained in indoor baths is commonly in the range 1 to 2 mg/l.

However, because of the dangers associated with the handling of chlorine gas in liquid form, the Department of the Environment recommended in 1978[9] that its use for swimming pool water disinfection should be phased out by the end of 1984. Instead sodium hypochlorite, calcium hypochlorite, electrolytically generated chlorine, compounds of bromine, or other chlorine releasing agents are recommended. Guidelines for the application of such chemicals have been published, 1979–81, by the DoE and a publication on the quality and treatment of swimming pool water is under preparation and to be published in 1985.

As an alternative to pressure filtration through sand beds, diatomaceous earth filters are useful (see Section 7.34). Diatomaceous earth is added to the water and this is deposited on to the surface of 'candles' which filter out impurities other than bacteria. The latter are dealt with by chlorination as with pressure filters. From time to time the flow through the candles is reversed, thus washing off the diatomaceous earth and collected impurities. The advantages of diatomaceous earth filters are that they take up less room than do pressure filters and that chemicals for coagulation are not necessary.

Taste and odour removal

8.23 Causes of tastes and odours

The source of a taste or odour in a water is often difficult to track down, but the following list includes the most likely causes.

(1) Decaying vegetation, such as algae, may give rise to grassy, fishy, or musty odours. Algae may cause offensive odours as they die off and some living algae cause taste and odour troubles.
(2) Moulds and actinomycetes may give rise to earthy, musty, or mouldy tastes and odours in a water which may be wrongly attributed to algal growths. In stagnant waters and especially water in long lengths of pipeline left standing in warm surroundings, such as the plumbing system of a large building, the moulds and actinomycetes have favourable conditions for growth and the first water drawn in the morning may have an unpleasant taste or odour of the kind mentioned.
(3) Iron and sulphur bacteria produce deposits which, on decomposition, release an offensive smell. The sulphur bacteria in particular give rise to the sulphuretted hydrogen (rotten egg) smell.
(4) Iron above a certain amount will impart a bitter taste to a water.
(5) Excessive sodium chloride will impart a brackish taste to a water.
(6) Industrial wastes are a prolific source of taste troubles and odour troubles of all kinds, of which those produced by phenols are the most frequently experienced. Phenols are present in tar compounds, and are found in gas undertaking wastes, and in water which has passed over tarred roads, but they can also be produced by decaying vegetable matter. In the presence of free chlorine the phenols form a 'chlorophenol' or medicinal taste which is quite pronounced. Even so small an amount as 0.001 mg/l phenol may react with chlorine to form an objectionable taste. A severe outbreak of taste troubles occurred in North West England in January 1984 when up to 0.01 mg/l phenol was detected in the River Dee due to an industrial waste discharge.
(7) Chlorine will not, by itself, produce a pronounced taste except in large doses, but many taste troubles accompany the injection of chlorine into a water because of the reactions which follow between chlorine and a number of substances. These tastes are usually described as 'chlorinous' and are fleeting, coming and going, and very difficult to track down to a particular cause. They are more noticeable in a pure water and give rise to many complaints from consumers when the water is put into supply with a residual free chlorine content. Many of these complaints are local and arise from the reaction of the chlorine with some grease on a tap or drinking utensil: 0.05 mg/l will cause such complaints to arise in some undertakings; elsewhere 0.2 mg/l may be left in the water without anyone noticing it.

8.24 Methods for the removal of tastes and odours

Of the methods suggested below each has its place according to the character of the water being supplied. Where chlorine is suspected as being the agent giving rise to the complaint the first attempt should consist of raising the initial dosage of chlorine to ensure that organic and vegetable matter in the water are partially oxidized, and this may then be followed by removing the excess chlorine as much as is desired by the addition of sulphur dioxide. Aeration is only useful in certain circumstances, but ozonization is more effective. The addition of activated carbon is a sound standby treatment to use in conjunction with filtration. It is wholly or partly effective against all sorts of taste and odour troubles and it can be used intermittently when these troubles arise. The methods in more detail are as follows.

Superchlorination This remedy is simple and cheap, if it works. It has been found by experience that if sufficient chlorine is added to a water to give an excess of free chlorine over and above that absorbed by organic and vegetable matter in the water then tastes and odours are reduced. If insufficient chlorine is added these tastes and odours may be quite severe. The excess of free chlorine breaks down the odour-producing substances by oxidation. The dosage required varies with the type of water: it will usually be between 0.2 and 2.0 mg/l. Dechlorination (see below) is usually adopted to remove some or all of the excess chlorine after sufficient contact time has been given.

Chloramine, which is a combination of ammonia and chlorine obtained by adding ammonia to the water in proportions varying from 1:2 to 1:4 of ammonia to chlorine according to the type of water, is a way of preventing taste troubles arising from the presence of free chlorine. It is a negative solution to the problem of taste prevention and it has the disadvantage that the chloramine residual left in the water is a less active disinfecting agent against pollution than free chlorine. However, it is more stable and therefore more effective in retaining a residual through a distribution system.

Chlorine dioxide, formed by adding sodium chlorite to chlorine solution is a stronger oxidizing agent than chlorine and is sometimes found to be more effective than superchlorination. It is especially useful for avoiding chlorophenol tastes. See also Section 9.16.

Dechlorination, by the addition of sodium thiosulphate for small supplies and sulphur dioxide (as a gas) for large supplies, is again a negative way of avoiding taste troubles by removing any free chlorine from a water. As a result there is no disinfecting agent left in the water.

Aeration will sometimes improve a water which is poor in palatability or stagnant. The particular use of aeration is to get rid of sulphuretted hydrogen smells found in deep well water. If iron is present in the water this may precipitate out as a result of the aeration and filtration will then be necessary. Aeration is seldom effective when tastes are caused by organic pollution.

Ozonization is highly effective, but costly. It is a more powerful oxidizing agent than chlorine and has no after-taste troubles. Ozone will improve the appearance of a water, remove traces of colour, and should remove odours if applied in sufficient quantity, but is not always effective against 'earthy' or 'mouldy' tastes and odours. It is an excellent finishing treatment to a poor quality water, its effects being wholly positive with no residual to cause trouble.

Activated carbon in powder or granular form can remove or reduce many kinds of tastes and odours by adsorption of the material producing them. It has been used in household filters since the nineteenth century in the form of charcoal. 'Activated' carbon is very finely divided carbon which presents a large surface area for adsorption. It can be added to the water in powder form. Good mixing with the water is essential; the point of addition is either before any mixing and sedimentation tanks or just before the water passes on to rapid gravity filters. The dosage can be intermittent in the range 8 to 20 mg/l or continuous at about 3 mg/l. When the water is filtered the carbon is left on the top of the beds and is removed by backwashing. For effective use the dose must be adequate, the mixing thorough, and the time of contact long enough for the material to carry out its work.

Activated carbon filters can be used for the removal of tastes and organic impurities. They consist of a bed of activated carbon in granular form and are used in a few waterworks after the rapid sand filters[10]. Compared with dosing with powdered carbon, the use of carbon filter beds has the advantage that the ratio of carbon to water whilst in the filter is obviously many times greater than with a dose of powder. Much more favourable adsorption equilibria should thus be obtained provided, of course, that the carbon remains active to the specific substances required to be removed. Carbon filters are generally the most effective of all methods of removing 'earthy' or 'mouldy' tastes or odours. They are also effective in removing a wide range of complex organic substances, for example pesticides and aromatic hydrocarbons, and their use will no doubt become more widespread under modern conditions of water supply.

Flushing of mains Many complaints of taste and of discoloration due to iron etc. come from water which has been left to stagnate in the ends of mains and the flushing out of dead ends of mains is one of the commonest methods of preventing such troubles. This should be a carefully controlled routine operation in a waterworks distribution system; if it is not, complaints will certainly arise.

Treatment of water for corrosion prevention

8.25 Nature of corrosion

In waterworks terminology water is said to be corrosive if it attacks any of the metals or materials commonly used in a waterworks system. The engineer is also faced with the prevention of external corrosion to water structures such as pipelines. Corrosion may be caused by physical processes, chemical processes, or electrochemical processes, and by bacterial action but, because these categories are interdependent, it is more convenient to divide corrosion into three other classes, distinguished by the different means adopted for their prevention. These classes are:

(1) corrosion primarily caused by motion of a water,
(2) corrosion primarily brought about by the presence of dissimilar metals in a water system,
(3) corrosion brought about by the chemical and bacteriological properties of a water.

8.26 Corrosion caused by water motion

Water flowing at high velocity may often be at sub-atmospheric pressure (its potential energy being converted to kinetic energy). It will then frequently be turbulent and therefore subject to rapid changes of pressure. At instants of low pressure very small vapour or gas bubbles may be released by the water, which collapse with implosive force the moment the pressure is again increased or (what amounts to the same thing) the bubble moves to an area of higher pressure. This phenomenon is particularly likely to happen when high velocity water impinges on to an obstacle where the repeated collapse of such bubbles upon a metallic or concrete surface will quickly cause deep pitting and erosion of that surface. This is the phenomenon known as *cavitation* and it is likely at any point where high turbulent velocities of water are followed by a sudden increase of pressure or by pulsating pressures. Corrosion of this kind may occur at the tips of the impeller blades of centrifugal pumps, at valve gates where valves are kept almost closed, at nozzles where high velocities are created, and at bends in pipes and conduits where high velocity flow reduces pressure to sub-atmospheric. The real remedy lies therefore in the redesign of the structure to avoid the high velocity, the sudden increase of pressure, or the pulsating pressures.

8.27 Corrosion caused by the presence of dissimilar metals

If water is placed in contact with two dissimilar metals having a high difference of electropotential, conditions favourable for severe electrolytic corrosion will be created. Two such materials commonly found in waterworks and having a high electropotential between them are copper and zinc. If the two metals are in contact the zinc becomes anodic and corrodes in the locality of the contact. Copper and aluminium will cause electrolytic corrosion, also to a lesser extent will copper and iron. In general, therefore, the use of copper in a water supply system must be undertaken with care. The worst corrosion of this type is caused by the water first taking copper into solution and then carrying this copper on towards zinc or aluminium, particularly in hot water systems. The means for preventing this type of pollution are not difficult to adopt: lengths of copper tubing should not be so long as to permit the water to take up large amounts of copper; galvanized tanks may be painted inside with a non-tainting bituminous paint; copper ball valves should have pressed seams not soldered ones and should preferably be painted; aluminium utensils should not be consistently filled from water which has stood overnight in copper tubing or cylinders. A complete household installation of copper to approved specification is generally very satisfactory.

8.28 Corrosion caused by properties of water

This type of corrosion has many causes and is best considered under several headings.

General considerations and 'Langelier' index Some waters have a distinct tendency to corrode metals due to a high content of chloride or other saline constituents. In some cases residual chlorine, particularly free residual chlorine, i.e. without ammonia being present, will tend to cause corrosion due to its effect on the oxidation–reduction potential. The extent to which corrosion occurs, however, depends in most cases largely upon the acidity or carbon dioxide content of the water, and conversely it can largely be controlled by the presence of calcium carbonate alkalinity with a sufficiently high pH. Another factor tending to reduce corrosion in many river derived waters is organic matter which probably has a contributory protective effect[11].

The amount of carbon dioxide in water in equilibrium with the atmosphere is about 0.5 mg/l, whereas amounts up to as much as 80 mg/l are found in some waters from underground sources. Anything in excess of 0.5 mg/l can be removed by aeration, although in practice most aeration plants only achieve partial removal, as already discussed in Section 8.16. From the point of view of corrosion a major consideration is whether or not the carbon dioxide in a water is sufficient to keep its dissolved calcium bicarbonate in equilibrium with solid calcium carbonate. If the amount of free carbon dioxide is more than sufficient it is termed 'aggressive' free carbon dioxide and the water will tend to dissolve any protective coating of solid calcium carbonate or any other solid calcium carbonate with which it may come in contact. Conversely, when a water is deficient in free carbon dioxide it will tend to deposit some of the calcium bicarbonate which it holds in solution.

The best way to ascertain whether or not a water is in equilibrium with calcium carbonate is to carry out the 'chalk test'. In this test a water is placed in contact with powdered chalk and the pH value is determined at which it no longer deposits or dissolves calcium carbonate. This pH value is designated 'pH_s' being the 'saturation' pH value of that water.

Langelier's index, I, is then denoted as the numerical difference between the actual pH of a water and its pH_s, i.e.

$$I = pH - pH_s$$

When I is negative the saturation pH value is higher than the actual pH value. Therefore the water is undersaturated with $CaCO_3$ and it will tend to dissolve existing sources of calcium carbonate so that it will not deposit any protective coating on metallic surfaces. Negative values of I therefore indicate a corrosive water; positive values a relatively non-corrosive water.

Equations and nomographs are available which give the saturation pH (or pH_s) of a water from its content of dissolved solids, temperature, and calcium carbonate alkalinity. There is some doubt, however, as to whether these equations and nomographs are based upon a sufficient foundation of practical work; if the chalk test is very carefully carried out this gives the most accurate value. A graph which is given by Cox[12] has, however, been found a very useful practical guide and the pH figures in Table 8.2 have been read from the graph as published. The table also includes the approximate figures for free carbon dioxide corresponding to each carbonate alkalinity with equilibrium pH.

Table 8.2

Calcium carbonate alkalinity (mg/l)	pH at equilibrium, i.e. pH_s	Free CO_2 at equilibrium (mg/l)
25	8.7	Absent
50	8.1	1
75	7.7	3
100	7.5	6
150	7.4	12
200	7.3	25
250	7.2	32
300	7.0	60

Following this introduction with particular reference to the carbon dioxide and calcium carbonate alkalinity of waters in general, reference will now be made to the occurrence of corrosive characteristics and their control in particular types of waters.

Hard surface derived waters with fairly high carbonate hardness A large proportion of lowland river waters associated with sedimentary geological formations, for example in the south-eastern parts of England, have fairly high carbonate, or temporary, hardness due to calcium bicarbonate. Although some of the water has originated from springs underground, it has been flowing exposed to the air and the pH is therefore usually fairly high, corresponding to a fairly low content of free carbon dioxide. Theoretically, therefore, such waters tend to deposit calcium carbonate, but in practice the calcium bicarbonate remains in solution under normal conditions and only very gradually deposits, even in such situations as on filter sand. In very many cases not only river waters but also stored and treated waters have a substantially positive Langelier index which helps to render them non-corrosive. In addition their residual content of organic matter may tend to give a protective effect on metals such as galvanized iron which otherwise would be subject to corrosion[11]. Generally, therefore, such waters are very satisfactory for the avoidance of corrosion with metals and they are not plumbosolvent. An example is London's water supply from the Thames, a typical analysis of which may include a pH of 7.9 which corresponds to a free carbon dioxide content of only about 5 mg/l in association with a carbonate hardness of about 200 mg/l.

Hard underground waters with substantial carbonate hardness The mineral composition of many underground waters from chalk, limestone, and other sedimentary sources may be similar to that of river water, but often the carbon dioxide content of the water pumped from underground is much higher. On the whole it is often quite near to the amount which is in equilibrium with solid calcium carbonate, and this might be expected if the water in an outcrop formation is in fact in close contact with chalk. Therefore the free carbon dioxide content of many underground waters can easily be reduced by aeration. If, sometimes, the amount is above the saturation amount, pH_s, the excess may be regarded as 'aggressive' free carbon dioxide and it is certainly advisable to remove that part. In fact to improve the characteristics of the water for use with a metal such as galvanized steel it is usually advantageous to reduce the content of free carbon dioxide below the saturation figure, i.e. to give the water a positive Langelier index. A water with a positive index is, of course, theoretically liable to deposit calcium carbonate, but in practice it will probably not deposit appreciably in water mains and tanks. Its anti-corrosion effect probably depends in part upon the rapid deposition of calcium carbonate on cathodic areas when electrochemical corrosion is initiated, the effect of this deposit being to stifle the incipient corrosion. Calcium carbonate is also, of course, precipitated from the carbonate, or temporary, hardness in waters when they are boiled or gradually, even when heated well below boiling, in domestic or other heating appliances.

It may be mentioned here that waters with fairly high carbonate hardness due to calcium bicarbonate are seldom appreciably plumbosolvent, even if they have a slightly negative Langelier index. Probably the main reason for this is the low solubility of lead carbonate whose solubility product is 3.3×10^{-14} compared with 1.0×10^{-8} for calcium carbonate.

Sodium hexametaphosphate (popularly known by the trade name 'Calgon') can be added to a water in small doses of 1 to 2 mg/l and is said to be a corrosion 'inhibitor'. A high initial dose is usually given to the water, followed by a continuing subsequent dose. Reduction of tuberculation and pitting and the prevention of precipitation of iron and manganese are other benefits attributed to the use of sodium hexametaphosphate. It is perhaps more widely used for the treatment of boiler feed waters than for the treatment of public water supplies.

Underground waters with low carbonate hardness but high free carbon dioxide Waters of this type constitute many small sources from some of the older geological formations and some from gravel wells and springs, although many of the gravel sources have high non-carbonate hardness.

With such waters it may be convenient and economical to remove part of the carbon dioxide by aeration. A further reduction may be achieved by adding lime and this at the same time increases the carbonate hardness according to the following equation:

$$2CO_2 + Ca(OH)_2 = Ca(HCO_3)_2$$

If the water already contains sufficient hardness then it is sometimes convenient to neutralize the free carbon dioxide with caustic soda because that reagent is very soluble and can be administered as a small volume of a strong solution. For the correction of corrosive characteristics of very soft waters passage over beds of limestone chips or through a slow filter containing granular limestone has been used. More than one range of improved and more active proprietary granular materials containing calcium carbonate (or lime) and magnesium oxide is now available for small- or large-scale use.

Soft waters from surface sources In this category there are a number of very pure lake waters, e.g. in the north-eastern parts of England and in Scotland, which have an organic quality almost approaching that of underground waters, but they do not contain any appreciable content of free carbon dioxide. Such waters are therefore almost neutral in reaction. Many of them have very low carbonate hardness, e.g. 10 mg/l, and from the point of view of corrosion in general and plumbosolvent action in particular they need very close consideration. In many cases they provide very large supplies and so the most realistic method of approach is to make observations on mains and fittings and to assess plumbosolvency by taking samples from consumers' taps both under normal running conditions and after standing in the pipes overnight. With some of these supplies an economical form of treatment is simply to add a very small dose of lime which can be found satisfactory. In cases where a greater amount of calcium carbonate hardness must be added, use of calcium sulphate with soda ash or sodium bicarbonate may be convenient. Another method is to use lime or finely divided calcium carbonate plus carbon dioxide formed from the burning of propane gas. The same methods of adding calcium bicarbonate are applicable in the case of distilled water which is mentioned in Section 8.35.

However, many very soft waters derived from moorland catchments are coloured brown by peaty matter and this contains organic humic and fulvic acids which give the water an acid reaction, with a pH of about 6 or lower. From the point of view of avoiding corrosive characteristics in general and plumbosolvent action in particular the organic acids do not merely require neutralization, but they require to be removed[13] and this is usually achieved with alum or with an iron coagulant, as mentioned in Section 7.22. The reaction of these coagulants needs further consideration in the context of avoiding corrosion. Reference to the equation given in Section 7.20 shows that the addition of aluminium sulphate actually removes carbonate hardness, the carbonate or bicarbonate being converted to free carbon dioxide. If lime is added after filtration this carbon dioxide can usually be largely converted again to calcium bicarbonate. However, if the lime is used to provide alkalinity to precipitate the alum no carbonate is introduced into the water. If the water contains a reasonable amount of hardness without carbonate the use of soda ash with the alum produces carbon dioxide and again most of this can be converted to calcium bicarbonate if lime is added after filtration. If the water is deficient in calcium then it may be very convenient to add calcium sulphate with the alum because the plant would provide ample opportunity for the calcium sulphate to dissolve.

Some soft waters have caused pitting of copper pipes due to manganese not being removed. An extreme type of water which flows from certain moorland catchments, e.g. in various places in Yorkshire, has a pH as low as 4.5 or even 4.0. This acidity is due to a trace of aluminium sulphate of natural origin and the content of carbonate or bicarbonate is zero so they must be added in the treatment. In contrast another type of surface water which is of interest is the very soft water found in many rivers and lakes in African countries, e.g. Lake Victoria, which contain a small excess of bicarbonate in the form of sodium bicarbonate. Any treatment required to minimize corrosion may be only a very small dose of lime. Many lime-softened waters are examples of soft waters which experience shows to be very satisfactory, even with galvanized iron tanks. Care must be taken, however, to avoid reducing the carbonate hardness too far, either in the softening or in the subsequent stabilization (see Section 8.5).

8.29 Dezincification

After unprotected iron or steel, one of the most widely used materials which is susceptible to corrosion is galvanized iron or steel. It is susceptible to corrosion by excess of free carbon dioxide dissolving the zinc and, at the other extreme, it is susceptible to any waters in which the pH is excessively high, especially above 10, due to unsatisfactory adjustment of the final stage in treatment. However, the term dezincification is usually applied to an effect on alloys containing copper and zinc by which the zinc is dissolved out. This effect can be seen by cutting and polishing a section of the metal when the dezincified part will show the typical colour of copper instead of the original brass colour.

An example of this effect which has received particular attention in recent years is called 'meringue' dezincification because of the voluminous white layer of corrosion product which appears. The effect of this is to cause failure in the action of fittings, mainly hot water fittings, when they are constructed of hot-pressed brass, but it does not occur with terminal fittings such as taps.

This effect has been investigated by Turner[14] and was found to occur with waters having a pH of over 8.2 and having the ratio of chloride to carbonate (temporary) hardness greater than is shown in Table 8.3 which is taken from Turner's curve.

Table 8.3

Chloride (Cl mg/l)	Carbonate hardness ($CaCO_3$ mg/l)
10	10
15	15
20	35
30	90
40	120
60	150
100	180

The findings summarized in this table reinforce those relating to the Langelier index (see Table 8.2) in indicating that many very soft waters need the addition of carbonate hardness to improve their behaviour towards metals. If a water contains an appreciable amount of chloride then simply raising its pH without increasing carbonate hardness may increase dezincification and fail to avoid other troubles with metals. In many cases it may be best for a water undertaking to introduce byelaws banning the use of fittings liable to meringue dezincification. However, the implications of the research summarized briefly in Table 8.3 do not seem to apply to all waters and further research is in progress.

An indication of methods of adding some calcium bicarbonate to waters has already been given in connection with corrosion in general by soft water in Section 8.28.

In the lime–soda softening of hard river derived waters the most economical procedure is to reduce the carbonate hardness to a low figure by using more lime than soda. Some river waters, however, have a chloride content of 60 mg/l or more and in some plants in recent years it is mostly the non-carbonate hardness which has been removed by using a high dose of soda ash in order to leave the ratio of carbonate to chloride satisfactory to avoid meringue dezincification.

8.30 Bacterial corrosion

Bacteria known as the 'sulphate-reducing bacteria' existing in anaerobic conditions, i.e. in the absence of oxygen, can cause iron to be corroded.

Many of these bacteria are capable of living on a mineral diet and they produce, as a result of their metabolism, hydrogen sulphide which attacks iron and steel and forms the end product ferrous sulphide. Steel thus becomes pitted and cast iron becomes 'graphitized'. Graphitization is a surprising, even astonishing, phenomenon. The cast iron pipe so attacked may have, to all external appearances, nothing wrong with it—yet it may be cut with a penknife.

Now we are here dealing with the external corrosion of a steel or iron main because the sulphate-reducing bacteria may be most numerous in waterlogged clay soils where no oxygen is present and where sulphur in the form of calcium sulphate is very likely to occur. In such clays the sulphate-reducing bacteria are one of the most virulent forms of attack on iron and steel; the most commonly used type of protection is to have an external protective coat to the pipeline. This external protection may take the form of a thick coating of bitumen, a 75 mm thick layer of rich concrete containing sulphate-resisting cement, or the pipe may be surrounded with chalk or gravel and sand which are free draining and thus permit the ingress of oxygen to the pipeline to prevent the continuance of anaerobic corrosion.

Inside a water main, not only may we find the sulphate-reducing bacteria in a water which lacks oxygen (as evidenced by the presence of sulphates in the water and also by the presence of hydrogen sulphide, the gas giving the rotten eggs smell), but we may also find 'iron bacteria', another and second cause of corrosion of iron. These iron bacteria are a group of organisms which have the power of absorbing oxygen, oxidizing iron, and storing it either from the water or from the iron of pipes. They are, unlike the sulphate-reducing bacteria, aerobic and their characteristic is that they form, under favourable conditions, into large deposits of slime which are highly objectionable in a public water supply, giving rise to objectionable odours, staining, and sometimes blocking filters. Their growth is an ever-present possibility when a water containing a high iron content is passed through water mains.

Within a main, in addition to the slime produced by iron bacteria, tuberculation may take place: this is a corrosion phenomenon which may eventually result in the failure of the main and will, even before that, result in serious diminution of the carrying capacity of the pipeline. This corrosion by tuberculation may in some cases be set off by the sulphate-reducing bacteria. The external surface of a tubercle, or nodule, consists of a hard crust of ferric hydroxide, often strengthened by calcium carbonate and manganese dioxide. Below this crust conditions tend to be anaerobic and it is here that the sulphate-reducing bacteria may be able to flourish creating further products of corrosion. The precise mechanism by which tubercles grow (and they may grow to be several millimetres in diameter and 20 mm or so proud of the pipe) can hardly be said to be completely understood, but the conditions in which they will grow are well known. Hence the two remedies for the prevention of internal tuberculation are the aeration of a sulphate- or hydrogen-sulphide-containing water, and the protection of the interior of the pipeline either with a thick coating of bitumen (a spun bituminous lining) or with a layer of mortar spun on to the interior of the pipe. The characteristics of the water as discussed in Section 8.28 are also important.

Desalination

8.31 Introduction

Desalination is a term used to describe processes used for the reduction of dissolved solids in water, usually referred to as total dissolved solids (TDS) and measured in mg/l. Sometimes conductivity is used as the measurement of total dissolved solids as it is easy to measure in the field with a conductivity meter. The SI units are millisiemens per m. For most purposes the TDS value of a water can be obtained by multiplying the conductivity value by 0.66, except in certain special cases, e.g. mine waters. Natural waters may be classified in broad terms according to their TDS values as follows[15].

Type of water	*TDS value (mg/l)*
Sweet waters	0–1000
Brackish waters	1000–5000
Moderately saline waters	5000–10000
Severely saline waters	10000–30000
Seawater	Above 30000

Several methods have been commercially developed for the desalination of high TDS waters and the selection of the right process, especially in the 500 to 5000 mg/l range, requires careful evaluation of process efficiency, plant capital, and running costs. As a general guide, the most frequent application of the various desalination processes has been in the following categories of TDS waters.

Process	*TDS value (mg/l)*
Ion exchange	500–1500
Electrodialysis (and reverse osmosis)	500–3000
Reverse osmosis with standard membranes	1000–5000
Reverse osmosis with high resistance membranes	5000 and over
Distillation	Above 30000

In terms of installed world capacity distillation accounts for about 75% of all desalination installations, with the largest concentration of plant in the Middle East and North Africa. In recent years, however, reverse osmosis systems have been developed for desalting seawater and the installed capacity of such plants is steadily increasing.

The use of freezing is theoretically capable of producing water free from saline constituents, but it has not been adopted on a commercial scale and so it will not be dealt with here. Reviews of the process have been given by Johnson[16] and Denton *et al.*[17].

8.32 Ion exchange

The use of ion-exchange processes for reducing mineral or saline constituents in water is discussed in Section 8.11 where it is referred to as demineralization instead of desalination. Whilst the two terms mean exactly the same, the word demineralization is used most often in industry to describe the particular application of ion-exchange resins to remove some or all of the mineral and saline salts in water. The use of ion exchange rarely finds application for waters containing more than 1500 mg/l of dissolved solids.

8.33 Electrodialysis

In this process a potential gradient is applied between electrodes and the ionic constituents in the water are thus caused to migrate through semipermeable membranes. The membranes are selective to cations and anions and demineralization occurs in a series of cells separated by pairs of cation and anion membranes. A number of membrane pairs can be installed in parallel between a pair of electrodes to form a stack and several stacks can be connected hydraulically together either in parallel or in series to increase output. The reduction in dissolved solids obtained by electrodialysis is directly related to the electrical energy input and therefore in practice to the cost of electricity. The location and cost of an electrical supply therefore play an important role in determining the viability of electrodialysis compared with other desalination methods. Electrodialysis membranes are prone to fouling, due primarily to bacteria, organic material, suspended solids, and scale-forming substances such as carbonates and sulphates of calcium in the feed water. In one development known as electrodialysis reversal the effects of scale are minimized by periodic reversal of the polarities of the electrodes, thus reducing or eliminating feed pretreatment costs.

8.34 Reverse osmosis

In this process the separation of dissolved salts from water is achieved by a hydrostatic pressure across a semipermeable membrane separating the feed water from the desalted water which is called a permeator. Commonly used membrane materials in desalination are cellulose acetate in spiral-wound membrane configuration and polyamide or cellulose triacetate in spiral-wound or hollow-fibre membrane configuration.

Generally the cellulose acetate membranes are relatively more sensitive to pH and temperature, whereas the polyamide membranes are more sensitive to chlorine.

The salt-rejection properties of a membrane can be varied by a membrane manufacturer so as to produce high salt-rejection and low flux rates or, alternatively, low salt-rejection and high flux rates. In practice salt rejections of up to 90% can be achieved with most commercial membranes; seawater desalination can also be carried out by reverse osmosis using special membranes to withstand high operating pressures and to give the high salt-rejection characteristics required. Typical operating pressure and salt-rejection properties of reverse osmosis membranes for such a duty are 55 to 70 bar pressure and better than 98.5% rejection respectively. Higher salt rejections can be achieved by using a multipass system where the permeate from the preceding pass is the feed to the subsequent pass.

The following equations give an approximate representation of the flow of water and dissolved solids through membranes.

$$Q_w = W_p(\Delta p - \Delta \pi)$$

and $$Q_s = K_p(\Delta C)$$

where Q_w is the water flux through the membrane (m/s), Q_s is the salt flux through the membrane (kg/(m^2 s)), W_p is the membrane permeability coefficient for water (m^3/(N s)), K_p is the membrane permeability coefficient for salt (m/s), Δp is the pressure differential across the membrane (N/m^2), $\Delta \pi$ is the osmotic pressure differential across the membrane (N/m^2), and ΔC is the concentration differential across the membrane (kg/m^3). Cellulose acetate membranes have water flux rates in the range 5×10^{-6} to 1×10^{-5} m/s and polyamide membrane rates are about one-tenth of this range.

The equations assist in understanding the basic criteria controlling the reverse osmosis process. It can be seen that as the feed water pressure is increased water flux increases but salt flux remains constant, i.e. both the quality and the quantity of the product increase with a higher driving force. It is also evident that at constant applied pressure the water flux decreases as the solute concentration of the feed is increased. This is due to the reduction in the driving force caused by the higher feed osmotic pressure. As more water is extracted (called water recovery) from a given feed, the salt concentration in the bulk solution increases and at constant applied pressure the water flux falls due to high local osmotic pressure. Higher water recovery also raises the salt flux due to higher salt concentration in the bulk solution.

The flux of a membrane is usually stated by a manufacturer for operating temperatures for about 25 °C, but temperature can affect performance: for example a 1 °C rise in temperature above the stated temperature for a membrane can increase the flux by about 2.5% to 3.0%; when the feed water temperature is below the rated temperature the flux decreases by a similar amount. Feed water temperatures consistently higher than the rated temperature can have a detrimental affect on the membrane. Although modern membranes are continually improving with longer life and robustness, nevertheless their performance can be significantly affected by fouling due to the presence of materials such as suspended solids, organics, silica, iron, manganese, and sparingly soluble salts in the feed water; membrane deterioration due to physical (temperature), chemical (pH and chlorine), and biological action; compaction due to excessive operating temperatures and pressures. Care is therefore always needed to ensure that the membranes are correctly applied in a suitable environment. Chemical cleaning is usually recommended by the manufacturer as a standard procedure, but full performance recovery is not obtained if the membrane is subjected to continual abuse.

Reverse osmosis is a concentration process and therefore water recovery is affected by the presence of sparingly soluble salts such as the carbonates and sulphates of calcium. With increasing water recovery the concentrations of sulphates and carbonates of calcium in the bulk solution also increase to such a level that they precipitate and block the membranes. Usually, with some form of pretreatment of the feed water, it is possible to recover about 75% of the feed water as product water, although recoveries of up to 90% can be achieved with multistage reverse osmosis in which the concentrate from each stage feeds successive stages.

In seawater desalination the high concentration of dissolved salts in the feed limits achieving high recovery rates which then rarely exceed 35% of the feed water.

Ion-exchange, electrodialysis, and reverse osmosis processes are all susceptible to fouling so that consideration of pretreatment requirements should never be overlooked. So-called 'conventional' stages of pretreatment consisting of chemical coagulation, sedimentation, and sand filtration may not be sufficient for many feed waters and it may be necessary to add stages of aeration and contact filtration for iron and manganese removal, activated carbon filtration for organic material, polishing sand filters for colloidal matter, lime–soda softening or base exchange for silica and sparingly soluble inorganic salts, and disinfection for bacteria and biological material, as well as pH adjustment with acid or alkali and removal of chlorine, if previously used for disinfection, by activated carbon or sodium bisulphite dosing.

8.35 Thermal processes

The earliest application of desalination was the distillation of seawater on ships and the necessary apparatus was being installed on steam ships as early as 1884. The total concentration of salts in normal seawater is approximately 3.5%, i.e. 35 000 mg/l. Distillation works on the principle that the vapour produced by evaporating seawater is free from salt and the condensation of that vapour yields pure water. The majority of plants use the multistage thermal flash process (MSF), but other processes used on a smaller scale are horizontal tube multiple effect plants (HTME), multiple effect distillation using vertical, enhanced surface heat exchange tubing (VTE), and vapour compression evaporators (VCE). More recently there has been some interest in the development of hybrid units using the best features of each process, and in the use of dual purpose stations for both desalination and the generation of power. The trend towards higher operating temperatures means that greater attention has to be paid to the reduction of corrosion and the use of cost-effective materials and chemicals to combat corrosion. In all desalination plants careful consideration has to be paid to the avoidance of pollution of the surroundings. Noise does not normally create any problems provided that blow-off silencers are used; oil spillages have to be guarded against if large quantities of fuel oil are to be brought by sea; acid handling equipment has to be designed to prevent accidental spillage. Safe disposal of the brine effluent (strength initially approaching double that of seawater, but subsequently diluted in distillation plant by return cooling water) may present a complicated problem. The effluent may have positive, negative, or neutral buoyancy at various times and the copper content has to be kept under surveillance.

One of the major design parameters of all thermal desalters is the performance ratio which is, in effect, a measure of the efficiency of energy utilization. The amount of energy required to desalt a given brine concentration varies according to the degree of sophistication of the plant installed, i.e. annual power costs reduce as capital costs increase. Other factors to be taken into account include size of units, load factor, growth of demand, interest rate on capital, and technical matters concerning the auxiliary services, repairs, and maintenance. In the detailed design something of the order of 70 design parameters have to be settled. Many of these are concerned with the safe or most economic limits for the temperatures, velocities, and concentrations of the coolants, brines, brine vapour, steam, steam condensate, and boiler feedwater, with particular reference to the prevention of scaling, corrosion, erosion, the purity of the distillate, the efficiency of heat exchangers, and the nature and cost of the auxiliary plant involved.

Distilled water approaches zero softness with an alkalinity not likely to exceed 2 mg/l. It is aggressive to metal and asbestos cement pipes and will take up calcium from mortar-lined pipes. It is unsuitable for distribution and is unpalatable, being flat and insipid. To remedy these deficiencies the distilled water should be treated with carbon dioxide and hydrated lime to produce an alkalinity of at least 35 mg/l as $CaCO_3$, as mentioned in Section 8.28 for the treatment of soft waters from surface sources. A positive Langelier index value needs to be achieved (see Section 8.28). The carbon dioxide must be added to the distilled water first and the most convenient method is to use liquid CO_2 injected by a V-notch vacuum operated carbonater, working on the same principle as a chlorinator (see Section 9.12). An alternative is to mix the distillate with another source of water having adequate alkalinity, but often such a source is not available. Seawater itself cannot be used for blending with distilled water as it does not contain a sufficiently high ratio of alkalinity to salt to permit this.

Major problems which can occur with seawater evaporators are: scale formation on heat transfer surfaces due to the presence of carbonates and sulphates of calcium; internal corrosion due to hot sodium chloride and the presence of dissolved gases such as oxygen, ammonia, and hydrogen sulphide; plant start-up problems and running at low capacity.

8.36 The costs of desalination

The subject of desalination costs can be discussed only cursorily here because of its diversity and complexity. The purpose of this brief review is therefore to indicate the principal items contributing to the production costs of desalinated water and to point out how these are likely to be influenced by local conditions. The major factors to be considered in any desalting application include: type and characteristics of the saline feed; type of desalination process to be used; local costs of primary energy source (e.g. oil or gas) or electricity; availability and costs of chemicals needed for feed pretreatment, product conditioning, and plant cleaning. Also important, in the case of distillation plants, is the choice of performance ratio and whether dual purpose operation is proposed (Section 8.35). Except for the smallest plants, staffing costs are not usually a major item. Table 8.4 summarizes the principal cost contributory factors of the three main desalination processes currently popular (1984) as applied to single purpose plants (i.e. producers of water only). In the case of brackish waters (as defined in Section 8.31) the primary energy and power demands are usually less than half of those for seawater desalination, whether electrodialysis or reverse osmosis is used. Thermal processes, including vapour compression, are not usually adopted for brackish water desalting because of their unfavourable energy demands. The feed water and chemical consumptions of brackish water desalination are also only about one-third of those for seawater reverse osmosis.

The costs of desalted water are very variable and site specific. Amortization of the initial capital investment, together with energy costs, usually accounts for up to 80% of the total water cost. The energy components can be calculated from the data in Table 8.4 if local primary

Table 8.4 Cost contributory factors for major types of seawater desalination processes.

Process	Usual maximum size of unit to date (m^3/d)	Total primary fuel energy demand (MJ/m^3)	Components of energy demand		Chemical consumption (kg/m^3)	Seawater consumption (m^3/m^3)	Annual cost of spares and replacements as % of initial capital cost of desalter (%)
			Power ($kW\,h/m^3$)	Heat (MJ/m^3)			
Multistage flash evaporation (MSF)	32000	200–400	0.5–1.0	185–370	0.05–0.30	5–10	3
Reverse osmosis (RO)	6000	50–100	5–10	Nil	0.50	3–4	8
Mechanical vapour compression (MVC)	1000	100–250	10–20	0–60*	0.06	2.0–2.5	2–4

Notes:
Data refer to stand-alone plants.
Heat and power assumed generated on site.
Thermal efficiencies assumed: 80% for steam boilers for MSF plant; 33% for diesel generators for RO and MVC plant.
Higher energy value of diesel fuel taken as 37.85 MJ per litre.
1 MJ = 0.278 kW h.
* Dependent upon heat recovered from jacket of diesel plant.

energy unit costs are known. Only the briefest indication of capital costs can be given in the present context for the average size of the three major types of desalination plant listed in Table 8.4. Plant costs, erected and commissioned, but excluding cost of civil works, local product storage, and all other site-specific costs (such as cost of intake, discharge of effluents, fuel storage and handling etc.) in late 1983 were as given below.

Type of plant	*Cost (£ per m^3/d output)*
Multistage flash evaporation plant	1000–1700
Reverse osmosis plant	1800
Mechanical vapour compression plant	2000

In the case of reverse osmosis and mechanical vapour compression plants, if diesel generators are not required because power can be taken from the grid supply or some other source, the plant costs would reduce by £300 and £500 per m^3/d respectively. For brackish water the capital costs of desalination plants are much lower, but it is not possible to give figures since brackish waters vary widely in type and salinity.

Radioactivity

8.37 Removal of radioactive substances

An introduction to radioactive substances in water was given in Section 6.36. Referring again to natural radioactivity, the removal of radon by aeration of a water supply having a very unusual amount of 15 000 pCi/l was described by Hoather and Rackham[18] and Raffety[19]. Experience of operation of the plant has shown that if the aeration is adjusted to reduce the radon content to 700 pCi/l then it reduces the free carbon dioxide from 60 to 10 mg/l, the treatment to avoid corrosion being completed by the addition of lime. Referring to beta activity from fallout, it was found that about half was removed from River Thames water by the storage and slow sand filtration plants[20]. The storage allows some time for radioactive decay and the filtration removed suspended matter. Water softening removed a substantial proportion of strontium 90 in solution in water not only because of its chemical identity with non-radioactive strontium, of which about 0.5 mg/l is present in many hard waters, but also because of its similarity to calcium. Demineralization of a water similarly would remove practically all radioactive elements from a water in the same ratio in which it removed the related non-radioactive elements. Iodine 131 is another important radioactive isotope formed in nuclear fission, but fortunately its half-life is only eight days. It was shown by Eden *et al.*[21] that when it was added to a water 90% of it could be removed by coagulation, sedimentation, and rapid filtration, provided that 0.08 mg/l of silver nitrate was also added. This removal depends upon the extremely low solubility of silver iodide.

Special decontamination filters of steel wool, clay, and activated carbon or resins have been found to remove 99.99% of radioactive materials. Whether any of these processes would produce a water safe for drinking under conditions of emergency or accidental contamination would depend upon the intensity of contamination because complete removal is not achieved.

For effluents from radiochemical laboratories or atomic energy or similar establishments, elaborate treatment processes are operated under very strict control. In the case of small amounts of short-lived isotopes such as may be discharged from hospitals, treatment may be unnecessary, but the quantities must not exceed an authorized amount. The Radioactive Substances Act 1960 brought the use of all radioactive substances in Britain under the control of the appropriate government department, and the authorization and control of any radioactive discharges accordingly come under the control of the Department of the Environment instead of the river authorities who control other types of discharges.

References

(1) Crawford M D, Gardner M J and Morris J N, Mortality and Hardness of Local Water Supplies, and Mortality and Hardness of Water, *Lancet*, **1,** 1968, pp. 827 and 1092.
(2) Crawford M D, Gardner M J and Morris J N, Cardiovascular Disease and the Mineral Content of Drinking Water, *Brit. Med. Bull.*, **27,** 1971, p. 21.
(3) Hilson M A and Law F, The Softening of Bunter Sandstone Waters and River Waters of Varying Qualities in Pellet Reactors, *JSWTE*, **19,** 1970, p. 32.
(4) Gledhill E G B and McCanlis A W H, Softening of Chalk Well Waters in a Pellet Reactor, *JSWTE*, **19,** 1970, p. 51.
(5) AWWA, *Water Quality and Treatment*, 3rd Ed., McGraw-Hill, 1971.
(6) Bolas P M, Some Experiences in Iron and Manganese Removal Using Catalytic Sludge Blankets, *JIWE*, **19,** 1965, p. 531.
(7) Waterton T, Manganese Deposits in Distribution Systems, *Proc. SWTE*, **3,** 1954, p. 117.
(8) Gameson A L H, Weirs and the Aeration of Rivers, *JIWE*, **11,** 1957, p. 477.
(9) DoE, *Statement on the Use of Chlorine Gas in the Treatment of Water of Swimming Pools*, Circ. 72/78, HMSO, 1978.
(10) Shinner J S and Davison A S, The Development of Bough Beech as a Source of Supply, *JIWE*, **25,** 1971, p. 243.
(11) Campbell H S, Corrosion, Water Composition and Water Treatment, *JSWTE*, **20,** 1971, p. 11.
(12) Cox C R, *Operation and Control of Water Treatment Processes*, WHO, Geneva, 1964, p. 203.
(13) Miles G D, *JSCI*, **67,** 1948, p. 10.

(14) Turner M E D, The Influence of Water Composition on Dezincification of Duplex Brass Fittings, *JSWTE*, **10,** 1961, p. 162, and **14,** 1965, p. 81.
(15) Fair C M, Geyer J C and Okun D A, *Water and Wastewater Engineering*, John Wiley, 1968.
(16) Johnson K D B, Desalination in Britain, *JSWTE*, **17,** 1968, p. 94.
(17) Denton W H, Hardwick W H and Johnson K D B, *Proc. 1970 Symposium: Water Treatment in the Seventies*, SWTE and WRA, p. 145.
(18) Hoather R C and Rackham R F, Some Observations on Radon in Water and its Removal by Aeration, *JIWE*, **17,** 1963, p. 13.
(19) Raffety R C, Some Notes on Plant for the Removal of Radon by Aeration, *JIWE*, **17,** 1963, p. 23.
(20) Windle Taylor E, Fortieth Report on the Results of the Bacteriological, Chemical and Biological Examination of the London Waters for the Years 1961–62, Metropolitan Water Board, p. 76.
(21) Eden G E, Downing A L and Wheatland A B, Observations on the Removal of Radio-isotopes During Purification of Domestic Water Supplies, *JIWE*, **VI,** 1952, p. 511.

9
Disinfection of water

9.1 Disinfectants available

The term 'disinfection' is used, meaning the reduction of organisms in water to such low levels that no infection of disease results when the water is used for domestic purposes including drinking. The term 'sterilization' is not strictly applicable because it implies the destruction of all organisms within a water and this may be neither achievable nor necessary. Nevertheless the word is often loosely used, as in 'domestic water sterilizers'.

On a plant scale the following disinfectants are in common use:

(1) chlorine,
(2) chloramine,
(3) chlorine dioxide,
(4) ozone.

Other disinfectants used, principally for the situations as given, are:

(1) ultraviolet radiation for small public supplies, railway refreshment vehicles etc.,
(2) silver for in-house 'water sterilizers',
(3) iodine for temporary, small-scale use in emergencies,
(4) potassium permanganate, possibly for predisinfection of a water before other treatment, also for iron and manganese removal,
(5) boiling for domestic drinking supplies, usually in an emergency but sometimes as a routine precaution.

The organisms in water which it may be necessary to kill by disinfection include bacteria, bacterial spores, viruses, cysts, and protozoa, the last being low forms of animal life such as worms and larvae. The resistance of these organisms to processes of disinfection varies according to the type and subtype of organism present, the type of disinfectant used and the form in which it is used, the number of organisms present, the physical and chemical characteristics of the water (principally temperature, amount of suspended solids and organic matter present, and pH), the amount of disinfectant applied, and the time period for which the disinfectant is applied. The disinfection of water is therefore a subject of some complexity and only an indication can be given below of the principal factors influencing the choice of a disinfectant and its efficiency.

Chlorination and the ammonia–chlorine process

9.2 Action of chlorine

The precise action by which chlorine kills bacteria in water is not known. It is believed that the chlorine compounds formed when chlorine is added to water interfere with certain enzymes in the bacterial cells which are vital for the support of life. It is not now believed that chlorine destroys bacteria by oxidation (by destruction) because the quantity used appears to be too low for this purpose. Nevertheless chlorine is a strong oxidizing agent which will break up organic matter, restrain algal growth, and convert iron and manganese in the water to their oxidized forms (see Section 8.14).

9.3 Compounds formed

When Cl_2 is added to water free from organic matter and ammonia, hypochlorous acid is formed.

$$Cl_2 + H_2O = HOCl + HCl$$

The very weak HOCl is further dissociated into H^+ and OCl^-, the extent of the dissociation being dependent upon the pH (see Section 9.4). The hypochlorous acid HOCl and hypochlorite ion OCl^- are together known as the 'free available chlorine'.

If ammonia is present in the water, or if ammonia is added to the water, then other compounds will be formed: monochloramine NH_2Cl, dichloramine $NHCl_2$, and trichloramine NCl_3 (nitrogen trichloride). Of these compounds, the monochloramine and the dichloramine together, in total, are known as the 'combined available chlorine'. In the investigation of these reactions in more detail the work of Palin has been prominent[1]. When the ratio of chlorine to ammonia is less than 5:1 the residual is mainly monochloramine. As the ratio by weight increases, dichloramine is produced until at 10:1 there are almost equal amounts of monochloramine and dichloramine. As the ratio further increases to 20:1, trichloramine or nitrogen trichloride begins to form, together with increased amounts of free available chlorine. These reactions are not precise: they are influenced by temperature, pH, and the time of contact.

The free available chlorine is many times more powerful than the combined available chlorine as a bactericide. Quoting from the work of Butterfield *et al.*[2], Whitlock[3] estimates that 25 times as much combined available chlorine must be used to achieve the same degree of kill as free chlorine in the same time. If similar doses of free and combined chlorine are used then the combined chlorine will take 100 times as long as the free chlorine to achieve the same degree of kill. Since ammonia is often naturally present in a water, it is therefore usual to add sufficient chlorine to react with the amount of natural ammonia present and a further dose of chlorine sufficient to create an excess of free chlorine for speedy disinfection. Free chlorine acts more rapidly in an acid or neutral water than under alkaline conditions and this indicates that hypochlorous acid is a more powerful bactericide than ionized hypochlorite. Of the combined chlorine, i.e. of the chloramine, the dichloramine is more powerful than the monochloramine. There follow from these findings a number of consequences which are set out in the next section.

9.4 Factors influencing the disinfection efficiency of chlorine

It is important to take into account the following factors when treating water with chlorine.

Effect of turbidity and organic matter It is sometimes found advantageous to apply chlorination at more than one stage in the treatment of a water, the first stage is then referred to as prechlorination, especially when applied before coagulation and filtration treatment. The effect of organic impurity and of turbidity as well as ammonia in the water is to make it difficult to obtain free residual chlorine. Also the penetration of chlorine and therefore the destruction of bacteria in particles of suspended matter may be very uncertain. It is always necessary, therefore, that disinfection by chlorine or other agents is completed as a final stage in water which is free from turbidity. Even in the chlorination of water under small-scale emergency conditions, e.g. for military use, it is accepted that some form of filtration must be applied unless the water is already free from suspended matter.

Absorption of chlorine by metallic compounds A substantial amount of chlorine may be used to convert iron and manganese in solution in the water into their higher states of oxidation which are insoluble in water (see Sections 8.14 and 8.15). Removal of the iron and manganese is therefore desirable.

Absorption of chlorine by ammonia compounds The ammonia compounds may exist in organic matter or, alternatively, ammonia may exist separately from organic matter (see Section 6.5) and in either case they will form combined available chlorine which, as we have seen above, is not so effective a bactericide as free available chlorine. Some remarks of E Windle Taylor in a report on chlorination in 1954[4] are relevant and interesting:

> 'A difficulty arose during very cold weather. The biochemical oxidation of ammonia in the water by organisms in the sand of the filter beds stops when the temperature falls to about 4 °C. The filtered water then contains ammonia, and some of the added chlorine is taken in to form chloramine. When this happens it becomes not only too costly, but impracticable to oxidise all the ammonia with chlorine, for even 2 hours is not sufficient for the reaction to finish at the low temperature. Unavoidably the treatment becomes chloramination, the ammonia being native to the water as a result of the cold conditions.'

Low temperature causes delay in disinfection The difference in kill rate of bacteria between the temperatures of 2 and 20 °C is noticeable both with free and combined chlorine. A very substantial decrease in killing power takes place with lowering of temperature. This must be borne in mind when fixing the contact period.

Time of contact is important Even the sterilizing effect of free chlorine is not instantaneous. The percentage kill of bacteria depends upon the time of contact between the chlorine and the bacteria. A contact period of less than 10 minutes would seldom be regarded as suitable for disinfection of a water supply, even with the highest doses applied in favourable circumstances. The time that must actually be allowed for contact depends upon a careful appraisal of all the factors listed here. For disinfection by free chlorine acting in a clear water a theoretical contact time of 30 minutes may be suitable, but the design of any tank or arrangement for time of contact or retention must be taken into account here because such tanks are liable to short-circuiting. Most tanks will only have an efficiency of about 33%, i.e. the actual contact time they provide will be found to be one-third of the theoretical contact time, the latter being the volume of water in the tank divided by the rate of throughput. Baffles do not necessarily increase the efficiency of a tank since they may only increase the 'streaming velocity' of the water through a longer path; weirs tend to be more effective.

Increasing pH reduces effectiveness of chlorine In both free chlorination and combined chlorination the more effective sterilizing compounds, i.e. hypochlorous acid and dichloramine respectively, are formed in greater quantities at low pH values than at high values. For pH values up to 6.7 at least 90% of the free available chlorine, when ammonia is not present, comes from the hypochlorous acid HOCl (see Section 9.3). As the pH value rises the dissociation favours the production of the hypochlorite ion OCl^- so that, at pH 9 and 0 °C, only 4.5% of the free available chlorine is present as hypochlorous acid. Again Whitlock may be quoted (reviewing Butterfield's experiments): 'There was a profound difference in the extent of bacterial kill in the case of chloramine and a significant difference in the case of free chlorine when the pH was lowered from 9.8 to 7.0, the temperature remaining constant at about 20 to 25 °C.'.

9.5 Efficiency of chlorine in relation to bacteria and viruses

Other things being equal, the greater the numbers of bacteria the longer will be the time necessary to reduce them below a given figure, e.g. per 100 ml of water. The chlorine dose required to meet sudden intermittent and unpredictable pollution is therefore often greater than that required to meet pollution of a steady known degree. In regard to the nature of the bacteria themselves an important matter to note is that the spores of bacteria are well known to be more resistant to the action of chlorine than are the bacteria; fortunately the bacteria causing most water-borne diseases are not spore formers, and the persistence of a small number of *Cl. perfringens* per 100 ml of a water supply is not considered significant for health. The work of Butterfield *et al.*[2] showed that under nearly all conditions the typhoid bacillus and other enteric pathogenic bacteria are at least as susceptible as *E. coli* to chlorination.

With respect to viruses, however, there is evidence that at least some viruses are more resistant to chlorine than *E. coli*. Poynter *et al.*[5] in 1972 reported that Russian experiments had indicated that higher levels of residual chlorine and longer periods of contact were required to free water from viruses than were required to destroy *E. coli*, and similar results were obtained by Scarpino *et al.*[6]. However, the degree of kill in a given time depends upon the pH of the water and also upon the proportions of hypochlorous acid (HOCl) and hypochlorite ion (OCl^-) formed[5] (see Section 9.3). The subject is complex and research findings are not yet conclusive. Among the complications is that tests for *E. coli* are normally made on 100 ml samples, whereas samples tested for viruses must be very much larger (15 to 20 litres) because of the greater scarcity of viruses. Furthermore *E. coli* is taken only as an indicator of pollution and its presence does not necessarily imply that pathogenic bacteria are also present, whereas the presence of one virus has been shown to be sufficient to cause infection[7].

9.6 Chlorination and the production of trihalomethanes

As discussed in Section 6.44 research commencing in 1974 has shown that when chlorine is applied to a water containing organic matter, derivatives of methane are formed, known as trihalomethanes. Some of these (notably chloroform) are known to be carcinogenic at high dose levels to rats and mice. Although they have not been shown to be carcinogenic to humans at the low levels found in drinking water, the assumption is made for safety purposes that their presence may pose a potential human health risk and therefore their production should be minimized or avoided. In 1979 the US Environmental Protection Agency (EPA) advised that the maximum permitted level for total trihalomethanes in water delivered to the consumer should be 0.10 mg/l: a survey of 113 cities in the USA in 1975 found levels up to 0.695 mg/l for total trihalomethanes and up to 0.54 mg/l for chloroform itself[8].

The consequence of these findings has been that emphasis is now being placed on the need to eliminate, or to reduce as much as possible, the organic content, or 'precursors', in a water before chlorine is applied. Whilst there are means of reducing trihalomethanes when formed (see Section 6.44), one way to reduce their formation is to avoid prechlorination of a raw or partially treated water which contains organic matter and to rely only upon final chlorination to disinfect a water after it has been clarified and filtered, when its organic content should be nil or very much reduced.

It must be emphasized that the present state of knowledge provides no evidence justifying the abandonment of proved and appropriate chlorination techniques for the disinfection of water. Vogt[8] sets out the complex 12-stage approach which the EPA advises should be followed before any change of chlorine disinfection procedure is contemplated. 'The basic operation principle,' he states, 'should be to maximize precursor removal prior to the addition of the oxidant, so as to minimize disinfection demand and byproduct formation.' However, regardless of the changes which are adopted, 'the microbiological quality of the drinking water must not be compromised'.

9.7 The ammonia–chlorine process

Combined available chlorine or 'chloramine' resulting from the reaction of chlorine with ammonia in water (see Section 9.3) is not commonly used as a primary disinfectant because of the acknowledged greater efficiency of free chlorine. However, ammonia is sometimes deliberately added to produce a chloramine residual in the final water which, passing into the distribution system, is longer lasting than free chlorine. This can assist in the control of 'aftergrowth' of bacteria or low forms of animal life in a distribution system, or may be used to control algae, or to reduce taste problems which arise from the application of chlorine alone. The ratio of ammonia to chlorine is usually in the range 1:3 to 1:4 and where this type of residual chloramination is adopted the ammonia is usually added after the free chlorine has had time to disinfect the water.

However, where a water already contains ammonia the production of chloramine is unavoidable (see Section 9.4) and Mackenzie[9] noted that in practice this ammonia–chlorine treatment, i.e. adding chlorine after ammonia is present, could act as a more effective disinfectant than was expected. Hoather[10] showed that this arose because the chlorine killed bacteria more quickly than it combined with ammonia, but Hoff and Geldreich[11] reported that further researches have indicated that the process is dependent upon a number of factors not yet fully investigated, and that it should only be approached with caution for primary disinfection purposes.

In general, chloramine is not in widespread use except where 'breakpoint chlorination' (see Section 9.8) proves impracticable or for the maintenance of a residual in the distribution system. However, Vogt[8] suggests that where some form of prechlorination must be adopted the addition of ammonia to produce chloramine can substantially reduce the production of trihalomethanes because the surplus free chlorine will tend to associate with the ammonia, thereby minimizing further formation of trihalomethanes. He quotes three cases where this treatment has been successful.

9.8 Breakpoint (or free residual) chlorination

If a water naturally contains ammonia (or organic matter) it has been found that the addition of chlorine first causes the formation of chloramine until the ratio of chlorine to ammonia compounds reaches about 5:1. At this point the addition of further chlorine causes a reduction of the chloramines because of their oxidation by the excess chlorine. When this reaction has been completed, at a ratio of chlorine to ammonia nearer 10:1, the addition of further chlorine results in the formation of free chlorine. The point at which this free chlorine begins to form is called the 'breakpoint' for the water. The effect on the ratio between chlorine applied and amount of free chlorine produced in a water is shown in Fig. 9.1. Quite frequently the breakpoint is shown not so much by a pronounced 'dip' in the curve of residuals as by a flattening of the curve, the characteristic depending upon the type of water being treated. Where there is no ammonia (or albuminoid ammonia) present in the water there will be no breakpoint since free chlorine will become available in increasing quantities according to the increase of the dose applied.

The principal advantage of breakpoint chlorination is that it ensures the production of free chlorine which is much more effective than chloramine as a disinfectant. Taste and odour problems can be reduced and there may be colour removal as well. The process requires good control to prevent the dosage from dropping below breakpoint level or tastes and odours may suddenly be intensified.

9.9 Superchlorination and dechlorination

By superchlorination is meant the dosing of a water with a heavy dose of chlorine, often much larger than the usual condition of the water demands. This normally means that free chlorine will be obtained and to that extent the treatment is the same as in Section 9.8. The advantage of this method is that the water may be prepared to meet any intermittent pollution which is possible or likely. Thus, in the case of a well water which remains free of contamination for long periods of the year, sudden serious pollution after heavy rainfall may be possible and the onset of such pollution can never be forecast. The normally pure water may only require a chlorine dose of the order of 0.2 mg/l. To wait for the pollution to occur and then to increase the dose might be impracticable, and therefore a continuous high dose rate is applied and an adequate time of contact is given for the worst possible conditions so that the supply may be made safe against any eventuality. After the contact period the water may be dechlorinated by the injection of

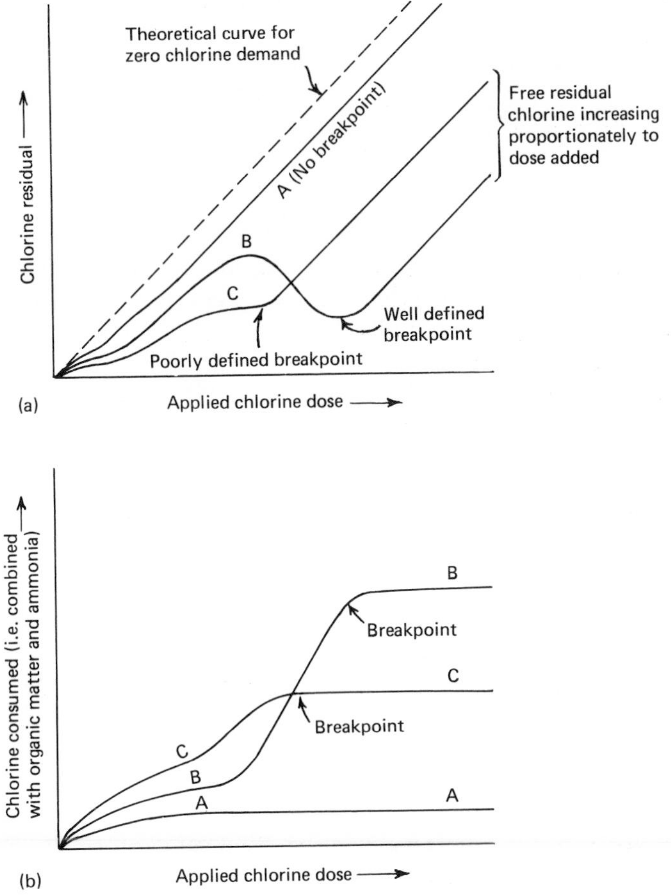

Fig. 9.1 (a) typical chlorine dose: residual curves. (b) typical chlorine consumption curves. (Note that the temperature, pH value, and contact time all affect the shape of the curves as well.)

sulphur dioxide, and this sulphur dioxide dose may be so controlled that either the whole of the chlorine is removed or only a part is removed, leaving a residual of the desired order to go into supply. With automatic methods of control and 'fail-safe' arrangements these processes have become more sophisticated.

9.10 Taste troubles

It is difficult to free pure chlorine as such in water, even at large doses of up to 1 mg/l. Tastes are caused by the reaction of chlorine with compounds in the water and then they are very noticeable and give rise to many complaints. The tastes are, however, fleeting and difficult to track down to their real source. These tastes are generally 'medicinal', 'iodoform', or 'chlorophenolic' and arise from very small quantities of oils, organic matter, or phenols in the water. One part of some phenols in 500 million parts of water can give rise to a detectable taste in the presence of chlorine. Algae may give rise to the oils causing the taste reaction and even an extremely pure water, such as a water from a deep boring in the chalk, can give rise to taste troubles and complaints when a residual of chlorine is carried into the distribution system. It is often difficult to ascertain whether the taste complained of by a consumer arises from the quality of the water itself as it leaves the works, or whether it is caused locally by an infinitely small amount of grease or other material on the utensil used for drinking. The fact that such complaints are then sporadic, isolated, and do not conform to any pattern of time or place seems to indicate that they are short lived, probably locally arising, and lie upon the threshold of detection by the average palate. Yet, when met, they are quite definite.

Such taste troubles indicate the value of superchlorination and chlorination above the breakpoint since these are two good methods of breaking down and oxidizing the taste- and odour-producing substances. It is noticeable that the most elusive problems of taste production occur most frequently in pure waters, where even as little as 0.02 mg/l residual left in the water going into supply may bring about complaints; residuals of ten times this amount, i.e. 0.2 mg/l, are regularly held in treated river and impounding reservoir waters with no complaints.

9.11 Chlorine dosage rates and contact times generally adopted

Chlorination is still the most widely used disinfection procedure. When used the primary requisites for adequate disinfection are:

(1) a clear water, free of suspended solids and organic matter,
(2) a high enough dosage rate to produce an adequate amount of free chlorine in the water,
(3) adequate contact time.

The dosage rates and contact times adopted in any particular case depend upon a careful appraisal of all the factors discussed in Sections 9.4 to 9.9 above in the light of adequate sample analyses of the water and as guided by specialist advice. In general chlorine dosage rates to final treated waters are usually in the range 0.5 to 2.0 mg/l: the lower rates tending to be those used with clear well waters; the higher rates relating to treated surface waters or to well supplies potentially liable to experience sudden pollution where superchlorination followed by dechlorination after adequate contact time may be advised.

Theoretical contact times, i.e. volume of stored water divided by throughput rate, should be not less than 30 minutes, preferably one hour should be given. These normally imply actual contact times in tanks of 10 minutes and 30 minutes respectively. Where protection against viruses in the final water is known, or suspected, to be necessary a larger chlorine dose and up to three times the contact time required for *E. coli* removal may be required. The WHO 1971 International Standards (see Section 6.1) state: 'In practice 0.5 mg/l of free chlorine for one hour is sufficient to inactivate virus, even in water that was originally polluted.'.

Application of chlorine

9.12 Injection of chlorine

Chlorine is available as a liquified gas, contained in cylinders or drums, and for small supplies it is available as hypochlorite solution. The gas is liquified and under pressure in these containers, the cylinders containing about 30 kg of liquid chlorine and the drums containing about 860 kg. There are other smaller or larger containers, but these two are the most usual sizes. In a few very large works the chlorine gas is stored in much larger pressure vessels which are refilled from tankers. The pressure within the containers is of the order of 0.275 N/mm^2 at 0 °C to about 0.550 N/mm^2 at 21 °C. When such bulk storage of chlorine is adopted special arrangements have to be made to handle chlorine gas blown off from the containers as they are filled, and temperature control is also important.

Chlorine should not be injected direct from containers into the water: the practice is dangerous and the dosage rate can be neither known accurately nor kept under control as it will reduce as pressure within the container lowers due to the withdrawal of gas or the lowering of temperature. For all normal operating circumstances it is necessary to use a chlorinating apparatus which both controls and measures the gas applied. Such an apparatus is shown in Fig. 9.2 and Plate 13A. The gas is withdrawn from the cylinder and is reduced in pressure and then made to pass through an orifice of exact dimensions, the rate of flow of the gas being controlled by the pressure applied across the orifice. If necessary this control of the rate of flow of the gas can be made to vary automatically so that a constant residual amount of chlorine is left in a varying flow of water (see Fig. 9.3). Once the flow of gas has been controlled it is mixed with a small flow of water, thus forming a concentrated chlorine solution which is passed on to the injection point.

If the pressure required to inject the chlorine solution is not too high this pressure may be applied to the solution water before it mixes with the chlorine gas. At the chlorinator this unchlorinated water is passed through an orifice causing its kinetic energy (thus its speed) to increase,

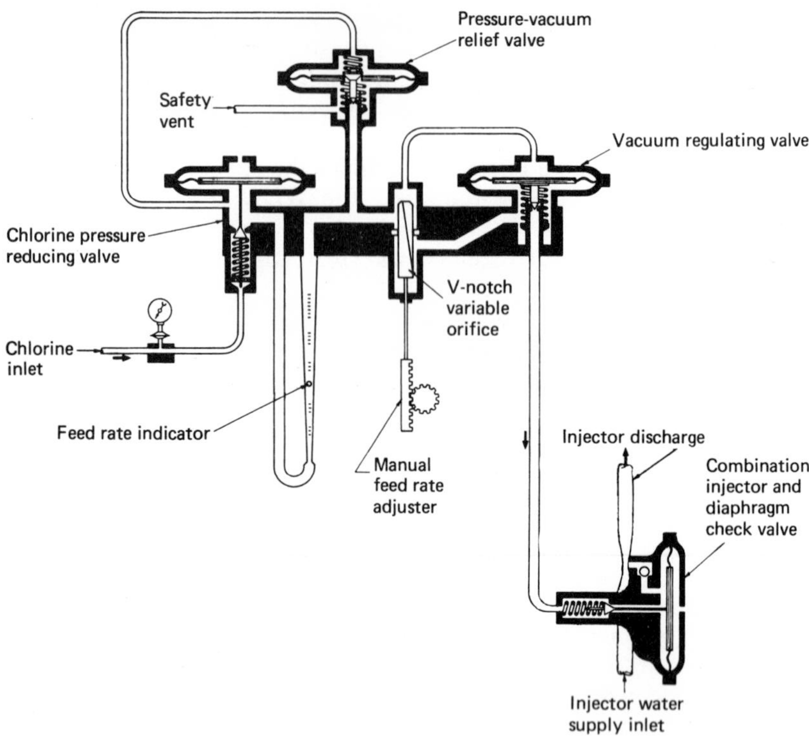

Fig. 9.2 Diagram showing the control of gas flow through a chlorinator. The chlorine gas enters on the left of the diagram, is controlled as to pressure and flow, and is then mixed with high pressure injection water as shown on the right of the diagram. (Wallace & Tiernan Ltd.)

and its pressure to fall, and so at that moment it can suck in the chlorine gas with which it mixes. Downstream of the orifice the pipe diverges so that the original pressure is nearly regained: this pressure may be sufficient to inject the solution at the point of application without further pumping. If, however, the pressure required to apply the solution is high it may be necessary to increase the pressure of the solution water after it has mixed with the chlorine; the pump used for this operation must be made of non-corroding metals because the chlorine solution is very corrosive.

When chlorine gas is 'dry', i.e. no water is present, the gas is not corrosive, but in the presence of moisture (which it absorbs) it is very corrosive. Hence most parts of a chlorinating apparatus must be made of non-corroding materials, e.g. glass or plastics. Even so, if there are any small leaks about a chlorinating apparatus, rapid corrosion of iron

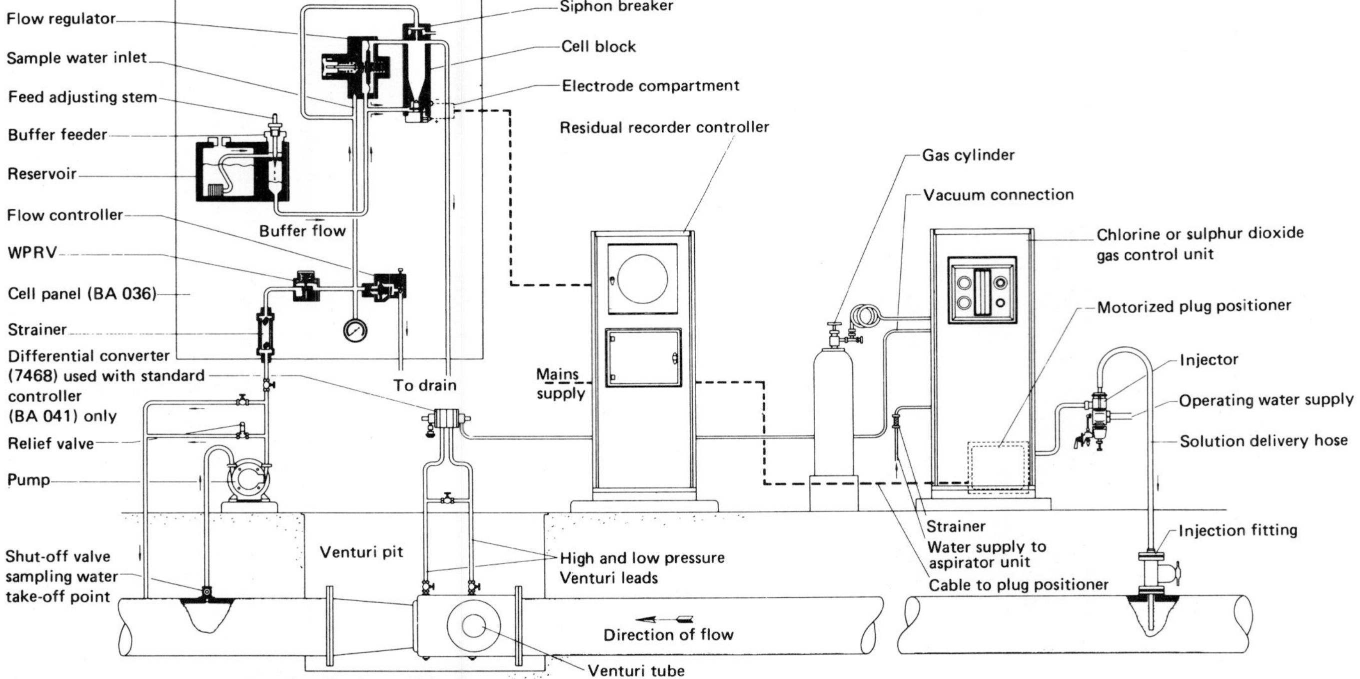

Note: In practice, the sample water take-off point and injection fitting must be positioned below the centreline of the main.

Fig. 9.3 Flow diagram for chlorination control. (Wallace & Tiernan Ltd.)

and steel nearby will be brought about. Pipelines for the delivery of the chlorine solution may be made of plastic-coated steel for high pressures or of rigid PVC for lesser pressures. Chlorinating apparatus is usually installed in duplicate and a full range of spare parts should be kept in store. The chlorine cylinders or drums should always be installed in a separate room from the chlorinating apparatus, having access from the open air. Doors should open outwards. A weighing machine should be provided and the amount of chlorine used daily should be recorded; this ensures a check that the dosage rate applied is correct. A respirator should be available outside the chlorine store or chlorinating room, away from any possible area of contamination. Chlorine gas is heavier than air and therefore ventilators equipped with exhaust fans discharging to the open air should be arranged at skirting level. Equipment can be supplied to detect chlorine gas leaks and to transmit an alarm state to a supervisor. In the past such equipment has not always been reliable, but manufacturers now claim an accuracy of detection down to 2 ml/m^3 of chlorine in air.

In cold weather the liquid chlorine in the cylinders or drums may not gasify readily, especially if the rate of withdrawal required is high. (The change of state from liquid to gaseous state absorbs heat.) Therefore the store room should be maintained at a reasonable warmth of about 10 to 15 °C and the chlorinating room should be about 3 °C higher to prevent reliquification of the gas within the pipelines. Secure apparatus should be provided for completely safe handling of drums and cylinders. With 70 kg cylinders and about 15 °C temperature 1.4 kg/h of chlorine gas can be withdrawn, or 4 kg/h on an intermittent basis. Using 860 kg drums the withdrawal rates are between 4.5 and 9.0 kg/h continuous, and 36 kg/h intermittent. When more than 18 kg/h is required continuously then liquid chlorine should be discharged to an evaporator.

9.13 Other apparatus used in connection with chlorination

Where dechlorination is adopted, a very similar apparatus to the chlorinator is used for injecting sulphur dioxide into the water. The amount of sulphur dioxide which needs to be applied to dechlorinate the water completely is 0.9 mg/l of sulphur dioxide to 1.0 mg/l of chlorine residual in the water. The dechlorinating action is almost immediate and no contact time needs to be specially provided. There is no resulting residual chemical in the water other than a small trace of acid which is quickly neutralized by the alkalinity of the water. Chlorine residual controllers can automatically control the rate of chlorine or of sulphur dioxide applied so that a constant amount of residual chlorine is left in the water. They can also be arranged to sound an alarm if the residual drops below a permitted value, and for remotely controlled works arrangements can be made for the pumps or works to be automatically shut down if there is a failure in chlorination. Automatic changeover devices are also available for bringing a new gas cylinder or drum into service when the one in service has become exhausted.

9.14 Point of application of chlorine

It is convenient to inject chlorine into water where the pressure is not too high, but the primary requirement is that any point of application must be such as to give adequate time of contact for the chlorine with the water before this leaves the works or before the water is dechlorinated. For preference, the point of application should be just upstream of a weir or Venturi tube or some other similar point where turbulence of the water will ensure thorough mixing of the chlorine. Aeration after the addition of chlorine is not objectionable as it will only reduce the chlorine content by a very small amount, but long exposure to sunlight will result in a decrease in the chlorine content.

Chlorine is sometimes added direct to a well or borehole, but this practice is not to be recommended. If it is adopted it is never possible to know either the degree of pollution or its incidence in the original raw water. In consequence the residual relative to the initial dose of chlorine can be the only controlling factor when this method is used. Other disadvantages which may follow are incomplete mixing of the chlorine with the water in the well since the chlorine dose may be applied to 'dead' water in the well, corrosion of the well or borehole lining, and corrosion of the pump impellers.

9.15 Testing for chlorine

The main test used for residual chlorine for many years has been that using orthotolidine and this test is still in use in some places outside the UK. However, orthotolidine is now recognized as being among those chemical substances capable of causing cancer. It is now a 'controlled substance' under the Carcinogenic Substances Regulations 1967 which apply to factories and thus to premises where water supplies are treated, and which require medical examinations of employed persons using the substance.

Following the introduction of these regulations, the use of orthotolidine for water testing has been largely superseded in Britain by the Palin test[12] using DPD* which was already established as a useful alternative and supplement to orthotolidine. Depending upon the information required, the method may be used simply for total residual chlorine, to differentiate between free and combined residual chlorine, or to estimate separately free chlorine, monochloramine, and dichloramine. (A supplementary procedure is available for nitrogen trichloride.)

The necessary reagents are supplied as tablets with detailed instructions. The red colours obtained are matched in the usual simple types of

* *N,N*-diethyl-paraphenylene diamine.

apparatus, e.g. 'Nesslerisers' or 'comparators', with ranges of standard colour discs from as low as 0.02 mg/l up to 10 mg/l of chlorine, as required. The user does not need to study the contents of the several tablet reagents, but it is of interest to note that the function of No. 3 tablet in allowing the determination of combined residual chlorine depends upon its contents of potassium iodide. Alternatively the necessary reagents can be prepared as solutions, but this is more complicated and the solutions have very limited stability for keeping.

The only substance likely to be present in waters to give a false reading is manganese (in its higher states of oxidation). Correction for this interference is applied by means of a separate test in which an arsenite solution is added before the DPD tablet (substantially the same correction was applied when orthotolidine was used).

The test procedure described above is for laboratory or plant use*. It should be regularly applied by the attendants at the treatment works. If the test procedures are properly organized in the first place and conscientiously carried out thereafter, they provide final evidence that a water has been properly chlorinated before it leaves the works. Hence a sampling point for such tests should be on the outgoing supply main from the works where it is intended that a residual should be carried into supply. The frequency of testing depends upon the quality of the water and its liability to pollution and how important chlorination is for the safety of the supply. In general the test is not so onerous that it cannot be carried out at least as frequently as once per shift, i.e. every eight hours. A greater frequency may be desirable if no automatic residual chlorine recorder has been installed, but where there is a residual recorder its results should be checked regularly by the chemical test.

It is usual to issue strict instructions concerning the application of chlorine to the water to the effect that no water at all may be put into supply during any failure of the chlorinating plant. If a chlorine residual is maintained in the distribution system it should be checked at least once per week, choosing different sampling points from time to time.

Chlorine dioxide

9.16 Disinfection using chlorine dioxide

Chlorine dioxide is most commonly used as a disinfectant in cases where problems of taste and odour arise, particularly those arising from the presence of phenols[13,14]. It can also be used for the oxidation of organics and the reduction of iron, manganese, and colour. In addition it provides a relatively stable disinfectant residual in distribution mains.

The usual production on site is achieved by adding chlorinated water from a normal chlorinator to a sodium chlorite solution ($NaClO_2$). Since

*For field testing see Section 9.20.

an excess of chlorine must be added to complete the reaction the product contains both free chlorine and chlorine dioxide (ClO_2). In a less frequently used alternative process concentrated solutions of sodium chlorite and hydrochloric acid are mixed to produce chlorine dioxide and there is no excess of free chlorine[15]. The process is more costly than simple chlorination.

Apart from its superior taste and odour prevention or reduction properties, chlorine dioxide has a bactericidal efficiency generally shown to be at least as great as that of free chlorine, with better disinfecting properties at high pH values. Its viricidal efficiency has not been extensively investigated. Chlorine dioxide has the advantage of not reacting with ammonia so that its use can be advantageous for waters having a high ammonia content (for which reason it is frequently used for disinfection of swimming pool waters). Vogt[8] suggests that its use instead of chlorine may be of value in limiting the production of trihalomethanes, provided that the amount of free chlorine associated with its production is small. If a large dose of chlorine dioxide is applied, sulphur dioxide can be used for dechlorination in the same manner as for chlorine alone.

Ozone

9.17 Action of ozone

Ozone is a gas, O_3, which has a powerful oxidizing effect causing rapid and effective disinfection of a clear water. It has the considerable advantage of reducing colour, although mostly by oxidizing organic matter rather than by completely removing it. It can improve palatability, although it may not be very effective against 'earthy' or 'musty' tastes or odours. In the popular view it is regarded as a 'natural' means of disinfecting a water and is therefore not objected to as chlorination sometimes is. Criticisms of the use of ozone are that it must be manufactured on site and this is costly and complicated; it leaves no persistent residual in a water; it may cause unwanted precipitation of iron and manganese; it is not suitable for a turbid water.

The effect of ozone is rapid, the usual contact time being between 5 and 15 minutes. Dosage is of the order of 1 to 2 mg/l. Excess ozone after treatment may be removed by cascading or giving a further short detention time: these processes are adopted where it is necessary to prevent corrosion effects arising. Tests have shown that an adequate dose of ozone will rapidly destroy bacteria, bacterial spores, and viruses. The ability to destroy spores within the usual contact period is an advantage since many of these are not destroyed by chlorine except when chlorine doses are high. The work of Evison[16] on the inactivation of viruses by ozone is of particular interest. The WHO 1971 International Standards advise that: '0.4 mg/l of free ozone for 4 minutes has been found to inactivate virus, but somewhat more rigorous treatment would perhaps be desirable because the resistance of hepatitis virus to ozone is unknown'. This advice is apparently based upon the work of Coin, 1964–67 (reported by Rice *et al.*[17]), which related to poliomyelitis virus only. The free ozone level required is the level after the initial ozone demand is satisfied. A higher residual ozone and longer contact time are usually adopted in practice.

Greenberg[18] states that it is generally agreed that ozonation does not produce trihalomethanes and Vogt[8] suggests that consequently the use of ozone to remove precursors may offer an advantage over chlorine, but this view is offset by the fact that the health significance of the many organic oxidation products of ozone is not yet known[18]. The lack of a persistent residual from the use of ozone also raises problems since this defect is overcome in many plants by the addition of a small amount of chlorine to the final treated water. Unfortunately Rice *et al.*[17] report that this has been found to cause an increase of trihalomethanes in the final water at some plants, sometimes in excess of that produced by chlorine alone. Sankey and Whatmough[19] reporting experience at the Watchgate treatment works in the UK in 1980 indicated doubts as to whether chlorine and ozone should be used conjunctively since they obtained some unexplained evidence that there might be a direct reaction between the two. However, ozone is widely used as a disinfectant, particularly in France, and its properties are at least as good as those of chlorine.

9.18 Production of ozone

Ozone is produced by passing a silent discharge of high voltage alternating current through dry air. A voltage of between 4000 and 20 000 volts is applied to dielectric plates about 6 mm apart, or to concentric tubes, through which the dry air is blown. While some ozone will be produced at the normal frequency of 50 cycles per second, it is usual to step up the frequency to 500 or 1000 cycles per second. High voltage and high frequency give higher efficiency of ozone production. The plates must be water cooled (small installations, air cooled) and the air passing between the plates must be dry and dust free. Drying is accomplished by passing the air through beds of silica gel or other desiccant or by refrigeration of the air.

The concentration of ozone produced by modern plant is of the order of 15 to 20 grammes per cubic metre of air. The ozone-containing air is then introduced into the water, either by an injector system which draws it in under reduced pressure or by forcing it under pressure through perforated pipes or ceramic material. The degree of absorption depends upon the depth of immersion of the injection apparatus below the water level in the contact tank and the fineness of the air bubbles introduced, and will vary from 60% to 90%. The volume of ozonized air to be handled is large. Ozone gas is highly toxic; whilst it is said to have a detectable odour at concentrations as low as 0.01 ppm[17], it does not have the distinct warning smell possessed by chlorine and may not occasion immediate discomfort when breathed. Careful safety measures have therefore to be adopted with its use. Any ozone in the vented gas from the absorption tower must be destroyed by carbon absorbers and monitoring equipment is usually placed nearby to ensure that this is achieved.

The efficiency of ozonization has been improved over the years, but capital and running costs are high. Sankey and Whatmough[19] reported that the total operational energy requirement at Watchgate in 1977 was equivalent to 27 watt-hours per gramme of O_3 produced, with the total cost of the order of three times the cost of equivalent chlorination. Their paper is useful in describing the problems encountered in relation to corrosion and the production of adequately dry air. However, plant development continues and higher efficiency and reliability can now be expected.

Other disinfection practices

9.19 Chlorination of small supplies

Chlorine gas injection plant is not suitable where local labour is insufficiently skilled to maintain it or the supply to be chlorinated is in too remote a location for the reliable procurement of chlorine gas and spare parts for the injection apparatus. Even the lack of an appropriate gasket can result in the plant becoming deficient because a chlorine gas leak is highly corrosive in a damp atmosphere. Also some supplies are too small to justify expensive gas chlorinating plant.

In such cases the following forms of chlorine are available for use:

(1) bleaching powder, i.e. chloride of lime,
(2) sodium hypochlorite solution,
(3) calcium hypochlorite granules,
(4) 'chlorine tablets',
(5) chlorine produced by electrolytic means.

9.20 Bleaching powder

Bleaching powder or chloride of lime is probably the cheapest form of chlorine available for the chlorination of small supplies. It contains 30% to 35% by weight of chlorine and its strength only slowly declines when the material is kept in sealed drums. The usual method is to make up a suspension of the powder in water, let the lime in it settle, and draw off the supernatant chlorine-water for dosing purposes. Assuming the bleaching powder releases 33.3% of its weight as chlorine, some typical solutions used might be as follows.

Chlorine solution strength	*Chlorine content*	*Amount of bleaching powder required at 33% w/w*
1.0% (10000 mg/l)	10 g/l	30 g/l
0.1% (1000 mg/l)	1 g/l	3 g/l
0.03% (300 mg/l)	0.3 g/l	1 g/l

Considerable problems attend the use of bleaching powder. If large quantities must be handled there is a dust nuisance so that exhauster fans should be installed and the men handling the powder should wear masks. There is a heavy sediment of lime when the bleaching powder is mixed with water; this sediment must be disposed of, but the supernatant still contains a large amount of suspended lime which tends to clog up any kind of drip-feeding device that can be devised. Many water authorities who use bleaching powder solution report a constant series of difficulties in trying to maintain dosing apparatus in working order. It is also difficult to measure out the correct dose of bleaching powder required for the chlorination of small supplies. The WHO[20] recommends first making up a 1% solution comprising 40 g of bleaching powder to one litre of water (this apparently assumes the bleaching powder has only 25% available chlorine). It is then recommended that three drops of this solution should be added to one litre of water for drinking, but this advice is not practicable for the disinfection of small community supplies. In the latter case a weaker solution is required that can be measured out in cupfuls (or Coca-Cola tinfuls) to the body of water to be disinfected.

If, say, 60 to 80 g* of bleaching powder (at 33% to 25% chlorine content) are added to 10 litres of water, and this is allowed to stand so that the lime sludge settles, the supernatant should contain about 2 g/l of chlorine (a 0.2% solution). A litre of this solution added to 1 m^3 of water will result in a dose of 2 g/m^3 or 2 mg/l which is a typical level of dosing. Either this diluted solution can be applied in the number of litres required or, for smaller quantities, the amount to be dosed can be stipulated in cupfuls or Coca-Cola tinfuls: there being five cupfuls or three Coca-Cola tinfuls to one litre (volumes 200 ml and 335 ml approximately respectively).

These strong solutions of chlorine are very unstable and will rapidly lose their chlorine content if exposed to the air or sunlight. Hence they must be stored in closed dark-walled containers. Even when so stored a 0.1% solution will lose a substantial amount of its chlorine in three to seven days. The chlorine content can be checked by a high range chlorine test based upon an iodine release method, using a comparator

* Bulk density of bleaching powder is about 400 g per 1000 ml, so a cupful (200 ml volume) will contain about 80 g.

calibrated up to 250 mg/l (0.25% solution). Also a tablet test is available for chlorine contents up to 50 mg/l*.

For the chlorination of wells various methods have been tried of suspending a pot (or a coconut) containing bleach, or bleach mixed with sand, in the well. The pot (or coconut) has a small hole drilled in it, or the pot is porous. By this means it is intended that the chlorine should slowly diffuse into the well water, but this may not always happen as the reaction between the lime and the water may create a sealing layer which prevents further diffusion of chlorine from the interior body of bleaching powder.

In general bleaching powder is an unsatisfactory medium for the application of controlled doses of chlorine on a daily basis. Its use should be avoided wherever possible because it so often results in the overdosing of a water with chlorine; consumers object to the strongly chlorinous taste of the water and in consequence will use other sources of water for drinking which are found more palatable but which may be highly polluted. However, this may be a 'counsel of perfection' as alternative sources of chlorine are more expensive and are often difficult to procure reliably in remote areas and in under-developed countries.

9.21 Sodium hypochlorite solution

Sodium hypochlorite solution is available under many brand names as an ordinary household disinfectant. It is a clear solution containing 1% to 15% w/w of available chlorine (density 1.27 kg/l at 14/15% w/w). It rapidly loses its strength when exposed to the atmosphere or sunlight, so it must be supplied in suitable small containers so that, once opened, the contents of a container are used up in about a week. This makes the use of sodium hypochlorite solution expensive as a substantial part of the cost lies in the packaging and freight. When large quantities are required, expensive carboys have to be used which must be returned to the maker for refilling or the cost is prohibitive. As a consequence sodium hypochlorite is too expensive to use for supplies in under-developed countries, although its freedom from suspended matter makes it suitable for drip-feed apparatus.

9.22 Calcium hypochlorite granules

Calcium hypochlorite granules contain 65% to 68% w/w chlorine (bulk density of granules about 0.9 kg/l). They can be supplied in 45/50 kg drums with plastic liners. They are readily soluble and granules can be dosed direct to a supply or a standard strength solution can be made up for dosing. They give a clearer solution than does bleaching powder but, after initial standing, the supernatant of a 0.5% chlorine solution will still contain up to 1% suspended matter. It is preferable to use a soft

* Information from Wilkinson & Simpson Ltd UK.

water, such as rainwater, for making up the solution as the use of a hard water results in some precipitation of its hardness. The storage life of the solution is similar to that quoted for solutions of bleaching powder above. Nevertheless the granules form one of the more useful chlorine compounds to use for small supplies in remote areas, the cost of using them being substantially less than the cost of using sodium hypochlorite solution delivered in carboys.

9.23 Chlorine tablets

Chlorine tablets are available in a range of sizes and chlorine content. They are normally too expensive to use on a large scale, but they are useful in certain circumstances for the chlorination of small supplies since the appropriate tablet size can be used for a given quantity of water. The smallest tablets contain about 0.12 g of chlorine and are used by travellers for the disinfection of drinking water. The largest tablets, containing 3 g or more of chlorine, are frequently used for disinfection of swimming pools and are suitable for use in disinfecting mains. They are available for rapid release of chlorine, or for slow release over several hours.

9.24 Electrolytic production of chlorine

Sodium hypochlorite solution can be produced by the electrolysis of a brine solution or seawater. A d.c. current is passed through the electrolyte, producing chlorine at the anode. Hydrogen gas is also evolved and must be safely vented. Solutions containing 8 g or more of chlorine per litre can be produced. Electrical power consumption is 5 to 6 kW h per kg of chlorine produced; a wide range of sizes of generators is produced. The apparatus is useful for the easy production of chlorine on site for the disinfection of small supplies, provided that an electricity supply is available. The use of a brine solution is sometimes preferable to the use of seawater because the latter will need straining before use and the manganese content must not be too high or the efficiency of the electrodes is impaired. Dyachkov[21] reports widespread use of the method in a number of countries, for example in the USSR units producing between 1 and 100 kg/d of chlorine are widely used in small town supplies and in rural areas.

9.25 Use of ultraviolet radiation

Ultraviolet (UV) radiation is only suitable for the disinfection of small supplies which are free of colour, turbidity, and suspended matter. Jepson[22] has provided a good review of the process. UV radiation lies between 15 and 400 nm* wavelength, but it is at about 260 nm that peak disinfection is achieved, falling to zero at 320 nm. The mechanism of kill

* nm = nanometre (or millimicron).

is not fully understood, and the degree of kill depends *inter alia* upon wavelength and the product of the intensity of radiation and the time of exposure. A rather wide range of lethal dosages is quoted for various percentages of kill and various organisms, so it is usual to design equipment for a high factor of safety. Minimum UV dosage specified by the US Department of Health is 16 mW s/cm^2 at 253.7 nm. The UV penetration depends upon the clarity of the water. Experimental work by Hoather[23] showed that 50% of radiation penetrated 15 cm of clear well water, but the same penetration occurred through only 5 cm of a typical treated surface water.

The UV radiation is produced by a low or medium pressure mercury arc discharge lamp in a quartz cylinder of about 50 mm diameter, sheathed in an outer metal tube of about 100 mm diameter. The water to be sterilized is passed through the annular space between the two tubes. A typical tube is about 0.5 m long, the lamp being 1.0 to 1.2 kW power with an output of 55 to 95 W in the range 230 to 330 nm. Lamps of this type described by Jepson give an exposure time of 0.74 to 1.11 seconds of radiation to the flowing water, total outputs being of the order of 60 to 100 m^3/d. Larger plants are reported from the USSR dealing with up to 3000 m^3/h, electrical consumption being quoted as averaging 10 to 16 watt-hours per m^3 for surface waters. Small UV apparatus is reported to be used on British Rail catering vehicles, the supply being mostly drawn from public supply systems so that the disinfection is an additional precautionary measure.

9.26 Other disinfectants and disinfecting procedures

Most other disinfectants are for use in special circumstances as described below.

Silver in the form of colloidal or 'katadyn' silver has a disinfecting effect. It has been used on a small scale for very small plants with outputs of only a few litres per day. Individual house supplies (for consumptive use) can be treated by passing the water through silver-impregnated 'candles' comprising hollow cylinders through which the flow is from outside to inside. The silver content of the treated water is too small to have any toxic effect on the consumer, even if such water is drunk for a lifetime. Portable units with miniature pumps are made by at least two firms and are valuable for travellers, but the process has never been adopted on a large scale. However, the sterilizing effect of silver vessels was known in antiquity.

Iodine or compounds based upon iodine, usually in tablet form, can be used as a disinfectant but only in an emergency, or temporarily, for disinfecting small quantities of water for drinking because of the possible toxic effect of iodine on susceptible individuals. Tablets are usually sized to treat one litre of water each; otherwise two drops of tincture of iodine of 2% strength will suffice for one litre of water.

Where severe pollution is probable the dose can be doubled. A contact time of at least 20 minutes should be given.

Potassium permanganate is an oxidant, but its use for disinfection is reported to be of doubtful efficacy against disease organisms and is not recommended. It is, however, widely used in treatment plants for the precipitation of iron and manganese and for taste and odour control. It is also now being considered as a substitute for prechlorination since this can result in a lesser production of trihalomethanes[24].

Boiling water is an extremely useful disinfection procedure and is reported as equally effective whether the water is clear or cloudy, relatively pure or highly contaminated[20]. For complete sterilization of a polluted water it is necessary to bring the water to 'rolling boil' and to keep it at this state for at least five minutes. At high elevations one minute extra should be given for every 1000 m above sea level because of the lower temperature at which boiling takes place. Where local supplies of water are clear but of doubtful bacteriological quality many expatriates boil all water for drinking and, after it has cooled, store it in a refrigerator. This is sound practice. The wise traveller, when offered a choice of drinks, will choose tea rather than a cold drink made from the local water supply on the basis that even if the water has not been brought to boil the process of heating will have reduced its living bacterial content.

Other disinfecting processes are normally only partial, but their value has not to be neglected either on a plant scale or for small volumes of water. Storage, clarification, softening, and filtration can all reduce the numbers of living organisms in a water. Provided that no recontamination takes place, storage can substantially reduce the numbers of pathogenic bacteria present in a raw water. The provision of large storage reservoirs such as those adopted for London's water supply is of great benefit. At the other end of the scale, in undeveloped rural areas, the two- or three-day storage of household water in clean covered earthenware pots will also give at least a partial benefit.

9.27 Disinfection of swimming pool water

The subject of swimming pool water disinfection is not dealt with here because it poses special problems arising from the heavy organic loading and continual recirculation of the water. Chlorine, ozone, and to a lesser extent bromine and other chemicals may be used for disinfection. Relatively high levels of free chlorine residual are advised, 1.5 to 2.0 mg/l, and appropriate pH control must be exercised. A series of publications giving desirable practices for treatment and disinfection are published by the DoE[25].

9.28 Disinfection of water mains and tanks

Before a new or repaired main is put into service it should be first thoroughly flushed out with clean water and then disinfected. Often the main is also swabbed initially with a foam swab to remove dirt, slime, and debris from the pipeline. Chlorine is universally used for disinfection, although the source may be chlorine gas or bleaching powder. Dosage rates are commonly not less than 20 mg/l and contact times at least 24 hours. Often much higher dosages and contact times are used.

It is frequently found difficult to achieve satisfactory coliform test results on water samples drawn from a new or repaired main which has been disinfected in the normal manner. Investigations have shown that the most common cause of this is the presence of vegetable-based lubricants and soil in the annular spaces of joints. These support bacterial growth and chlorinating the body of the water in the pipeline fails to disinfect these annular spaces. The problem can be avoided by using appropriate inert jointing materials and lubricants and ensuring that joints are as clean as possible. Sometimes disinfecting tablets are additionally left in joint spaces.

Considerable aftergrowths may occur in distribution mains, such as slimes, iron and manganese deposits and iron bacteria, corrosion products etc. In such systems it is often difficult to retain any residual chlorine in the water. Swabbing followed by slug-dosing with a heavy dose of chlorine passed slowly along the mains is the immediate cure, but the permanent cure lies in improving the quality of the water distributed. No practicable level of residual chlorine can prevent these troubles arising if algae, alum, iron, manganese, suspended solids, or organic matter etc. are permitted to enter and precipitate in the distribution system.

Tanks are normally hosed down with strong jets of clean water initially and then brushed down with a strong chlorine solution. The initial filling is made with water containing a relatively high dose of chlorine which is allowed to stand in the tank for at least 24 hours. This water is discharged to waste, the tank refilled with clean water, and after again being allowed to stand for a period a sample of the water is tested for the presence of coliform organisms.

Prevention of pollution

9.29 Basic principles

It is better to prevent pollution of the source than to attempt to remove it at the treatment works. The knowledge gained during the past ten years has reinforced the old principle that the best source of water is that which is least polluted. Recent research work has raised health questions concerning:

(1) the practice of softening,

(2) the possible production of carcinogenic substances in water from processes of disinfection,

(3) the inadvisability of a high salt intake from water,
(4) the possible long-term significance of trace amounts of organic and inorganic substances in water,
(5) the efficacy of traditional methods of treatment and disinfection against viruses,
(6) the value of *E. coli* as an indicator of pollution.

It is the responsibility of the professional water engineer to handle this new and often incomplete information rationally and to safeguard the public from gaining unreasonable fears concerning it. The information must be viewed as a whole and in context. Much of it lies below a measurable level of significance and some of it cannot pose foreseeable risks to the individual that in any way compare with other dangers of living universally accepted.

Taken as a whole the new evidence confirms the essential validity of long-standing practices in public water supply, namely:

(1) the source should be as remote as possible from pollution,
(2) the catchment to the source should be kept under regular surveillance against pollution,
(3) the quality of the raw water should be monitored,
(4) industrial wastes may contain more unknowns than domestic wastes and are the more to be avoided,
(5) storage of water gives valuable protection,
(6) the clarification of a water is a pre-eminent requirement before it is disinfected,
(7) traditional methods of disinfection using chlorine or ozone are still the best practicable barriers against the transmission of disease by water,
(8) detection of *E. coli* in water is still the best practicable method for indicating the presence of pollution.

9.30 Source protection

If the supply is drawn directly from an impounding reservoir then the methods used for sewage disposal from every property in the catchment area must be kept under observation and altered where necessary. Large effluents, or effluents which can enter the reservoir near the drawoff point, or which are of a nature presenting special hazard, should be piped below the reservoir. For ordinary domestic sewage from single properties, if the effluent is discharged some distance upstream of the reservoir, the natural purification which takes place in streams and rivers may be sufficient for the effluent to be tolerated. Each case must be examined on its merits. Properly designed septic tanks or small treatment works which will cause some degree of bacterial purification are preferable to cesspools, since emptying of the latter is always liable to be neglected. The manuring of fields and also the dipping of sheep may have to be controlled.

A river supply must necessarily be treated with even greater watchfulness than an impounding supply since there is a constant risk of new or altered effluents of a dangerous nature being discharged. Many cases of sudden pollution do arise and, of these, a large proportion are due to accidental discharges from industrial plants, the mishandling of chemicals on farms, or from road accidents involving tankers or other lorries carrying substances of a polluting nature. Oil pollution is a relatively frequent occurrence. To guard the supply it is best to arrange for continuous monitoring of the raw water both well upstream of the intake and at the intake itself. This, however, needs to be supplemented by regular inspection of the river. Liaison with other water authorities abstracting from points upstream, and with the police and local fishermen, aids the swift reporting of unusual incidence of pollution. In the positioning of the intake care must be taken not to site it too near any discharge of sewage effluent or storm-water discharges. Raw water storage, sufficient even for a few days, is usually very desirable, if only to even out variations in quality and to allow time for adjustments in treatment. At least 24 hours should be given, but preferably three or more days should be allowed.

The complete protection of an underground source against entry of pollution is frequently impossible when it is on or near the outcrop of the formation from which it draws water, or when the formation is fissured chalk or limestone. The 'catchment area' to such a source can seldom be well defined. While every visible source of pollution upon the supposed catchment should be removed or controlled as far as practicable, it is unwise to assume that, having done so, the source is necessarily 'protected'.

With this type of underground source it is constantly necessary to bear in mind that, although the source may have given pure water for many past years, pollution can occur at any time in the future. The onset of such pollution can be sudden, can occur without any warning, and may be severe. Such extreme instances are fortunately rare, but they do occur, especially when sources are in or near populated areas with building or road works in progress. As an example of what may occur a large water authority in the north of England had to close down one of its major underground sources for several months whilst an intensive search was made for the cause of the sudden pollution of a well which, up to then, had given pure water for seventy years.

Probably the safest sources of water are deep boreholes drawing from a water-bearing formation which is covered over a sufficient area by an impermeable stratum so that, as long as the borehole lining remains sound, it is almost impossible for pollution to gain access. If, however, the water-bearing formation is fissured, particular attention to the condition of any disused boreholes in the neighbourhood is necessary. These should not only be covered at the top; they should also be filled in, unless they are retained under the supervision of the water authority either for emergency use or for observation. Chlorination or other disinfection treatment is usually considered desirable as a precautionary measure to maintain a wide margin of safety, even for a water pumped from under an impermeable stratum.

9.31 Design precautions

In the design of structures, among the practices which should be adopted to minimize pollution are the following.

(1) Wells and boreholes should have an adequate depth of watertight lining to seal out surface water. No well or borehole should be chlorinated direct because it is important to test the natural condition of the water.
(2) Service reservoirs should be covered over. Air vents should be covered with durable fine meshed screens. Exterior drains should be laid to prevent high levels of groundwater outside.
(3) Pipes should be clean when laid and debris from the trench should be prevented from entering the pipeline by keeping a temporary stopper in the exposed end of the pipe. For large mains the interior should be cleaned out by hand after every two or three pipes are laid and after all interior jointing work is completed.
(4) Air valves on mains should be above the highest possible groundwater level that can occur in the pit or chamber in which the air valve is sited.
(5) Ball hydrants should not be used. (These were an old type of hydrant in which a ball was depressed to allow flow through the hydrant.)
(6) Filter beds and any other tank for the holding of treated water should be watertight and proof against the entry of groundwater.
(7) Water mains should not be laid in close proximity to sewers, but if this is unavoidable extra care must be taken to protect both the main and the sewer from leaking or fracturing.
(8) The supply mains should be kept under pressure by means of a full 24-hour supply: no pumping of supplies from the mains should be allowed which causes the pressure to reduce below atmospheric.

9.32 Administrative procedures

The production and transmission to consumers of a continuously wholesome supply of water is the water engineer's most important duty. He does not necessarily act alone in this matter: not only should he have the expert advice of a chemist and bacteriologist available in a full-time or consultative capacity (according to circumstances), but he should also have contact with a medical adviser. In Britain the local Medical.Officer of Health of the local government authority has independent statutory responsibilities to ensure that a wholesome supply of water is maintained within his area and this acts as a useful supplement to the surveillance of the supply. In practice this Medical Officer of Health is kept informed, by the water engineer, of the standards of quality maintained in the supply and is available for consultation where matters of public health are concerned.

In selecting men for employment in any situation in which they could present any risk of pollution of the water, a water authority should arrange the necessary tests to exclude any carrier of typhoid bacteria. Such a person would represent a hazard to the undertaking since these bacteria may be excreted for the remainder of a person's life, even though he may have no symptoms of the disease. All works such as the extension of wells or adits or the cleaning out of service reservoirs, filters, tanks, conduits etc. receiving treated water must be strictly supervised to see that no contamination is taken into the structure. It is usual for men working in such areas to be provided with clean gumboots which are disinfected by walking through a solution of chloride of lime or hypochlorite before entering the structure.

The day-to-day procedures for maintaining a satisfactory quality of supply without interruption consist mostly of delegating clearly defined responsibilities and of ensuring, from the top, that these responsibilities are efficiently executed. Often it is the routine duties of checking that the water is properly chlorinated, and so on, that are of paramount importance, and the wise engineer will set up systems for ensuring that such routine duties are not neglected. He may well check these routines personally from time to time to show, by his actions, that they are still of fundamental importance.

The authoritative booklet in the UK on this subject was published in 1978 by the DoE[26].

References

(1) Palin A T, Chlorine Dioxide in Water Treatment, *JIWE*, **2,** 1948, p. 61; 1950, p. 565; *WWE*, 1950, pp. 151, 189 and 248.

(2) Butterfield C H *et al.*, US Public Health Reports 58, 59 and 61, US Government Printing Office, Washington, 1943–46, also *JAWWA*, **40,** 1948, p. 1305.

(3) Whitlock E A, The Application of Chlorine in the Treatment of Water, *WWE*, January 1953, p. 12.

(4) Windle Taylor E, London Metropolitan Water Board Annual Report, 1954.

(5) Poynter S F B, Slade J S and Jones H H, The Disinfection of Water with Special Reference to Viruses, *JSWTE*, **22,** 1972, p. 194.

(6) Scarpino P V, Berg G, Chang S L, Dahling D and Lucas M, A Comparative Study of the Inactivation of Viruses in Water by Chlorine, *Water Research*, August 1972, p. 959.

(7) Sobsey M D, Enteric Viruses and Drinking Water Supplies, *JAWWA*, August 1975, p. 414.
(8) Vogt C and Regli S, Controlling Trihalomethanes while Attaining Disinfection, *JAWWA*, January 1981, p. 33.
(9) Mackenzie E F W, *J Inst. San. Eng.*, **39,** 1940, p. 507.
(10) Hoather R C, The Bactericidal Effect of Ammonia–Chlorine Treatment, *JIWE,* **3,** 1949, p. 507.
(11) Hoff J C and Geldreich E E, A Comparison of the Biocidal Efficiency of Alternative Disinfectants, *JAWWA*, January 1981, p. 40.
(12) Palin A T, Methods for Determination in Water of Free and Combined Available Chlorine, etc., *JIWE*, **21,** 1967, p. 537.
(13) Atkinson J W, The Control of Taste in the River Dee Supply of the West Cheshire Water Board, *WWE*, 1962, p. 146.
(14) Wallwork J F *et al.*, Identification of Phenols in the River Test and their Treatment with Chlorine Dioxide, *JSWTE*, **18,** 1969, p. 203.
(15) Malpas J F, Disinfection of Water Using Chlorine Dioxide, *JSWTE*, **22,** Part 3, 1973, p. 209.
(16) Evison L, *JBWA*, **9,** 1972, p. 14.
(17) Rice R G, Robson C M, Miller G W and Hill A G, Uses of Ozone in Drinking Water Treatment, *JAWWA*, January 1981, p. 44.
(18) Greenberg A E, Public Health Aspects of Alternative Water Disinfectants, *JAWWA*, January 1981, p. 31.
(19) Sankey K A and Whatmough P, Experience in the Use of Ozone, *JIWES*, September 1980, p. 435.
(20) *The Purification of Water on a Small Scale*, Technical Paper No. 3, WHO, International Reference Centre, The Hague, March 1973.
(21) Dyachkov A V, Recent Advances in Water Disinfection, Special Subject No. 5, *Proc. IWSA Congress*, 1976.
(22) Jepson J D, Disinfection of Water Supplies by Ultra-violet Radiation, *JSWTE*, **22,** Part 3, 1973, p. 175.
(23) Hoather R C, The Penetration of Ultra-violet Radiation and its Effect in Water, *JIWE*, **9,** 1955, p. 191.
(24) Ficek K J and Boll J E, Potassium Permanganate: An Alternative to Prechlorination, *Aqua.: JIWSA*, 1980.
(25) DoE, *Swimming Pool Disinfection Systems*, Five booklets dealing with different disinfectants, HMSO, 1979–82.
(26) DoE, *Safeguards to be Adopted in the Operation and Management of Waterworks*, HMSO, 1978.

10
Hydraulics

10.1 Limitations of use of ready-made tables

In this chapter particular attention will be paid to explaining some of the basic theories of hydraulics which must be used in practical problems. To solve hydraulic problems in the field many of the formulae used in practice are not strictly true; they are empirical formulae which work well for the practical use for which they were intended. However, there are a number of occasions where the incorrect use of an empirical formula may lead to gross error in the calculations, and there the only true answer can be obtained by returning to fundamental considerations. There are two reasons why this warning has to be given. Firstly, there are many tables and diagrams in existence which give great assistance in the solving of hydraulic problems by empirical formulae and the engineer who has not been forewarned of their limitations may easily slip into the error of using them at the wrong time because they are so readily to hand. Secondly, the use of empirical tables and nomograms erroneously, when in fact an approach by way of basic theory should have been made, can result in serious errors which affect the safety and stability of a waterworks structure. Hence this chapter is more particularly concerned with elucidating those subjects of design in hydraulics which are important to the water engineer from a practical point of view, and which are liable to be misunderstood.

10.2 Bernoulli's equation

For the water engineer Bernoulli's equation is the most useful of basic equations, to which he will find it necessary to return many times. It is usually stated as

$$\frac{u^2}{2g} + p + z = \text{constant} \qquad (10.1)$$

$$\text{kinetic energy} + \text{pressure energy} + \text{potential energy} = \text{constant}$$

where u is the velocity of a unit mass of water, p is its pressure, and z is its height above some given datum. The limitations of this equation must be carefully noticed. It applies only to steady water flow within a 'streamline' and to flow where no energy is lost through friction. A streamline is a very small cross-section of the total flow of water, so small that flow within it may be taken as smooth and changing only gradually. This formula does *not* apply to flow within a commercially sized pipe, and if the reader should see the formula applied to pipe flow he should realize that, in that case, it is a modified version of Bernoulli's equation which he sees before him. Such modification will be later explained. In the form above the equation expresses what we can readily grasp in words, namely that the total energy possessed by a unit mass of water is comprised of three parts: its kinetic energy of movement, its pressure energy, and its position energy above some given datum in a gravitational field. This equation also expresses the fact that for an ideal frictionless fluid the sum total of the energy does not change as the unit mass of water moves from one position to another, so long as no energy is taken away by friction and none added (as by a pump).

If we now consider the more practical case where energy is lost through friction, water not being an 'ideal' fluid, we shall obtain a modified form of Bernoulli's equation which relates the total energy possessed by unit mass of water at position A to that possessed at position B, measured above a common datum. This equation is

$$\frac{u_A^2}{2g} + p_A + z_A = \frac{u_B^2}{2g} + p_B + z_B + E \tag{10.2}$$

where E is the energy lost through friction as the unit mass of water moves from A to B. This may be expressed pictorially as Fig. 10.1.

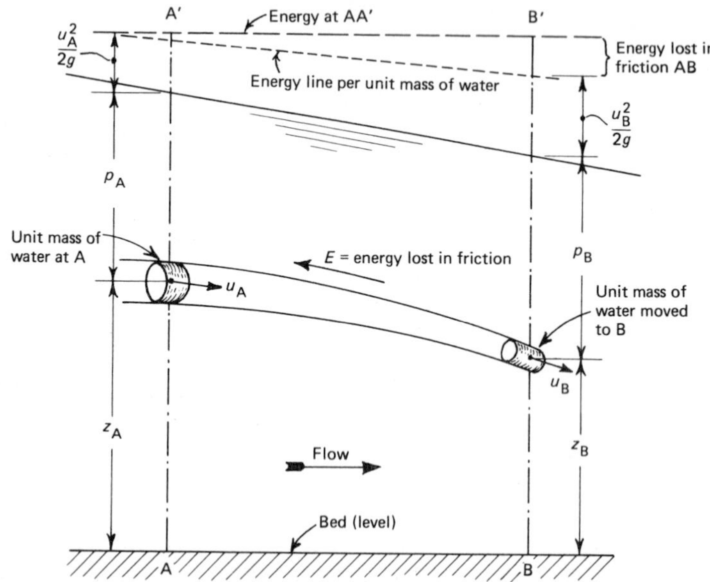

Fig. 10.1 Variation in energy as unit mass of water moves from position A to position B.

Flow in pipes

10.3 Bernoulli's equation applied to pipe flow

The flow of water within a pipe is not uniform. The walls of the pipe exert a drag on the water flow so that the speed of the water is greater at the centre of the pipe than around the perimeter. Microscopically close to the walls of the pipe the flow of the water may be said to be practically zero. At the centre of the pipe the speed of travel of the water is greatest and is about 1.2 times the average speed. Figure 10.2(a) illustrates this.

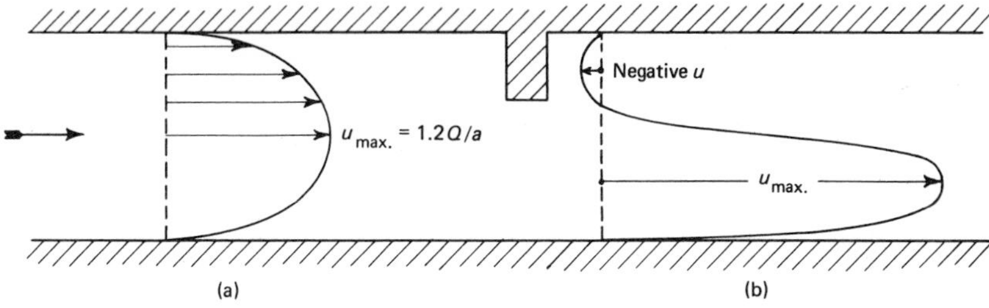

Fig. 10.2 (a) distribution of velocity of flow in a pipe and (b) effect of an obstacle in the pipe.

The ratio of the maximum speed of the water within the pipe to the average speed varies according to the pipe diameter and the roughness of the walls of the pipe and, of course, it varies very greatly if an obstacle occurs within the pipe, as Fig. 10.2(b) shows. The average speed of water within the pipe is easy to compute: it is simply the rate of discharge divided by the cross-sectional area of the pipe, i.e. Q/a where Q is the discharge in, say, cubic metres per second and a is the cross-sectional area of the pipe bore in square metres. Let us denote this velocity Q/a as V to distinguish it from individual streamline flow velocities u_1, u_2, u_3 etc.

We know that Bernoulli's equation can apply to any streamline flow, but to apply the same equation to flow within the pipe as a whole we see that we must sum the energy equations for all the separate streamlines because their velocities are not all the same. The true summation of the kinetic energy of the water at any cross-section of the pipe would be

$$\sum \frac{u_1^2}{2g} + \frac{u_2^2}{2g} + \frac{u_3^2}{2g} + \ldots + \frac{u_n^2}{2g}$$

where the values of $u_1, u_2, \ldots, u_n$ would range from zero to approximately $1.2V$. Also the summation of the pressure on unit masses is $\Sigma p_1 + p_2 + \ldots + p_n$ and the sum of the position energy $\Sigma z_1 + z_2 + \ldots + z_n$.

If we now intend to write Bernoulli's equation for the pipeline flow as a whole and we write

$$\frac{V_A^2}{2g} + P_A + Z_A = \frac{V_B^2}{2g} + P_B + Z_B + E \qquad (10.3)$$

we have a relationship which is not strictly true because V_A and V_B are the mean velocities, i.e. the Q/a values at sections A and B of the pipeline, and $V^2/2g$ is not equal to the summation of the individual $u^2/2g$ values for the streamlines. In practice the error incurred by this assumption is so small as to be negligible, the ratio of $V^2/2g:\Sigma u^2/2g$ being 0.98:1, and the errors of measurement in the field are such that a difference of this order does not affect the practical value of the calculation. We have therefore made the assumption that the ratio is 1:1 and that the total kinetic energy flow across a section of the pipeline is $V^2/2g$ where V is Q/a or the mean velocity of flow.

Similarly we should note that we have assumed P_A, P_B and Z_A, Z_B are values measured to the centreline of the pipe, i.e. we have assumed that all the individual unit mass elements of water are concentrated at the centreline. The former again introduces negligible error and the latter no error, so long as the pipe is flowing full. However, we must realize that once the pipeline is partly empty, and there is a free water surface, then the centre of mass and the point of application of the average pressure on a cross-section may depart widely from the geometric centre of the pipe. This becomes of importance when we consider channel flow; in the meantime it must be realized that even the modified form of Bernoulli's equation above does not necessarily apply to a pipe which flows partly empty.

Now that this problem has been examined in detail we can see that, without practical error, we have an equation which connects flow at any section of a pipeline with the flow at any other section of the same line.

10.4 Units

The most common practice is to express all terms of the above equation in 'metres head of water', this being found the most practicable. This system of units has, in fact, been presupposed by writing Bernoulli's equation in the form above, although instead of metres we could measure in inches or feet head of water or any other height of water column. Assuming we call P 'metres head of water', then Z must also be measured in metres and the kinetic energy is comprised of $(V\,\text{m/s})^2/9.81\,\text{m/s}^2 = \text{metres}$, i.e. we must keep V in m/s and use 9.81 as the value for g.

10.5 Practical application to flow in pipes

If the pipeline has a uniform diameter through the length considered then $V_A = V_B$ and the kinetic energy term is constant. Therefore Bernoulli's equation reduces to

$$P_A + Z_A = P_B + Z_B + E$$

If, furthermore, the pipeline has the same level at A as it has at B then

$$P_A - P_B = E$$

which means that the pressure lost is equal to the energy lost through friction. If, as occurs in practice, the pipe is of constant diameter, but undulates in level between various sections, then we can measure P_A and P_B at any two points and determine the heights Z_A and Z_B above a given datum at these points. Then

$$\text{friction loss } E = (P_A + Z_A) - (P_B + Z_B)$$

The main interest of the engineer, faced with problems in the field, is to have ready means of computing the value of E for a given pipeline. In general two types of formulae are available, namely the dimensionally correct and the empirical.

10.6 The Colebrook–White formula

A dimensionally correct expression for the head lost by friction when water flows through a circular pipe is Darcy's formula:

$$\text{friction loss } H \text{ (m head of water)} = \frac{fLV^2}{2gd} \qquad (10.4a)$$

where f is Darcy's coefficient and is a dimensionless number, L is the length of the pipeline in metres, V is the average velocity of flow (m/s), d is the internal diameter of the pipe (m), and g is 9.81 m/s^2.

Values of f vary according to the degree of turbulence of flow. Colebrook and White showed that it was related to the roughness and diameter of the pipe, the velocity of flow, and the viscosity of the water. The consequent Colebrook–White formula for circular pipes of commercial size flowing full is therefore:

$$V = -2(2gdi)^{1/2} \log_{10}\left(\frac{k}{3.7d} + \frac{2.51\nu}{d(2gdi)^{1/2}}\right) \qquad (10.4b)$$

where i is the hydraulic gradient H/L, k is the roughness of the pipe (m), ν is the kinematic viscosity of water (m^2/s), and V, d, H and L are as before. The minus sign expresses the negative gradient of H/L in the direction of V and is cancelled by the negative value of the logarithm. The kinematic viscosity of clean water is 1.310×10^{-6} m^2/s at 10 °C and 1.011×10^{-6} m^2/s at 20 °C. Water temperatures in the UK are about 10 °C, but in tropical climates may be up to 20 °C. The difficulties of solving equation 10.4b for i or d can be overcome by using an iterative method of successive approximations on a programmable calculator; otherwise values of V for values of i can be graphed out for a given pipe diameter and interpolation can be used.

Suitably conservative design values of k for use in network analysis of distribution systems are as follows.

For new or recent installations	$k = 0.000\,50$ m
For systems up to 30 years old in fair condition	$k = 0.001\,25$ m
For systems 50 years old and over with some deterioration	$k = 0.003\,00$ m

The equivalence of these to values of C in the Hazen–Williams equation (see Section 10.7) is given in Fig. 10.3 for a 1 m/s flow rate.

For design of individual pipelines it is more difficult to quote values of k. The value is meant to represent the roughness of the pipe in terms of equivalent sand grain size, but this is a notional conception which is often difficult to apply in practice, even if a pipeline can be opened for inspection. Also the roughness of the wall of the pipe is not the only cause of losses within a pipeline; there will be losses also from discontinuities at joints, tees, and valves, and from turbulence at bends. Quoted values of k for various types of pipe can be of some assistance, but it is advisable to check that the k value used does not imply too high a C value in the Hazen–Williams formula as shown in Fig. 10.3. It is important to remember that measurements of pressures and flows are not more accurate than ±5% in practice, with errors often greater.

10.7 Empirical formulae

The field difficulties of measurement are such that, in practice, the engineer is free to choose for most of his calculations that empirical formula which it is most convenient to handle, having regard to the problem he needs to solve. Rarely is it necessary to choose the formula which close examination of experimental results indicates will be likely to give most precise results.

Hazen–Williams formula For many practical conditions the Hazen–Williams formula is deservedly still popular for several reasons: it is well documented and easy to use; it has been in use for many years and there are numerous experimental results which indicate fairly reliable values of the coefficient C to take in most circumstances; it is reasonably accurate over a range of pipeline sizes and flows which are widely experienced. This formula can be expressed as

$$H = \frac{6.78L}{d^{1.165}}\left(\frac{V}{C}\right)^{1.85} \qquad (10.5a)$$

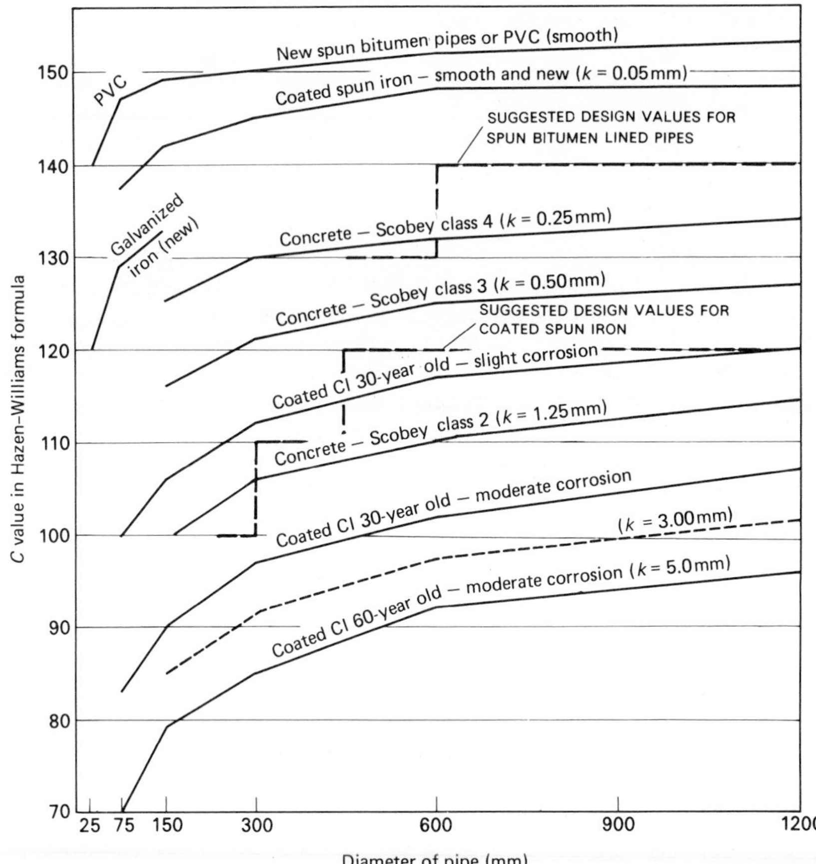

Fig. 10.3 *C* values in the Hazen–Williams formula and *k* values in the Colebrook–White formula according to P Lamont (*WWE*, January 1969).
Notes:
(1) Suggested design values are for a non-aggressive and non-sliming water.
(2) The curves apply to a 1 m/s flow rate. For 2 m/s reduce *C* values by 5% below 100, 3% below 130, and 1% below 140. For 0.5 m/s increase *C* values by the same amounts.
(3) Scobey classes are: *class 4*—first class interior finish with all joint irregularities removed; *class 3*—good interior finish with joints filled and concrete made on steel forms; *class 2*—imperfect interior finish, and as tunnel linings; *class 1*—old concrete pipes with mortar not wiped from joints ($k = 5.00$ mm).
(4) For asbestos cement pipes use values for spun bitumen lined pipes.

or $$V = 0.355Cd^{0.63}\left(\frac{H}{L}\right)^{0.54} \qquad (10.5b)$$

where C is a coefficient and L, V, and d are as defined for equation 10.4a above.

It will be noticed that in equation 10.5b $V = kC$ for a given set of conditions, so that we may draw a convenient diagram of flow against head lost for a range of pipes, all for the value $C = 100$, and adjust the answer readily for any other value of C. Figure 10.4 has been drawn on this basis. The formula is sufficiently accurate for pipe sizes of 150 mm upwards and for values of C not substantially below 100. It is more accurate for the larger diameter of pipes and for flows of the order of 1 m/s. Values of C which may be taken are as set out in Fig. 10.3.

Manning's formula The Manning formula has the advantage that $H \propto V^2$, the coefficient n being constant for a given type of pipe. Since the friction loss caused by fittings, e.g. bends, tees, tapers etc., is also usually expressed as kV^2 the Manning formula is conveniently used for lengths of pipeline involving many fittings whose effect is appreciable and must be added in to the calculations. The formula is more accurate than the Hazen–Williams formula for estimating high flows, or flows in old rough-surfaced pipes where the C value in the Hazen–Williams formula is well below 100. Manning's formula (for circular pipes) is in metric units

$$H = \frac{n^2}{0.397^2} \frac{LV^2}{d^{4/3}} \qquad (10.6a)$$

or

$$V = \frac{0.397}{n} d^{2/3} \left(\frac{H}{L}\right)^{1/2} \qquad (10.6b)$$

Values of Manning's coefficient n are set out in Table 10.1 which also includes values of n for open channels (a matter dealt with in Section 10.12).

Estimation of head lost through fittings etc. In such cases the head lost is usually expressed as $H = kV^2/2g$, i.e. it is expressed as a proportion, k, of the kinetic energy. The value of k is determined by

Table 10.1 Values of Manning's coefficient *n* for various types of surfaces.

Surface	Manning's coefficient *n*
Smooth metallic	0.010
Large welded steel pipes with coal-tar lining	0.011
Smooth concrete or small steel pipes	0.012
Riveted steel or flush-jointed brick	0.015–0.017
Rough concrete	0.017
Rubble (fairly regular)	0.020
Old rough or tuberculated cast iron pipes	0.020–0.035
Cut earth (gravelly bottom)	*0.025
Natural watercourse in earth	*0.030–0.040
Natural watercourse in earth with bank growths	*0.050

* Values for half-bankside depth of flow; at bankfull discharge values may be 20% less.

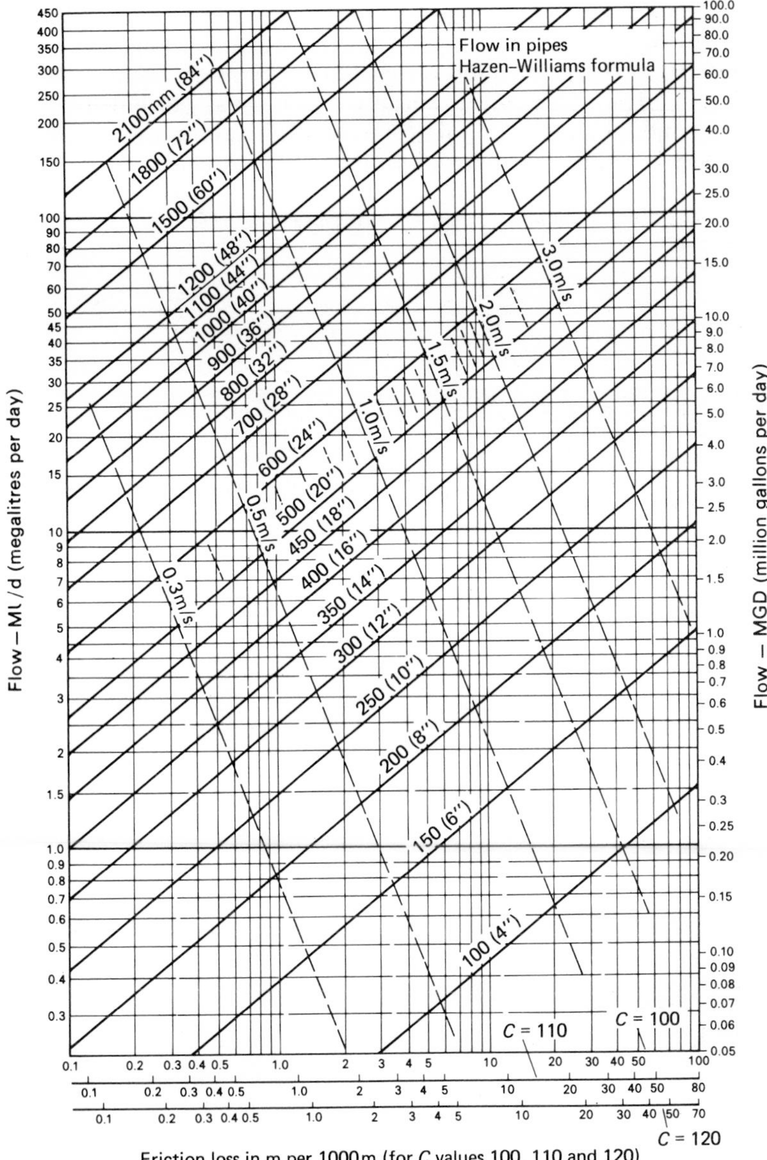

Fig. 10.4 Flow and friction loss in pipes based upon the Hazen–Williams formula.

experiment and many such experiments in the past enable us to draw up values of k for particular types of obstructions, some of which are set out in Table 10.2. It will be noticed that this relationship assumes $H \propto V^2$ which is nearly enough correct because flow is usually turbulent around obstacles and, in most practical problems, the head lost through fittings in a pipeline is small compared with the total frictional loss.

Table 10.2 Friction loss through constrictions in pipelines and channels, expressed as k in the formula $H = kV^2/2g$.

	k values		
Item	Theoretical	Allow*	Notes
Entrances			
Standard bellmouth pipe	0.05	0.10	V = velocity in pipe
Pipe flush with entrance	0.50	1.00	V = velocity in pipe
Pipe protruding	0.80	1.50	V = velocity in pipe
Sluice-gated or square edged entrance	—	1.50	V = velocity through opening (value includes expansion loss)
Bends—90° (45° half values given)			
Medium radius (R/D = 2 or 3)	0.40	0.50	
Medium radius (R/D = 2 or 3) *mitred*	0.50	0.80	
Elbow, or sharp angled as in a channel or mitre	1.25	1.50	
Tees—90°			
In-line flow	0.35	0.40	Assumes equal diameters
Branch to line, or reverse	1.20	1.50	Assumes equal diameters
Exits			
Sudden enlargement, ratio 1:2	0.60	1.00	
Bellmouth exit	0.20	0.50	Must be well tapered
Sudden contractions			
Loss on contraction *and* subsequent expansion where ratio of area of opening to area of pipe or channel is:			
1:2	1.00	1.50	V = velocity through opening
2:3	0.65	1.00	V = velocity through opening
3:4	0.40	1.00	V = velocity through opening
Contraction or enlargement only	—	1.00	
Gate valve fully open	0.12	0.25	

* The values under the column 'Allow' are conservative and intended for field use, i.e. not laboratory use. These values are recommended for short pipeline or conduit loss calculations. It is particularly important to allow losses for all sharp changes in direction or cross-sectional area of flow.

Flow in channels

10.8 Energy of flow

The free surface of water in a channel is a direct measure of the potential energy, or position energy in a gravitational field, that the water possesses. The remainder of the energy possessed by a unit mass of water is its kinetic energy. Since the sum total of these two types of energy—in the absence of an external applied force such as friction—is constant, it follows that there can be no *increase* of kinetic energy except at the expense of an equivalent *decrease* of potential energy, and vice versa. Hence an increase of the speed of flow of the water must be accompanied by a drop in level of the free water surface. A retardation of the speed of flow must be accompanied by a rise in level of the free water surface. One of the most important practical consequences of this is that if water from a still pond or pool enters a culvert (flowing into it) then there must be a drop in level of the free water surface at the entrance to the culvert in order that the water can accelerate to flow into and along the culvert. This phenomenon is of the greatest practical importance in deducing water flow in channels.

The kinetic energy of a mass m moving at v m/s is $\frac{1}{2}mv^2$. For a mass of water which weighs 1 kg in the earth's gravitational field the kinetic energy then becomes

$$\frac{1}{2}\frac{1\,\text{kg weight}}{g}v^2 = \frac{v^2}{2g}$$

the units of this expression, as we have seen in Section 10.4, are 'metres head of water'. Thus the kinetic energy term $v^2/2g$ may be said to represent the equivalent potential energy of $v^2/2g$ m height of water column above a given datum. Therefore, if a unit mass of water accelerates from standing (0 m/s) to moving at V m/s, its free water surface must fall a distance of $V^2/2g$ m. Neglect of this necessary stage through which water must pass before it can flow down a channel is an error which can have dangerous consequences when estimating the discharge capacity of a reservoir overflow.

Suppose we have a channel connected to a reservoir as shown in Fig. 10.5. We wish to compute the flow that will occur down that channel under a given reservoir level.

Now an attempt may erroneously be made to compute the discharge down the channel A–B by using ready-made tables of flow for channels. However, in order to do so we shall find that it is necessary to know the depth of flow in the channel. If we assume that the depth of flow in the channel is the distance of the reservoir level above the base of the channel at the entrance, distance H in Fig. 10.5, we may seriously overestimate the possible flow. We shall have omitted to bring into account the reduction in the level of the free water surface, $V^2/2g$,

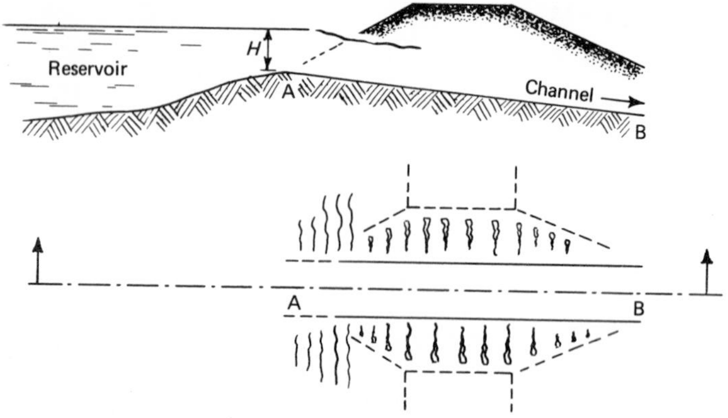

Fig. 10.5 Channel fed from a reservoir.

which is necessary in order that the water shall actually move into the channel and pass down it with velocity V.

It is instructive to examine the problem more closely to see whether we can find an expression for V. It is clear that the water level cannot and will not drop below the level of the base of the channel at A. Consequently the maximum velocity that could under any circumstances occur in the channel would be given by $H = V^2/2g$, i.e. $V = \sqrt{(2gH)}$, but at that drop in water level the depth of flow over the channel entrance would be nil, and the total flow would therefore also be nil. At the other extreme the depth of water in the channel could be H, but in this case the drop in water level would be nil and $V = 0$, so that again no flow would occur down the channel. We therefore suspect—and this is quite true—that there is some value of d, the depth of flow in the channel, which lies between $d = H$ and $d = 0$ such that the quantity of flow is a maximum. This depth d is called the 'critical depth of flow', d_c, and the accompanying value of the velocity is called the 'critical velocity', V_c, and these, as we shall show below, can be computed.

10.9 Critical depth of flow

Consider any cross-section XX′ of a *horizontal* channel and let us neglect friction energy lost for the moment (Fig. 10.6).

Let us apply Bernoulli's equation for a unit mass of water at position A, which is h_A m below the water surface. Then, measuring energy from the base of the level channel, we have for this individual unit mass of water

$$\frac{u_A^2}{2g} + p_A + (d - h_A) = \text{constant} = H \text{ (say)}$$

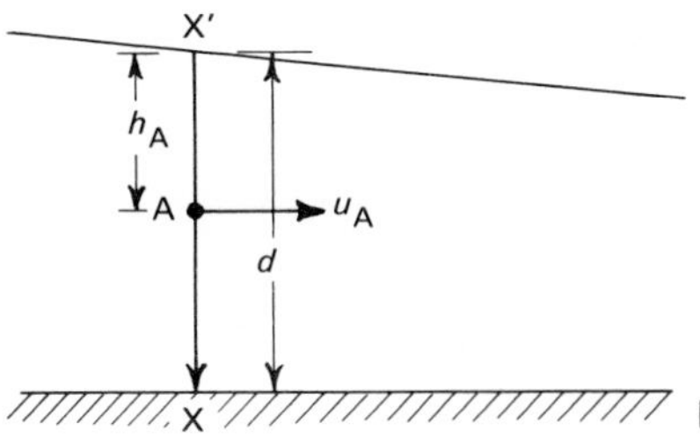

Fig. 10.6 Flow in a horizontal channel.

Now since our units are in metres head of water the pressure p_A on the unit mass of water at A is simply h_A, its depth below the free surface. If we therefore substitute h_A for p_A the equation reduces to

$$\frac{u_A^2}{2g} + d = H \tag{10.7}$$

for that particular streamline at A. For any other streamlines at points B, C... etc. the same relationship will hold and if we furthermore assume, with negligible practical error, that the velocity of flow through the section XX' is uniform, i.e. $u_A = u_B = u_C = \ldots = V$, the mean velocity of flow through the channel at section XX', we can then obtain for total flows through the whole section XX'

$$\frac{V^2}{2g} + d = H \text{ (a constant)} \tag{10.8}$$

Now the discharge per metre width of channel is $Vd = q$, say, where q is in m^3/s, d is in m, and V is in m/s. Substituting q/d for V in equation 10.8 we obtain

$$\frac{q^2}{2gd^2} + d = H$$

i.e. $$q^2 = (H - d)2gd^2 \tag{10.9}$$

Differentiating q with respect to d in order to find a maximum value for q we obtain

$$2q\frac{\mathrm{d}q}{\mathrm{d}d} = 4gdH - 6gd^2$$

For a maximum $\mathrm{d}q/\mathrm{d}d = 0$. Therefore $4gdH = 6gd^2$. Whence

$$d = \tfrac{2}{3}H \tag{10.10}$$

Therefore, for a rectangular channel (since we took a unit width of channel) approached by water having a limited total energy head H, maximum discharge will occur when the depth of flow in the channel is two-thirds of H. This we call the 'critical depth of flow' and label d_c. H represents the total height of the still water in the reservoir above the bed of the channel at the entrance, i.e. its total energy head.

For other depths greater or less than d_c the flow will be less than the maximum possible, as we can find out if we draw the whole graph of q against d expressed by equation 10.9, either by taking H as unity or by expressing d as a proportion of H (see Fig. 10.7).

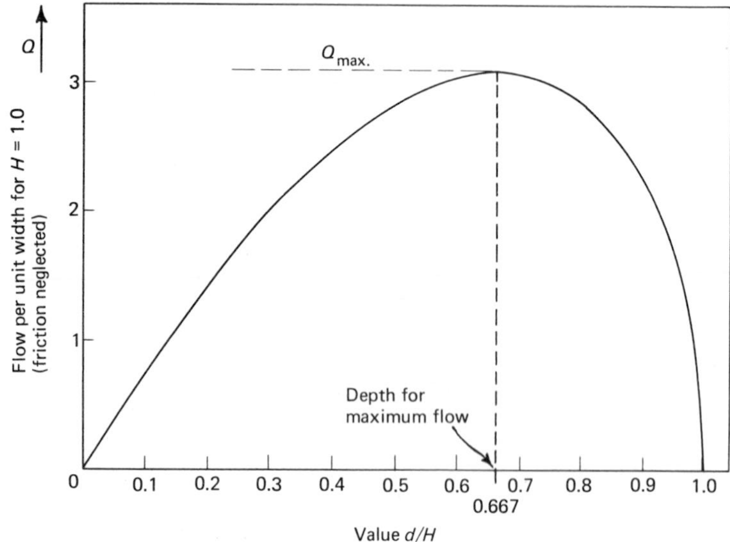

Fig. 10.7 Relationship (in rectangular channel) between quantity and depth of flow for a given total energy head *H*.

For other shapes of channel similar calculations will give different values of d_c relative to H and a different curve of q against d. Since the relationship of d_c/H is rather more complicated where the channel is trapezoidal, semi-circular, or of some other shape, the reader is advised to consult other hydraulic textbooks and specially prepared tables for the evaluation of d_c in those cases. Nevertheless the principle applied for deducing the value of the critical depth, whatever the shape of the channel, is the same. For many practical purposes channels may be assumed to be rectangular despite some slope to their sides, provided that they are not too narrow, and the use of the simple relationship for a rectangular channel is always of assistance for a first run through of the calculations where exact accuracy is not intended or needed in the initial stages.

10.10 Critical depth occurrence in a channel

We can now see, referring back to Section 10.8, how we can determine the depth of flow and thus the drop in water level, and consequently V and Q, at the entrance to a channel. Applying our knowledge concerning critical depth and the principle of conservation of energy we are able to state that, providing no external forces come into play immediately downstream of the entrance to the channel, the flow there will be critical, as Fig. 10.8. This is a most important factor because it is often

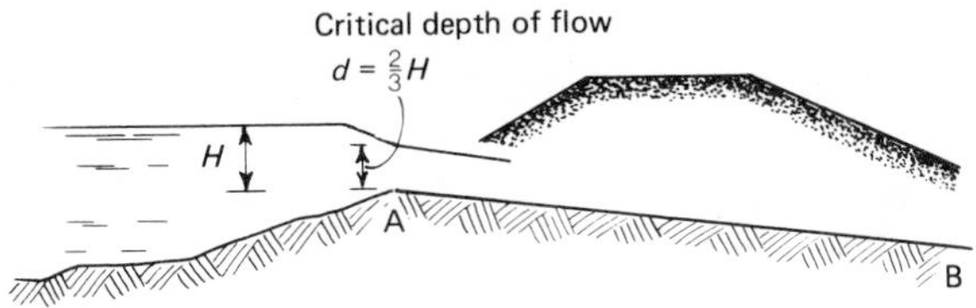

Fig. 10.8 Critical depth of flow at a channel entrance, with the condition that the slope AB is greater than the critical slope.

the limiting condition to flow in a channel. To put the matter colloquially, it does not matter how steeply the channel falls subsequently to the entrance, there can be no more flow down that channel than 'can get into the entrance' as determined by the conditions for critical depth. Immediately downstream of the entrance to a rectangular channel the flow will be critical, the depth of water will be two-thirds of H, and therefore the velocity of the flow will be $V = \surd(2gH/3)$ where $H/3$ is the drop in water level. Hence the discharge can be calculated.

The qualification 'providing no external forces come into play' is important: we have so far neglected friction. If friction causes loss of energy between the reservoir and the point where critical flow occurs then the full value of H may not be available to create critical depth. The actual critical depth will then be $\frac{2}{3}H'$ where $H' = H -$ the energy lost in friction, which is denoted by E. Usually the value of E for the case of a reservoir discharging to a channel is so small that it can be neglected.

The next and equally important matter to investigate is to consider whether the slope of the channel downstream of the entrance is sufficient to 'take the water away' from the entrance. If the slope of the channel is very small the release of potential energy, i.e. the drop in bed level, may not be sufficient to equate or surpass the loss of energy through friction which occurs as the water moves through the channel at critical velocity V_c. This means that the speed of the water must slow down, its water surface must rise, and it will in fact 'back up' and drown out the smaller depth d_c occurring at the entrance. In fact d_c never forms and the flow of water at the entrance is not critical and we are faced with a new problem, the essence of which is that flow is not limited by the critical depth which can take place but is limited to a lesser quantity by the flow capacity of the channel, having regard to its shape, its roughness of bed and walls, and its slope. This again is an important aspect of channel flow, particularly in relation to the spillways of impounding reservoirs where, quite often, the overflow spillway has to be rather flat for some distance across the crest of the dam.

10.11 Flow in a channel where critical depth is not reached

Let us assume that we have a length of channel fed from a reservoir as shown in Fig. 10.8.

As a first attempt we compute the critical depth, velocity, and flow at the entrance A of the channel. We now need to find out whether this discharge can be maintained in the length of channel AB; if it cannot then flow will not be critical at A.

Consider a short length S_1S_2 of the channel as shown in Fig. 10.9. We know that the total energy at S_1 must equal the total energy at S_2 plus the friction energy E dissipated, i.e.

$$\frac{V_1^2}{2g} + d_1 + z_1 = \frac{V_2^2}{2g} + d_2 + z_2 + E \qquad (10.11)$$

where we should note that, in the first instance, we have assumed the depth of the water to vary as it moves from S_1 to S_2. We now place into this equation, on the left-hand side, the critical flow values V_c for V_1 and d_c for d_1 which we have already found. On the right-hand side of the equation we express V_2 in terms of known quantities V_1 and d_1 and the

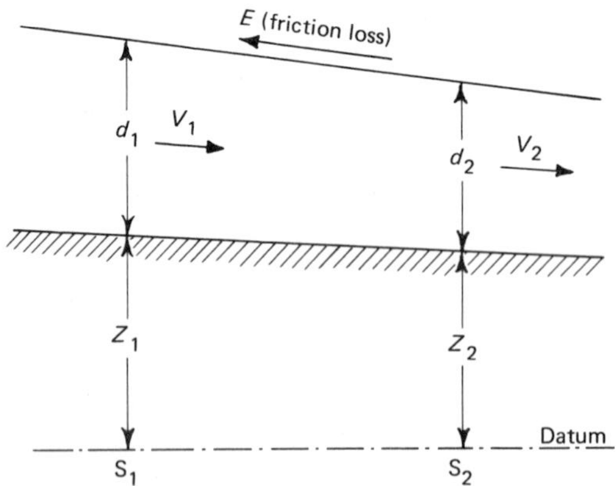

Fig. 10.9 Flow between any two sections S_1 and S_2 of a uniform channel.

unknown quantity d_2 by using the relationship $V_1d_1 = V_2d_2$. We make an estimate for E using some appropriate friction formula, such as the Chézy formula (see Section 10.12) which is $V = C\sqrt{(mi)}$ where C is a constant appropriate for the type of channel we have. For V and m in the Chézy formula we have to make an 'intelligent guess', which we may later have to revise when we have gone through the calculation once. They should represent mean values for the section S_1S_2. They may, however, be taken as the initial values for section S_1 without introducing much error in problems involving gradually changing flow.

In this equation we now have known values for all terms excepting d_2 and this we can proceed to compute. The equation is, of course, a quadratic in d_2 and we shall have to dispense with that value of d_2 which appears, by inspection, to be unreal or extremely unlikely. (Some care is required in this.) We shall now find whether d_2 is less than d_c, equal to d_c, or more than d_c. If it should happen that d_2 is exactly equal to d_c, this means that the depth of flow and the velocity are constant, the free water surface remaining parallel with the bed, and the fall of the bed is just sufficient to overcome the frictional resistance to flow. In that case critical flow would be maintained throughout S_1S_2, which is unusual. If d_2 is less than d_c then the depth of flow is decreasing and the velocity of flow is therefore increasing. This means that the potential energy released by the fall of the bed is greater than the energy used in overcoming friction, so that the balance is used in accelerating the flow of the water. When d_2 is greater than d_c the depth is increasing and therefore d_c could not, in fact, occur at the entrance to the channel, section S_1. We must then compute the flow by methods shown below.

10.12 Steady-state flow (the Chézy formula)

In steady-state flow, as we have seen, the energy lost through friction equates with the potential energy given up by the fall in bed level. Thus, in the last of the foregoing cases where we saw that flow could not be critical at the entrance because the channel was flat, or only slightly sloping, if we use a formula which will give us the steady-state flow for a given channel slope for a range of water depths d, we can solve the problem by choosing a suitable value of d such that $V^2/2g + d = H$ where H is the head of the still water in the reservoir above the bed level at the entrance to the channel.

A suitable formula to use is the Chézy formula, which is

$$V \text{ (m/s)} = C(mi)^{1/2} \qquad (10.12)$$

where m is the hydraulic mean radius (equal to the cross-sectional area of flow divided by the wetted perimeter), i is the hydraulic gradient, i.e. H/L where H is the head lost in friction in length L, and C is a coefficient (not to be confused with the Hazen–Williams pipeline coefficient C).

Values of the Chézy coefficient C may be calculated using values for Manning's n, Table 10.1, for if the Chézy formula and the Manning formula are compared* it will be seen that $C = 1.0m^{1/6}/n$. We are then enabled to proceed as follows.

In the Chézy formula $V = C(mi)^{1/2}$ we have an expression connecting velocity with energy loss through friction and, by definition, we know that for steady-state flow the energy loss through friction is equal to the fall in bed level. Therefore for unit length of channel

$$V = Cm^{1/2}\left(\frac{z_1 - z_2}{L}\right)^{1/2} \tag{10.13}$$

where $(z_1 - z_2)/L$ is the slope of the channel expressed as fall per unit of length. Hence, by a tentative choice of d, which enables us to compute m, we can compute V from equation 10.13 and can then substitute this value of V in $V^2/2g + d$, the sum of which must equal H. If it does not, we choose another value of d until we get that steady-state flow whose kinetic energy plus depth of flow is just equal to the available head at entrance H. The computation is not so laborious as it sounds because, in practice, m varies only slightly with d. Alternatively tables are available (and this is the use of such tables in particular) relating results of the Chézy formula to definite values of H for various types and shapes of channel.

10.13 General and particular conditions of channel flow

So far we have paid particular attention to the development of critical flow at the entrance to a channel (Section 10.10); we have discussed the problem which arises if conditions downstream do not permit this critical flow to occur at the entrance, and we have shown (Section 10.12) how in that case to determine the discharge by reference to steady-state flow. These are the two primary limiting conditions of channel flow, and in every channel length (however irregular are its fluctuations of slope) one or other of these limitations, or each one in turn, may come into play to govern the conditions of flow. As a final clarification of the subject we may illustrate one or two common cases likely to be experienced.

In Fig. 10.10 is shown the profile of a slope down an impounding reservoir spillway. The channel, whose bed is represented by the line ABCD, is assumed to be of uniform width. In this case the slope of AB is insufficient to produce critical flow at A (see Section 10.10). Also the length AB is too short for steady-state conditions to develop in AB (see

* We have from equation 10.6b above: $V = (0.397/n)d^{2/3}i^{1/2}$. For a circular pipe $m = \pi d^2/4\pi d = d/4$. Therefore $d^{2/3} = (4m)^{2/3}$. Substituting this gives $V = 0.397 \times 4^{2/3}(m^{2/3}/n)i^{1/2} = 1.0(m^{2/3}/n)i^{1/2}$. Comparing with the Chézy formula $V = Cm^{1/2}i^{1/2}$ it is seen that $C = 1.0m^{1/6}/n$ where n is the Manning's n.

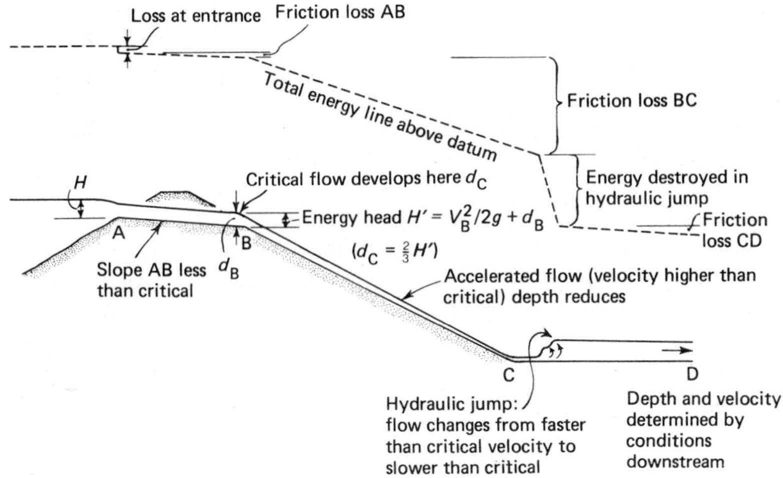

Fig. 10.10 Flow down a reservoir spillway channel.

Section 10.12). Instead the critical depth of flow occurs at B and it is therefore at B that we must start our calculations. We make a first tentative assumption that the total energy head H at A is also available at B and this permits us to calculate the critical depth of flow at B. We now have a first approach to the quantity flowing down AB and can make an assessment, as a result, of the friction energy loss in this short section of channel. This permits us to revise the critical depth calculation at B where the energy head is now taken as H', which is equal to the total energy head (H) minus the energy lost in channel AB. Once we have adjusted the calculation to the desired accuracy, this gives the maximum discharge of the channel under the given reservoir level.

Downstream of the point B the channel BC is steep. Here the excess potential energy released by the fall of bed accelerates the flow of the water and, applying equation 10.11 and the type of calculation described in Section 10.11, we may, by taking short lengths of the channel at a time, trace the speeding-up of the water and its reduction in depth of flow as it approaches C. At C the water enters a channel CD which has only a gentle slope to it. Hence at some point D downstream steady-state-flow conditions will develop (see Section 10.12). Calculating velocity and depth backwards from D we realize that at C, at the bottom of the spillway, there must be a change of depth. This change of depth is quite sudden and it is called a 'standing wave' or 'hydraulic jump'.

10.14 Hydraulic jump

A hydraulic jump or standing wave is formed when water at a velocity greater than the critical velocity, and therefore flowing at a depth *less* than the critical depth, is suddenly forced to slow to a velocity less than the critical velocity, and therefore flows at a depth *greater* than the critical depth. This is simply a typical instance of equation 10.9, which was graphed out in Fig. 10.7 where we saw that for any fixed quantity q there can be either a high value or a low value of d, the depth of flow. Whereas formerly we were considering the maximum value of q for a given amount of energy, i.e. we were dealing only with the 'peak' of the curve in Fig. 10.7, the energy added to the water consequent upon its fall down the steep slope BC has been increased and with this extra energy the water has a choice of depths of flow and velocity. Which one it takes is dependent upon the channel conditions.

The hydraulic jump therefore represents a change of regime of flow from fast speed at shallow depth to slow speed at larger depth, there being also a dissipation of energy in the turbulence of water formed at the hydraulic jump. Such a hydraulic jump is a cheap device for destroying surplus energy of the water without causing large erosion of the bed and sides of the channel, for much of the energy is dissipated by the water impacting upon itself causing great turbulence and boiling.

To sum up, in dealing with channel flow problems we first have to consider where the restrictive conditions apply, such as at obstructions across the river or at sudden changes of bed slope which cause critical flow conditions to develop, or in long lengths of channel where steady-state conditions of flow appertain. From these 'control' points we must then work upstream and downstream from the known conditions by the method set out in Section 10.11 and can note the interference of any restrictive condition upon another and so finally arrive at the whole regime of flow for the whole channel.

10.15 Flow through a narrowing and falling channel

So far we have dealt with flow through channels of uniform width. Cases are often met in practice where a channel narrows as it falls. Assuming the channel may be regarded as rectangular in shape, we have by Bernoulli's equation (modified for flow as a whole):

$$H = z + D + \frac{V^2}{2g}$$

where z is the distance above datum, D is the depth of flow, and V is the mean velocity of flow.

Differentiating with respect to distance x (see Fig. 10.11) from starting point we obtain

$$\frac{dH}{dx} = \frac{dz}{dx} + \frac{dD}{dx} + \frac{d(V^2/2g)}{dx} \qquad (10.14)$$

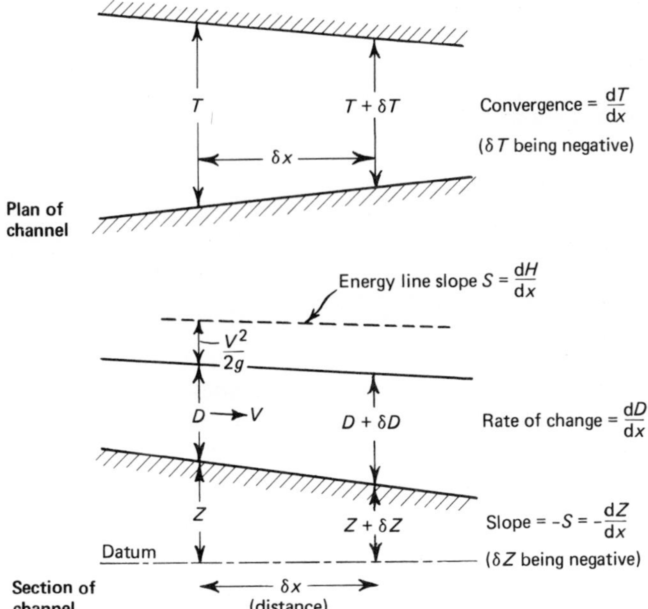

Fig. 10.11 Flow in a narrowing and falling channel.

Now V is dependent upon a, the cross-sectional area of flow, which for a rectangular channel is TD where T is the top width of water. Therefore

$$V^2 = (Q/TD)^2$$

and substituting for V in equation 10.14 we obtain

$$\frac{\mathrm{d}H}{\mathrm{d}x} = \frac{\mathrm{d}z}{\mathrm{d}x} + \frac{\mathrm{d}D}{\mathrm{d}x} + \frac{\mathrm{d}(Q^2/T^2D^2)}{2g\,\mathrm{d}x}$$

i.e. $$\frac{\mathrm{d}H}{\mathrm{d}x} = \frac{\mathrm{d}z}{\mathrm{d}x} + \frac{\mathrm{d}D}{\mathrm{d}x} - \frac{Q^2}{gT^3D^2}\frac{\mathrm{d}T}{\mathrm{d}x} - \frac{Q^2}{gT^2D^3}\frac{\mathrm{d}D}{\mathrm{d}x}$$

Q being constant, but both T and D varying with distance x.

Now $Q^2/T^2D^2 = V^2$ so

$$\frac{\mathrm{d}H}{\mathrm{d}x} = \frac{\mathrm{d}z}{\mathrm{d}x} + \frac{\mathrm{d}D}{\mathrm{d}x} - \frac{V^2}{gT}\frac{\mathrm{d}T}{\mathrm{d}x} - \frac{V^2}{gD}\frac{\mathrm{d}D}{\mathrm{d}x} \qquad (10.15)$$

Putting $\mathrm{d}H/\mathrm{d}x = -s$ (slope of energy line) and $\mathrm{d}z/\mathrm{d}x = -s_0$ (slope of base),

$$-s = -s_0 + \frac{\mathrm{d}D}{\mathrm{d}x}\left(1 - \frac{V^2}{gD}\right) - \frac{V^2}{gT}\frac{\mathrm{d}T}{\mathrm{d}x}$$

Whence

$$\frac{\mathrm{d}D}{\mathrm{d}x}=\frac{s_0-s+\dfrac{V^2}{gT}\dfrac{\mathrm{d}T}{\mathrm{d}x}}{1-(V^2/gD)}$$

So that for any short length of channel δx in which the width of the channel narrows by an amount δT we have

$$\text{change of depth } \delta D=\frac{(s_0-s)\delta x+\dfrac{V^2}{gT}\delta T}{1-(V^2/gD)}$$

where, over the length δx, D is the mean depth over the section, T is the mean width over the section, V is the mean velocity of flow through the section, s is the slope of the energy line, or fall in hydraulic gradient, which is the total energy less the loss of energy caused by friction (calculated from some formula such as the Chézy formula), and s_0 is the fall in the bed.

By this means we can find the change of depth of water and velocity, knowing the initial depth and flow, step by step along the channel taking short lengths such as 5 or 10 m according to the rate of change of the width of the channel. We have, of course, to take trial values for the means D, T and V in the first instance, which we correct having found a first value for δD.

Weirs, flumes, and Venturi meters

10.16 Broad crested weirs

Consider a weir across a stream as shown in Fig. 10.12(a).

From a consideration of critical depth we know that the depth of flow over the cill of the weir will adjust itself so that the maximum flow takes place for the amount of energy available. Therefore the height of flow over the weir will be d_c, the critical depth for flow q. Therefore if the total energy head upstream of the weir is H then d_c will be, as for a rectangular channel, equal to $\frac{2}{3}H$. We also have

$$H=\frac{V^2}{2g}+d$$

i.e. $\quad H=q^2/2gd^2+d$

where q is equal to Vd for unit width of channel.

For maximum or minimum

$$\frac{\mathrm{d}H}{\mathrm{d}d}=-\frac{q^2}{2g}\frac{2}{d_c^3}+1=0$$

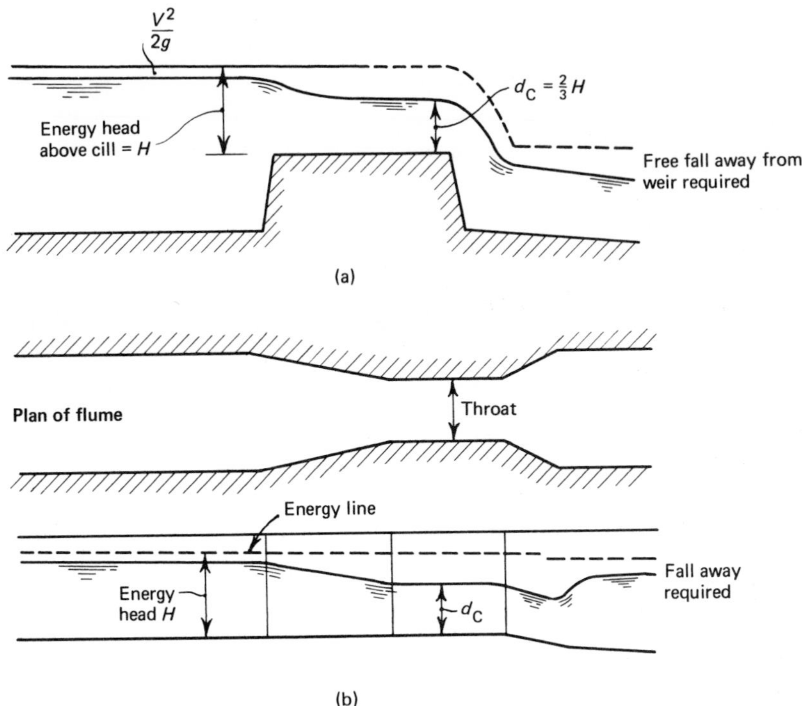

Fig. 10.12 (a) flow over a broad crested weir. (b) flow through a flume.

where d_c is the critical depth, i.e.

$$\frac{q^2}{gd_c^3} = 1$$

i.e. $$q = g^{1/2}d_c^{3/2}$$

We can therefore measure the discharge (per unit width) by a single observation of d_c. Alternatively, since $d_c = \frac{2}{3}H$,

$$q = g^{1/2}(2H/3)^{3/2}$$

i.e. $$q = 1.705H^{3/2}$$

where q is in m^3/s and H is in m for a 1 m unit width.

We can then obtain the discharge by measuring the height of the water above the cill level of the weir, upstream of it, noting, however, that the value of H includes the $V^2/2g$ term for the kinetic energy head of the water, although this figure will usually be found to be so small in practice that it may be ignored.

An important point is that there must be a free fall of water after the weir; the weir must not be 'drowned out' by a return of water level to H, or thereabouts, after the weir.

10.17 Flumes

A flume is a constriction in the water channel, such that critical flow in the 'throat' of the flume is formed (Fig. 10.12(b)). At the throat the discharge per unit width of channel is increased so that, as with a weir,

$$q = g^{1/2} d_c^{3/2}$$

In order to force this flow through, the total energy height upstream of the weir must be H where $H = 1.5d_c$. Therefore, as with a broad crested weir,

$$q \text{ (m}^3\text{/s)} = 1.705H^{3/2}$$

H being measured in m, where q is the discharge *per unit width of throat* and H is the total energy height upstream of the throat. Downstream of the throat the channel reverts to normal size and a small standing wave or hydraulic jump is formed. This causes dissipation of energy and the free surface level of water flowing away is usually sufficiently low to prevent 'backing up' into the flume. If it is not, a fall must be given to the bed of the channel downstream of the flume.

For a flume to BS 3680, as shown in Fig. 10.13, the length of the throat must be not less than 1.5 times the maximum total head to be measured, surfaces of the approach channel must be of smooth

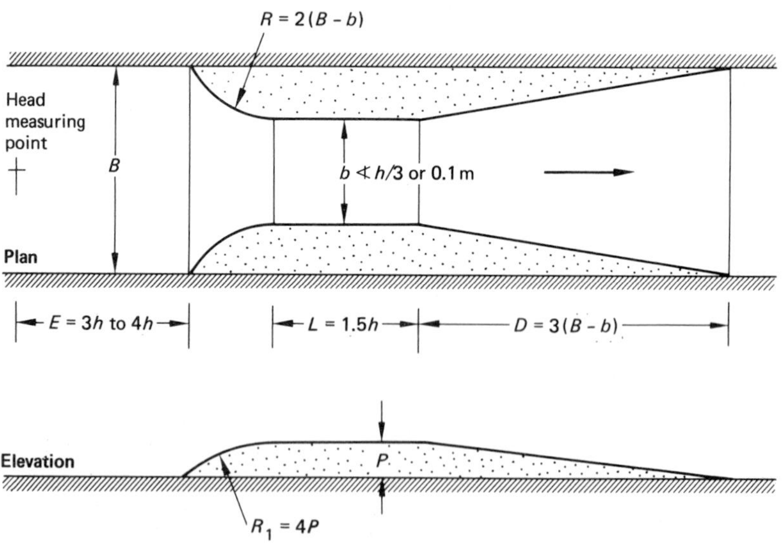

Fig. 10.13 Flume to BS 3680.

concrete, the invert of the approach channel must be level for 1.3 m upstream of the point of measurement, the divergence downstream must not exceed 1 in 6, and the flow must be subcritical upstream.

For a flume to the dimensions shown in the figure the discharge is given by

$$Q\,(\mathrm{m}^3/\mathrm{s}) = \tfrac{2}{3}(2g/3)^{1/2} C_V C_D b h^{3/2} \tag{10.16}$$

where h is the head measured upstream above the base of the flume if flat bottomed, or above the surface of the hump where the invert is raised, and C_V and C_D have values as given in the tables below.

C_V values for *flat invert* flumes depend upon the ratio b/B (see Fig. 10.13) as follows.

b/B	0.10	0.20	0.30	0.40	0.50	0.60	0.70
C_V	1.002	1.009	1.021	1.039	1.063	1.098	1.146

C_V values for throated and raised invert flumes depend upon the ratios b/B and $h/(h+P)$ as follows.

$h/(h+P)$	1.0	0.9	0.8	0.7	0.6	0.5	0.4	0.3
b/B								
0.3	1.02	1.02	1.01	1.01	1.01	1.00	1.00	1.00
0.4	1.04	1.03	1.02	1.02	1.01	1.01	1.01	1.00
0.5	1.06	1.05	1.04	1.03	1.02	1.01	1.01	1.00
0.6	1.10	1.08	1.06	1.04	1.03	1.02	1.01	1.01

C_D values for both flat invert and humped invert flumes depend upon the ratios L/b and h/L as follows.

L/b	0.4–1.0	2.0	3.0	4.0	5.0
h/L					
0.1	0.95	0.94	0.94	0.93	0.93
0.2	0.97	0.97	0.96	0.95	0.95
0.4	0.98	0.98	0.97	0.97	0.96
0.6	0.99	0.98	0.97	0.97	0.96

10.18 Sharp crested weirs

A formula may be deduced on theoretical grounds for the discharge of a sharp crested weir, the discharge through any small element of the flow through the weir being $\delta A(2gH)^{1/2}$ where H is the depth of the element below the free water surface. Formulae of the type $Q = CLH^{3/2}$ will be derived for a rectangular notch weir, and of the type $Q = C\tan(\tfrac{1}{2}\theta)H^{5/2}$ for a V-notch weir whose central angle is θ. In both these cases H is the height of the water above the lowest point of the weir crest and C is a constant found by experiment. In practice the discharge of a weir is affected by the end conditions, i.e. whether a rectangular notch is narrower than or equal to the channel approach to it, hence empirical formulae are used.

BS 3680 Part 4A gives recommendations for the measurement of flows by weirs, notches, and flumes. For thin plate rectangular weirs and triangular notches the upstream face must be a smooth plane at right-angles to the bed and sides of a vertically-sided horizontal channel, the weir being placed centrally thereto. The weir crest must be 1 to 2 mm thick as shown in Fig. 10.14 and its edge must be free of burrs and scratches. No rounding of arrises must occur as may follow from filing or cleaning with abrasive paper. The following equations then apply.

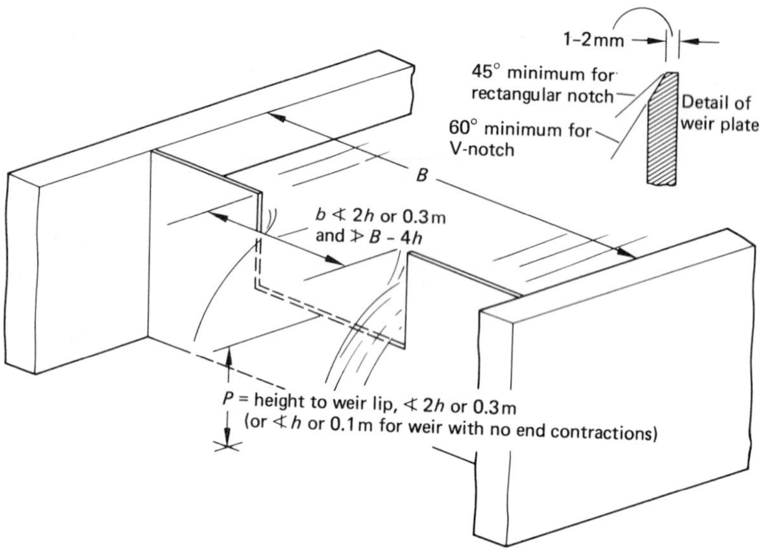

Fig. 10.14 Rectangular weir with fully developed end contractions.

For a rectangular weir with no end contractions, i.e. the weir being the full width of the channel,

$$Q\,(\mathrm{m^3/s}) = \tfrac{2}{3}(2g)^{1/2}C_D b h_e^{3/2} \tag{10.17}$$

where $C_D = 0.602 + 0.083h/P$, $h_e = h + 0.0012$, h = measured head (m) over weir a distance 3 h to 4 h upstream, P = height (m) of cill of weir above channel floor, b = breadth (m) of weir, and $g = 9.81\,\mathrm{m/s^2}$.

For a rectangular thin plate weir with fully developed end contractions

$$Q\,(\mathrm{m^3/s}) = \tfrac{2}{3}(2g)^{1/2}C_D b h^{3/2} \tag{10.18}$$

where $C_D = 0.616(1 - 0.1h/P)$ and h, P and b are as defined for equation 10.17 above.

Where the area of waterway in the channel is less than ten times the area of waterway over the weir then h must be replaced by $h' = h + 1.4v_a^2/2g$ where v_a is the velocity of approach in the channel.

For a 90° thin plate weir with fully developed end contractions

$$Q\,(\mathrm{m^3/s}) = \tfrac{8}{15}(2g)^{1/2}\, C_D h^{5/2} \tag{10.19}$$

where h (m) is the head referred to the vertex of the notch and is measured $3h$ to $4h$ upstream and C_D is given by the following table.

Head h (m)	0.050	0.075	0.100	0.125	0.150	0.200	0.300
Value of C_D	0.608	0.598	0.592	0.588	0.586	0.585	0.585

The height of the notch of the weir above the base of the channel must be not less than $2.5h$ and the width of the approach channel not less than $5h$. The nappe (or underside) of the weir must in each case be fully aerated. Where a rectangular weir is the full width of the channel whose walls extend downstream of the weir, a pipe must be used to aerate the underside of the nappe.

A defect of all sharp edged weirs is that they are easily damaged by debris brought down by the river or stream, since the metal edge (which should be of brass) is necessarily thin for accurate measurement. Deposition of silt occurs upstream of any weir and is another source of error. The advantage of a flume is that debris and silt will normally be swept through and the flume, although somewhat less accurate than a sharp edged weir, therefore remains more consistently accurate for long periods of time. Flumes can be constructed to take very large flows of water. For reservoir overflows the principles of a broad crested weir measurement have to be adopted at the spillway lip. The construction of the spillway is massive, since durability is required under all circumstances, and the crest or lip of any spillway or bellmouth shaft is therefore of thick stone. Provided that the crest profile is regular and even, flow in large quantities can be measured to the degree of accuracy required, although in most cases it is necessary to have a model test to confirm the relationship between Q and H. At small flows these large spillways cannot be expected to be accurate, especially when wave action causes intermittent discharges, but these smaller flows can probably be measured in the stream below the dam.

10.19 Venturi and orifice meters

These types of water meters work on the principle of causing water to flow through a constriction in a pipeline (see Fig. 10.15). At this constriction there is a reduction in the pressure of the water and the amount of the reduction measured together with a knowledge of the cross-sectional areas of pipe and constriction permit measurement of the flow as follows.

Assuming the pipeline is level, and neglecting friction losses between the two sections A and B, as shown in Fig. 10.15(c), we have

$$p_A + V_A^2/2g = p_B + V_B^2/2g$$

Hence the pressure difference, H, between the two sections A and B is given by

$$H = p_A - p_B = \frac{V_B^2 - V_A^2}{2g}$$

Rewriting this in terms of total flow through the pipe, $Q\,(\mathrm{m^3/s})$, the diameter of the pipeline, $D\,(\mathrm{m})$, and the diameter of the constriction, $d\,(\mathrm{m})$, we obtain

$$Q = \frac{\pi d^2}{4}\left(\frac{2gH}{1-(d/D)^4}\right)^{1/2}$$

For any given Venturi tube or orifice d and D are fixed, and we also know the throat area, $a\,(\mathrm{m^2})$. We therefore have, theoretically,

$$Q = akH^{1/2}$$

where $k = \{2g/[1-(d/D)^4]\}^{1/2}$ which is a constant for a particular meter.

In practice a further constant C must be introduced, i.e.

$$Q = CakH^{1/2}$$

where C varies for a Venturi tube from 0.95 to 0.99 (usually 0.98).

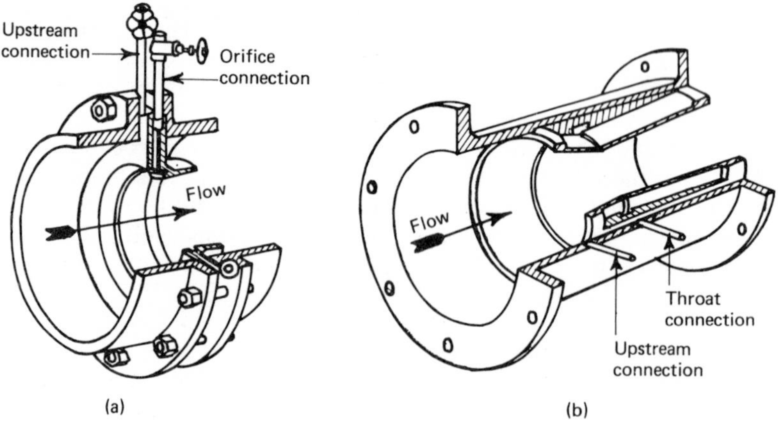

Fig. 10.15 (a) and (d) orifice meters; (b) Dall type Venturi tube; (c) cross-section of a Venturi tube.

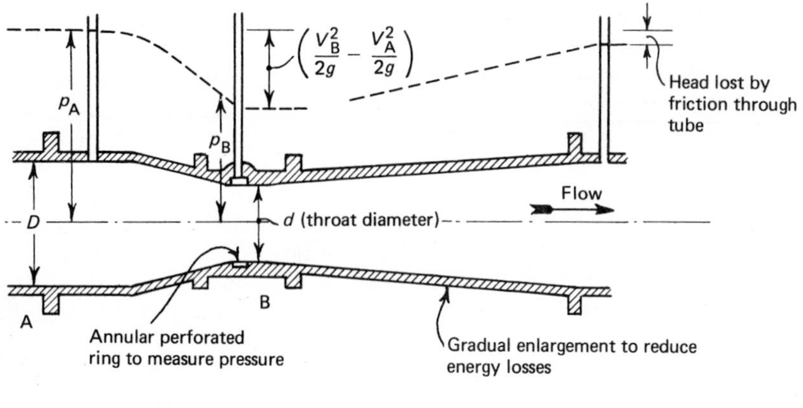

(c)

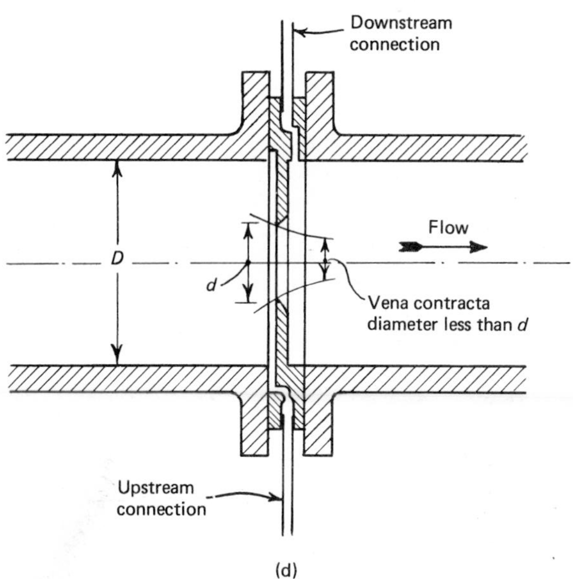

(d)

Fig. 10.15 (c) and (d).

An orifice meter has exactly the same formula for discharge, but C varies much more widely than for a Venturi tube, being dependent upon the diameter of the pipe, D, the velocity of flow, and the ratio d/D. Where the latter is in the range 0.4 to 0.6 and for pipe diameters over 200 mm, the value of C will be in the range 0.60 to 0.61 for the usual velocities of flow experienced in a distribution system.

The accuracy of measurement of both Venturi tubes and orifice meters is severely affected by flow disturbances created by fittings in a pipe. Detailed conditions for accurate measurement are laid down in BS 1042 Part 1: these can generally be met by ensuring that for d/D ratios not exceeding 0.6 there must be at least twenty diameters of straight pipe without a fitting upstream of the meter and seven diameters of straight pipe below.

The loss of head through a Venturi tube can be made very small because of the good recovery of pressure head in the long downstream taper, but the loss of head through a disc orifice in a pipeline is substantially higher because of the sudden expansion of the diameter downstream of the orifice. Hence Venturi meters, or some form of them, such as the Dall tube illustrated, are preferred when it is desired to measure a flow with the minimum loss of pressure through the meter.

Other flow meters

10.20 Ultrasonic and magnetic flow meters

There are two main types of ultrasonic flow meter—the Doppler effect meter and the transit-time meter. The Doppler effect meter requires reflecting particles or gas bubbles in the water and readings therefore depend upon many factors, such as the flow profile and the number, size, and distribution of particles. Although they comprise the majority of ultrasonic meters made, they are not satisfactory for the measurement of clean water and can be inaccurate unless calibrated for a given situation. Transit-time meters measure the difference in transit time of ultrasonic beams sent diagonally across a pipe, one directed downstream in the direction of flow and the other against it. Sophisticated electronic techniques are necessary to measure the small time difference which is of the order of 90 nanoseconds in a 100 mm pipe. The sensor heads are usually set at 45° to the direction of flow, and the time difference is measured by frequency tuning. Difficulties can arise from the presence of air bubbles or suspended solids, but the meters are otherwise suitable for the measurement of clean water. Lengths of straight pipe are required for 10–15 diameters upstream and 2–5 diameters downstream. An accuracy of ±1% is claimed provided that the flow is turbulent. The meter is not suitable for measurement of laminar flows, but these are seldom experienced in water supply systems. Sensors are being developed for insertion through tappings made on existing mains so that portable transmitters and recorders can be used for measuring flows in various mains of different diameters to an expected accuracy of about ±2.5%.

Magnetic flow meters work by creating a magnetic field across a pipe. Sensors on either side measure the induced emf in the flowing water which is proportional to the rate of flow. An accuracy of ±0.5% is possible provided that the apparatus is set up by a skilled instrument engineer who will need to ensure that no other electrical apparatus can interfere with the accuracy of measurement. The magnetic flow meter is expensive and is meant for permanent installation, but it has a high accuracy and can measure flow in either direction.

11

Service reservoirs

Function and capacity

11.1 Functions

A service reservoir has two main functions:

(1) to balance the fluctuating demand from the distribution system against the output from the source,
(2) to act as a safeguard for the continuance of the supply should there be any breakdown at the source or on the main trunk pipelines.

While it may be possible for the source itself to give a fluctuating output in step with the demand, it is not usually economic to follow this policy. Supposing the peak flow rate is twice the average, then the source must be capable of delivering twice the average demand and similarly the delivery pipeline must also be capable of delivering twice the average demand. Therefore there will be only 50% usage of the capital expended on these works and this would represent a serious wastage of capital expenditure. The aim must be to keep the sourceworks and the main trunk line working as much as possible at maximum capacity and to balance the changes in demand by a service reservoir.

The variation in rate of consumption for a typical industrial area over two days is shown in Fig. 11.1. There the night flow is shown to be some 10% less than the average for the whole day and the peak flow, occurring about midday, is some 18% greater than the average. In a small non-industrial area there is a larger fluctuation (see Fig. 11.2). In this case the night-time flow is about 30% less than the average and the maximum flow is between 40% and 60% in excess. In general, the smaller the area fed, the greater is the range of fluctuation of the demand. In a large industrial area the night-time rate of flow can be quite high because of continuous round-the-clock working of factories and because many large consumers may have storage tanks which fill up at night. In small urban areas containing very little industry, fluctuations in the rate of flow may be most marked because of the mass activity of the population served.

11.2 Minimum storage to even out hourly demands

If we look again at Fig. 11.2 we shall see that we can conveniently divide the day's consumption into three 8-hour periods. From midnight until

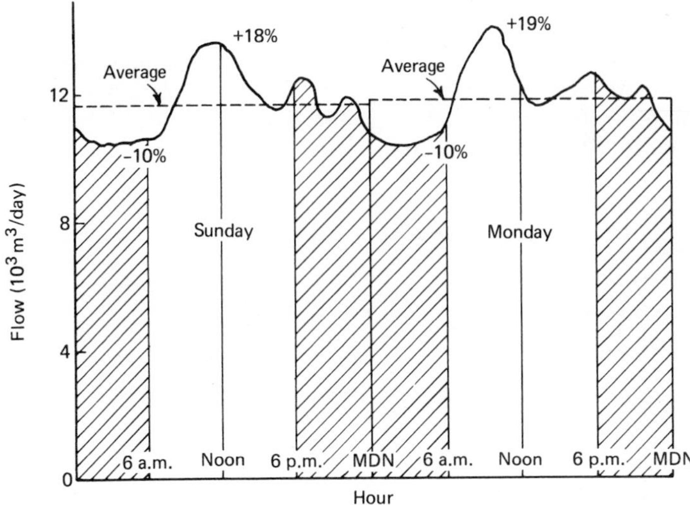

Fig. 11.1 Variation of flow in trunk mains feeding a large industrial area; for smaller areas there would be a much greater flow variation.

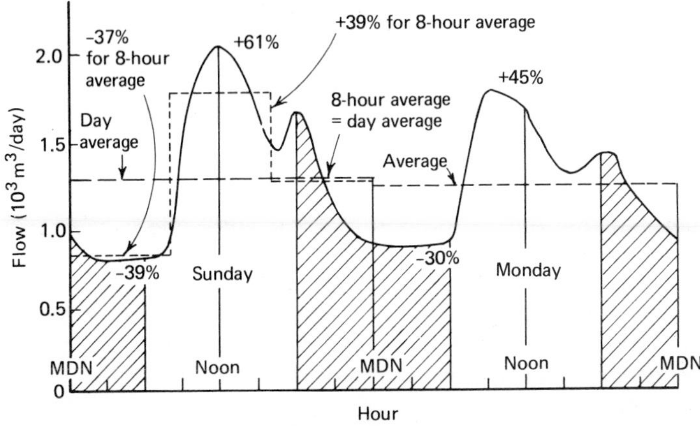

Fig. 11.2 Typical variation in consumption rates for a small non-industrial area of about 10 000 population.

8 a.m. flow is less than average (37% less in the example shown), from 8 a.m. until 4 p.m. flow is greater than average (39% greater in the example shown), and from 4 p.m. until midnight the flow is about average. On this simple basis we can readily deduce that during the period 8 a.m. until 4 p.m. excess flow will have to be drawn from store and the amount of this excess flow will be $8/24 \times 39\% \times$ average for day.

This amount of water will have to be put back into the reservoir during the following night in order to fill the reservoir ready for the next day. Turning now to Fig. 11.3 we see plotted there two lines. The full line shows an hour-by-hour plotting of the demand in relation to the average consumption for the day. The dotted line shows the same thing, but 'rationalized' for three 8-hour periods. We see that the 8-hour period

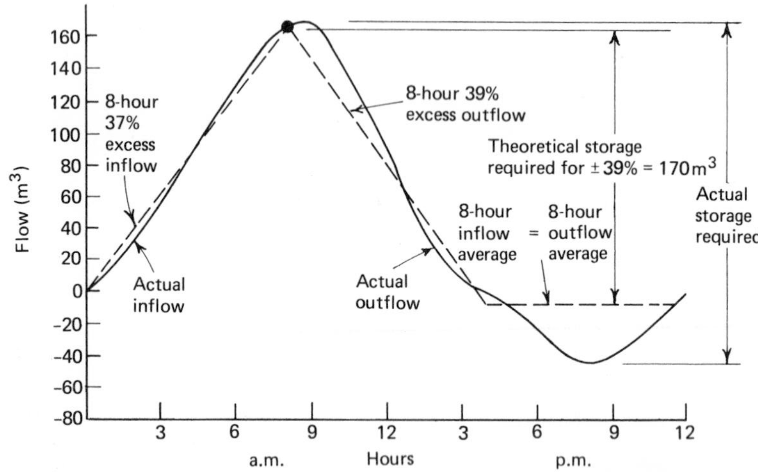

Fig. 11.3 An inflow–outflow graph prepared from the first day's consumption shown in Fig. 11.2.

line is a good approximation to the hourly line for the first 16 hours of the day, but that during the period 4 p.m. until midnight an excess of consumption during the hours 6 p.m. until 8 p.m. causes a dip in the hourly line below the 8-hour line. The storage deduced as being required from the 8-hour line is 170 m^3, i.e. one-third of 38% of the average consumption for the day, but the storage shown to be required by the hourly line is 40 m^3 more, i.e. 210 m^3. Thus the 8-hour line underestimates the amount of storage required by about one-fifth. Nevertheless, in the absence of detailed knowledge concerning the hour-by-hour variation of flow, we can make a very good judgement of what is the minimum amount of storage required by making an estimate of the maximum demand 8 a.m. until 4 p.m., calculating the consequent storage, and adding about 25% to this figure. The maximum demand figure to use will vary from 15% to 40% in excess of the average for the whole day depending upon the size of the area considered. The lower percentage would be taken as applicable to large industrial areas of 100 000 population upwards; the higher percentage would apply to a small urban area of under 20 000 population or to holiday towns. It is not possible to lay down fixed rules, but much can be done simply by considering the nature of the area served.

The foregoing calculation shows the minimum storage required to average out the flow of one particular day. In practice we shall need more than this storage: firstly because there will be variations from one day to the next and secondly because it may not always be convenient, or possible, to have the service reservoir full every morning at 8 a.m. To try to comply with this requirement would be to impose a rigid routine of operation upon the sourceworks which would be neither convenient nor economical. If we give double the theoretical requirement, however, we can safely fluctuate the contents of the service reservoir above and below the 50% full line, so giving much greater flexibility. Thus supposing we have an 8-hour maximum flow rate which is 30% in excess of the average flow for the day (this excess being a usual figure) then we shall need the following amount of storage:

$$2 \times 0.30 \times \tfrac{1}{3}(\text{day's total consumption}) \times 125\%$$

which equals *one-quarter* of the day's consumption. We may take this figure as giving a good guide to the minimum practicable storage which must be provided on a distribution system solely for the purpose of levelling out the hourly fluctuations of flow. If this is all that is required we see that it is a relatively modest amount of storage which should be cheaper to provide than going to the cost of increasing the capacity of the sourceworks and the trunk line by 30%. It must not, however, be regarded as sufficient to safeguard the continuance of the supply against all contingencies.

11.3 Contingency storage

Contingency storage is required to meet breakdowns at sources, repair of bursts on major mains, loss of water from bursts, and major fire demands. The storage required to meet these contingencies cannot be arbitrarily fixed: it depends upon the nature of the source, the layout of mains, and what safety precautions are possible. The following indicates the types of breakdowns commonly experienced and the consequent storage required to safeguard the supply.

Borehole pumping stations or boosting stations can have four to five hours interruption of supply due to electricity supply failure, but in any case such stations should be designed for not more than 22 hours working out of 24, or routine maintenance is difficult. Water treatment plants also need the ability to shut down for four to five hours as a minimum in order to effect any necessary repairs or alterations. River intake sources present particular hazards due to sudden pollution, but this is best met by providing *raw* water storage at the intake since the polluted water should not pass to the treatment works (see Section 7.1).

For the repair of major mains, six to eight hours minimum time for repair should be allowed *plus* two to four hours for refilling the system when the main has been repaired. The loss of water from the burst may also be substantial. Major fire consumptions can be 5000–15 000 m^3, but usually the water authority will receive notice of the extra demand and should be able to increase output accordingly.

The breakdown times quoted above are what might be termed 'normal': if things go wrong, double the time may elapse before supply can be restored and every water engineer can think of an instance when this has happened. Also a breakdown will not conveniently occur just when the reservoir is full; it may well occur during the maximum demand period when the overdraw may rise to 25% of the day's supply (see Section 11.2). In addition some allowance must be made for 'bottom water' in the reservoir to prevent its complete emptying.

Taking all these factors together at least 24 hours' service reservoir storage is necessary to cope with variations of demand, trunk main bursts, and sourceworks plant breakdowns. *This is the usual minimum service storage required and installed*: the fact that it is not provided on all undertakings is irrelevant, the reason usually being lack of money to undertake the necessary construction.

11.4 Three-day storage

While one-day storage may be sufficient for a particular day's fluctuations, certain days during a whole week will experience higher consumption than others. In some areas Monday, the traditional wash day of the week, proves to have the highest consumption, but since World War II habits have changed and Sunday, no longer the traditional day of rest, often proves to be the day of highest consumption, particularly in urban areas. At seaside holiday towns the influx of visitors at the weekends can push the day's consumption up by as much as 50% over the average for the time of the year. With one-day storage provided, therefore, there can only be a short time lag before the day's consumption must be matched by the output from the source. However, it may not be economic or easy to keep the output of the source closely in step with the consumption: there will be a time lag between knowing what the consumption for the day is likely to be and changing the output of the source; it may not be economic to change the output immediately because of the incidence of maximum demand or peak demand electricity tariffs, or because of the need to bring out extra staff on duty or to employ them on overtime working. If three-day storage is available these difficulties are minimized. The day's demand can be known and steps can then be taken to make maximum use of cheap off-peak electricity supplies and to keep the maximum demand charge to a minimum. Overtime working for staff can be kept to a minimum and maximum use made of normal working hours. Three-day storage is therefore the amount of storage nearly every water engineer would like to see for his undertaking because it is beneficial to the smooth running of the undertaking. Whether it is essential is another matter depending upon the nature of the works.

11.5 Summary in regard to storage

Purely for balancing flows to an average sized distribution system, about one-quarter of a day's supply should be stored, but this is not sufficient to safeguard the supply against breakdowns. The minimum storage for safeguarding the continuance of a supply during breakdowns is one day's supply. Where daily fluctuations are large, trunk mains are not duplicated, or the sourceworks relies upon pumping, and in all cases where it is desirable to adopt the best practice in relation to service reservoir capacity, then three days' storage should be provided.

11.6 Position and elevation of reservoirs

If the service storage is to be of maximum value as a safeguard to the undertaking against breakdown then it should be positioned as near as possible to the area of demand. From the service storage the distribution system should spread directly, with such ramification of mains that no single breakage could cause a severe interruption to the continuity of the supply. There should be sufficient interconnection between the distribution mains so that, should a breakdown of any one main occur, a supply may be maintained by rerouting the water. It is, of course, not always possible to find a high point which is in the centre of the distribution area and the best must be done in the circumstances. If the high point is remote from the area of demand the aim should be to feed the demand area by two major mains from the service reservoir. If there is some high ground which is not quite high enough then a water tower, or several water towers, may meet the need. It is also usually necessary to site the service reservoir at such elevation that a steady pressure is maintained at all points of the distribution system, sufficient to reach the topmost storey of three- or four-storey buildings, together with sufficient additional (or residual) pressure to enable a good flow to be maintained to those topmost points.

The elevation at which it is desirable to position a service reservoir depends upon both the distance of the reservoir from the distribution area and the elevation of the highest buildings to be supplied. If the distribution area varies widely in elevation it may be necessary to use two or more service reservoirs at different levels so that the lower areas do not receive an unduly high pressure. Wherever possible the use of non-standard pipes for high pressures should be avoided as such pipes are expensive. Also it has to be borne in mind that prior to the introduction of ductile iron pipes (see Table 13.1) the majority of cast iron distribution mains were laid in class B, class C, or (more rarely) class D pipes, suitable for maximum working pressures of 61, 91, and 122 m respectively. Generally, 45 to 75 m static pressure is that which best suits the domestic distribution system. It should result in distribution pipes of moderate size and thickness, and wastage from leaks should not be excessive at this figure. Pressures below 45 m will be likely to cause trouble in supplying extensive distribution areas; pressures above 120 m, although by no means unknown, may be considered too high, tending to result in excessive leakage losses.

Pressure reducing valves are sometimes installed in pipelines from service reservoirs in order to lower the pressure to low lying areas, but this is a policy which should not be resorted to unless no alternative exists. If the pressure valve fails to operate at any time, which it will do if it is not properly maintained, then the low main would be subjected to a sudden increase of pressure and may burst, while excessive wastage of water would also take place through ball valves unaccustomed to working at a high head. Break-pressure tanks are preferable to pressure reducing valves since they give better protection to a low pressure zone. In connection with pumping schemes break-pressure devices represent a direct waste of pumping energy and they should not be used unless no more economic solution exists.

Design of service reservoirs

11.7 Introduction

The most convenient way to approach this subject is to illustrate some actual examples of service reservoir design. These are shown in Figs 11.4 and 11.5. The following are some notes on the salient features of service reservoirs and the alternatives that may be adopted. The subject is a wide one and for detailed design other books should be consulted.

11.8 Depth and shape

There is an economic depth of service reservoir for any given site. For a given quantity of water either a shallow reservoir having long walls and a large floor area may be constructed or, alternatively, a deep reservoir constructed with high retaining walls and a smaller floor area. Depths most usually used are as follows.

Size (m^3)	*Depth of water (m)*
Up to 3500	2.5–3.5
3500–15 000	3.5–5.0
Over 15 000	5.0–7.0

These figures do not apply to water towers or prestressed reinforced concrete tanks. Factors influencing depth for a given storage are:

(1) depth at which suitable foundations conditions are encountered,
(2) depth at which outlet main must be laid,

(3) slope of ground, nature and type of backfill,
(4) the need to make the quantity of excavated material approximately equal to the amount required for banking, so as to reduce unnecessary carting of surplus material to tip,
(5) the shape and size of land available.

A circular tank is geometrically the most economic shape, giving the least amount of walling for a given volume and depth. It is not always adopted, however, either because it does not make the best use of available land or because the slope of the ground contours shows that a rectangular tank would be more economical. The shape of the land available is often the determining factor in built-up areas and it is for this reason that we most often find circular reservoirs on high, open moorlands and similar places, rather than in built-up areas. A rectangular reservoir, with ratio of sides about 1.2 : 1.5, will show a benefit over a square tank when an internal wall is required to divide the reservoir into two compartments. The design leads to greater simplicity of roof design and inlet and outlet arrangements.

11.9 Roofing

The water in service reservoirs is treated water ready for consumption and therefore it ought not to be open to pollution from the atmosphere. Seagulls, which feed at sewage farms and come to wash themselves in reservoirs, are a noted source of pollution, and these and other birds will not be prevented from settling on the water simply by extending wires across.

Concrete roofs can be designed as flat slabs on columns, as cast-in-situ slabs resting on beams, or as a series of precast and prestressed beams laid side by side and grouted together. If ordinary reinforced concrete slabbing is used, spans of 3.5 to 4.5 m, with thicknesses in the range 150 to 200 mm, will be found to be the usual economic dimensions. The use of precast prestressed beams permits the use of spans of up to 7.5 m, thus reducing the number of columns which have to be cast. It is most important with large roof areas to provide for shrinkage and expansion of the roof caused by temperature changes, and the roof must not be fixed on to or against the walls, but must be free to slide over them. Concrete roofs are traditionally 'earthed' over, partly as a means of insulation, but sometimes possibly as camouflage required by the planning authorities. There is no reason why a concrete roof should not just be left covered with a layer of gravel for insulation if the appearance is not objected to; the disadvantage of adding earth on top of the gravel is that the dead load on the roof is thereby increased and the cost made greater. However, the gravel layer must not be omitted if the roof is earthed over because drainage is important so as to prevent leakage of contaminated water into the reservoir. A roof, despite being termed

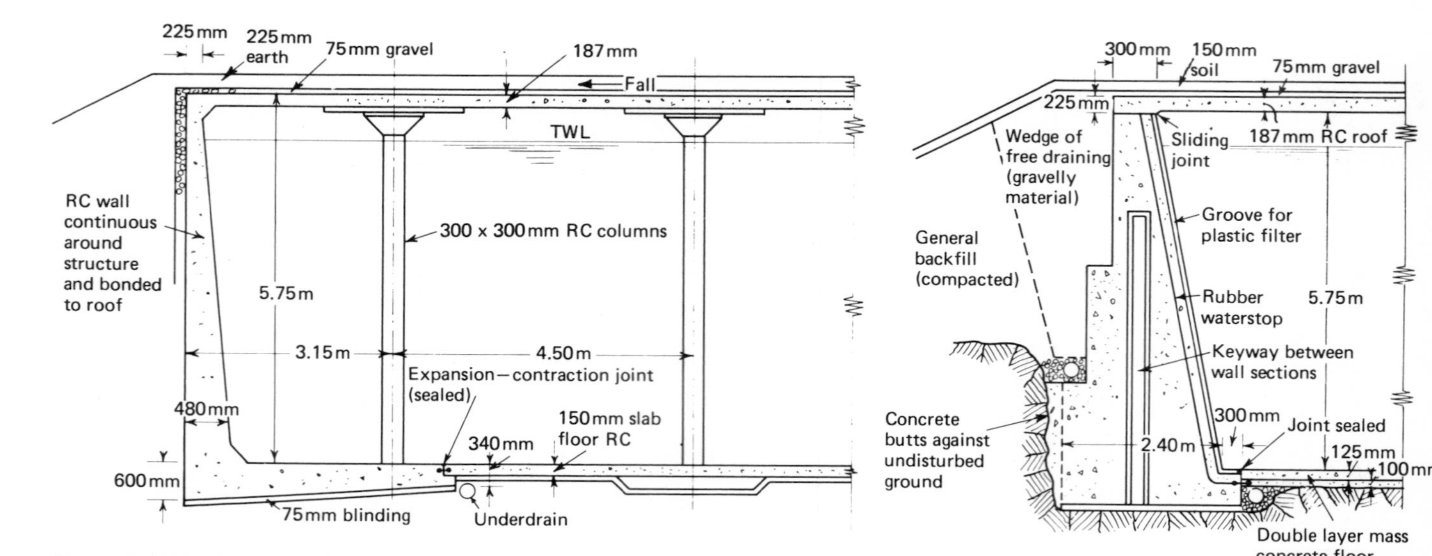

Typical RC design for a small service reservoir (Note: Do not omit check for flotation.)

Typical mass concrete wall to service reservoir

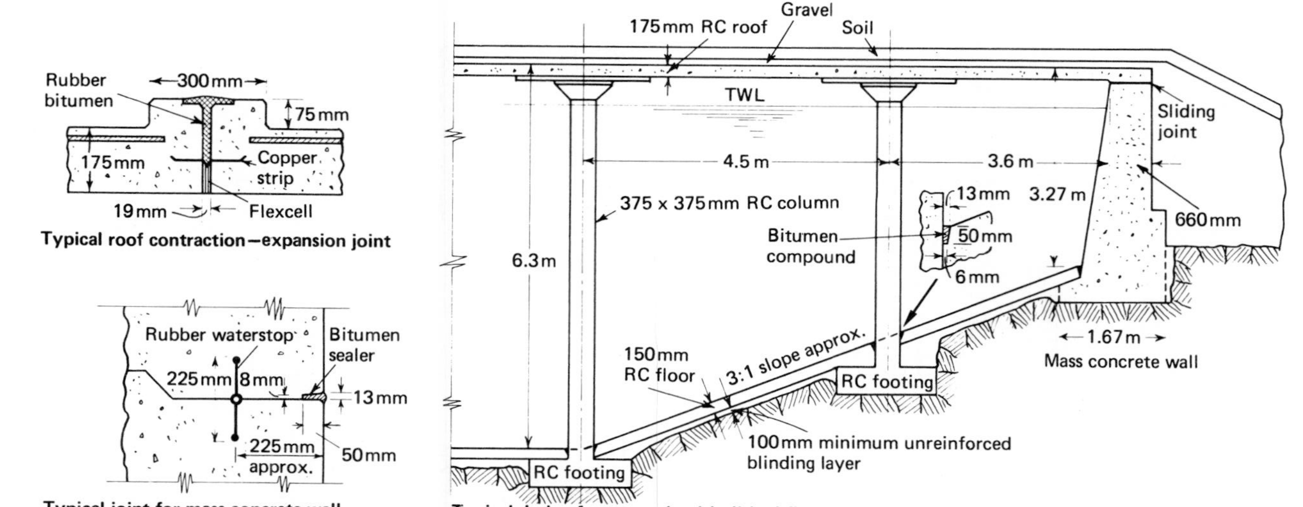

Fig. 11.4 Typical designs for service reservoirs.

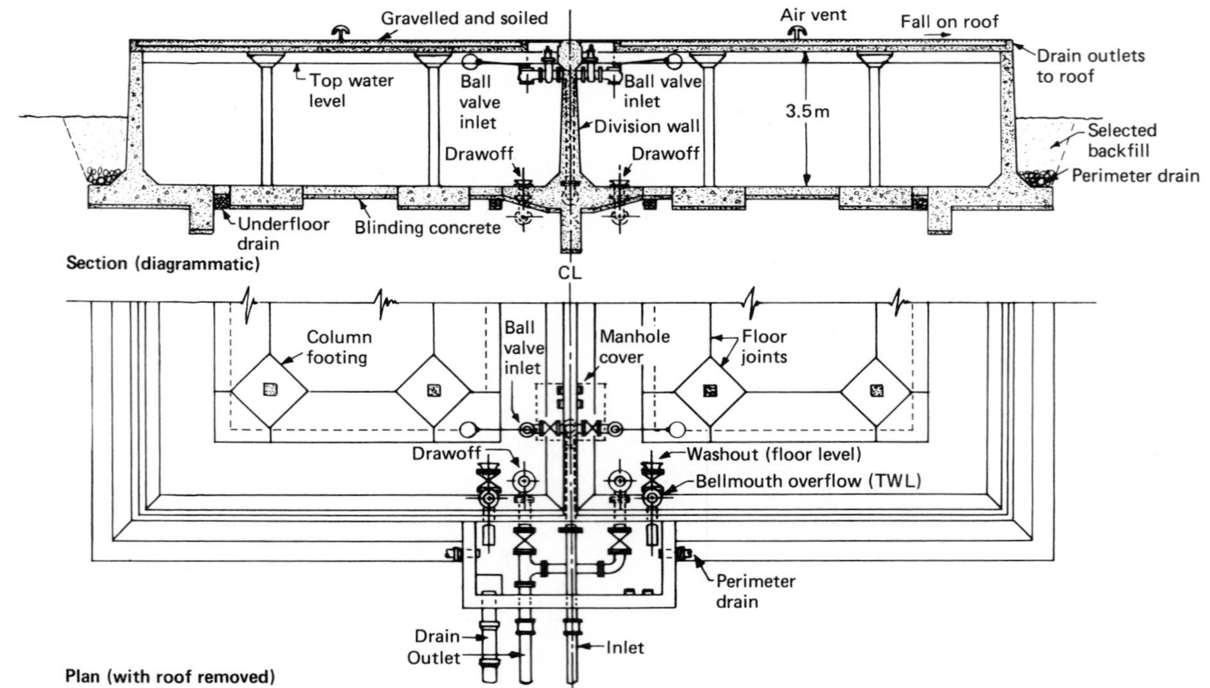

Fig. 11.5 A small service reservoir with reinforced concrete walls showing inlet and outlet arrangements. (Binnie & Partners.)

'flat', should, of course, have a fall to assist drainage. Arched roofs are sometimes used and, although this seems the correct design for a material like concrete, they are not likely to prove cheaper than the flat slab unless very large areas are to be covered, permitting reuse of formwork many times. Arched roofs do, of course, exert some extra thrust on the walls, necessitating stronger wall design.

Aluminium domed roofs are cheap for the covering of circular tanks and have special advantage for the covering of existing reservoirs because their clear span enables them to be erected without the addition of columns inside the reservoir and without interfering with the use of the reservoir.

Columns, when required to support roofs, are most economically made of ordinary reinforced concrete cast-in-situ. Casting columns inside permanent shuttering composed of asbestos cement pipes may seem attractive, but this is more expensive than the use of wood shuttering. Some contractors complain that they do not like circular columns; some contractors complain that they do not like square columns. All contractors find sloping column footings difficult and expensive to construct and column 'heads' slightly less so. The designer may ignore these various opinions and use what seems to him best in the circumstances, so long as he makes sure to give adequate cover to his bar reinforcement.

11.10 Walls of concrete reservoirs

Walls of concrete reservoirs are constructed to many different types of design, but generally three main types may be distinguished: mass concrete gravity walls, reinforced concrete walls, and prestressed walls. For normal sizes of tank and good ground conditions the mass concrete gravity wall has much to recommend it (see Plate 14A). Its simplicity invites a low price for construction; its weight and mass result in durable construction suitable for many different types of foundation and backfill; the joints, which should be no more than 10 m apart, are easily made watertight; there is no difficulty in making the walls impermeable. The major precaution required is to see that the corners of the tank are well designed and, if necessary, supported by a rough concrete backfilling against undisturbed ground. Whether the waterface is vertical or sloping depends primarily upon the results of the stability analyses for various trial designs, but a slope to the inner face or waterface permits a rake to the access ladders, thus making them much safer to use.

Mass concrete reservoir walls are not suitable for ground having a low bearing or shear strength. In this case a reinforced concrete wall must be used, cantilevering from a broad base or raft which is also reinforced. The projection inside the reservoir of the wall base is called the 'toe' and the projection outside is called the 'heel'. Both the wall and the toe are usually tapered, but this tapering should not be too severe or there will be difficulties in placing the concrete. The length and thickness of the toe and heel have to be adjusted so as to obtain the most economic design having regard to the stresses on the wall and the maximum permissible stresses on the foundations. The toe is a relatively economic part of the structure since it replaces a portion of the floor which would have to be constructed of concrete anyway. The heel is more expensive since it enlarges the amount of excavation and backfill required, but it may be essential, either for the reduction of maximum bearing pressure on the foundation under reservoir full conditions or for the development of adequate shear strength against sliding of the wall under reservoir empty or full conditions. In small rectangular reservoirs the reinforced concrete walls may be continuous and also joined to the floor and roof, making a box-like structure. The minimum thickness for reinforced concrete walls should be not less than 225 mm, partly because with less thickness there is some danger of leakage, but mostly because of the need to place four layers of steel in the wall and to maintain a minimum cover to the outer layer of at least 40 mm.

In designing all reservoir walls, whether reinforced or not, the designer must beware of making unwarrantedly precise assumptions concerning the loading on the walls from the backfill. It is not good engineering practice to design wall thicknesses to the last centimetre and reinforcement to the last square millimetre when the magnitudes of the forces exerted by the backfill are known only to plus or minus 25%. Earth pressures are not, as yet, precisely measurable and, whatever theory is used, the final design ought not to show an unusual departure from the kind of design which has proved satisfactory in the past.

Prestressed concrete tanks are usually circular, with the prestressing being applied only in the horizontal plane by winding the prestressing wires around the outside of the tank wall. When the wires have been placed and tensioned, a 50 to 100 mm thickness of pneumatic mortar is 'gunited' on to the outside of the tank, thus covering the wires. A firm foundation to the walls is necessary and provision for movement of the walls relative to the solid foundation concrete is also needed. Prestressed tanks can be made deep because the extra tension induced in the walls by greater water depth can be met by adding more strands of prestressing wires, the thickness of the wall not being greatly increased. Roofs can be domed, the thrust from the dome being taken by a thrust ring incorporated at the top of the tank. The chief defect of the prestressed circular tank is the same as that for a reinforced concrete circular tank, namely that it is difficult to design for unbalanced external forces from backfill. Thus such tanks must have uniform backfill conditions, which presupposes that they are on relatively flat ground, not ground which slopes steeply so that the tank is surcharged one side and has very little backfill against it on the other. The price of a prestressed tank is not usually found to be less than the price of a mass

Plate 13 (A) Chlorinators and recorders (Wallace & Tiernan Ltd.) (B) Chlorine comparator and (C) comparator test kit featuring Lovibond colour discs and Palin test tablets. (Manufacturers: Tintometer Limited, and Wilkinson & Simpson Ltd.) (D) Chlorine residual meter 'Residometer'. (Manufacturers: Wallace & Tiernan Ltd.)

Plate 14 (A) Mass concrete walled service reservoir under construction for Luton Water Company. (B) and (C) A reinforced concrete walled service reservoir under construction for Bristol Waterworks Company. (Engineers: Binnie & Partners.)

Plate 15 (A), (B) and (C) An unusual 30 000 m^3 reservoir constructed for Bournemouth and District Water Company, 1980. Precast concrete units, 9.5 m high, erected and circumferentially stressed by external prestressed wire, with gunited mortar layers inside and outside wiring. (Contractors: Preload Ltd.) (D) An earth-embanked raw water reservoir: interior concrete lining yet to be placed.

Plate 16 (A) and (B) Vertical spindle mixed flow pumps at Wixoe Pumping Station, Essex River Authority. The first photograph shows two 1000 kW fixed speed driving motors and one 500 kW variable speed a.c. motor; the second shows the pumps below rated at 1.05 and 0.52 m^3/s respectively. (C) Vertical spindle, double entry split-casing centrifugal pumps (motors above not shown), Langford Pumping Station, Southend Waterworks Company. (D) Horizontal spindle, double entry split-casing pumps at Grafham Pumping Station. (Engineers: Binnie & Partners.)

concrete walled tank or a normally reinforced concrete tank because the savings in concrete and steel are offset to a large extent by the cost of the special methods of construction.

11.11 Design procedures

It is now usual to design concrete liquid-retaining structures by limit state theory in the same way as that adopted for normal reinforced concrete building structures to CP 110. The British Code of Practice BS 5337 (1976): the structural use of concrete for retaining aqueous liquids, which is based upon CP 110, gives two main criteria when using the limit state method. The first is to ensure the safety of the structure against failure or 'collapse' at the ultimate loads and the second is aimed at ensuring that the structure remains serviceable at working loads throughout its expected life. The Code does, however, still preserve what is now termed the 'Alternative Method' of design, which is the more familiar elastic 'modular ratio' method which formed the basis of the earlier code CP 2007.

The use of limit state design will generally result in reduced thicknesses of the members, particularly walls, and will also in many cases show a saving in the quantity of reinforcement. It should therefore be adopted in preference to the 'Alternative Method', although the latter, being so well known, is still useful as a simple method of design*.

One of the important features introduced into the latest Code is the greater attention paid to serviceability conditions: methods are included to maintain satisfactory control over the early thermal cracking as well as those cracks which result from the structural loading. The procedure involves calculating the probable maximum widths of cracks and restricting these to limiting values depending upon the degree of exposure. The provision of movement joints and their spacing is treated in some detail and the designer is left to choose from a number of options the method of controlling early thermal effects and restraining shrinkage which he considers is best for any particular structure.

The Code now uses the term 'characteristic strength' instead of '28-day works cube strength' for concrete and the same term is applied to the 'yield stress' for the steel. The characteristic strength is defined as the value of concrete strength or yield stress in the reinforcement below which not more than 5% of the test results may be expected to fall.

The grades of concrete usually adopted for liquid-retaining structures are 25 and 30†. These numbers represent the 28-day characteristic

* The 'Alternative Method' is also a safe method because the vast majority of all concrete service reservoirs in existence were designed according to the elastic theory—and are still standing.

† The grade 25 concrete can be taken as approximately equivalent to the earlier designated 1:2:4 (cement:sand:large aggregate) mix and the grade 30 concrete as equivalent to the earlier 1:1.5:3 mix.

strength of the concrete expressed in newtons per square millimetre. Since the difference in cost of high yield deformed bars over mild steel is small, and economies result from the use of the former, all reinforcement except that used for links or binders in beams and columns should be of a high yield variety with a minimum characteristic strength of 410 N/mm^2.

One of the most important practical recommendations of the Code is that the cover to reinforcing steel should be not less than 40 mm, and this should apply to the subsidiary steel, such as binders etc., as well as to main steel when no subsidiary steel is used.

Cracking in mass or reinforced concrete walls may be reduced by adopting the following measures:

(1) using a concrete grade not over-rich in cement, e.g. using grade 25 concrete rather than grade 30, even though this may mean a slightly greater thickness of section due to the lower strength,
(2) using a 40 mm coarse aggregate size rather than the more frequently adopted 20 mm size,
(3) keeping the water content of the mix as low as possible and adding an air-entraining agent to assist in producing adequate workability,
(4) using steel shutters (not plywood or other wood shutters) so as to reduce the maximum temperature reached by the concrete during setting.

By these means reinforced concrete walls, 9 m long by 7 m high, concreted in one lift to their full height, have been constructed without cracking.

11.12 Floors

In all but the smallest tanks the floor is cast as a separate structure from the walls. Underfloor drains are therefore important to prevent uplift and these underdrains must have a free fall to an open outlet. If underdrains are not possible then the whole floor must be of rigid construction, reinforced against uplift and held down by the walls and column bases. For this the total weight of the structure when empty must be sufficient to prevent flotation caused by any possible outside groundwater level. On a firm foundation, where underfloor drains are possible, it may be quite unnecessary to reinforce the floor. It should then be cast in squares of about 3.0 to 3.5 m side, alternate squares being cast first before casting the infilling squares so that shrinkage at the joints will be minimized. Sometimes floors are laid in two layers, the joints in the upper layer being staggered in relation to the joints in the lower. There is, however, probably little to be gained by two-layer construction provided that a blinding layer of concrete 100 to 125 mm thick is used to seal the surface of the natural ground and also that proper care is paid to sealing the joints between the slabs. These joints consist of a groove into which a plastic compound, which adheres to the concrete, is poured.

11.13 Division walls and piping

Full height division walls add greatly to the cost of a service reservoir and a more economical practice is to install a half height division wall. This should be quite sufficient to enable the supply to be maintained through one-half of the reservoir while cleaning of the other half is undertaken. However, if the flow of water can be maintained without difficulty simply by constructing a bypass to the tank, this will prove cheaper altogether because not only is a division wall saved but also the cost of one set of inlet and outlet piping.

The inlet to a service reservoir may be at any elevation, but if taken into the bottom of the tank there should be a non-return valve on the inlet pipework so that, should the pipe burst, the contents of the reservoir are not discharged back through the inlet main. A bottom inlet pipe can be advantageous when the pressure on the inlet pipeline is very small because the head and therefore the flow into the tank will be increased when the reservoir water level is low. The outlet pipe must, of course, be taken out at low level and can be arranged with its invert at floor level provided that a sump is constructed just in front of the outlet into which debris will be trapped instead of being swept into the main. A 'trip valve' is useful on the outlet main to shut down automatically should there be a burst on the main which causes a flow in excess of anything normally experienced. Once such a valve has operated it must not be opened hurriedly, but the whole of the distribution system must be carefully filled again (a procedure which may take some hours) so as to avoid entrapped air and surge shocks developing therefrom. Sudden opening of the valve causing water to be discharged at high pressure into partially empty mains would result in bursts.

The overflow to any service reservoir is, of course, an important safety feature. It must be capable of taking the maximum inflow to the tank under any circumstances without the water level inside the tank rising so high as to cause an uplift pressure on the roof or to exert an excess pressure on the walls which has not been taken into account in the design. A bellmouth pipe overflow is not so effective as the incorporation of an overflow chamber within the tank, from the bottom of which the overflow pipe is led off. Use of an overflow chamber permits a longer weir for the overflow of the water than can be obtained by the use of a bellmouth and this can result in a substantial saving in the cost of the tank as a whole because the height of the tank may then be kept to an absolute minimum.

Stop valves must be inserted on both inlet and outlet mains, but these may, with very little loss of head, be made smaller in diameter than the pipelines by some 25%. By this means a cheaper installation is achieved and the valves themselves will be easier to operate. There is no reason why valves cannot be inserted inside the tank, providing the operating spindles are made of a non-corroding metal such as phosphor bronze. If a dry chamber for the valves is built outside the tank this may involve some expensive constructional work unless the natural fall of the ground permits a shallow chamber. Precautions must be taken to lay all pipes beneath the surrounding embankment on solid ground so that differential settlement is minimized. Where a pipe passes under the embankment and then through the wall of the reservoir there should be at least two flexible joints incorporated in the line at the back of the wall to permit settlement of the wall without fracture of the pipeline. Pipes should on no account be laid around the perimeter of the tank beneath the embankment, but should be kept well clear of the structure of the tank and any ground disturbed by the construction operations.

Inlet control valves may take a variety of different forms. Ball valves may be used for shutting off the supply when the reservoir is full, but the defect of ball valves is that they may cause a large loss of head at entry even when wide open. Float-controlled valves will cause less loss of head when wide open, but they must be made of durable materials and be reliable in design so that they give trouble-free operation for long periods of time. It is better not to rely upon inlet controllers but to incorporate some water level measuring device which transmits information back to the source of supply so that restriction of flow can take place there. This is only possible, however, when the reservoir is at higher elevation than the source and receives pumped water.

11.14 Access manholes

Two manholes should be provided for each separate compartment of a service reservoir, one of which should be large enough (about 1.2 m square) to make an adequate entry through which to lower scaffolding, buckets etc. for any necessary repairs. An upstand wall about 300 to 450 mm high should be cast integrally with the roof concrete around the opening so as to prevent seepage of rainwater into the reservoir from the roof. A sturdy handhold on the edge of the manhole is a great comfort for ease of entry. Access ladders should never be vertical because they will be used by men who are carrying equipment in one hand and ladders should therefore be to a convenient rake. Covers should be locked down to prevent unauthorized persons from prising them up and no single piece of cover should weigh more than 50 kg—preferably its weight should be less than this. A large entrance needs to be barred over with loose removable bars beneath the cover so that, if the latter slips when being lifted off, it will not fall into the reservoir.

11.15 Special types of service reservoirs

Where a flat site is underlain by suitable material, the reservoir may be formed by using the excavated soil to form an embankment around the perimeter. Such a reservoir can be covered over with a concrete roof of flat slab and column construction. The surface of the excavation needs to be concreted or asphalted to make it sufficiently watertight. The concrete may be reinforced or mass depending upon the type of foundation. The asphalt will have to be laid on an undercarpet of sand or ashes overlying a hardcore base. In favourable circumstances earth banked reservoirs can show a large saving over concrete walled reservoirs, but the site conditions need to be right. A flat area is, of course, essential and the soil should be suitable for compacting in the earth embankments. Membranes of butyl rubber, reinforced bitumen, or high density polyethylene can also be used but they need a suitable bed and must be carefully laid, jointed, and protected if they are to be successful. In all reservoirs of this type it is vital to include underdrains and to prevent any uplift occurring.

In certain areas coal or salt mining may be expected to cause extensive settlement of the reservoir and special design must therefore be adopted to permit this settlement to take place without leakage occurring from the reservoir. If the reservoir is small it may be designed as a single rigid box, but usually the size of reservoir required does not permit of this type of construction. Moderate settlements may be met by making all members of the reservoir, e.g. the wall units, the columns, the floor segments, and the roof beams, independent units which are capable of moving relative to one another, but which are connected by flexible watertight seals. Larger settlements require the construction of the reservoir in rigid units of sections of wall, floor, and roof, each of which can settle relative to the adjacent units without leakage being caused.

11.16 Joints in reservoirs

There is an increasing risk of shrinkage cracking occurring with concrete unit lengths greater than 10 m, whether reinforced or not. It is better to work to shorter lengths of 6.7 to 7.5 m if one hopes to be sure of avoiding shrinkage cracking. Joints are therefore necessary in walls and they usually consist of a flexible membrane of rubber or plastic (PVC) compound with a central bulb to it which permits considerable differential movement. Concrete adjacent to the membrane must be dense and well placed. In addition to this the waterface of the joint should be grooved to receive a plastic sealing compound which adheres to the concrete, thus giving an additional safeguard. For floors the plastic-filled groove is usually used alone without the rubber or PVC waterstop. The groove must be dried, cleaned, and primed before the plastic is inserted. Special care is necessary in cleaning out the floor joints which get very dirty during construction. The termination of a waterstop at the base of a wall must be carefully considered so that no path of leakage remains.

For preference the waterstop should be carried down the wall and turned into the floor.

Where little settlement is expected, after the initial settlement during construction, a satisfactory joint may be made by caulking a preformed joint on the waterface between sections with lead wool. The advantage of this method is that lead is a durable material of proven reliability and, with sealing arranged on the surface, it can always be examined and recaulked if not properly driven home the first time or if further shrinkage has taken place. The waterstop embedded in the centre of the concrete suffers from the disadvantage that once inserted it cannot be repaired or renovated.

In all cases where contraction joints are constructed the concrete face should be bitumen painted so that a definite separation at the required point occurs between successive pours of concrete. Reinforcement may be carried across contraction joints, but in this case it is necessary to seal both inner and outer faces of the concrete by inserting plastic compound in grooves. The depth of such grooves cannot be greater than the depth of cover to the steel reinforcement so that this is limited to 25 or 38 mm.

11.17 Testing and achieving watertightness

When a concrete reservoir is first filled it should be left to stand for at least three days to allow the concrete to absorb water. The measurement of water level for testing may then commence. BS 5337 requires that the total drop in water level over the next seven days shall not exceed one-thousandth of the average water depth of the full tank for a satisfactory test. If the walls have been designed so that they are stable under water pressure without the assistance of the backfill, inspection of the exterior of the joints to the tank is possible and advisable. However, it is not usually economic to have such thick retaining walls that they are stable under water load solely for the purpose of testing the tank. Neither will the construction contractor find it convenient to delay any backfilling until he has completed the tank because this will involve stacking the backfill material and rehandling it. Therefore many tanks must be tested without the chance of examining the exterior of the walls. In order to save water it is best to put 0.6 to 0.9 m of water in the tank first so that the floor can be tested. The underdrains need to be watched and the outflow from these before, during, and after the test should be carefully measured, by a temporary gauging weir, and logged down. Judging by eye is not at all satisfactory—a simple V-notch weir should be installed. If the floor appears watertight the reservoir can be filled to half depth and then to full depth, allowing the tank to stand to permit absorption at each level before taking the final acceptance test.

11.18 Searching for leaks

It is not easy to make a reservoir fully watertight and the following notes as to what steps might be taken to track down the point of leakage may be found helpful.

(1) Flows from underdrains should be examined. These drains should preferably have been so designed that they can give some idea as to the location of the leakage; they should not all join to one common outlet point before being measurable.

(2) The inlet and outlet valves and the scour valve must be tested to ensure that they are not passing water out of the tank. The only secure way of knowing this is to have short removable sections of the mains on the outer side of the valves so that these sections can be removed and the outflow, if any, through the valve observed and measured. If testing can be done before making all the pipe connections this will avoid the need for all the pipelines to have removable sections in them. Valves ought not to leak but many do, not necessarily through faulty design but most commonly because the gate is not shut properly because of dirt in the gate groove.

(3) The rate of leakage at full depth, half depth, and with about 0.6 m of water in the tank should be measured so that some idea is obtained as to the possible height of the leakage point within the tank. It is not likely that any revealing mathematical relationship between rate of leakage and depth of water in the tank will be established because any 'crack' through which water is leaking may be vertical or horizontal, long or short, and there may be several such cracks. It is usual to find that leakage is less when the depth of water is less, but occasionally one may find that there is no leakage at all below a certain level and this is a useful piece of knowledge.

(4) If the leakage is not traced by the above methods then the tank must be emptied and subjected to the most careful internal inspection. Emphasis must be placed on careful inspection. Good lights, adequate ladders, plenty of time, and a consistent pattern of examination should be adopted. It is extremely easy to miss a faint crack in the wall or floor. Walls (particularly the joints next to the corners) should receive special attention for it is here that there is most likelihood of movement having occurred. After emptying a tank the walls should be kept under observation when drying off because there is a certain stage of drying when leakage is evidenced by a dark patch on the surface of the concrete, even though this is the waterface. This patch will be short lived, but it may give a clue as to the whereabouts of a poor area of concrete through which leakage is taking place.

(5) The floor joints should be inspected. Jointing material should be examined to see if it has sunk, has holes in it, or has come away from or failed to bond to the concrete of the sealing grooves. The majority of leakages arise from defects of this kind. Wall joints should similarly be examined.

(6) If failure still results, about 0.6 m of water should be put into the

reservoir and be left to stand until the water is quite still. Then crystals of potassium permanganate may be dropped into the tank, widely spaced, and left for a considerable time. Then, descending into the tank with a good light and walking over a pre-arranged walkway so as not to disturb the water, streaks of colour may be noticed from the permanganate crystals showing some definite flow towards a point of leakage.

(7) As an alternative to method (5), about 150 mm of water can be put into the tank, a hole or several holes bored through the floor, and compressed air can be blown under the floor. In certain conditions of foundation air bubbling upwards through the water may indicate where faulty floor joints occur.

(8) If, despite all these attempts, the cause of leakage is still unaccounted for then more drastic measures may have to be undertaken, such as digging pits in the bank to inspect the rear of the wall joints, placing further sealing strips (such as glass fibre embedded in bitumen) over joints, or even rendering wall face areas. Sealing of leakage through a reservoir floor has been achieved by gravity grouting. About 450 mm of thin grout mix is put into the reservoir and the cement is kept in suspension by continually sweeping the floor and disturbing the water with squeegees for two successive days. Thus grout passes into the unknown paths of leakage and the cement sets. It should not be necessary, however, to adopt these measures unless poor construction has taken place. Where ordinary care has been taken with construction, failure to find the cause of leakage should be taken as an indication that the interior inspection has not been carried out carefully enough and this should be repeated. In the long run most troublesome leaks are discovered to be in some rather obvious place which has not been thoroughly examined in a first or even a second examination.

12
Pumping plant

Pumps

12.1 Reciprocating pumps

While the majority of pumps used for water supply purposes are centrifugal, with a rotating impeller, the reciprocating pump still has its uses and it was, of course, the earliest type of pump invented.

The ram pump is the most common form of reciprocating pump and it consists of a piston reciprocating within a cylinder, the latter being provided with inlet and outlet valves for the water. Water is sucked in by one stroke of the pump and forced out by the next stroke of the pump. If there is only one such cylinder operating then the output of water must necessarily be fluctuating, but if three or more cylinders are operated cyclicly a sensibly constant rate of flow is maintained. This type of pump, with three cylinders, is termed a 'triple throw ram pump' (as shown in Plate 17C), the pistons being connected 120° apart to the operating crank. Ram pumps were in common use before the introduction of centrifugal pumps. Their slow speed was admirably suited to the speed of reciprocating steam engines to which they were connected either directly or through gearing.

The triple throw ram pump is still used today, although it is usually driven by an electric motor. It is used where exceptionally high heads must be produced since no special design is necessary for this other than making the working parts sufficiently strong to sustain the head. Efficiencies of up to 94% or 95% can be obtained, falling off with piston and cylinder wear.

Small ram pumps driven by fractional horsepower variable speed electric motors are in widespread use for the injection of chemical solutions and suspensions in water treatment. The speed of the pump can be varied proportionately to the flow of water to be treated and the stroke of the pump can be made adjustable: thus any desired rate of chemical injection per unit volume of water to be treated can be maintained.

The bucket pump is another form of the ram pump, but is arranged to operate vertically for drawing water from a well or borehole. Instead of a piston a 'bucket' is given a reciprocating motion up and down in the rising main. This bucket has a valve on it which opens as the bucket descends, allowing the bucket to descend below the water line. When the bucket rises the valve closes so that water is lifted upwards. An alternative arrangement permits the descent of the bucket to force the water up the rising main by a series of valves, this type of pump being called a 'lift and force' pump. The village pump is a single throw bucket pump operated manually and it illustrates one of the chief advantages of the bucket pump, namely that it is always ready for instant use and will function even if it is operated erratically either at slow or fast speed.

In the past very large triple throw bucket pumps were used to draw water from underground, driven by triple expansion steam engines. Maintenance and repairs were infrequent because of the slow speed of operation and robustness of construction. Many continued in use for three-quarters of a century, but became uneconomic because of the cost of steam raising.

Small single throw bucket pumps are used with modern wind vanes as motivating power, again illustrating the reliability of these pumps under erratic operating conditions.

The hydrostat is, in essence, a water-operated ram pump. A large volume of water flowing in one pipe is used to operate a ram which, in turn, is connected to a smaller ram pump which pumps part of the water to a higher elevation through a branch pipe. It is not possible to raise the pressure of all of the water, of course, and the portion of water which continues down the major pipeline suffers a loss of head. Hydraulic rams are not convenient in public water supply systems because any increase of demand on either the high or low pressure side will quickly invalidate the set-up, and the ram pump will then become worthless and will have to be replaced by a booster pumping set.

12.2 Centrifugal pumps

All pumps which operate by a rotary action are called rotodynamic pumps and, of these, the centrifugal pump is the first type to be considered.

The principal part of a centrifugal pump comprises an impeller which is rotated at high speed. The impeller consists of two discs having a number of spiral blades between them. A pump with one impeller is shown in Plate 17B and one having six impellers is shown in Plate 17D. The impeller is often made of bronze and is one casting. One of the discs is fixed to the shaft of the pump, the other has a central hole in it which creates an annular space around the shaft: this is the 'eye' of the impeller. When the impeller is rotated at high speed, water is sucked into it via the eye, drawn through the impeller, and flung off the tips of the vanes, much as a 'catherine wheel'. This gives the water around the periphery of the impeller a high kinetic energy and, in the diffuser chamber around the impeller, part of this energy is converted to pressure energy, part to forward movement of the water through the connected pipeline, and part is lost in eddy formation. The conversion to useful pressure energy and forward movement of the water with minimum energy loss is accomplished by careful design of the impeller and diffuser chamber. With good design the efficiency of the pump can normally be in the range 78% to 82%, including all energy lost in bearing friction as well as the hydraulic loss within the pump. Efficiencies of up to 90% are possible with special design, but these are obtained by, amongst other things, fine clearances between the moving and static parts of the pump so that this efficiency may fall off with wear.

12.3 Types of centrifugal pumps

While all centrifugal pumps work on the principle set out above, their construction may vary considerably according to the type of duty required from the pump.

Multistage pumps consist of several impellers and diffuser chambers clamped together in series, the impellers being fixed to one shaft. Plates 17A and 17D show a typical multistage pump. The water from the first diffuser chamber is led to the second impeller, from the second diffuser to the third impeller, and so on. By this means the pressure developed by the pump is increased. For general waterworks duties the maximum pressure normally developed by one impeller is between 80 and 100 m head. Higher heads can be produced by higher speeds of rotation and larger impellers, but these measures increase the cost of manufacture.

A 'split casing' centrifugal pump is shown in Plates 16D and 17B. The advantage of this design is that it is easy to remove the upper half of the casing, thereby enabling access to the impellers and diffuser chamber for cleaning, if the quality of the water to be pumped makes this desirable.

A special advantage of the multistage pump is that it is possible to have a 'dummy stage' for one or more of the stages. This is simply a diffuser chamber without an impeller in it, thus allowing for the addition of impellers if the pump is later required to develop more pressure. The efficiency of a pump is not materially altered by the dummy stage and its introduction at the time of installation when future increases in head can be foreseen can prolong the life of a pump. It is important to remember that the driving motor must have enough power to drive the pump when the impeller is added to the dummy stage.

One of the problems of the design of a centrifugal pump is how to overcome end thrust. The pressure of the water over the eye of the impeller is low relative to that exerted on the back of the impeller and axial thrust develops. Several means are adopted for balancing this thrust which would otherwise quickly cause wear on the pump and shorten its life. Small pumps can absorb the end thrust by the use of ball or roller bearings. For larger pumps a double impeller may be used with the water dividing and entering the impeller from both sides; in this case there is no appreciable unbalanced thrust to counteract. A multistage pump cannot have double entry arranged for each impeller, moreover the total end thrust is high. A common device for overcoming end thrust on this type of pump is to incorporate a balancing disc on the shaft, high pressure water being led to one side of it so that most of the end thrust is taken by the balancing disc.

The vertical spindle pump is frequently used for pumping water from a well. The driving motor (Fig. 12.1) is at the surface, but the pump (Fig. 12.2) is immersed in the water perhaps 20 to 30 m below and it must therefore be driven by a vertical spindle. This spindle rotates within a tube or 'sleeve' of about 75 to 125 mm diameter which is held centrally in the rising main by spider bearings; the pumped water is delivered to the surface via the annular space between the sleeving and the rising main. A typical arrangement would be a 250 mm diameter rising main in 3 m lengths bolted together with flanged joints, the sleeve tube being 100 to 125 mm diameter, with spider bearings carrying lignum vitae bushes for the spindle at every joint in the rising main. The bearings to the spindle are water lubricated, lignum vitae being a wood, and this water is fed down through the sleeving, the water being taken off the pumping main and filtered before it is passed down to the bearings. The whole weight of the spindle, and the pump impellers and the hydraulic thrust created, is taken upon a Michell thrust bearing positioned at the top of the shafting, just below the coupling to the motor. A Michell thrust bearing is based upon the principle of taking the thrust on a thin film of oil trapped between a number of tilting blocks radially positioned on one face of the bearing which rotates, in a bath of oil, on a flat disc forming the other half of the bearing. It is an effective thrust bearing, but the power lost in it can be high.

Vertical spindle machines are very reliable being robust and designed to meet continuous heavy duty running. They are expensive in first cost and take a considerable amount of time and skill to dismantle or erect when repairs are necessary. Their capital cost may be double that for a horizontal spindle pump.

Submersible pumps (see Plates 17E and 17F) should strictly be termed 'submersed-motor' pumps because this is their particular characteristic. The pump is directly connected to an electric motor sited immediately below it and this motor is capable of running under water. The modern practice is to allow the windings to be surrounded by water from the well or borehole so that the water can act as a coolant and also as a bearing lubricant. Some designs seal the whole motor off from the water. The motor must of necessity be a squirrel cage a.c. motor running at a fixed speed unless a variable speed frequency changer is interposed on the line. Power to the motor is fed by waterproofed cables clipped to the side of the rising main.

Submersible pumps are quickly and easily installed and they can be of small diameter. Also no space is taken up within the rising main by a spindle and sleeving as with the vertical spindle pump. They need not be

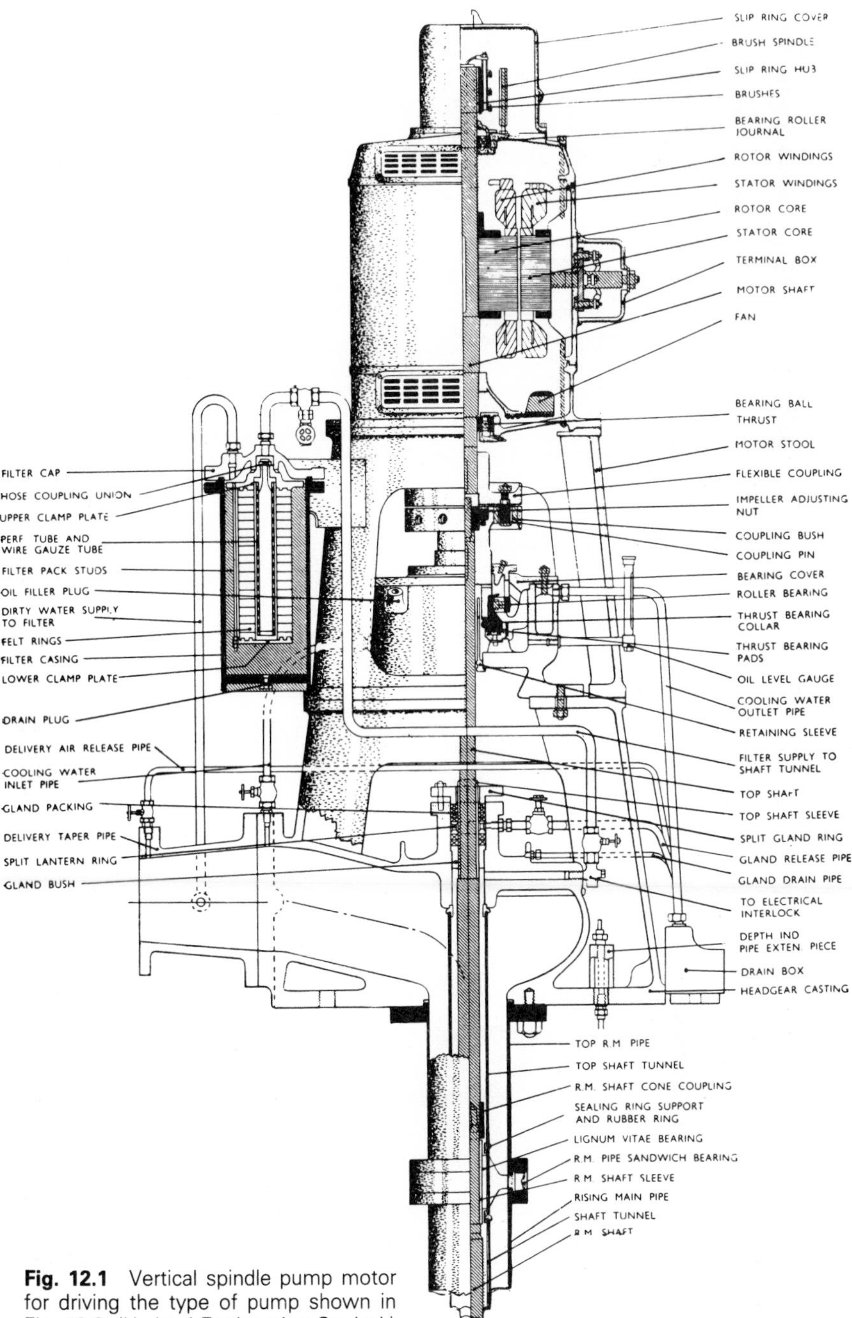

Fig. 12.1 Vertical spindle pump motor for driving the type of pump shown in Fig. 12.2. (Harland Engineering Co. Ltd.)

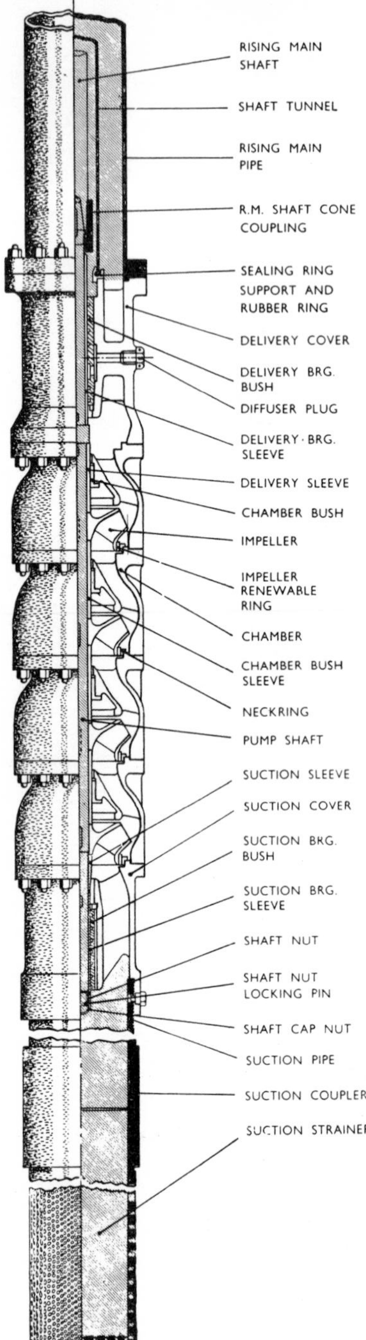

Fig. 12.2 Shaft-driven multistage bore-hole pump which would be driven by the motor shown in Fig. 12.1. (Harland Engineering Co. Ltd.)

installed truly vertical and are in fact being used horizontally as booster pumps in distribution mains. Their reliability in non-corrosive waters has been proved over the years, and even in corrosive waters they have the distinct advantage that when they need attention they can easily be withdrawn and reinstalled in a boring. They are cheaper than vertical spindle pumps and do not need any housing over them, but they are generally less efficient because of the special design of the motor.

12.4 Characteristics of centrifugal pumps

The characteristic curves for a typical true centrifugal pump are shown in Fig. 12.3. It will be noted that the head curve is relatively flat up to

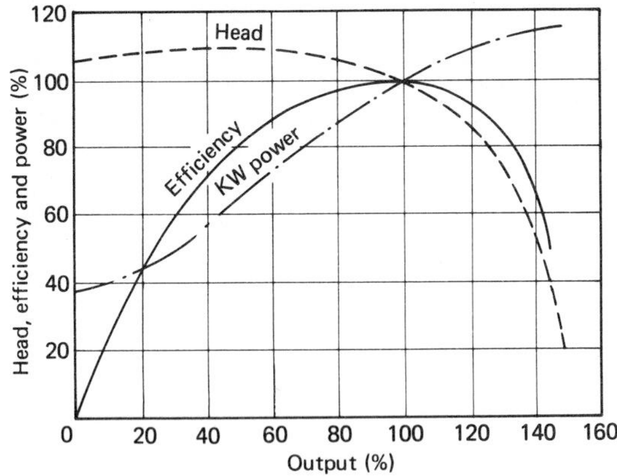

Fig. 12.3 Characteristic curves for a centrifugal pump at constant speed. Note that the head developed shows unstable characteristics up to about 90% output and a continuously falling head curve would be preferable and in some instances necessary (see main text).

the design duty point and that the power at zero flow is only about 40% of that required at design duty. This illustrates two useful characteristics of the centrifugal pump: it can be started against a closed valve and the power required to start it is somewhat under half of that required at the duty point. It is common practice to start a centrifugal pump against a closed valve and to close the valve before the pump is shut down. The pump comes to no harm provided that the process is not so long continued that the pump becomes heated. The reduced power required for start-up is also beneficial in reducing the starting current when an electric motor is used as the prime mover.

It will be noticed that the maximum head produced by the pump, for its given speed, is not greatly in excess of the design head. However, in the particular instance shown this head–output curve is unsatisfactory since for heads above the duty point there are two possible outputs. The pump is therefore unstable and this could cause trouble if it were operated in parallel with another pump. For preference the head–output curve should continuously decline for increasing output so that the pump is stable and the maximum head at no flow should not be excessive. The efficiency curve should be reasonably flat about the design duty point so that there is no great reduction in efficiency if the actual pumping head is slightly different from the duty.

The head developed by a centrifugal pump is sensitive to small changes of speed. Hence if a pump is rotated at slightly less than its design speed its output against a given head may be substantially reduced. This can arise in the case where the supply voltage to the electric driving motor is not up to standard and the motor is an induction motor. With reduced voltage the 'slip' of the rotor (see Section 12.8) will increase, thus reducing the speed of rotation of the pump and having a substantial effect upon its head or output.

Air in a pump causes its efficiency to reduce substantially and may also induce severe corrosion from cavitation (see Section 12.20). Air within a pump is evidenced by a hard crackling noise heard within the pump, almost as if the pump had some gravel left inside it. If such a sound does not disappear a short while after starting up, the cause should be investigated and eradicated.

12.5 Axial flow and mixed flow pumps

Axial flow pumps are of the propeller type, the rotation of which forces the water forward. (Water turbines can be of exactly the same design, the flow of the water turning the propeller.) Mixed flow pumps act partly by centrifugal action and partly by propeller action, the blades of the impeller being given some degree of 'twist'. There is no precise dividing line between centrifugal, mixed flow, and axial flow pumps.

In general axial and mixed flow pumps are primarily suited for pumping large quantities of water against low heads, whilst centrifugal pumps are best suited for pumping moderate outputs against high heads. Axial flow pumps must be submerged for starting and are most often used for drainage purposes or for transferring large quantities of water from a river to some nearby ground-level storage. Mixed flow pumps are shown in Plates 16A and 16B.

Characteristic curves for typical mixed flow and axial flow pumps are given in Fig. 12.4. It will be noted that the starting power for the mixed flow pump in this particular case is about the same as the duty power required, but for the axial flow pump the starting power is substantially in excess of the duty power. Axial flow pumps are therefore not started against a closed valve, but against an open valve so as to minimize the starting current required.

Even the smallest of pumps may be mixed flow rather than truly centrifugal. Thus the borehole pump shown in Fig. 12.2 is really a mixed

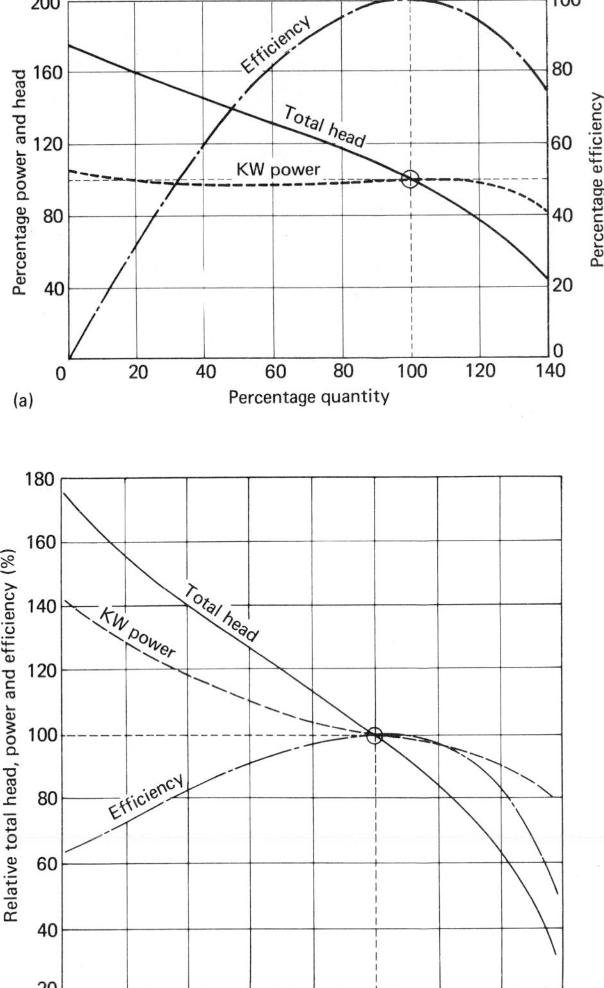

Fig. 12.4 Characteristic curves for (a) a mixed flow pump and (b) an axial flow pump. (Allen Gwynnes Pumps Ltd.)

flow multistage pump. The designer has had to design a pump which is restricted in impeller diameter so that the pump can be inserted in small borings; the quantity required has forced him to adopt a mixed flow impeller design and, because one impeller does not give sufficient head, the designer has been forced to add several such impellers in series.

12.6 Choice of pumps for waterworks purposes

The horizontal centrifugal pump is eminently suitable for all waterworks duties except those of handling very large volumes of water against low heads and pumping from wells and boreholes. The main advantage of the horizontal pump is that it is cheap and a great variety of designs are available to meet a wide range of pumping conditions. For a single unit the output can range from 50 Ml/d by 60 m head to 10 Ml/d by 200 m head. The general 'run of the mill' waterworks duty is from 10 to 25 Ml/d per unit by 30 to 120 m head and in this range the horizontal pump is cheapest. The pump should have stable characteristics and should be 'non-overloading', i.e. the power absorbed should not be excessive if the head drops. This is not always possible to arrange and electrical overload trips must protect the motor. A low head of this kind could develop if the delivery main from the pump were to burst near to the pumping station. The efficiency curve must also indicate no severe dropping of efficiency for moderate variations of flow or head about the duty point. When intending to purchase a pump it is most important to inform the manufacturer of the complete range of duties the pump is likely to be called to meet; to specify the duty point only, i.e. the theoretical normal conditions under which the pump is to operate, could lead to large loss of efficiency if the actual running conditions vary from the theoretically calculated duty.

For wells and boreholes the choice lies between vertical spindle pumps and submersible pumps. We have already mentioned the characteristics of these two types of pumps and the special usefulness of the submersible pump and its cheapness. The vertical spindle pump may be regarded as a 'heavy duty' pump and one advantage it has over the submersible pump, despite its extra cost, is that it may be driven by a synchronous electric motor. Thus the power factor can, by convenient control facilities, be brought to unity and the consequent cost of the electricity supply is reduced, although the motor is more expensive and maintenance is increased. Large outputs are often handled by vertical spindle pumps driven by synchronous electric motors. Variable speed is another advantage that can be more easily obtained with the vertical spindle pump than with the submersible pump which requires a frequency changer for variable speed. Variation in speed is often necessary when pumping from a well or borehole because the output from the well may have to be kept in step with the supply demand on the same pumping station, irrespective of the seasonal fluctuation in water levels in the well. A thorough appraisal of all the operating conditions likely to apply must be made before choosing the best type of pump for the duties.

For pumping large quantities of water against low heads, again the vertical spindle motor driving a mixed flow pump is suitable, with the pump immersed in the water and the motor sited above the highest water level likely to be experienced and being variable speed if necessary. It is not always necessary to use the vertical spindle pump for taking water from a tank or main and pumping it to a high level because the cheaper horizontal centrifugal pump can be sited in a dry well at low water level of the tank so that the pump is automatically primed. It is preferable not to site centrifugal pumps so that they must exert some suction lift on the water because their efficiency drops and cavitation troubles may be experienced. However, depending upon the design, up to 4 m lift can be arranged without difficulty and up to 5.5 m is possible. Similarly, because in most instances there is no reason why a horizontal centrifugal pump should not be used for boosting a piped supply, there is little point in having a vertical spindle pump do the same duty since the vertical spindle pump is more expensive than the horizontal type.

For high lift pumping the most economical pump is the fixed speed horizontal multistage centrifugal pump. Submersible pumps are also used for this duty, either for pumping direct from a well to a high level tank or for inserting as boosters in a main in a pit below ground level. This is advantageous when it would otherwise be impossible or expensive to erect a building to house the normal type of horizontal pump and motor.

For the full output of a large pumping station a single pump is seldom relied upon. Adequate standby is essential to ensure a continuity of supply and if the full duty can be handled by only one pump then a duplicate of this must nevertheless be installed. If, however, pumps are installed of 50% total required output capacity then the installation of three such pumps will ensure 100% output on the breakdown of any one machine, and 50% output in the more unlikely event of two pumps failing at the same time. The cost of the standby plant is then less than when two pumps each of 100% capacity are installed.

Prime movers

12.7 Steam and diesel engines

Triple expansion steam engines are no longer working in British waterworks. Their high capital and running costs and the size of building required for them make them uneconomic as compared with diesel or electric prime movers. Some of the larger water undertakings installed steam turbines for very large pumping units and some used gas turbines for heavy duty standby purposes. However, steam plant is no longer used.

The use of the diesel engine as a prime mover in the UK also declined substantially since the 1950s, although many installed in earlier years remain in operation. Most frequently they are used for a.c. or d.c. generation as this gives a greater flexibility of operation as compared with diesel engine direct drive through gearing to pumps. Except in remote situations where an electricity grid line may not be available or may be very expensive, or in oil-producing countries where fuel oil may be plentiful and cheap, diesel engines are mostly used for the purpose of providing standby power generation in the event of a grid line failure or for the purpose of 'peak-lopping'. The latter consists of the avoidance of high electricity tariffs at peak hour rates, or the minimization of maximum demand charges for electricity, by changing to diesel power at the appropriate times. However, proposals of this sort have to be considered most carefully because it may be uneconomic to run diesel engines for short periods of time and they tend not to give of their best when run on an intermittent basis. Where the cost of diesel engine power generation is little more than the cost of a grid supply of electricity, many water engineers would favour the diesel engine because of its reliability when properly maintained and the fact that three to six months supply of fuel oil may be stored on site. For remote locations where long grid lines may be subject to failure, or where grid supplies are erratic, the diesel engine with a good supply of fuel oil in store is difficult to beat for reliability.

The installation of even one diesel engine, however, will also mean the need to install all the costly ancillary apparatus that must go with it: the cooling water system, fuel storage tanks, starting air bottle and compressor, a suitable crane for dismantling, and a proper workshop and set of spares for repair and maintenance. Normally two engines should be installed since if one develops a major fault, such as the need for new bearings or cylinder linings, considerable time will be required to repair it.

12.8 Electric motors

Most of the electric motors used for driving waterworks pumps are a.c. fixed speed induction motors and most of the differences arise from the type of starting required or whether or not speed control is required.

(a) *A.c. induction motors* are usually either of 'squirrel cage' or 'slip ring' design. The squirrel cage motor is the more simple: it is one in which the rotor windings are not brought outside the machine, i.e. the windings are 'caged'. In the slip ring motor the rotor windings are brought out to slip rings on the outside of the machine, on these ride carbon brushes so that it is possible, through them, to put resistance in series with the rotor windings. By this means the starting current taken by the motor can be controlled and the speed can also be controlled, although at the expense of losses in the resistances. Induction motors are relatively cheap and of standard design. Their speed is related to the numbers of pair of poles provided for the field winding, the frequency of the supply, and the amount of 'slip'. Slip, which is the lag of the rotor behind the speed of rotation of the flux, varies from 5% for small machines to 2% for large machines.

Starters If a standard squirrel cage motor is started by switching it direct on line a large starting torque is developed (about 120% full load torque) and this is accompanied by a starting current which is about six times the full load current. This heavy starting current will cause trouble to the electricity supply and can also damage the field windings of the motor. If 20% to 40% full load torque is applied to the ordinary centrifugal pump this will be quite sufficient to bring it up to speed against a closed delivery valve*, which can then be opened, and the power consumed will increase as the pump starts to deliver water. The problem of starting is an important one and four methods, set out below, are in normal use with a.c. induction motors.

(1) Direct-on starting is normally permissible only for small, medium voltage motors up to 3 kW, but if the machines are high voltage then direct-on starting may be permitted by the electricity authority for much larger machines. It depends upon the starting torque requirements of the pump and the consequent starting current required as to whether or not direct-on starting is permissible. Machines with multicage windings may permit direct-on starting with initial currents of the order of 3.5 to 4 times the running current.

(2) Star-delta starting can be used for small squirrel cage motors. The stator (or field) windings are first connected 'star', i.e. phase to neutral, so that the voltage first applied to the field windings is about 240 volts for a normal grid supply. The line current taken at start is then not more than twice the full load current and the starting torque is 33% full load torque. When the motor has reached its maximum speed at this connection a switch is thrown so that the windings are connected 'delta', i.e. the full line to line voltage of, say, 415 volts is applied to them. This causes the motor to pick up speed to its maximum without further increase of current above the maximum starting current. Star-delta starting is simple to install and operate and is extremely useful for starting machines which are slightly too large for direct-on starting. It cannot be used for large machines.

(3) Transformer starting can also be applied to squirrel cage motors. The line voltage applied to the field windings is first put through an autotransformer. Usually three steps of voltage are provided: 40%, 60%, and 75% of full line to line voltage, the tapping selected being that most favourable to the particular duty, thus we speak of this kind of starting as 'autotransformer starting'. The starting current is normally about 2.5 times the full load current and the starting torque is about 36% full load torque with 60% tapping. If this starting torque is not sufficient a higher tapping of the transformer must be used for the first step. The method is used for starting large motors. Autotransformers are usually worked in conjunction with press-button starting so that the

* The axial flow or propeller pump must be started against an open or partially opened valve.

whole operation is gone through automatically once the main breaker has been set and the start button pressed. With normal autotransformer starters there is an undesirable pause created in the supply at change-over and a special starter known as the Korndorfer starter is used to obviate this difficulty.

(4) Rotor resistance starting can only be used with slip ring motors since the method consists of putting a resistance in series with the rotor windings via the slip rings. This increases the impedance of the motor so that the rotor and stator currents are reduced, the line current is thereby more nearly brought into phase with the line voltage so that a proportionately larger starting torque is available for a given current flow. With metal grid resistances the resistance is cut out step by step, and the number and sizing of these steps can be so arranged that a starting torque in excess of 100% full load torque can be made available without the line current exceeding 150% of the full load current. When all the resistance is cut out the motor connections are exactly the same as if the machine were a squirrel cage motor, and it will continue to run in the same fashion. In order to prevent continual wear of the brushes on the slip rings the brushes can be arranged to lift off the rings once the rotor is up to maximum speed and the final resistance is cut out. It may occur to the reader that resistances could be put in series with the field windings, but this would result in a large loss of torque for a small reduction of the starting current.

(b) *A.c. synchronous motors* have the rotor windings fed by a d.c. exciter, i.e. the windings carry an imposed current and do not have current induced in them by the lag effect caused by the rotor speed being less than the speed of rotation of the magnetic flux caused by the stator. Hence the synchronous a.c. motor is in reality an induction motor whose slip is zero and it rotates at the same speed as the magnetic field, i.e. it is synchronous with the magnetic field. Thus a two-pole stator fed by a 50 Hz supply would rotate at $1 \times 50 \times 60 = 3000$ revolutions per minute and a four-pole machine would rotate at half this speed, i.e.

$$\text{speed} = \frac{1}{\text{number of } \textit{pairs} \text{ of poles}} \times \text{number of cycles}$$

Although the synchronous motor is more expensive than the induction motor it offers the distinct monetary advantage of procuring unity power factor, i.e. the line current is taken in phase with the voltage so that

$$\begin{aligned}&\text{total power absorbed (watts)}\\&\quad = \sqrt{3} \times \text{line volts} \times \text{line amperes} \times 1.0\end{aligned}$$

where 1.0 represents the unity power factor. The power factor of an induction motor, where the current must inevitably lag behind the voltage, is in the region of 0.85 unless power factor correction capacitors are installed, as they sometimes are. Hence there is a distinct advantage in installing synchronous motors, especially when most electricity tariffs are designed to penalize users of electricity at lower power factor. Starting of synchronous motors is usually achieved by connecting them into line as an induction motor first, using autotransformer starting, and when the motor is near to maximum speed the exciter d.c. current is fed into the rotor circuit and the motor pulls into synchronous speed. The exciter is a small d.c. generator running on an extension of the main motor spindle.

(c) *D.c. motors*, although tending to be costly, are suitable for use where large or frequent adjustment of speed is required, particularly if power is generated on site by d.c. generators. If a grid supply is used then it must be rectified or it is possible to have variable speed a.c. motors. The starting of a d.c. motor is achieved by gradually cutting out a resistance which is connected in series with the field. The development of silicon and germanium rectifiers has made d.c. machines more popular.

(d) *Variable speed motors* Before deciding to use a variable speed drive for a pump every effort must be made to see if the necessity can be avoided because it will certainly involve greater cost and decreased efficiency over a fixed speed drive. Pumps may be obtained which will give a fairly flat curve, i.e. they will need only a little throttling to achieve correction of the rate of flow desired and this will not be accompanied by much reduction in efficiency. Dummy stages can be inserted in multistage pumps so that they may later be equipped to meet more onerous duties by the insertion of an additional stage. A balancing tank may represent a cheaper method of running two pumps in series and give some advantages of layout also.

If none of these devices can solve the problem then direct drive from a diesel engine is one way of obtaining variable speed, but this is expensive and speed-increasing gears may have to be used which can prove troublesome. If variable speed electric motor drive is desired then the choice may be made between:

(1) d.c. commutator motors,
(2) a.c. slip ring motors with rotor resistance,
(3) a.c. commutator motors,
(4) other types of a.c. motors.

Direct current motors will, of course, necessitate either site production of direct current or rectification of the grid supply. A limited speed range is possible by the use of a variable resistance in series with the shunt field windings. Alternating current slip ring motors may be reduced below their maximum speed by retaining a resistance in series with the rotor windings. Provided that the speed reduction required is not great this method has the advantage of simplicity, although some power must inevitably be wasted. Liquid resistances will permit infinitely variable speed within the motor's range. Alternating current commutator motors are efficient variable speed motors, but maintenance of the brushes is required. Regulation of speed is obtained either by shifting the position of the brushgear on the commutator for rotor-fed machines or by an induction regulator on stator-fed machines. Other variable speed a.c. machines may work on the basis of changing the frequency of the supply by various means.

Automatic control of pumping

12.9 Savings of automatic control

The extra capital expenditure required to give automatic control may be repaid many times over by the saving of labour attendance costs. A single pumping station attendant will be a charge of upwards of £6000 per annum on a water undertaking and his 'capitalized value' will be upwards of £60 000. Automatic control is usually applied to grid-fed electric motors, not to diesel driven pumps. Control may be completely automatic, i.e. both starting and stopping controlled, or semi-automatic, when it is usual to arrange only for shutdown to be automatic. The latter system is cheap and advantageous when, as is usual, single shift attendance must be provided for station cleaning and maintenance duties.

12.10 Protective devices

Among the various protective devices which are available the following functions may be mentioned.

(1) *For protection of electric motors*
(a) no volt protection,
(b) phase failure protection,
(c) overloading protection,
(d) earth leakage protection.

(2) *For protection of pumps and pumping system*
(a) low suction pressure,
(b) high discharge rate, e.g. burst main,
(c) prolonged high delivery pressure,
(d) lack of chlorine in supply,
(e) surge control,
(f) automatic slow closure of valves.

(3) *For control of pumping*
(a) high or low water level in reservoir,
(b) high or low pressure in delivery main,
(c) high or low flow,

(d) time switching,
(e) speed variation or bringing in of pumps to maintain a given flow or pressure.

12.11 Push-button starting

Push-button starting permits starting of electric motors by unskilled personnel. A series of interlocking and properly sequenced electrical contactor switches, with appropriate time lags, are set into operation upon the closing of the main breaker and the pressing of the starter button. These contacts then take an induction or synchronous motor through all the necessary steps of the starting routine and the attendant has only to open the main delivery valve on the pump at the appropriate time.

Most controlling devices which translate information from the water supply system to the electrical input side of the motors are either float- or pressure-controlled devices. A float can either make or break an electrical contact when reaching a preset level; a series of such contacts can be arranged in cumulative steps to speed up a variable speed motor step by step or to bring in one pump after another. When flow control is desired the sensing of flow is frequently carried out by a differential head producer, such as a Venturi tube, so that the sensing device is in reality a pressure-operated mechanism. Pressure switches may directly convert pressure information into electrical information, but because pressure is often subject to surges it is normal to adopt some damping device, such as an air vessel on the line, to smooth out sudden pressure fluctuations so that 'hunting', i.e. constant cutting in and out of machinery, is avoided. Frequent stopping and starting of an electric motor must be avoided because starters will give trouble if used more frequently than their rated capacity of 'starts per hour'.

Boosting

12.12 Types of boosting arrangements

Boosting is the term used for any pumping arrangement which augments the pressure or quantity of water delivered through a prior existing mains system. (The term is loosely and erroneously used to mean simple pumping.) There are many possible arrangements, but chief among them we may distinguish the following:

(1) the addition of a fixed extra flow to an existing supply,
(2) the addition of a fixed extra pressure to an existing supply,
(3) the maintenance of a given pressure, irrespective of the flow.

12.13 Addition of fixed extra flow or pressure

This operation is comparatively simple. If we wish to double flow we may connect two similar pumps in parallel; if we wish to double pressure we may connect two similar pumps in series. For lesser increases the pump added in parallel or in series will be of a suitably smaller size. In either case, however, it is not sufficient to consider only the pump characteristics because we must also consider the characteristics of the system into which the pumps are to deliver. Figure 12.5 shows the characteristic flow–pressure curve for a pump A. For series working of

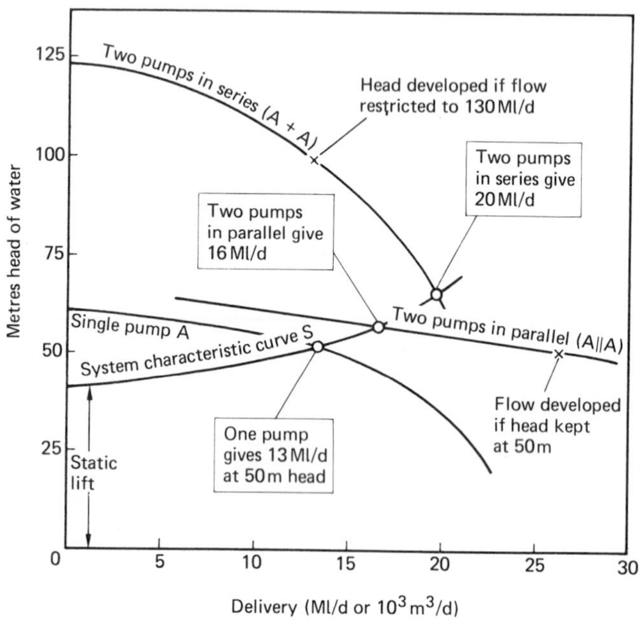

Fig. 12.5 Output of two pumps connected in series or in parallel.

two such pumps the joint characteristic curve (A + A) is obtained. We must then draw the characteristic curve S which indicates the head–flow relationship for the system into which the pumps are delivering water, because as the flow increases so frictional losses will increase also. The point of intersection of the curves (A + A) and S indicates the joint output of the two pumps A in series and we notice that the joint output is less than the sum of the individual nominal outputs. If the pumps are connected in parallel we obtain a pump characteristic curve (A parallel A) by adding flows for a given head and, again, the point where this curve cuts curve S indicates the joint output of the pumps when working in parallel.

Thus far so good, but this is still not sufficient to indicate whether it is a practicable proposition to connect the two pumps in series or in parallel. A hydraulic gradient must now be drawn for the system in which the pumps are to work so as to confirm that no suction troubles will occur and that the siting of the added pump is correct. Referring to Fig. 12.6, suppose we have pump A initially drawing water from reservoir R_1 and pumping it to reservoir R_2. When pump A is delivering

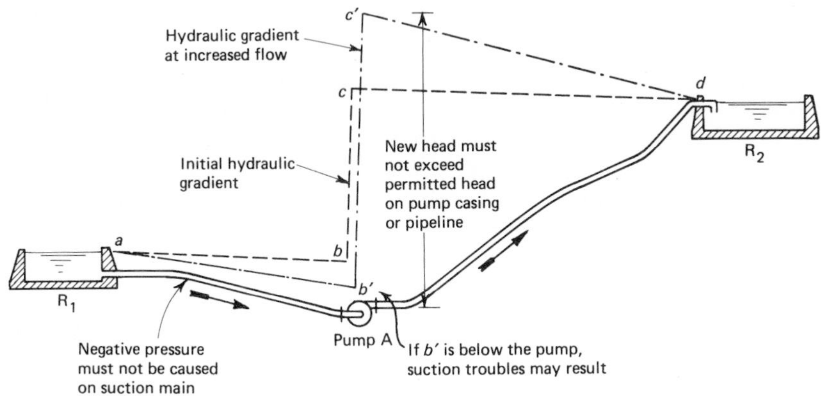

Fig. 12.6 Effect of boosting flow in a rising main to a reservoir.

water the hydraulic gradient for that pump's output will be the line *abcd*. If we increase the flow by an amount Q then the hydraulic gradient must change to some line $ab'c'd$ where b' is lower in elevation than b and c' is higher in elevation than c. The difference cb was the pumping head which had to be developed by pump A alone. The difference $c'b'$ is the new pumping head which must be developed by pumps (A + A). We are now in a position to check the following points, all of which must be satisfied.

(1) The position of b' must not be so low in relation to the elevation of the pumps that suction troubles will result. In fact it is better to have no negative pressure on the suctions because negative pressure may cause trouble with cavitation and will also reduce the efficiencies of the pumps.
(2) The enhanced pressure represented by c' must not be beyond the safe rated working pressure on the body of pump A, neither must it exceed the safe working pressure of the delivery main up to reservoir R_2.
(3) If the pumps are to be operated in series the joint duty must not cause pump A to be overloaded (if this is so then one might be able to overcome the difficulty by changing its impellers).
(4) Following upon (3), provided that pump A is not overloaded, we must investigate whether or not any serious loss of efficiency will result

by reason of the new duty for pump A (under joint operation) coming off its design duty, which should have represented its peak efficiency also. The joint efficiency of the two pumps working together must be examined and the conclusion may possibly be drawn that it would be better to scrap pump A altogether and have an entirely new pump capable of managing the whole of the enhanced duty.
(5) If the pumps are to be run in parallel they must have stable running characteristics.

If consideration (2) indicates that an additional pump cannot be sited alongside A then a possible solution is to place pump A′ somewhere along the line between the pumping station and R_2. If suction conditions are liable to cause trouble then all or part of the delivery main from R_1 to the pumping station may have to be duplicated to reduce the friction losses at the augmented flow.

12.14 Maintenance of a given pressure

One of the most frequent uses of a booster is to increase the pressure of water over a distribution system at times of high drawoff. It should be noted that the flow rate is variable. At low drawoff the pressure at the ends of the mains system may be adequate to meet all needs, but when the drawoff is high the terminal pressures may be too low for the water to be delivered to buildings situated at high elevation. Instead of laying additional feeder mains into the distribution area it may be more economical to boost the pressure at times of high flow. Figure 12.7 shows the hydraulic gradients that may apply before and after boosting.

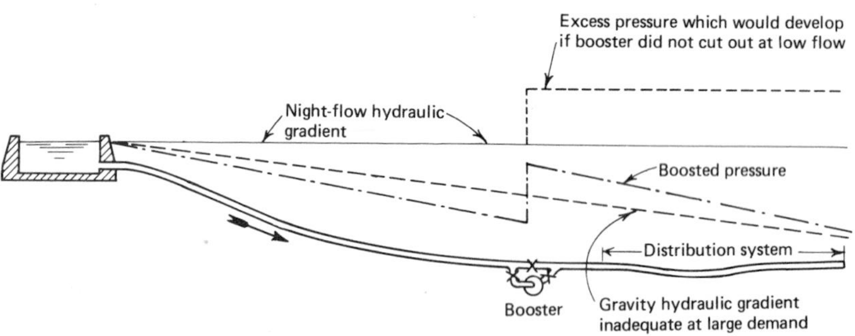

Fig. 12.7 Effect of boosting the pressure of water direct into a distribution system.

An immediate consequence of the boosting of the pressure will be that the flow is thereby increased also, and the amount of this increased flow can only be fore-estimated, not fore-measured. Since the area is already starved of water the existing records will not indicate the value of the true maximum demand at adequate pressure. The design duty required from the pump must therefore be judged by the engineer who will take into account that some margin will have to be kept in hand for further demands coming upon the system. The value of the pressure which is adequate to give the flow required must also be investigated and from these considerations the engineer will be able to judge the characteristics of the pump he must install.

This kind of booster pump will probably be arranged to cut in automatically when the pressure downstream of the pump reaches a certain low value or when the flow reaches a certain high value. A number of safeguards will have to be inserted into the pump controls to prevent hunting and ensure correct interpretation of events: such things as a burst main near to the pump delivery, which would increase flow and reduce pressure, must not be misinterpreted as a normal demand for more water. The duties required from a booster pump of this kind may be too wide for a fixed speed machine. For instance, if the drawoff demand only slightly exceeds the supply given by the unboosted supply then the pump is called upon to raise the head of the whole of the water by only a few metres, i.e. flow is large against a small head. The flow may later increase substantially so that the duty required from the pump is large flow against moderate to substantial head. If these conditions are more than a fixed speed pump can cope with (and they are likely to be) then a variable speed motor may be necessary, or a range of machines should be made available so that these can be brought into use one by one. Pressure surges must be prevented by slow opening and closing of valves and the control system may build up to one of substantial complexity. A more favourable solution to the problem is considered in Section 12.16.

12.15 Need for reliable booster controls

Elaborate automatic boosting systems incorporating a complexity of automatic controls are not always the best; simplicity is desirable. While pumping, repumping, and boosting systems can be built up into a complex automatic integrated system, all controlled from one central office, the cost of labour thus allegedly saved may be exceeded by the cost of having to set up and maintain a highly skilled corps of labour to service and maintain the complicated control apparatus. Reliability of supply is of the utmost importance to a water undertaking. It is of little use installing pumps and motors of the greatest reliability, i.e. to waterworks standards, and then to control their operation by some flimsy make-and-break switch which has a useful life of a year or so. All automatic control gear must be as robust and reliable as the equipment it controls and it should not be so complicated that hours are taken to find an instrument fault that has been the cause of a premature shutdown.

12.16 Use of balancing tank

A booster which is automatically controlled to maintain a given pressure when pumping into supply may involve very complicated automatic control which is liable to be upset by pressure surges or sudden drawoffs. It is also an expensive arrangement. Much of the complication can be avoided if a balancing tank, or water tower, can be built into the system at an appropriate point and this will also add certain advantages conducive to the safety and economy of the supply. Once such a tank is connected to the system the booster pumps need no longer be pressure controlled, but may be operated by the more simple device of water level control switches in the balancing tank. When peak drawoffs occur the water level in the balancing reservoir will fall and this may be arranged to cause a pump to start up in the booster station. With further lowering of the water level in the reservoir, further level switches may bring additional boosters into operation or the speed of the first pump may be increased. This system, having the single disadvantage of increased cost, has nevertheless many valuable characteristics which may be listed as follows.

(1) The peak drawoffs are smoothed out because the balancing tank will take the initial onset of load.
(2) Sudden large increases of drawoff which last only for a short time may be handled by the balancing reservoir without causing the booster pumps to start up at all. Such sudden increases can be caused by opening of washouts or by fire-fighting.
(3) The maximum capacity required from the booster pumps is reduced.
(4) The head range against which the pumps have to work is diminished.
(5) Pumps, once started, can continue to run until the balancing reservoir is again full and the load factor and efficiency of the pumps is thereby increased.
(6) Control of the pumps is simple and positive and the possibility of hunting and repeated stopping and starting of the pumps is almost entirely eliminated.
(7) Future extensions to the boosting system are much easier to arrange because a number of possible ways of further development will exist once there is balancing storage on the delivery line.

The size of balancing tank or water tower required is often small because the periods of peak drawoff on a water distribution system are quite short. Thus if a 5 Ml/d supply suffers from a peak drawoff of 50% over normal for a period of three hours the theoretical size of tank required to meet this demand, without boosting, would be about 312 m^3, but a tank of one-half or even one-third of this size would greatly reduce the maximum duty required from the booster pumps.

Plant layouts

12.17 Buildings and services required

There are a legion of different possible layouts for a pumping station, but some commonly adopted principles may be mentioned. *High tension switchgear* and transformer are usually necessary with a grid supply and they should be sited in a separate locked room from the main rooms. A transformer can be placed outside the building, where it will look ugly; if it is placed inside a building use can be made of the heat which it generates to provide some background heating to the station. *Main switchgear* and control panels should have adequate room both in front and behind. About 1 m should be regarded as the minimum at the rear, 2 m in the front. All switchgear should as far as possible be in line so that the main bus bars are as short as possible. *Instrument panelling* should likewise have plenty of room for access at the back and should not be excessively far away from the pumps and pipework, otherwise long lengths of tubing will be required for pressure gauges if conventional transmission methods are used. A *chlorination room* is usually an essential part of pumping station layout. Two rooms need to be provided, one to house the gas cylinders or drums and one to house the chlorinating apparatus. The former room should have direct access to the open air and both rooms should be provided with an exhauster fan at skirting level controlled by a switch outside the room. Moderate heating of both rooms is necessary, the chlorinating room being at slightly higher temperature than the chlorine store. If *air vessels* are necessary these can be placed outside the pumping station if the climate will permit this or inside in an unheated part of the station. Outside vessels will probably spoil the appearance of the station, but if they are kept inside in too warm a temperature heavy condensation will occur on the vessels causing staining of floors and rusting of the vessel. A *booking room* is necessary for the station attendant; a small *workshop* or at least a room for working in is useful and may conveniently also contain a sink and means for cooking light meals if the station is a fully tended one. *Sanitation* of the highest standard must be provided together with hot water and adequate washing facilities. The *main engine room* should leave reasonable working space about the machinery. Adequate room should be allowed for operating valves and good standing and passage room should be allowed between units where possible; there should always be a clear empty space in some part of the room sufficient for complete stripping down of one of the units. A *crane* is essential for all but the smallest pumping units, but it can be mobile. All *cabling, piping etc.* should be in conduits of ample size and runs should be carefully laid out. When designing cable ducts it is important to find out the minimum radius cables can be bent to. When designing pipe ducts for water supply every effort should be made to avoid T-junctions or 90°-bends in the line of flow. Radiused junctions and large radius bends will substantially reduce friction losses. All moving machinery must be securely bolted down to completely adequate foundation blocks. If pumps are sited over a well or tank they should preferably be fixed upon steel joists which are themselves adequate, rather than bedding pumps upon reinforced concrete slabs. Means of heating are essential to prevent damp lowering the resistance of electrical insulations; some background heat may be obtained from a transformer when current is being supplied. Electric motors are frequently provided with built-in heaters which are switched on when the motor is off duty. These are very effective and take little power, i.e. they take about 0.25 kW for small motors under 50 kW, 0.5 kW up to 100 kW, and up to 1.5 kW for very large motors.

Miscellaneous points

12.18 Water hammer

If a valve in a pipeline is suddenly closed, a large positive shock pressure will build up against the valve; this is called water hammer. If a pump suddenly ceases to pump, a large negative shock wave will be developed in the delivery pipeline just ahead of the pump. Such shock waves will travel up and down a pipeline at approximately the speed of sound in water. When any shock wave reaches the closed end of a pipe it will be reflected with the same sign; when it reaches an open end, such as a reservoir, it will be reflected with reversed sign. Hence in a delivery pipeline of length L the negative shock wave will travel to the open end of the pipeline and be reflected as a positive shock wave back on to the pump in time $2L/a$ where a is the velocity of the wave (usually between 600 and 900 m/s when cast iron pipes are used). At time $2L/a$ after cessation of pumping, therefore, a pressure rise will occur at the pump, particularly if the pump has ceased delivering water and the reflux valve has already shut. The pressure created will theoretically be aV/g where a is as above, g is acceleration due to gravity (m/s), and V is the flow velocity reduction (m/s). If the reflux valve is not quite shut when the returning positive wave r reaches it, it will then be slammed shut and pressures higher than aV/g can develop. If the pump is still delivering water when the initial wave returns the process is more gradual and a shock wave of less intensity will be produced. If the pipeline has a high frictional characteristic then, again, the intensity of the pressure rise may be diminished.

The first precaution to take to prevent excessive surge pressure developing is to see that the reflux valve shuts immediately forward flow ceases. An ordinary flap valve may not achieve this; a quick acting 'recoil' flap valve may be necessary. A second precaution is to reduce the value of the initial negative wave by feeding water into the pipeline just downstream of the pump when the latter ceases to operate. There are two methods of doing this: the first is to connect an air vessel to the delivery main; the second is to connect the suction pipe of the pump to the delivery pipe by a branch connection which incorporates a quick opening one-way valve in it.

The onset of the negative wave, or reduced pressure, will cause water to be drawn into the delivery main from the air vessel or suction pipe, thus diminishing the magnitude of this negative wave and consequently reducing the size of the return positive wave. Connection to the suction is cheap, but it is effective only in certain circumstances. An air vessel with an air compressor to go with it is expensive, but it is an effective means of reducing surge pressures. A third method of reducing water hammer is to permit water to escape from the pipeline on the return positive pressure, via spring loaded valves placed in the delivery main just ahead of the reflux valve.

Estimating the exact amount of water hammer likely to be experienced in any given system is difficult, but this must not lead the engineer into ignoring a phenomenon which presents a very real risk, especially when all-electric pumps are installed. A current failure, causing sudden stoppage of pumps, may create large water hammer blows which occur so quickly that their intensity may not be revealed by ordinary Bourdon pressure gauges. The worst effects can be avoided by installing a suitable non-return valve on the pump and even a moderate sized air vessel can bring about much diminution of pressure surges.

12.19 Pump suctions

It is preferable to site a pump so that negative pressures do not develop on the suction side. This negative pressure is liable to cause a falling off in the performance of the pump and it also prevents the pump from being automatically primed. If negative pressure is unavoidable a 'self-priming' pump must be specified. The suction of a pump must be properly designed and Fig. 12.8 shows a bad design and a good design for a suction. The aim is to keep all reductions of pressure to a minimum so as to prevent air being released from water when the pressure is reduced below atmospheric. *Vortexing* is another fault that can develop below a suction pipe and it can be diminished by inserting a straight vane across the inlet.

Most suctions have a strainer incorporated in them, but this must be kept in good condition. Foot valves, i.e. non-return valves, are a potential source of trouble and should be avoided where possible. If installed they must be of high quality and be kept well maintained or they may tend to stick. They are sometimes installed for the purpose of keeping a pump primed when it is idle, but they may not be very effective at this over a long period and it is better to arrange other methods of priming. They are also used, on occasion, to prevent back-rotation of a vertical spindle pump when the pump is stopped and

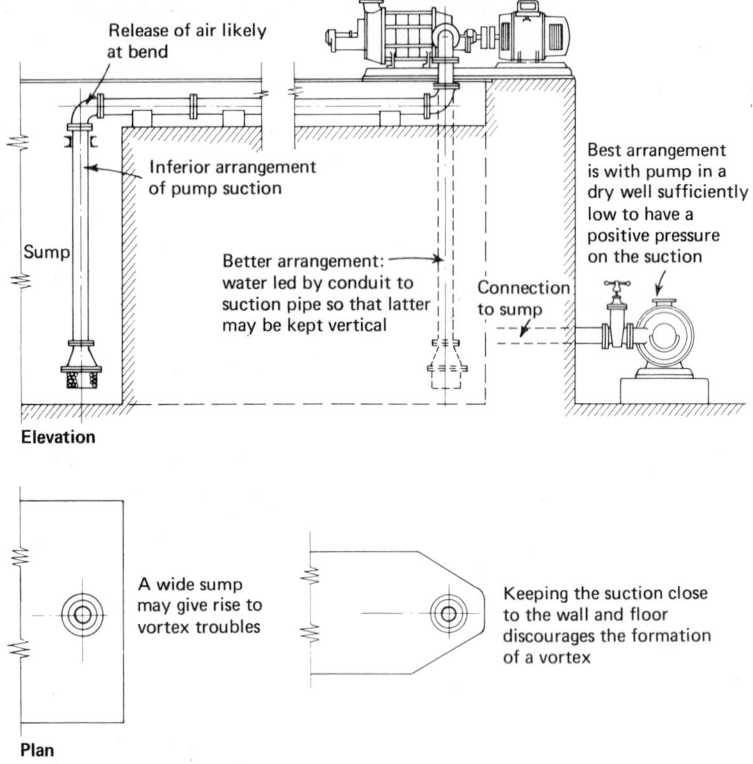

Fig. 12.8 Arrangement of pump suctions.

the vertical rising main is of some length. A high speed of back-rotation can be caused by the falling column of water when the pump is stopped and to restart a pump when it is back rotating may cause breakage of the shaft. Instead of a foot valve a pump may be specified as suitable for being turbined under reverse flow and a time delay switch is then incorporated in the switchgear to prevent restarting before the reverse rotation has stopped. The diameter of pump suction pipes is usually larger than the delivery pipe diameter to reduce friction losses on the suction side. Strainers must be of ample size and should be specified as having a total area of opening at least double that of the suction pipe. When submersible pumps are inserted in a narrow borehole it may be necessary to specify 'shrouds' to the pump intake strainers, which are circumferential shields around the body of the pump between the motor and the pump. These shrouds prevent excessive drawing of water from the borehole formation opposite the strainers; this is a matter of some importance if the borehole penetrates soft strata.

12.20 Cavitation

Cavitation can occur if suction conditions are poor. This is a phenomenon where air or water vapour may be released from water when the pressure is reduced. The entrained bubbles, which may be of minute size, are carried through into the pump, where entering a higher pressure zone they collapse. A stream of such bubbles continuously collapsing on the tip of an impeller blade causes the metal of the blade to be eroded. The blade becomes pitted and will be worn away in a very short time. Cavitation is prevented by good design of pump and by maintaining a positive pressure on the suction pipeline of the pump.

12.21 Costs and efficiencies of pumping plant

There is always difficulty in quoting costs and efficiencies because they vary so widely and every individual case has something special about it. The figures in Tables 12.1 and 12.2 are intended to give a rough guide as to the efficiencies normally achieved and UK costs in 1983. There are wide variations according to the power rating and type of motor, whether fixed or variable speed, whether squirrel cage or synchronous etc.

Table 12.1 Efficiencies.

Pumps	
Horizontal centrifugal	Medium size 80% to 82%, perhaps 85% large size. Even higher with special construction but at higher price
Vertical spindle shaft driven	Tending towards about 3% less than the horizontal centrifugal
Submersible	75% to 81% and can be lower, to about 70% for small sizes. Generally about 3% less again than the vertical spindle pump, the reason being that the pump is restricted in diameter
Electric motors	
For horizontal pumps	93% to 95%. Fixed speed a.c. induction
For vertical pumps	90% to 94%. Fixed speed a.c. induction
For submersibles	85% to 89%, less than the above because of the restrictions imposed on the design
Variable speed	About 3% to 5% less than with a squirrel cage a.c. motor

12.22 Effect of electricity tariffs

The charges made by Electricity Boards in the UK are designed to benefit the consumer who is able to take a good steady load of electricity at high power factor. The consumer who wants large power intermittently is penalized. Most of such tariffs comprise two, three, or even four separate charges as listed on p. 448.

Table 12.2 Overall fuel consumptions and costs.

Fuel consumptions	
Electrical driven pumps	About 1.0 kW for every 0.75 kW of water power output, this implies an overall efficiency of about 75% which would be usual. Up to 1.3 kW per 0.75 kW water power output or higher for small pumps or variable speed pumps
Diesel engines	0.21 kg of diesel fuel oil consumed per kW h of engine power exerted would be considered good, 0.28 kg per kW h being not unusually high. For lubricating oil add 5% to fuel oil cost
Costs	
Capital costs for complete electric driven pumping sets. (These figures include switchgear and station pipework.)	About £300 per kW installed would be about the lowest price now possible and would only be possible with fixed speed horizontal pumps in medium to large installations. Can range up to £750 per kW when there is a fair degree of instrumentation or for small duties
Pumping costs generally	Pumping generally in waterworks costs between £1.75 and £3.00 per cubic metre for average lifts; the figure includes labour, repairs, maintenance buildings, but no loan charges. Power represents about two-thirds of this cost

(1) A 'maximum demand charge' per kW of maximum demand in that month above a given figure (and sometimes, more onerously, per kW of maximum demand for the last twelve months)—the scale of charge is on a sliding basis and may also differentiate between summer and winter maximum demand.
(2) A 'unit charge' which is a monthly charge for the number of units of electricity consumed, the charge per unit being substantially higher for daytime consumption.
(3) A 'coal clause' or 'fuel clause' which increases the unit price of electricity according to the increase of basic fuel cost.
(4) A 'power factor clause' which increases the maximum demand charge if the power factor drops below a certain figure (usually 0.90).

The effect of the maximum demand charge (especially when based upon twelve months) can be onerous. The larger the gap between the maximum demand and the average demand, the more penal the maximum demand charge becomes and the more costly are the charges per unit. Hence it is a matter of great economic importance for pumping stations to be run at a high 'load factor', i.e. the average power consumed should be as close as possible to the maximum power required at any time during a month (or year), and pumps should be run for long periods rather than short, so that the units consumed are as cheap as possible because their price reduces as more units are consumed per kW of maximum demand during the month (or year).

Pumping for short and intermittent periods at high outputs is therefore an expensive way of pumping water. Similarly, the occasional bringing of pumps up to full load when they normally need run only at part load or bringing in additional pumps for parallel operating for short times are also expensive matters. Even the testing of pumps when first installed—perhaps running several together as a 'try out'—can bring a heavy financial charge running into several hundreds of pounds for so brief a run as 30 minutes. Choosing the right duty for pumps, the right hours of working, and the right amount of standby to be provided for any pumping station are matters that must be most thoroughly investigated. The reader should note particularly the benefits due to adequate service reservoir storage on the pumping line, from the threefold aspect of safeguarding continuity of the supply, levelling out pumping rates thereby reducing friction losses and lessening maximum power demanded of the motors, and permitting long steady running at a high load factor and so minimizing electricity charges.

13

Pipes, pipeline construction and valves

13.1 Types of pipes

Pipes used for waterworks distribution systems are made from the following materials:

(1) cast or 'grey' iron (now largely superseded by ductile iron),
(2) ductile iron,
(3) asbestos cement,
(4) steel,
(5) prestressed concrete, cylinder or non-cylinder,
(6) reinforced concrete cylinder,
(7) glass fibre reinforced concrete (a new type of pipe),
(8) uPVC (unplasticized polyvinyl chloride),
(9) GRP (glass reinforced plastic) or RPM (reinforced plastic matrix),
(10) polyethylene of low, medium, or high density.

There are still other types of pipes made from aluminium, copper, stainless steel, and various thermoplastics such as polypropylene, polybutylene, and ABS (acrylonitrile butadiene styrene). Except for copper pipe which is widely used for plumbing systems, none of these pipes is in general use for water distribution purposes.

Cast iron pipes

13.2 Cast or 'grey' iron pipes

These pipes were originally made in the nineteenth century by casting molten iron in vertical sand moulds. Many waterworks still have some of these pipes in use. Their walls were relatively thick and not always of uniform thickness because of occasional misplacement of the central core mould.

'Spun' grey iron pipes were first produced in the UK in the 1930s, the molten cast iron being poured into cylindrical moulds rotating at high speed so that the pipe walls were formed by centrifugal force. After a few moments of spinning the metal solidifies and the pipe contracts slightly and can then be drawn in a red hot state from the mould. It is then sent for reheating and slow cooling so as to reduce stresses arising from cooling. By this process a denser iron resulted and the pipes were of uniform wall thickness. These 'spun' iron pipes were in widespread use from the 1930s to the 1960s and now comprise a large proportion of the distribution mains of UK water undertakings. Three classes of such pipe were in common use: B, C, and D for working pressures of 60, 90, and 120 m respectively. Classes B and C were predominantly used, but because their outside diameters in the larger sizes were different, and also to reduce stocks and possible confusion in use, many water authorities used only class C pipes rather than a mixture of classes B and C. The manufacture of grey iron pipes has now been discontinued in most countries, except for the production of non-pressure pipes for rainwater and soil pipes.

13.3 Ductile iron pipes

Ductile iron pipes have now superseded grey iron pipes. They are centrifugally cast, i.e. they are 'spun' ductile iron pipes. The ductile iron is made by adding small quantities of magnesium (or cerium) to molten grey iron. This has the effect of transforming the lammelar or flaky form of graphite carbon into spheroidal form, thereby increasing the tensile strength and ductility of the iron. The resulting pipe is stronger and less liable to fracturing than grey iron pipe. Classes of ductile iron pipes and fittings and their recommended test and pressure rating are given in BS 4772, ISO 2531, and the British Code of Practice CP 2010 Part 3. Details are given in Table 13.1. Ductile iron pipes in the UK are available up to 2000 mm diameter, but in Japan up to 2600 mm. Standard lengths in the UK are 5.5 m for spigot and socket pipes, but pipes up to 8 m long are manufactured elsewhere.

13.4 Standard linings and coatings

For both internal lining and external coating the standard treatment comprises cold applied black bitumen to BS 3416 or hot applied bitumen-based coating to BS 4147. These are best sprayed or brushed on. Coal tar pitch has also been used and gives a smoother surface. For extra protection pipes can be either lined and sheathed as described for steel pipes (see Section 13.7) or mortar lined (see Section 13.9).

13.5 Joints for cast and spun iron pipes

Types of joints in use are shown in Fig. 13.1.

The run lead joint for use with spigot and socket pipes to former BS 78 is now mostly superseded by the various proprietary mechanical joints, but many mains still in use have been laid with this joint. Skilled workmanship is required to make the joint properly. The lead has to be heated to 400 °C at which temperature the molten lead shows strong rainbow colours when the surface scum is drawn aside. A clip is placed around the pipe against the annular space of the socket and the lead must be poured in one continuous pour through an opening left at the top of the clip. The operation is a dangerous one if not undertaken with care and the socket space must be quite dry to avoid blowback of the

Table 13.1 Standard ductile iron pipes.

Definition of class (BS 4772 and ISO 2531)
Class is defined by the value of K in the formula:

$$e = K(0.5 + 0.001DN)$$

where e is the pipe wall thickness (mm), DN is the nominal internal diameter (mm), and K (standing for 'Klass') can take any whole value . . .8, 9, 10 . . .etc.

Standard classes
The standard classes proposed in BS 4772 and ISO 2531 are:

$K = 9$ for socket and spigot pipes
$K = 12$ for fittings other than tees
$K = 12$ for flanged pipes (BS 4772 only)
$K = 14$ for tees

Works test pressures to BS 4772 and ISO 2531, together with maximum operating and hydrostatic pipeline pressures stipulated in CP 2010, are as set out below.

Diameter (mm)	Works hydraulic test pressure for pipes: ISO 2531, BS 4772 (bars)	Works hydraulic test pressure for fittings[1]: BS 4722 only (bars)	'Pressure rating', i.e. maximum sustained operating pressure on pipeline[2]: CP 2010 (bars)	Maximum field hydrostatic pressure on pipeline[3]: CP 2010 (bars)
80–300	50	25	40	45
350–600	40	16	25	30
700–1000	32	10	16	21
1100–1200	25	10	16	21
1300–2000	25	—	—	—

Notes:
(1) Figures for works test pressures of fittings are lower than for pipes because of the need to avoid distortion during test.
(2) As Table 2 of CP 2010 Part 3. Also maximum sustained operating pressure plus surge must not exceed 1.1 times pressure rating.
(3) As Table 6 of CP 2010 Part 3. Pressure is measured at the lowest point of the pipeline. The *actual* test pressure applied can be 5 bar in excess of the actual sustained operating pressure, if this is less than the figures shown.
It will be noted that BS 4772 covers pipe only up to 1200 mm size; ISO 2531 extends the range up to 2000 mm, but does not specify pressure rating and field test pressure.
1 bar = 10.17 m head of water.

lead by steam. The lead solidifies almost immediately after pouring and is then caulked up using a series of chisels. The joint is a rigid one and slight movement brings a tendency for such joints to weep; however, countless mains have been laid with this type of joint satisfactorily for the past three-quarters of a century.

Flanged joints, now to BS 4504 (metric), must be carefully aligned before the bolts are inserted and the flanges pulled together. The alignment must be almost as precise as that adopted for aligning motor

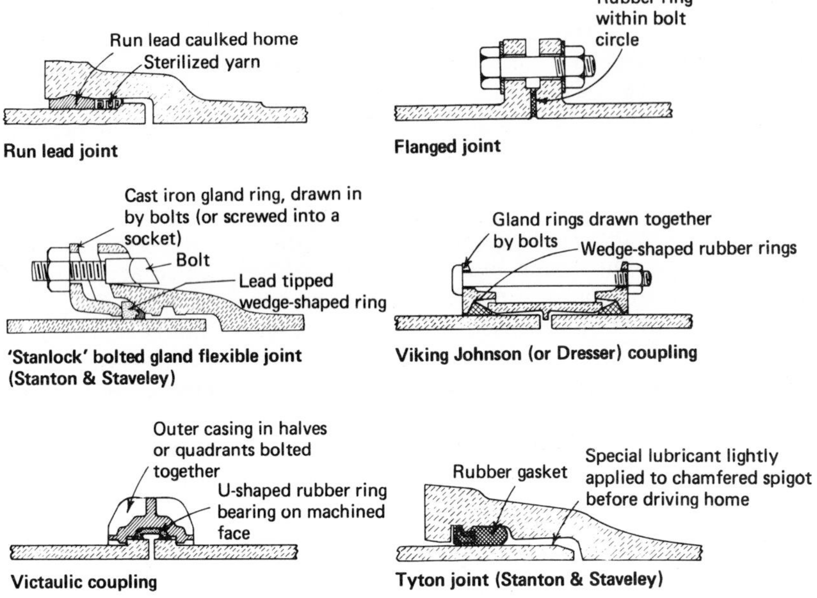

Fig. 13.1 Joints for cast iron and spun iron pipes.

couplings. To pull up malaligned flanges is likely to cause fracture of the pipe or flange. A rubber ring is inserted between the flanges, of such diameter that it lies inside the bolt circle but does not intrude into the pipe bore. The faces of flanges and the rubber ring must be perfectly clean before assembly and the bolts must be tightened up in sequence little by little so that an even pressure is maintained all round. No grease, bitumastic paint, oil, dirt, grit, or water should be permitted on the flange or rubber ring faces. The contact should be between clean dry metal and clean dry rubber. When making joint faces which are vertical, some difficulty may be experienced in keeping the rubber ring flat against the vertical flange face and, to counteract this, a little clear rubber solution may be used to tack the rubber ring on to the metal face. This is the only material whose use can be permitted in connection with rubber rings. If a greasy material, such as a bitumen-based adhesive, is used then tightening the flanges will cause the rubber ring to extrude into the pipeline and, however much further the flanges are squeezed together, an imperfect joint will result and be liable to leak. Rubber rings are 3.2 to 4.8 mm in thickness (material to BS 2494) and they must lie perfectly flat against the face of the flange without distortion.

Compressed gasket joints, i.e. bolted or screwed gland joints, work on the principle of forcing a lead tipped rubber ring into an annular space formed between a specially shaped socket and the plain spigot of the pipe. The rubber ring is forced into the annular space by a cast iron pressure gland which is drawn by bolts or screwed into the socket. The pipe barrel, the socket, and the rubber ring must all be scrupulously clean before erection and the pressure gland must be tightened up uniformly. Joints of this type are principally used for gas mains and are being superseded by the push-in type of joint for water pipelines in the smaller diameters.

Viking Johnson couplings are another form of patent rubber ring joint used for connecting together lengths of pipe which are straight-barrelled, i.e. no socket is used and no spigot beads are necessary. This joint is only occasionally used with cast iron piping, but is very widely used with steel pipes. A cross-section of such a joint is shown in Fig. 13.1 and it will be seen that it acts on the same principle as the patent ring type of joint. Slight angular deviation at joints is possible with these couplings; also, if the coupling is specified as without a central register it can be moved along the barrel of the pipe, thus permitting removal of a section of main.

The Victaulic coupling is used in conjunction with shouldered ends of pipes, thus holding them together longitudinally. The joint can be unmade and remade without difficulty. Hence this type of joint is most often used for temporary pipelines laid above ground. It is also sometimes used to connect meters in pipelines, thus facilitating removal of a meter.

The 'O'-ring joint (Tyton joint) is a specially shaped rubber ring fitted into the inside of the socket of a pipe and the spigot of the next pipe is then forced home. A little lubricant may be used on the inside of the gasket or on the outside of the pipe spigot before forcing the latter home. This is all that is required. Complete cleanliness of the socket, gasket, and spigot is of course essential. Pipes jointed in this manner can be deflected 5° up to 300 mm diameter; 4° for sizes 350 to 400 mm diameter; 3° for sizes 450 to 600 mm diameter. The simplicity of these joints and the ease and speed with which they may be made have brought them very much into favour.

The rubber for O-rings should be synthetic because natural rubber rings have been found liable to biodeterioration[1]. The synthetic rubber which appears most likely to be satisfactory is ethylene propylene (EPDM and EPM), but styrene butadiene (SBR) is also likely to be acceptable.

Asbestos cement and steel pipes

13.6 Asbestos cement pipes

Asbestos cement pipes are made of Portland Cement and asbestos fibre mixed into a slurry and deposited layer upon layer upon a cylindrical mandrel. When the required thickness has been built up the pipe is steam or water cured, cleaned, the ends are turned down to an accurate diameter for some 150 mm, and then (in the UK) the pipe is dipped in cold bitumen.

Asbestos cement pipes have the valuable property of being resistant to corrosive conditions, except in the case of sulphated soils which attack the cement. However, the pipes tend to be brittle and liable to be easily damaged by shock, hence they require careful handling and laying, the backfill against them should contain no large stones, and they should preferably not be laid beneath roads carrying heavy traffic and subject to vibration. If the pipes have to be rehandled several times (as for instance when they are exported overseas), breakages tend to be high. Service connections can be made to them but, because the screw-thread tapped into the asbestos cement is fragile, it is usual to clamp a permanent metal saddle around the pipe and to drill and tap through this for the insertion of a service pipe ferrule as shown in Fig. 13.2(a). Asbestos cement pipes for water supply are made in accordance with BS 486 and to CP 2010 Part 4 requirements suitable for working water pressures of 75, 100, and 125 m according to class. The works test pressure is specified as double the working pressure; the field test pressure on the line should be 1.5 times the working pressure. The latter should be the maximum static working pressure or the maximum operational pressure including an allowance for surge. Fittings for asbestos cement pipes, other than bends, are usually of grey or ductile iron, but sometimes enamelled steel or aluminium is used. In the smaller diameters and for the lower pressures asbestos cement pipes can be directly connected to iron fittings; in the larger sizes (300 mm and above) modified fittings must be used.

Joints Asbestos cement pipes are plain ended so that they are connected by means of collars and compressed O-ring rubber or synthetic rubber joints. The cast iron detachable joint compresses circular cross-section rubber rings in the annular space between collar and outside of the pipe. The Widnes coupling (see Fig. 13.2(b)) comprises a push-fit rubber ring which has to be lubricated. The Viking Johnson (or Dresser) coupling shown in Fig. 13.1 can also be used. Great care needs to be taken to ensure that the rubber O-rings as placed are not twisted, otherwise leakage will occur. A leak under pressure from a joint can sometimes cut right through the pipe wall. A gap must be left between the spigot ends of the pipes in hot climates to allow for expansion.

13.7 Steel pipes

Steel pipes of the larger sizes are made from steel plate bent to circular form, the edges of the plate being either lap welded or butt welded. If the edges are lap welded the pipe is reheated and rerolled so that its wall

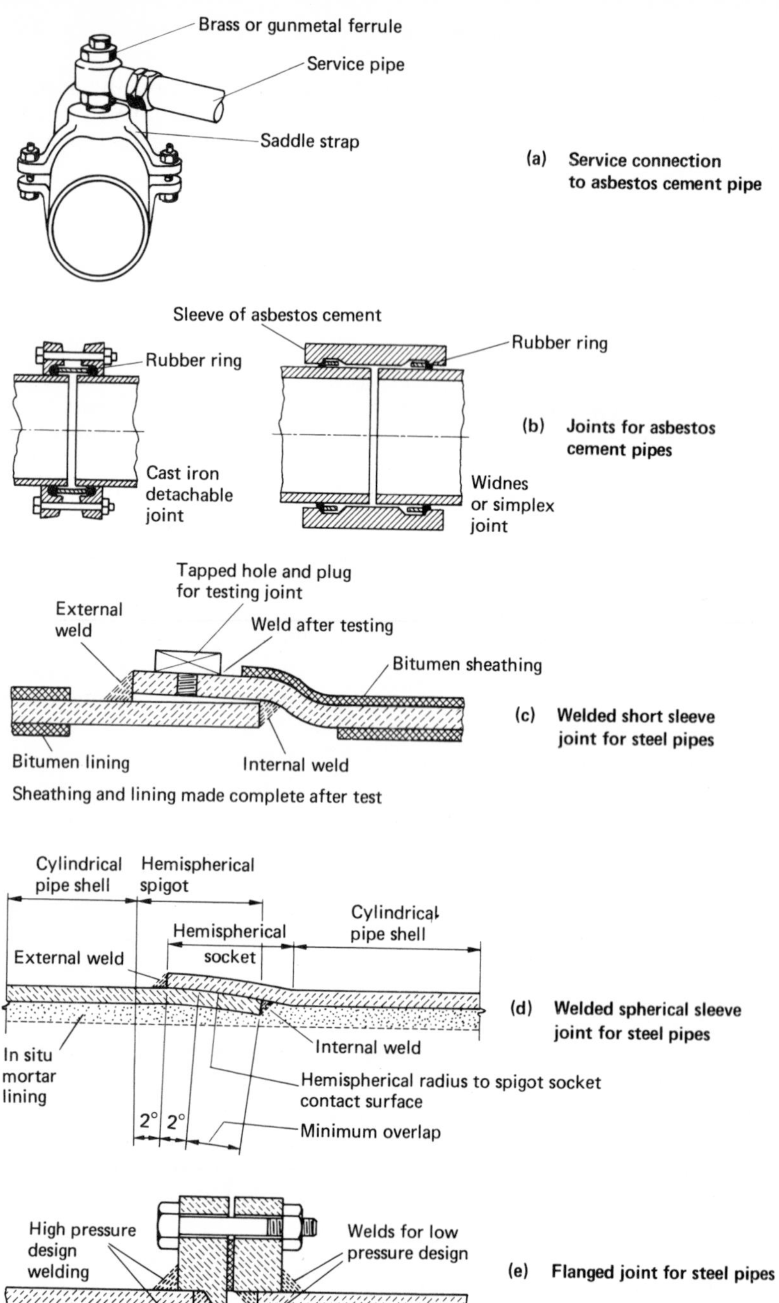

Fig. 13.2 (a) service pipe connection to an asbestos cement pipe; (b) joints for an asbestos cement pipe; (c), (d) and (e) joints for steel pipes.

thickness is uniform. The smaller sizes of pipes are made from billet bars or ingots of hot steel which are pierced and rolled into a cylinder of the required dimensions. The latter pipes are known as 'seamless' pipes. There are no standard classes for steel pipes because of the need to design the wall thickness for the specific internal pressure proposed, for prevention of distortion when buried, and for adequate reserve of thickness against corrosion. BS 534 lists a range of nominal sizes for steel pipes giving minimum recommended wall thickness for pipes laid underground as follows.

Diameter (mm)	*Thickness (mm)*
300–600	6.3
675	7.1
750–900	8.0
1050–1200	9.5

The working pressure for steel pipes therefore depends upon their diameter and thickness, the latter being increased from the minimum thickness if this does not give enough strength. CP 2010 Part 2 recommends the minimum thickness t (mm) should be derived from the formula:

$$t = \frac{pd}{2afe}$$

where p is the internal pressure (N/mm^2), d is the external diameter (mm), a is the design or safety factor (usually taken as 0.5), f is the minimum yield stress for the steel (N/mm^2), and e is the joint factor which can be taken as 1.0, except in the cases of electric fusion welding ($e = 0.9$), or furnace butt welding or where the weld zone is not normalized ($e = 0.85$). The negative manufacturing tolerance must be added to the thickness so derived (10% for welded pipes and 12.5% to 15% according to diameter for seamless pipes). Pipes are usually 9 to 10 m long, but can be cut to any length required when produced from plate which is spirally bent and continuously welded. Special pipes, e.g. bends, tees etc., can be made to any dimensions required, bends being made by cutting and welding together mitred sections of pipe.

Joints for steel pipes in common use are the Viking Johnson coupling (Fig. 13.1), the welded sleeve joint (Figs 13.2(c) and (d)), and the butt welded joint. Johnson couplings permit slight angular deviation of pipes at joints and a certain amount of longitudinal movement sufficient to avoid the need for contraction–expansion joints. The lining needs to be made good inside the pipe when Johnson couplings are used and this can be done from inside the pipe when its diameter is 600 mm or over, and from outside when the diameter is smaller. When completion of the lining must be done from outside a former is pushed into the pipe on a rod and expanded to form a temporary inside collar at the joint. Hot bitumen is then poured through a hole in the Johnson coupling, thus filling all the annular space at the ends of the pipe. The outside coating is made continuous by putting a mould over the coupling and filling this with bitumen and allowing to cool before stripping off the mould. Pipes for Johnson couplings must have true end diameters and if it is intended to cut pipes these must be ordered 'true diameter throughout', otherwise they will only be trued at their ends to a circle. Some difficulty may therefore be experienced when inserting a branch by Johnson couplings into an existing main at a point where its diameter may not be true.

Two types of welded spigot and socket joint are shown in Fig. 13.2. The tapered short sleeve socket (Fig. 13.2(c)) permits up to 1° deviation on larger sizes of pipes, but when any deflection occurs the external weld is large over part of the circumference. It is usual to weld both inside and outside the joint on pipes which are large enough to be entered for welding. An air test can then be applied to the annular space between the two welds, which permits pipe joints to be tested before backfilling the trench. This is a great advantage of this type of joint. The welded spherical joint (Fig. 13.2(d)) was developed to permit deviation without unduly increasing the size of external weld and, after some resistance by manufacturers, has been found practicable and more satisfactory than the short sleeve joint because of the smaller welds required. Butt welding of plain ended pipes is usually only adopted when the pipe wall thickness is substantial.

Standard linings and coatings for steel pipes are made of bitumen. Internally a spun bituminous compound is used having 1.6 mm thickness for the smaller diameters, increasing to 6.4 mm minimum thickness for the larger diameters. Externally the standard protection, known as 'sheathing', comprises a first coat of bituminous compound, followed by a second coat of hot bitumen containing fibres, to a total minimum thickness of 6.4 mm for pipes of 300 mm diameter and over. If pipes are to be 'wrapped' they receive a first coat of hot bitumen and then a second coat over which is wrapped bitumen-impregnated glass fibre. All pipes, whether sheathed or wrapped, are limewashed on the outside to prevent absorption of heat and stickiness of coating. To ensure the continuity of the bitumen linings and coatings a 'Holiday' detector is used. This comprises passing a brush carrying an electrical charge at 15 000 volts over the lining and coating; any pinholes or breakages are disclosed by an electrical discharge to the steel of the pipe and this can cause a bell or buzzer to sound. It is usual to test all pipes in this manner before they are lowered into the trench so that defects can be made good.

Mortar lining of steel pipes is frequently adopted and is dealt with in Section 13.9.

13.8 Thin walled steel pipes

At diameters above about 1000 mm the theoretical pipe thickness to meet usual internal pressures is frequently less than the thickness required to prevent the pipe distorting under backfill load. To save adding extra steel thickness for stiffening the pipe against backfill load the pipe may be laid under 'controlled backfill' conditions which, in practice, means bedding the pipe on to a preformed circular invert (60° width) in sand or fine gravel and carefully tamping the backfill in shallow equal layers either side of the pipe to achieve 95% Proctor compaction of the fill (the standard compaction test to BS 1377). By this means a thin walled pipe may be held from distorting more than 2% on its vertical diameter when under full backload condition[2]. The pipe has temporary stays inserted in it on the vertical diameter when first laid to elongate it in this direction and, after backfill, the stays are removed and the deflection is measured. If it exceeds 2% the backfill is removed and replaced to obtain the necessary circularity.

13.9 Mortar lining

Steel and ductile iron pipes can be mortar lined with cement–sand mortar and this forms one of the best types of interior protection. The bare metal pipe is spun at high speed on rollers and the mortar is poured as a slurry into the interior. The lining builds up by centrifugal force and when the required quantity has been poured the speed of rotation of the pipe is increased. This causes the mortar to compact further and surplus water, rising to the surface of the lining, then runs off as the pipe is tilted very slightly. The mortar mix comprises sand or fine broken rock aggregate mixed with Portland or sulphate-resisting cement in the ratio of 2.5 : 1 or 3 : 1 by weight. The lining is usually so well compacted after spinning that the pipe may immediately be taken off the spinning bed and then placed in a damp warm atmosphere for curing. The stipulated lining thickness may vary and some specifications are given in Table 13.2.

The mortar lining thickness for steel is generally thicker than for ductile iron pipes of the same diameter because of the flexibility of steel pipes as compared with ductile iron pipes. With regard to ductile iron pipes there has been some discussion as to the value of the minimum acceptable thickness and the choice should depend upon the quality of the water to be conveyed. After mortar lining, the interior of the pipe should be spray or brush painted with bitumen since this adds a substantial further measure of protection against aggressive acid waters. All fittings associated with mortar lined pipes are also mortar lined, but this has to be trowelled on. The discontinuity of the lining at pipe joints should also be filled if the pipe is of large enough diameter to give inside access.

Table 13.2 Mortar lining thicknesses (mm).

Nominal pipe diameter	Steel pipes		Ductile iron pipes		
	BS 534	AWWA C205	NWC	AWWA C104	Stock commercial
100–150	6	6.4(−0.8)	4.5	1.6	2.5(−1.0)
200–250	10	6.4(−0.8)	4.5	1.6	2.5(−1.0)
300	10	7.9(−1.6)	4.5	1.6	2.5(−1.0)
350–550	13	7.9(−1.6)	4.5	2.4	4.5(−2.0)
600	13	9.5(−1.6)	4.5	2.4	4.5(−2.0)
650–900	19	9.5(−1.6)	5.5	3.6	5.5(−2.5)
900–1200	19	12.7(−1.6)	5.5	3.6	5.5(−2.5)
Over 1200	25	12.7(−1.6)	8.0	3.6	8.0(−4.0)

Notes:
The negative tolerance is shown in brackets.
BS 534—the minimum thickness is stated (negative tolerance zero).
NWC Standard Specification—minimum mean of four readings: single reading tolerance = 1.0 mm.
AWWA C104—figures are for 'standard thickness' (negative tolerance zero), but 'double standard thickness' may be specified.
Stock commercial—as offered in the UK.

Concrete pipes

13.10 Prestressed concrete pipes

Prestressed concrete pipes are to be designed to BS 4625 (metric). They are made by tensioning high tensile wire wound spirally around a cylindrical core. This core may consist either of concrete which is prestressed longitudinally or of a thin steel cylinder which has a thick spun concrete lining to the interior as shown in Fig. 13.3. Pipes with metal cylinders to the core are called 'cylinder prestressed concrete pipes', the others 'non-cylinder'. When the wires have been wound on to the core, stressed, and anchored, a thick cover coat of concrete is applied externally to the pipe. This coating is applied pneumatically or by machine in even layers as the pipe is slowly rotated and acts as a cover to the prestressing wire. The chief advantage of prestressed concrete pipes is that they can offer a cost advantage over other pipes in certain sizes above 300 mm, particularly for higher pressures. Prestressed pipes can be made to withstand high pressures simply by increasing the number of turns of prestressing wire per unit run of pipe or by doubling up the layer of prestressing wire. A second advantage of prestressed concrete pipes is that they are proof against certain corrosive conditions that would attack iron and steel, although they do need special protection if the groundwater is saline or otherwise aggressive to concrete. In some cases the pipes are cathodically protected which

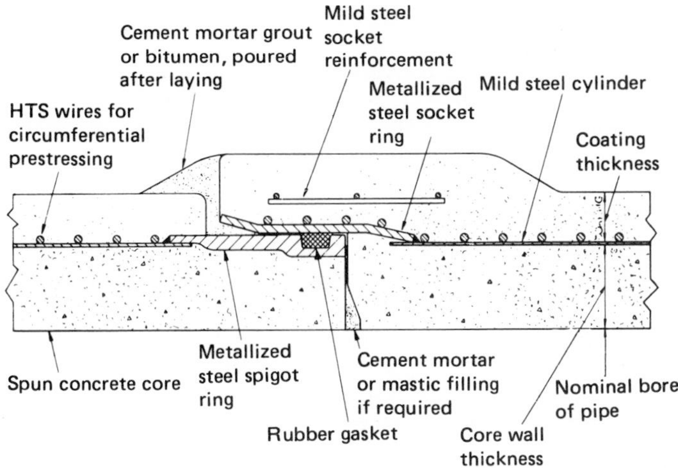

Fig. 13.3 Prestressed concrete pipe, with metal cylindrical core. (Stanton & Stavely Co. Ltd.)

involves the provision of electrical contacts to the prestressing steel and the cylinder at each end of the pipe so that these can be bonded together electrically after the pipe is laid. Alternatively, or in addition, the pipes can be coated with bitumen, coal tar enamel, or coal tar epoxy. The pipes can also be made using sulphate-resisting cement.

There are some disadvantages to prestressed concrete pipes. They have very limited flexibility at joints and it is preferable to limit deviation to 0.5° at any joint. In the larger diameters the pipes are very heavy and difficulty will occur in aligning them in soft ground unless they are placed on a prepared bed of granular material of adequate thickness to prevent uneven settlement. Connections are not easy to make after the pipeline has been laid and, to avoid the need for this, tees have to be incorporated in the line as it is laid. A second objection taken by some engineers to these pipes is that the outside coating of concrete may be liable to cracking because of a tendency for the pipe to expand when put under internal pressure. It should be noted that this coating cannot be prestressed. It is true some failures of external cracking did occur in the earlier years of usage of this type of pipe, but the difficulties appear to have been overcome and cases of failure are now very rare.

Joints for prestressed concrete pipes are usually of the socket and spigot O-ring push-in type. The socket is sometimes afterwards mortared up, but the joint then becomes rigid and the mortar infilling may be cracked if the pipes later expand due to exposure to the sun in a hot climate.

13.11 Reinforced concrete cylinder pipes

Reinforced concrete cylinder pipes are similar to prestressed concrete pipes, except that instead of using high tensile steel wire for circumferential reinforcement mild steel rod is used. The rod is wound spirally under tension on to a rotating metal cylinder, being welded to the cylinder at both ends. The reinforced cylinder is then covered internally and externally with concrete, the internal lining being spun on and the external coating being applied by impact and smoothed in layers. In the design of the pipe both the steel cylinder and the reinforcement are assumed to resist hoop tension stresses. The principal advantages of reinforced concrete cylinder pipes are: they can be designed for relatively high heads; they require less sophisticated manufacturing techniques than do prestressed concrete pipes; the concrete acts as a good protection to the steel and so permits some saving over the amount of steel that would be used in an equivalent steel pipe; the pipes are rigid and have a reasonable degree of resistance against rough handling and poor backfilling techniques. Disadvantages are as mentioned for prestressed concrete pipes.

13.12 Glass fibre reinforced concrete pipes

Glass fibre reinforced concrete pipes are a relatively new product. They use alkali-resistant glass fibre as circumferential reinforcement placed near the outer and inner surfaces of the pipe wall. This reinforcement is stronger than steel and considerably more resistant to the usual corrosion agencies so that a lesser pipe wall thickness can be used. The pipes have not yet come into widespread use, but they appear to have a considerable future.

Plastic pipes

13.13 Unplasticized polyvinyl chloride (uPVC) pipes

uPVC pipes are available in diameters up to 600 mm and sometimes for larger diameters. They are rigid, light pipes in 6 or 9 m lengths with socket and spigot joints either for use with solvent cement or a rubber ring. Classes B, C, D, and E (equivalent to ISO classes NP 6, 9, 12, and 15 respectively) are stipulated in BS 3505 and details are given in Table 13.3. The pipes are not fully suitable for use in hot climates and should not be used for hot waters. CP 312 Part 2 recommends uPVC pipes should not be used at temperatures exceeding 60 °C and for ambient temperatures in the range 20 to 60 °C a reduction of 2% of the allowable working pressure per 1 °C rise above 20 °C should be allowed. Maximum pressure surges due to water hammer must be included within the permissible working pressure because uPVC pipes can suffer from

Table 13.3 Unplasticized PVC pipes to BS 3505.

Nominal bore (mm)	Minimum OD (mm)	Maximum average wall thickness (at six points around circumference) (mm)			
		Class B 60 m working pressure	Class C 90 m working pressure	Class D 120 m working pressure	Class E 150 m working pressure
50	60.2	—	3.0	3.7	4.5
80	88.7	3.4	4.1	5.3	6.5
100	114.1	4.0	5.2	6.8	8.3
150	168.0	5.2	7.5	9.9	12.1
225	244.1	6.7	9.8	12.9	15.8
300	323.4	8.8	12.9	17.0	20.8
450	456.7	12.3	18.2	23.8	—
600	609.1	16.3	24.1	—	—

Note: Tolerances on OD range from +0.3 to +1.0 mm, and on wall thicknesses from −0.5 to −2.4 mm (negative) and +0.0 to +1.0 mm (positive). The largest tolerances apply to the highest class of the largest diameter shown.

fatigue problems. Typical uPVC pipe material has a relative density of 1.42, a softening point of about 80 °C, and a coefficient of linear expansion of between 5.0×10^{-5} and 8.0×10^{-5} per °C.

The main advantage offered by uPVC piping is its resistance to corrosion, hence its use for chemical transfer lines in a water treatment works and also its use for service pipes in cool climates. It is also light in weight, flexible, and has easily made joints. When laying uPVC pipes the trench bottom should be free of sharp stones and the fill should be brought up evenly either side to prevent distortion exceeding 5% of the diameter. Pipes should not be laid during freezing conditions because they tend to become brittle at low temperatures and cannot withstand freezing water.

Test pressures are usually 1.5 times working pressure, applied not earlier than 24 hours after making solvent cement joints. Air pressure tests should not be used and, when testing, some allowance should be made for pipe expansion. CP 312 suggests a satisfactory test should be a loss not exceeding one litre per day, per 1 km length, per 2.5 cm of nominal bore, per 30 m of test pressure, the pipeline having been allowed to stand filled with water for 24 hours previously. (This is three times the allowable loss for steel and ductile iron pipes.)

Fittings can be made in uPVC or metal: for the latter special connectors are available.

uPVC pipes have not been found entirely satisfactory in use, especially in the larger diameters above about 200 mm. In these larger diameters the pipe can distort from circularity if the backfill is not evenly compacted in equal layers on either side of the pipe and this can cause distortion at the joints leading to leakage. Also there seems to be a tendency towards embrittlement of the pipes with age. Solvent joints have not always proved satisfactory: if the joint is not carefully made the high coefficient of expansion and contraction of the pipe with change of temperature may cause it to break and, once broken, a solvent joint will leak substantially and cannot be remade. uPVC pipes are also degraded by ultraviolet light, the effect increasing with temperature so that the pipes must not be exposed to sunlight in hot climates for more than a day or two. Sometimes pipes are found split longitudinally.

13.14 Glass reinforced plastic (GRP) pipes

GRP pipes for water supply are made of various types of resins—the isophtalic and bisphenol polyester resins to BS 3532 or the epoxy resins to BS 3534—reinforced with glass fibres. The fibres may comprise continuous filaments wound on to the pipe as it is formed to create a crossply of reinforcement or they may comprise chopped strands mixed in the resin. Pipes can be centrifugally cast, but are more often progressively built up on a rotating mandrel. The pipe may comprise layers of different resins, one layer being cured before another is applied, or may be made from one type of resin throughout. These resins are thermosetting plastics which soften with heating, but which become irreversably hard with chemical change and cooling. Where different resins are used the inner layer comprises the most chemically-resistant resin, e.g. bisphenol, and the outer layer is a cheaper, stronger resin, e.g. isophtalic. Similarly the glass fibres used in each layer may be of different quality and make-up. The resins are hardened by the addition of a catalyst, accelerator, and hardener and by being brought to the right temperature for the particular resin used. Residual volatiles after curing should be less than 0.1%, otherwise taste problems may arise; residual amines or other toxic components must all be used up in the mixing and curing processes. Manufacture is a complicated process which must come under skilled control and should preferably be automated. The pipes are light, flexible, and corrosion-resistant, but their long-term life in service is not yet known since they have only been manufactured since the late 1960s. Potential problems can arise in the form of delamination and cracking due to imperfect manufacture and from strain corrosion if pipes are unduly stressed by deflection. There is also a possibility that the pipes may promote microbiological growth which can soften the resin. Pipes must be laid under strictly controlled backfill conditions to prevent unacceptable distortion of the pipe wall or bending.

Joints can be socket and spigot for use with O-ring synthetic rubber rings or push-in gaskets, or pipes can be rigid jointed using a resin adhesive. Fittings can be made of GRP pipe which is cut and bonded together using overlays of glass reinforced plastic matting which is bonded on; alternatively metal fittings can be used.

Reinforced plastic matrix (RPM) pipes are made of polyester resin, glass fibres, and silica sand. The sand increases the wall thickness and therefore the rigidity of pipes which can be an advantage for laying purposes.

The predominant use of GRP and RPM pipes is for conveyance of aggressive waters, or for laying in exceptionally aggressive ground conditions where both iron and concrete pipes would be severely attacked. GRP pipes are light in weight and can therefore be produced in long lengths of up to 12 m. Diameters of pipes commonly produced are in the range 400 to 2500 mm, but larger sizes can be made. Pipes can also be designed for very high internal pressures. BS 5480 gives standard pressures and stiffness classifications.

13.15 Polyethylene pipes

Polyethylene is a thermoplastic material which softens with heat. In its natural state it is translucent, but when used for pipes a black or blue pigment is added to reduce the degrading effect of ultraviolet light. Polyethylene pipes are light in weight and flexible, resistant to abrasion and corrosion, and have a better impact resistance at low temperatures than do uPVC pipes. They can be used in freezing temperatures. Three types of pipes are available for water supply purposes: low density polyethylene (LDPE) of relative density 0.90 to 0.93 to BS 1972, high density polyethylene (HDPE) of relative density up to 0.96 to BS 3284, and, more recently, medium density or modified high density polyethylene (MDPE) of intermediate relative density. Low density polyethylene piping is supplied in coils; medium and high density can be supplied in coils or ‘sticks’, i.e. straight lengths, dependent upon diameter. Table 13.4 gives some characteristics of the three types.

Polyethylene piping is only suitable for cold water mains laid underground in climates which are not exceptionally hot because of the substantial reduction of strength with increase of temperature and the need to give continuous support to the pipes. LDPE piping is normally only used for the smaller pipe diameters, particularly for service pipes; MDPE piping is now superseding HDPE for the larger diameters. The flexibility of all three types of pipes is an advantage when laying underground as the use of many bends is avoided.

Joints for LDPE pipes are made using metal or plastic compression couplings. These couplings are not so suitable for MDPE and HDPE pipes of larger diameter since there may be difficulty in making the joint tight enough to achieve adequate compression. Butt or sleeve welding of the thicker MDPE pipes is now widely adopted. An electrically heated plate is placed between the butt ends. When these have been brought to the right temperature the plate is removed and the butt ends

Table 13.4 Polyethylene piping characteristics.

		Low density (Type 32) to BS 1972			High density (Type 50) to BS 3284		
Long-term stress rating at 20 °C		3.2 N/mm^2			5.0 N/mm^2		
Relative density		0.90–0.93			Up to 0.96		
Classes		B	C	D	B	C	D
Pressure rating, i.e. maximum sustained working pressure at 20 °C (bar)		6	9	12	6	9	12
CP 312 reduced pressure rating							
at 30 °C		80%			65%		
at 40 °C		60%			40%		
at 50 °C		45%			25%		
Nominal diameter (mm)	Minimum OD (mm)	Minimum wall thickness (mm)					
50	60.1	5.3	7.6	—	3.5	5.1	6.6
80	88.6	7.8	11.2	—	5.2	7.5	9.7
100	113.9	10.0	—	—	6.6	9.6	—
150	167.8	—	—	—	9.8	—	—

Note: Medium or modified high density polyethylene commercially available is quoted as having a long-term stress rating of 5.0 N/mm^2, relative density of 0.94, and pipes are manufactured up to 600 mm diameter for up to 10 bar pressure rating at 20 °C. Reduced pressure ratings are quoted as 80% at 30 °C, 65% at 40 °C, and 50% at 50 °C. The coefficient of linear expansion is quoted as 15×10^{-5} per °C.

are pushed together until the fusion is complete. The operation has been made much more successful by the use of semi-automated jointing machines, the fusion temperature being controlled to a preset value with the butt ends being moved and brought together with the correct pressure for fusion. Sleeve joints can be similarly fusion welded, and tee-branches of the split collar type can be fusion welded on pipes, the hole in the sidewall of the pipe being afterwards reamered out.

Pipeline construction

13.16 Choice of pipes

Only an indication can be given here of general patterns in the choice of pipes.

For the middle range of pipe sizes used in water distribution systems ductile iron pipes are the type most widely used because of their rigidity, strength, toughness, and durability in many kinds of ground which are not exceptionally aggressive to iron. In addition a complete range of compatable, standard dimensioned ductile iron fittings and valves are available to make up a homogeneous pipeline, which simplifies both design and construction processes. Service pipe connections can also be tapped directly on to the main under pressure if necessary. In this same middle range of sizes asbestos cement pipes form the main alternative to ductile iron pipes, their advantages being that they are usually cheaper (especially if ductile iron must be imported), and they are more suitable for laying in aggressive ground conditions than are ductile iron pipes. However, asbestos cement pipes are more liable to breakage problems than are ductile iron pipes, both in transport and when laid, fittings must be mainly of ductile iron, and gunmetal or cast iron saddles have to be strapped to the pipes before service pipe connections can be made. Deterioration of these saddles can bring about severe leakage problems. Nevertheless the use of asbestos cement is widespread and will probably continue, provided that the manufacturing hazards involving asbestos fibre are not found unacceptable and also that it is clear that their use for the conveyance of drinking water creates no health risk. At present, although there is evidence that fibres do enter the water, especially when a main is tapped, there is no evidence that this represents a danger to health.

For the smaller pipe sizes in water distribution systems, 80 to 200 mm, it is likely that the use of medium density polyethylene (MDPE) tubing will increase in countries with cold or temperate climates now that easily made and satisfactory jointing and tee-connection systems are becoming available. The tubing is tough and durable, its flexibility does away with the need for many bends, and it is not damaged by low temperature conditions. In hot climates uPVC tubing may continue to be used for small diameter mains in the range 50 to 150 mm because of its better strength characteristics at raised temperatures, compared to those of polyethylene tubing, and also because it is easier to lay and possibly cheaper than ductile iron or asbestos cement. However, it suffers the defects mentioned at the end of Section 13.13 and its use in water distribution systems in temperate climates has reduced and may do so also in hot climates. Notwithstanding the foregoing remarks, both ductile iron pipes and, to a lesser extent, asbestos cement pipes are strong competitors for small diameter mains.

For trunk mains and large diameter pipes no general rules of choice can be laid down since steel, ductile iron, prestressed concrete, concrete cylinder, and GRP pipes may each be used in any particular case according to circumstances applying. Steel pipes are predominantly used for trunk mains or high pressure mains, the welded joints offering a distinct advantage in the latter case. Ductile iron pipes in the largest diameters tend only to be used instead of steel if their price is competitive or if it is expected that there will be difficulty in getting the skilled welders necessary to weld steel pipes. The alternatives of prestressed concrete, concrete cylinder, or GRP pipes tend only to be used because of special circumstances applying, such as price competitiveness, in-country manufacture as opposed to importing, aggressive ground conditions or aggressive water to be conveyed, or (in the case of concrete pipes only) where a greater margin of safety is required against rough handling and backfilling. GRP pipes have the principal advantage that they are not attacked by ground conditions or by waters, such as desalinated water, which are severely aggressive to both iron and concrete.

13.17 Laying of pipes

Strict control needs to be exercised over the laying of pipes because of the high capital cost of pipelines and their swift deterioration if not properly laid. In this section will be considered the key factors to which attention should be paid to produce a well laid pipeline.

The stock dumps for pipes should be properly planned and pipes should only be stacked one above the other if they are properly provided with timber supports. Pipes should not be stacked in areas where long grass may grow: in a dry period this grass can catch fire, thus ruining the exterior protection of the pipes. All pipes should be handled by using purpose made lifting slings of a wide fabric material so that the external coating is not damaged. The practice of lifting pipes by means of chains or wire ropes 'packed off' the barrel of the pipe by pieces of wood should be forbidden. This not only damages the pipe, but also can be a dangerous practice. When pipes are delivered, and again just before they are lowered into the trench, they should be inspected for flaws. The Holiday detector should be passed over steel pipes. Any coating or lining flaws detected should be made good. The interior of all pipes should be inspected as the pipe is lifted and any debris must be brushed out.

It is vital to make an even bed for pipes, with joint holes previously excavated in the positions required. All large stones must be removed from the bed and no hard bands of rock should straddle the trench. The use of boning rods and sight rails for every pipe is essential and work should be stopped until these are provided. What constitutes a 'large stone' which should be removed in the base of the trench depends upon the nature of the material forming the base of the trench. Generally any large stones over 75 mm size should be removed over one-third of the width of the trench, i.e. along the central band of the trench where the pipe will bed, but smaller stones than this may need to be removed for small pipes laid on hardish ground. In all cases where rock is encountered, and rock includes chalk with hard bands in it, the base of the trench must be filled with 150 mm of concrete on which the pipe is bedded whilst the concrete is still plastic. Sand can be used as an alternative, but concrete is better. An alternative method is to site the pipes on preformed concrete saddles carefully lined up and then to place the concrete below the pipe, working it well up beneath the underside of the pipe. Where rock is expected there should be liberal provision for concrete bedding in the bills of quantities. If this is not adequately provided for, the supervising engineer on site may be reluctant to order the proper bedding the site conditions demand because of the additional cost involved. He should not be put in such a position. The provision of concrete for bedding or fully surrounding pipes can solve many pipe-laying problems in a sound manner. The concrete not only gives full support to the pipe, protecting it from settlement on rock points and from excessive overburden pressure or traffic loading, but it also gives protection against corrosion in aggressive soils.

Lengths of pipeline should be laid to even grades: CP 2010 suggests not less than 1:500 on a rising grade (in the direction of flow) and not less than 1:300 on a falling grade. This, however, is a counsel of perfection and is frequently impossible when laying pipelines in flat ground. For distribution pipes which have service connections a flat grade does not particularly matter as air will be drawn off via the service connections. On trunk pipelines, however, it is important to arrange even rises to air valves.

Considerable trouble is experienced when laying pipelines through urban districts where many other services, e.g. gas, electricity etc., will have to be negotiated. No amount of preplanning work will ever reveal all the problems that will be encountered since records of these other services are seldom perfect or to the accuracy necessary. One might just as well insist that the foreman mainlayer keeps prospecting ahead of his pipelaying with trial pits to find such services, and it is preferable that he should get his trench excavated well ahead of where he is laying pipes, if the road authority will permit this. Stocks of $11\frac{1}{4}°$ bends ('$\frac{1}{32}$nd bends') should be held available as they are the ones most frequently used when negotiating the pipeline around obstacles.

Backfilling to pipes should be placed in even layers either side of the pipe up to soffit level and, in addition, for non-rigid pipes the backfill must be very carefully compacted to keep the pipe in a true cylindrical form. The backfill material adjacent to the pipe and for 150 mm above its crown must be free of large sharp stones that could puncture the sheathing of the pipe. When the material from the trench is being excavated it should be inspected and instructions should be given for setting aside material that should not be used against the pipe. This may perhaps be used in refilling the trench once the pipe has been properly covered with soft material.

A principal requirement for satisfactory pipelaying is care in making the joints. Achieving cleanliness in a muddy trench is far from easy and the men should be provided with the facilities required, such as clean water and buckets, plenty of wiping rags, and enough room to work and time to make the joint properly. The reward for taking care with each joint is a pipeline which passes the test requirements at the first test: this can save weeks of extra work.

Cover to pipes should normally be not less than 0.9 m, with 1.2 to 1.3 m under heavily trafficked roads. When pipes have more than 2.0 m cover it is necessary to check that they are strong enough to withstand the soil loading; if not, they should be fully surrounded with concrete. Thrust blocks must be designed having regard to both the pipeline thrust developed under water hammer or pipeline testing conditions and to the soil resistance available. Thrust blocks do not resist pipeline thrust by themselves: they transfer the thrust to the ground and the soil resistance that can be mobilized to resist this must be carefully analysed. All such blocks must bear against undisturbed ground.

13.18 Testing pipelines

It is best to test pipelines in reasonable lengths leaving the joints exposed to view. This is not always possible. The test usually consists of filling the pipeline with water, applying a pressure to the water either of about 50% in excess of the design or working pressure or of some fixed amount, such as 30 or 50 m water head, in excess of the maximum working pressure, and then measuring the amount of fall of pressure in a given time. After this time has elapsed (usually between 20 minutes and 1 hour) water is again pumped into the line under test, sufficient to bring the pressure back to its initial value, and the amount of water so pumped in is measured.

CP 2010 Part 3 for iron pipes gives the following standard for field testing:

> '1 litre of water per 10 mm diameter of pipe per km of length of pipeline, for each 24 hours and for every 30 m head.'

This standard, which is a fairly rigorous one, may also be applied to steel mains, but up to four times the allowance may be made for new mains of concrete and asbestos cement because of their absorption of water. American standards for testing appear to be far less onerous, but a properly laid pipeline should conform to the British Standard test requirements without difficulty if the workmanship is good.

Air should not be used for testing water mains. The test should be hydraulic and take place between blank flanges, bolted or welded to pipe ends, or caps may be used if fully supported by anchor blocks. Where pipes have flexible joints the end pipe must be fully anchored. Testing should not take place between closed valves because if the valve is already inserted in the line it will not be possible to detect any leakage past the valve, and if the valve is exposed at the end of a section of the line it will be in the 'open end condition' and will leak because it is designed for the 'closed end condition' (see Section 13.22).

Pipelines should be tested in reasonable lengths and, if possible, joints in the length under test should be left exposed for visual examination. The latter is, however, nearly always impracticable and also tends to result in imperfect backfilling because the spoil may be partly filled over the body of the pipes. Fluctuating test pressure results are likely to be caused by air locks in the pipe. To avoid air locks there must be suitable air valves on the pipeline and it must be slowly filled with water.

When a pipeline fails its test and it has been backfilled, searching for leaks can be troublesome. It is best to leave the pipeline under pressure for a day so that, possibly, a wet patch on the surface of the ground will indicate the whereabouts of failure. The pipe may also be sounded for leakage using normal waste detection methods (see Section 15.15) and it is possible to use some kind of tracer element in the water, although this is a skilled matter requiring a specialist. In practice, therefore, it is usual to dig down to the joints until the leakage point is found. Usually this is not as time-consuming as one might expect because wet ground is easily found if the pipeline is left under pressure.

13.19 Sleeving and cathodic protection

The external protections available for iron and steel pipes have been mentioned in Sections 13.4 and 13.7. Two additional protections are available: sheathing with polyethylene sheeting and cathodic protection. Clays, sulphate-bearing soils, moorland acid waters, and saline groundwaters are the principal causes of aggressive ground conditions.

In recent years the practice of loose sheathing all pipes and fittings with 1000 gauge, i.e. 0.25 mm thick, polyethylene sheeting has increased because of its simplicity and effectiveness. Pipes and fittings are laid on these sheets which are then wrapped over the pipeline and sealed with adhesive tape. Minor damage and puncturing of the sheet does not appear to impair the efficiency of the protection, although any such tear may be easily patched up with sticky tape. The effectiveness of the wrapping appears to be due to the fact that even though the sheeting is not watertight it holds a virtually static body of water against the pipe, thus reducing corrosion of the pipe metal.

Cathodic protection uses the principal of electrolytic chemical action to protect pipes by providing the pipeline at intervals with sacrificial anodes or, alternatively, by impressing a (negative) voltage on the pipeline so that it becomes cathodic with respect to buried anodes in the medium of the wet soil which acts as a weak electrolyte. Corrosion takes place at the anode and not at the cathode. Sacrificial anodes are usually made of zinc or magnesium and, electrically connected to the pipe, they cause a weak current to flow in the right direction since they are 'anodic' with respect to iron. They are buried a few feet away from the pipeline at 50 to 100 m intervals and are connected to it by an insulated cable. If impressed current is used a d.c. negative voltage is applied to the pipeline and the anodes wired to it can be made of iron as the impressed voltage is sufficient to drive current in the right direction. This type of cathodic protection is used where the soil resistivity is high since better protection can then be afforded than by the weaker electrical potential developed by using sacrificial anodes only. The lengths of pipe protected by a given anode must be electrically connected together. If the pipe joints use rubber rings then each pipe must be electrically connected across its joints by means of an insulated conductor welded on to two adjacent pipes.

Cathodic protection is not by itself complete protection of a pipeline against corrosion: it is an additional defence. Also it cannot be used in urban surroundings where there are many other services since it may initiate or increase corrosion of these other services by making them anodic to the pipeline.

13.20 Making connections

If a socket and spigot tee has to be inserted into an existing pipeline the length cut out of the latter must be slightly greater than the overall length of the tee. The socket of the tee is pushed up to fit one end of the cut pipe and the resulting gap between the two spigots at the other end is joined by using a collar. If a double-socketed tee is used this must be inserted using a plain piece of pipe on one side, again joined by a collar to the pipeline.

An alternative is to use an 'under-pressure connection'. This is shown in Plates 19C and 19D. A split collar is clamped on to the main, the collar having a flanged branch on it to which is bolted a valve. A cutting machine is attached to the valve, the latter is then opened, and the cutter is moved forward through the valve and trepans a hole in the side of the pipe. The cutter is withdrawn with the trepanned piece of pipe wall and the valve is closed. The cutting machine is then removed and the branch connection can be made. Steel and iron pipes can be cut in situ using a rotating cutting tool which is clamped on to the main. A manually operated wheel cutter can be used on small diameter cast iron mains of 80 or 100 mm size. Oxyacetylene cutting of steel and iron pipes can be used, but the cut is ragged and difficult to make exactly at right-angles to the axis of the pipe. Cast and ductile iron pipes can be cut above ground using a hammer and chisel. The pipe is placed on a timber baulk below the line of cut and is rolled back and forth as the chiselling proceeds: first to 'mark' the cutting line and then to deepen the chiselled groove. At a certain stage the pipe will come apart at the chiselling line. When asbestos cement pipes are cut, or sawn, a water jet must be directed on to the cutting point to prevent asbestos fibre being raised in the air.

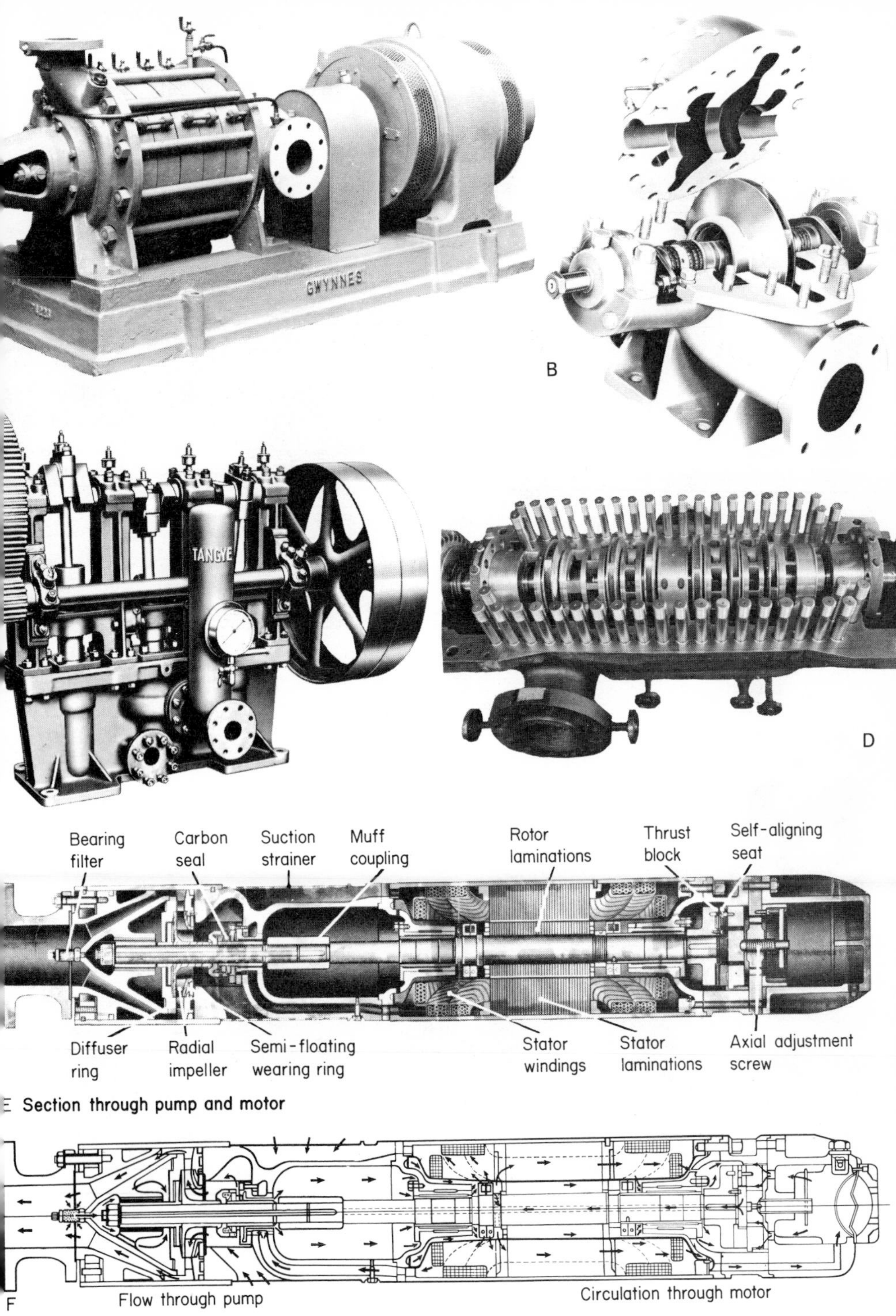

Plate 17 (A) Horizontal multistage centrifugal pump. (Gwynnes Pumps Ltd.) (B) Double entry split-casing centrifugal pump. (Worthington Simpson Ltd.) (C) Three-throw ram pump. (Tangye Ltd.) (D) High head six-stage centrifugal pump for duties up to 200 bar. (Worthington-Nord SpA Italy.) (E) and (F) Sections through submersible pump unit—single-impeller type. (Hayward Tyler & Co. Ltd.)

Plate 18 (A) Twin 300 mm diameter pipe arches of 50 m span over River Avon for Coventry Corporation water supply. (B) Unsatisfactory steel pipelaying in rocky ground in the 1960s: modern excavating machinery now permits a better and safer approach. (C) laying a 1.7 m diameter steel pipeline within a 50 m easement: topsoil stacked on the right; main excavation on the left.

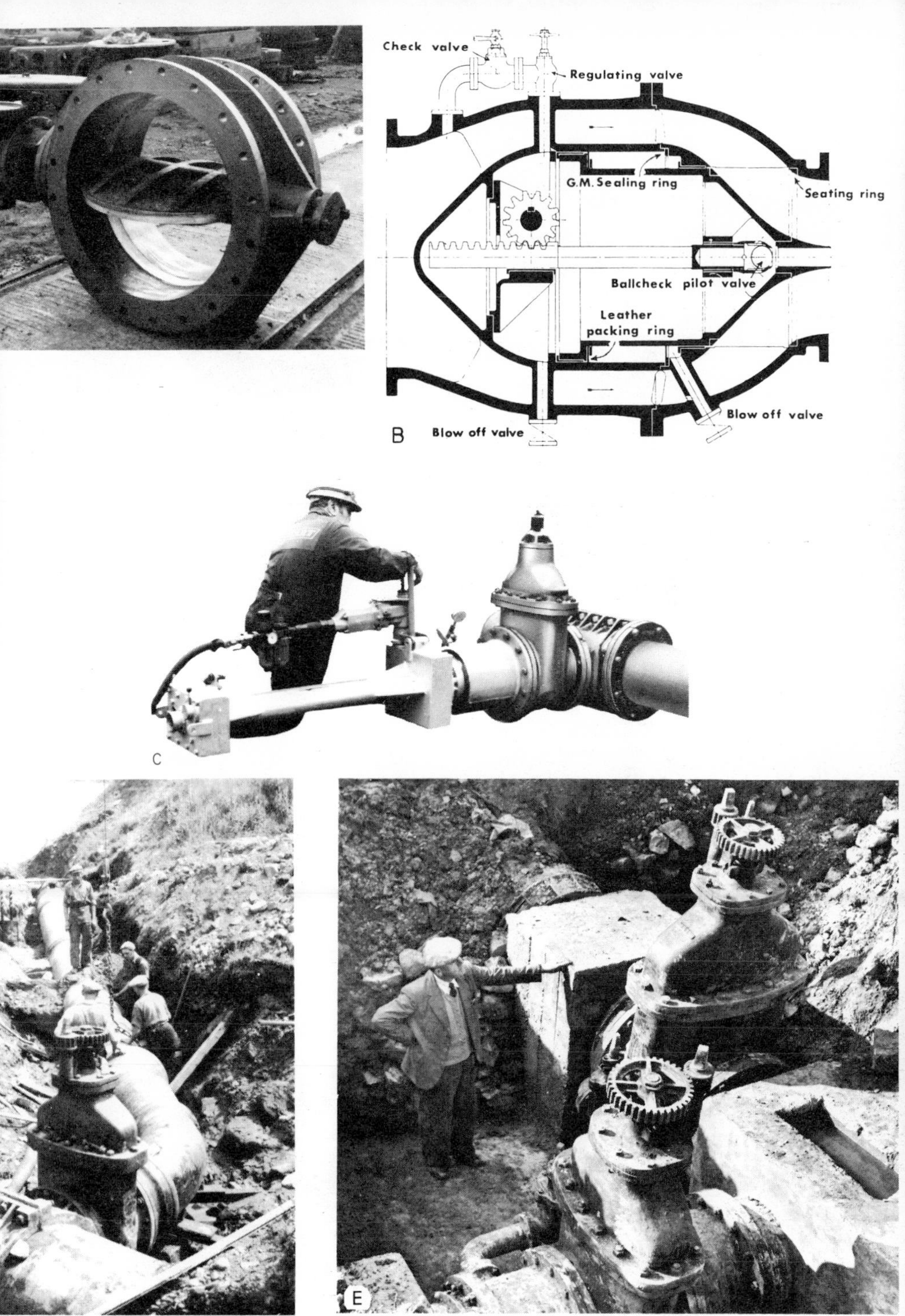

Plate 19 (A) Metal seated hand-operated 840 mm butterfly valve. (Glenfield & Kennedy Ltd.) (B) Section through Larner–Johnson control valve. (J. Blakeborough & Sons Ltd.) (C) Making an under-pressure branch connection to an existing main. The split collar is clamped on the main, the valve is opened for the drilling machine to trepan out a hole in the side of the main, the piece cut out is withdrawn, and the valve closed. (E. Peart & Co. Ltd.) (D) A 600 mm under-pressure connection on a 900 mm main, and a typical pipe 'trench' that is seldom like anything shown on the drawings. (E) Geared sluice valve with bypasses, for manual operation.

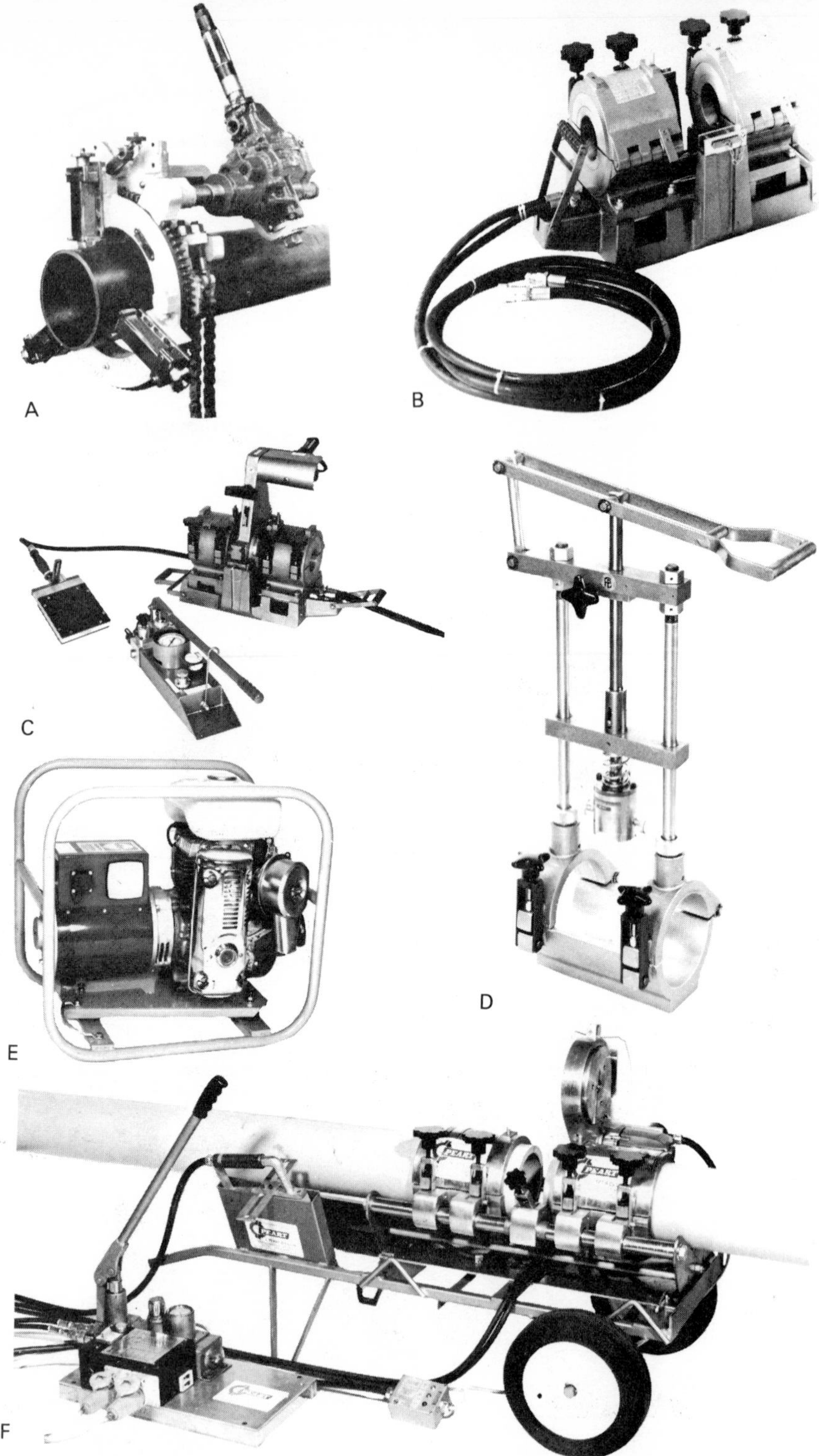

Plate 20 (A) A pipe cutting machine: the driving shaft on the right can be powered by an air motor. (E. Pass & Co. Ltd.) (B) and (C) Automatic butt fusion welding equipment for MDPE pipes: the second photograph shows the pipe-end trimmer inserted, with the heating plate separate and the handpump used for bringing the ends of the pipe together under controlled pressure. (Fusion Equipment Ltd.) (D) Equipment for welding a ferrule tee on to an MDPE pipe under controlled heating and pressure. (Fusion Equipment Ltd.) (E) A 110-volt portable generator for using with fusion welding equipment. (F) Trolley-mounted butt fusion welding equipment for MDPE pipes. (E. Peart & Co. Ltd.)

Stop valves, air valves, and hydrants

13.21 Valve types

Valves used in waterworks may be listed as stop valves, non-return valves, flow control valves, pressure control valves, energy dissipators, and air valves.

Stop valves are of several different varieties, namely:

(1) sluice valves which consist of a gate shut down into the pipeline,
(2) butterfly valves which consist of a pivoted disc in the pipeline, turned through 90° to block the flow of water,
(3) streamline valves which consist of a cone, moved longitudinally or expanded in diameter so as to stop annular flow around the cone,
(4) screwdown plug valves which consist of a plug or diaphragm which is forced on to a circular seating through which the water is flowing (used only for small pipes or low heads).

13.22 Sluice valves

Sluice valves were invented over one hundred years ago and have not

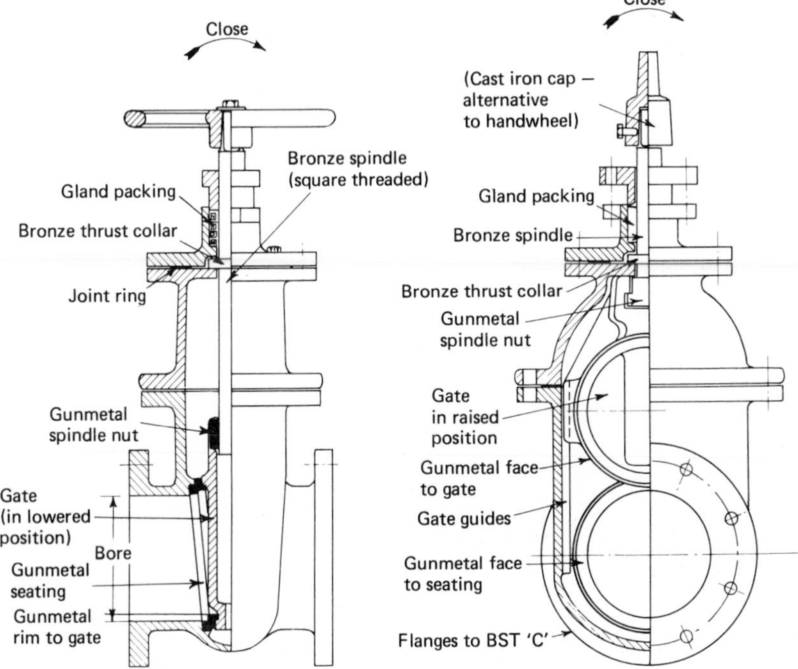

Fig. 13.4 Screwdown sluice valve for 300 mm diameter pipe. (Glenfield & Kennedy Ltd.)

materially altered in design. Figure 13.4 shows a section through a large valve. The gate is wedge shaped and is lowered into a groove cast in the body of the valve, the meeting faces being made of gunmetal. The design is simple and relies for watertightness upon the gunmetal faces being forced together to make a watertight joint. The sluice valve has a number of troublesome characteristics.

(1) When closed an unbalanced head of water will press the gate heavily against the seating; to re-open the valve against any large unbalanced head requires a great force. Above certain unbalanced heads for the larger sizes of valve it is impossible to open such valves manually, even if gearing is used. Where high unbalanced heads must be met, anti-friction devices, such as ball-bearing thrust collars and anti-friction rollers, must be used. Table 13.5 gives the maximum unbalanced heads that valves can be manually operated against without special anti-friction devices.

Table 13.5 Standard waterworks valves: maximum unbalanced pressure in metres head of water for manual operation.

Valve diameter (mm)	Hand wheel diameter (mm)	Maximum unbalanced pressure in metres head of water with:			
		Hand wheel	Key	Spur gearing and key	Worm gearing and key
75	265	115	135	—	—
100	300	80	135	—	—
150	340	40	110	135	—
225	420	20	45	85	—
300	450	10	24	55	—
450	675	6	9	20	50
600	750	2	3	12	30
750	750	1	1	6	16
900	900	—	—	4	10
1200	900	—	—	2	5

Notes:
Data by Glenfield & Kennedy Ltd, converted from Imperial units.
Force applied assumed to be 12.7 kg (28 lb) push and pull at ends of 1 m key or on rim of handwheel.

It will be seen that there are quite severe limitations to the manual operation of valves. While greater unbalanced heads may be met by providing ball-bearing collars to the valves and by providing external rising screws which can be greased, the most effective method of meeting high unbalanced heads is to provide a bypass to the valve.
(2) The amount of work required for full opening or closing of a large valve is also great. The time taken by two men to open one of the larger valves of 750 mm or more diameter may well be more than one hour.

This is not a defect hydraulically since large flows of water should not be suddenly stopped or severe water hammer will result, but it does mean that in waterworks shutdowns, as for instance for repairs of mains, considerable time must be allowed for operating valves.
(3) Sluice valves which are left shut for a long time tend to stick shut and require even greater force than that tabulated to get the gate off the seating. Similarly sluice valves which have been left open for a long time may not close properly because of collection of dirt in the gate groove which prevents proper closure of the gate.
(4) A sluice valve is only designed to close drop tight at a given unbalanced pressure; it may not close drop tight at greater or lesser unbalanced pressures.
(5) Sluice valves must be specified as designed for the 'open end test' or the 'closed end test'. A valve designed for drop tight closure under open end conditions may not close drop tight for closed end conditions. A closed end siting of a valve is when that valve is so fixed that it cannot expand when pressure is put on it, as in the case of a double-flanged valve connected into rigidly held pipework. An open end situated valve is one where the valve is free to expand, as for instance where it is held to the pipeline by one flange only or where the joint on the downstream side is a slip joint, such as Johnson coupling, which permits expansion of the valve body along the line of pipework.
(6) A sluice valve is not the proper device for controlling the rate of flow of water through a pipe. Figure 13.5 shows how the waterway on a sluice valve decreases with travel of the gate. Only the last 10% travel of the gate towards closure has any substantial effect on the flow rate, depending upon the pressure of the water in the pipeline, and this means that when controlling flow the water will be passing under the gate at high velocity. The gate may develop vibration as a consequence and cavitation resulting from the reduction in pressure of the high velocity water flowing under the gate may cause pitting and erosion.
(7) Repairs to a sluice valve (other than repacking of glands) involve emptying the main in which the valve is situated.
(8) Sluice valves are expensive and, in the larger sizes, they are heavy single items of equipment to handle (see Plate 19E).

Some of these difficulties can be minimized by making the valve about $\frac{5}{8}$ to $\frac{3}{4}$ the size of the pipeline. If the valve is connected in line by properly designed tapers, particularly if a long taper is given on the downstream side, very little headloss will occur when the valve is fully open. The amount of head lost will be a few cm of water pressure only. The cost of the valve will be less (by as much as 50% in the larger sizes), its insertion into line will be made easier because it will have less weight, and the force required to open it will be substantially reduced. This practice should always be followed where possible. Difficulty with sticking valves

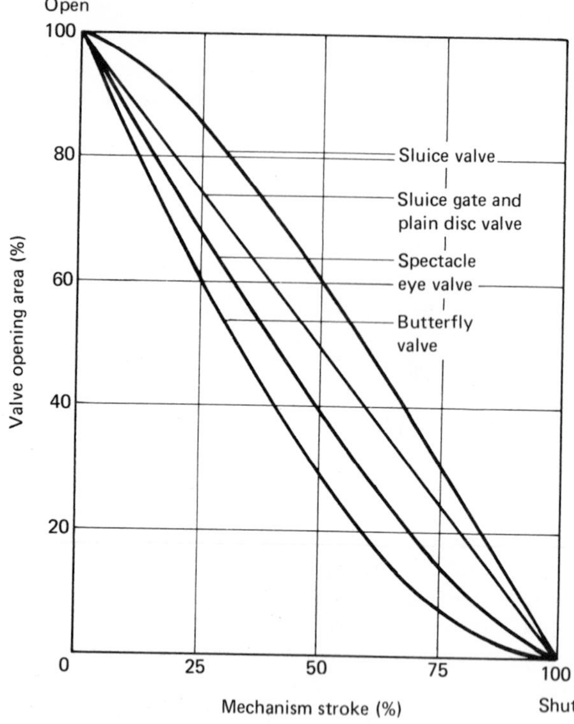

Fig. 13.5 Valve waterway areas in relation to travel of the gate. (Harland Engineering Co. Ltd.)

and dirt on the gate groove can be greatly reduced by operating valves. To leave valves unoperated for years on end is an unwise policy as they may not then close in some emergency. The problem of opening large valves against unbalanced heads is greatly eased by installing a bypass to the valve, on which bypass a second smaller valve is fixed as shown in Fig. 13.13. Bypasses need not be very large in diameter, a 75 or 100 mm bypass would normally be suitable for valves between 450 and 900 mm diameter. The effect of the bypass depends upon the downstream conditions of the pipeline. In general, however, even if the pressure is not substantially raised downstream of the main valve when the bypass is opened, pressure is reduced on the upstream side of the main valve and, accordingly, the force required to unseat the gate of the main valve is reduced. A bypass is also useful for filling mains since this can be opened first and the flow into the empty downstream section more carefully controlled than if the main valve were opened. Electrical or hydraulic power may be used to operate valves.

13.23 Butterfly valves

The use of the butterfly valve in water distribution systems has increased: it is easy to operate such a valve against unbalanced water pressures because of the partially balanced pressure against the disc. Butterfly valves can be metal seated or resilient seated: in the latter case the seat is usually made of natural or synthetic rubber and can be fixed to the disc or to the body of the valve.

Resilient seated valves can be specified as 'tight shut-off', meaning virtually watertight when shut against unbalanced pressures up to the design head. Hence such valves are the type usually specified for isolating valves in distribution systems. They do, however, prevent the passage of foam swabs used for cleaning mains, but this does not usually pose a problem if the valves are spaced reasonably far apart at 2 km or so intervals since the main can then be cleaned in sections. Short lengths of pipe of between 1.5 and 2.0 m on either side of the valve are made removable so that the cleaning apparatus can be inserted and removed.

Metal seated butterfly valves do not have tight shut-off characteristics and are mainly intended for flow control purposes where they need to be held in the partially open position. (Resilient seated valves are used for this purpose also.) Metal seated valves are not necessarily more durable than resilient seated valves since scouring of the seating can occur if silty water flows at high velocity past a disc which is held only slightly open. Solid rubber seatings are the type of resilient seating most usually adopted. Inflatable seals have been used on very large valves, but not always with success. One advantage of the butterfly valve is that closure of the valve in flowing water causes loose debris below the disc to be swept away. Disc position indicators are useful and strong disc stops integral with the body should be specified so that the operator can feel with certainty when the disc is fully closed or fully open.

Butterfly valves have been made to very large diameters (10 m or more) operating under very high heads and at high water velocities (20 m/s or more) and have proved successful in use. Permissible leakage under test should not exceed about 0.06 litre per hour per 100 mm of nominal diameter for resilient seated valves under pressures of up to 100 m.

13.24 Streamline valves

Streamline valves are efficient, easy to operate, give very good control of water flow, and have a low frictional resistance when wide open. Plate 19B shows a cross-section through such a valve. Where precise control of flow, manual or automatic, is required these are the best type of valve available, but their cost is high.

13.25 Screwdown valves

Screwdown plug valves are normally made only in the smaller sizes, of which the bibtap is a typical example. The body of the valve is so cast that the water must pass through an orifice which is normally arranged in the horizontal plane. A plug or diaphragm, or in the case of a bibtap or stopcock a 'jumper', can then be forced down on to this orifice by a screwed handle, shutting the water flow off as shown in Fig. 13.6. This principle is used in all sorts of valves for shutting off or controlling flow.

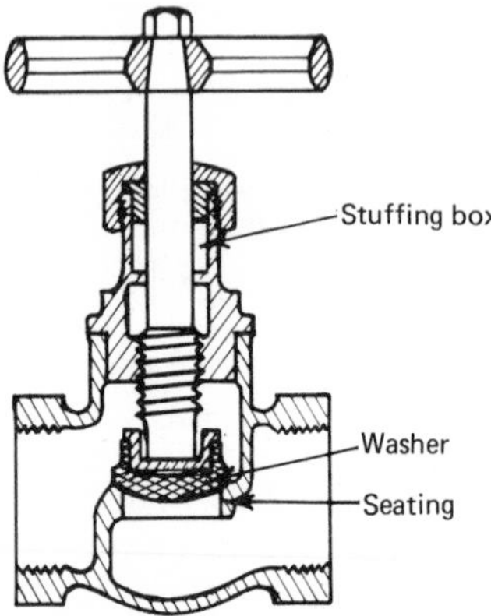

Fig. 13.6 Screwdown stopcock for service pipe.

The same principle applies to ball valves, to pressure or flow control valves, to hydrant valves, and to many other types of valves. Where the pressure is low the size of valve can be quite large, as for instance with simple plugs for stopping flow into a drain. When the size of pipe, and therefore of orifice, is small then high pressures can be controlled, as in the case of the ordinary domestic tap. The defects of these particular types of stop valves are that their seatings need renewal from time to time if they are in frequent use and that, even when wide open, they cause a considerable loss of pressure.

13.26 Non-return valves

The usual type of non-return valve consists of a flat disc within the pipeline pivoted so that it is forced open when the flow of water is in one direction and forced shut against a seating when the flow tries to reverse (see Fig. 13.7). The seating is arranged slightly out of perpendicular when the valve is to be inserted into a horizontal pipe so that the flap will close by gravity when there is no flow. The speed with which the flap acts is often important, especially on pump delivery lines where the valve is required to be fully shut the moment forward flow of the water ceases. The valve can be assisted to shut quickly at the appropriate time by bringing its pivoting spindle out through the body of the valve and putting a counterbalancing weight upon this. The defect of non-return valves is that they cause a substantial headloss because the weight of the gate must be lifted and kept suspended by the pressure of the flowing

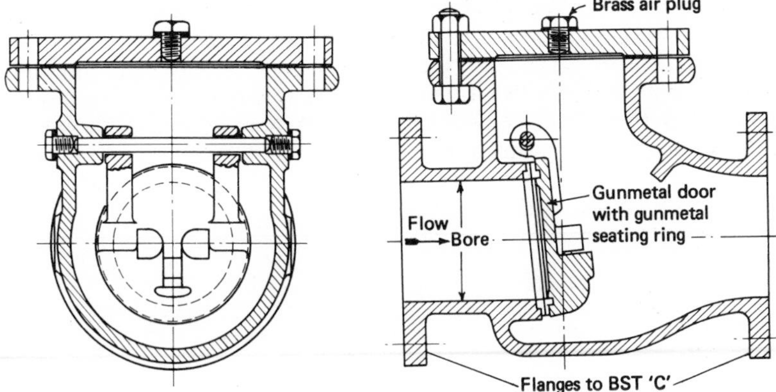

Fig. 13.7 Non-return valve. (Glenfield & Kennedy Ltd.)

water. If the gate is very large and heavy the headloss created can be large and this objection can sometimes be met by providing several smaller and lighter flaps in a single bulkhead.

13.27 Control valves

The operation of a typical control valve is best illustrated by taking as a first example the standard *pressure reducing valve* shown in Fig. 13.8.

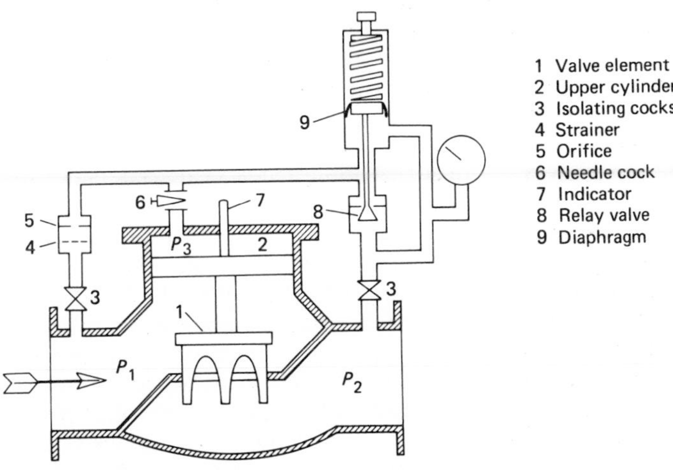

Fig. 13.8 Pressure reducing valve.
Operation: If the outlet pressure P_2 becomes too high to balance the spring load the relay valve 8 tends to close, allowing pressure P_3 to increase, causing the main valve 1 to tend to close, thus reducing P_2 to the preset value. Similarly, if P_2 becomes too low the relay valve opens, P_3 decreases, the main valve opens further, and P_2 rises again to the preset value.

This maintains a preset pressure in the main downstream of the valve irrespective of the pressure upstream. Of course, if the pressure upstream is insufficient the valve simply opens wide and no more pressure is available downstream than the upstream pressure less the loss through the valve, which is about $5v^2/g$ where v is the mean velocity at the valve inlet. Pressure reducing valves of this sort should be sized so that their full open capacity, as shown in Table 13.6, is more than adequate for the desired maximum flow.

Table 13.6 Full open capacities for pressure reducing valves.

Valve size (mm)	Maximum discharge	
	l/s	Ml/d
80	11	1.0
100	20	1.7
150	40	3.4
200	80	6.9
250	125	10.8
300	180	15.5
450	400	34.5
600	710	61.2

Note: Data by Glenfield & Kennedy Ltd.

A *pressure sustaining valve* is shown in Fig. 13.9 and operates to keep the pressure upstream of the valve to a given amount. This sort of valve is used, for instance, where a distribution area A feeds a second distribution area B. With a sustaining valve in the feedline, adequate pressure can be maintained in the distribution area A without excessive demand in B pulling the pressure down in A. (In effect A's demand takes precedence over B's demand.) When the valve is full open, however, the pressure upstream passes through to the downstream side, less the loss of head in the valve ($5v^2/g$). This valve can also be used as a pressure relief valve if connected on a branch to a main where the branch, downstream of the valve, discharges to waste.

A *constant flow valve* is shown in Fig. 13.10. An orifice plate in the main upstream of the valve measures the flow and creates a differential pressure. The valve adjusts so as to maintain a constant differential pressure across the orifice, thus maintaining a constant flow.

Variable flow control valves work on the same principle as the constant flow valve illustrated, except that the differential pressure set across the orifice is adjusted manually, or altered to follow some signal received from a controller, so that the flow in consequence is altered.

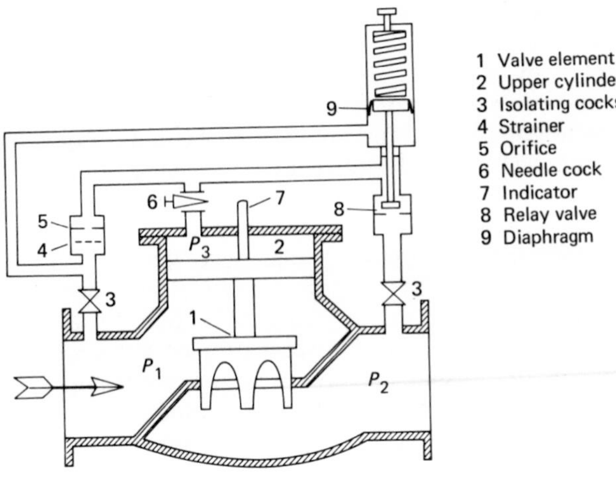

Fig. 13.9 Pressure sustaining valve.
Operation: When the inlet pressure P_1 becomes too high to balance the spring load the relay valve 8 tends to open, allowing pressure P_3 to diminish so that the main valve 1 tends to open, thus reducing P_1 to the preset value. Similarly, if P_1 becomes too low to balance the spring load the relay valve closes, P_3 increases, the main valve closes, and P_1 rises again to the preset value.

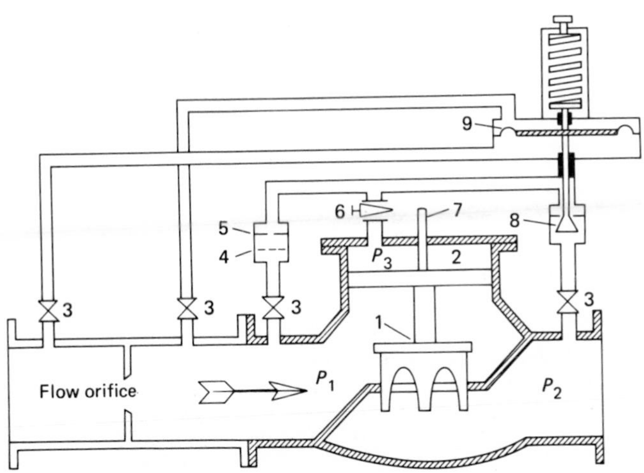

Fig. 13.10 Constant flow valve.
Operation: When the pressure difference across the orifice increases due to an increase of flow the relay valve 8 closes slightly and pressure P_3 is increased so that the main valve 1 tends to shut, thus reducing the flow in the pipe until the preset flow is reached when the main valve comes to rest. Similarly, when the flow decreases, reducing the differential pressure across the orifice, the relay valve opens slightly, P_3 reduces, the main valve tends to open, and flow increases until the preset value is reached.

13.28 Energy destroying valves

All control valves are energy destroying devices, but there is a particular kind of orifice device whose sole aim is to destroy the energy of fast flowing water before it falls or is ejected into a river or a basin. These are known as 'jet dispersers' and are used at the ends of pipelines under high head where water must be given a free discharge. If the full bore jet of water were to be left to fall upon the soft bed of a river then a large hole would be quickly scoured in the river bed. The aim of the jet disperser is therefore to break up the solid flow of the water and disperse it into numerous fine jets or droplets. This is accomplished by giving the water a 'twist' in the jet disperser so that it sprays out over a wide area and the velocities of the water drops are reduced by friction through the air. Jet dispersers are usual on scour pipes, including washouts from mains, when the head on the pipe is likely to be high.

13.29 Air valves

Before a pipeline can be filled with water, means must be provided for releasing air from it. Once the pipe is full of water, however, any aperture for release of air must close so that no water is lost. An air release valve is designed to meet this condition. The orifice through which the air is to pass is circular and below it is a ball of special composition which will float in water (Fig. 13.11). So long as air exists below the orifice the ball will be floating on the water surface below, but as soon as water begins to be released the ball will rise and close the orifice. The design must, however, take several other factors into

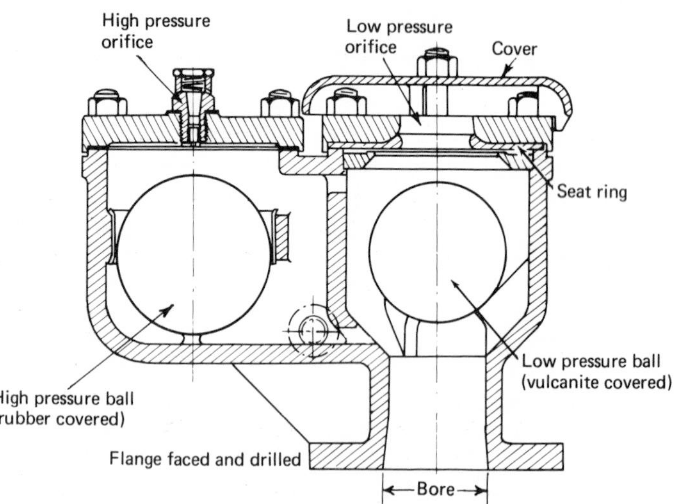

Fig. 13.11 Double orifice air valve. (Glenfield & Kennedy Ltd.)

account, for example the ball must not be sucked on to the orifice by high velocity air and nor must the ball vibrate up and down. It is usual to provide any large pipeline with two sorts of air release valve, one being a 'large orifice' air valve designed to release large quantities of air when a pipeline is being filled with water and the other having a smaller orifice which is designed for continuous operation, releasing small quantities of air as they collect in the main.

Apart from air which must be discharged to permit the filling of a pipeline, air can be introduced from pumps and from the release of air in solution in the water as its pressure reduces. The latter amount depends upon the pressure reduction and other factors, but water at standard conditions contains 2% dissolved air by volume. Whilst it is obvious that some air will collect at high points, there is also always a likelihood of air collecting at changes of gradient in the pipeline (see Fig. 13.12). The air is not necessarily shifted by the water flow, especially on a downhill gradient where air at the soffit of the pipe may

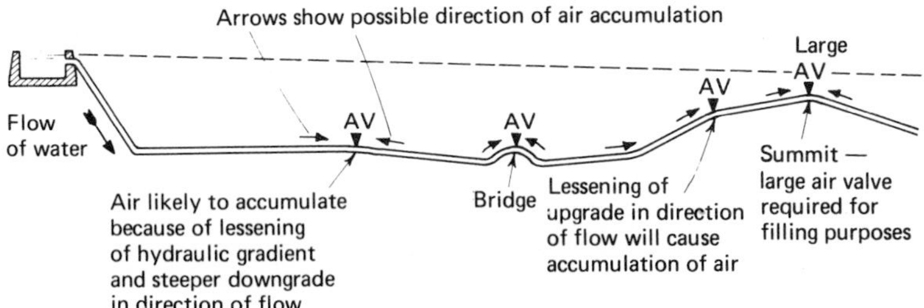

Fig. 13.12 Diagram showing desirable positions for air valves on a length of pipeline.

not be able to travel backwards against the flow. Movement of air in a forward direction may also be impeded in the case of a relatively flat main. Both these cases are worsened if there is continual release of air from solution or if air is taken in through some pump. Any movement of air along a pipeline is slow: when filling a 2 km length of main, for instance, although the flow from some downstream hydrant may soon be steady it will be interrupted by occasional bursts of air for perhaps half an hour or more.

When filling a main it is common practice to open all fire hydrants along the length of line to be filled, shutting them only when they cease to discharge air. Whilst large orifice air valves could be used instead, the fire hydrants permit the discharge of dirty water to waste and also permit sampling of water quality. Therefore it is a debatable point as to whether hydrants or large orifice valves should be provided for filling a main.

It is seldom necessary to put small orifice air valves on distribution mains since air will be removed via the service pipe connections which should be soffit connected. It is only necessary to put small orifice valves on distribution mains where there is a sudden hump in the pipeline, such as when it is laid over a bridge.

Bearing the above comments in mind, positions for air valves might be summarized as follows.

(1) *Double orifice air valves are required*
(a) at high points where air must emerge to permit filling and at high points relative to the slope of the hydraulic gradient,
(b) where a falling pipeline steepens its gradient,
(c) where a rising pipeline flattens its gradient,
(d) at intervals on long lengths of pipeline (with no fire hydrants) where necessary to permit the release of air for filling the pipeline in convenient sections, generally at 2 km intervals or thereabouts.

(2) *Single orifice valves may be required additionally*
(e) on access manhole covers and other local 'humps' where air may collect,
(f) at some suitable point a short way downstream of any pump,
(g) after a pressure reducing valve if there is a substantial reduction of pressure,
(h) (possibly) at 0.5 km intervals on downward legs of pumping mains if the pump may have entrained air and on substantially horizontal mains at 1 to 2 km intervals, especially when the flow velocity tends to be low, account being taken of small orifice valves provided under (1) and the presence or absence of service pipes connected to the soffit.

Not so many small orifice valves are required towards the ends of a system as over the beginning lengths where the aim is to ensure early release of air. Large orifice valves are usually sized according to the estimated rate of filling or emptying of a main, but fast filling is inadvisable. For a large main over 10 km long a filling time of three hours would be reasonable, with shorter lengths filled in not less than one hour. The flow capacity of an air valve needs to be taken from manufacturers' catalogues, bearing in mind that the air pressure within the main will be low, being only a fraction of a bar*. Small orifice valves have an orifice size which is related to the operating pressure in the main: the higher the pressure the smaller must be the diameter. It would be usual to ensure that the total small orifice capacity provided on the system is sufficient to release air at 2% by volume of the water flow. Again capacities of valves should be taken from manufacturers' cata-

* Blakesborough's catalogue takes 0.33 bar as a basis, but states that lower pressures 'are normal when filling a main'.

logues, but in this case the pressure taken is the working pressure in the main at the location where the air valve is fixed.

A sluice valve or stopcock should be sited below the air valve, thus making it possible to remove the air valve for repairs without shutting down the main. Care should be taken to site all air valves above the highest possible groundwater level that can occur in any pit; if this is not done polluted water may enter the main via the air valve when the main is emptied. On steel pipelines anti-vacuum valves may be essential to prevent the pipeline collapsing when water is drawn off by emptying the main from a lower point. It is important to check the need for this on large diameter steel pipes. An air valve of the large orifice kind can be used as an anti-vacuum valve.

13.30 Valve operating spindles

A valve must often be placed so deep that it cannot be reached with the normal length of key so that an extension spindle must be arranged for the valve, running in brackets rigidly attached to the chamber walls. These extension spindles may be obtained for the exact length required. It is worthwhile making spindles which are to be immersed in water, such as those for operating valves inside a reservoir, of manganese bronze so that they are proof against corrosion. There is, unfortunately, no standard direction of rotation for operating valves, but the most usual method is given in Table 13.7.

Table 13.7 Valve spindle rotation.

Type of valve	Direction of rotation of main spindle in order to close valve
Valves operated by keys or extension spindles	Anticlockwise
Valves operated by handwheels	Clockwise

13.31 Washouts

Despite the name, a washout is seldom used for scouring or 'washing out' a main because its diameter is usually too small to create sufficient flow velocity in the main to wash out debris. Washouts are usually 100, 150, or 225 mm diameter and they are principally used for emptying a main or for removing stagnant or dirty water. In open country it would be usual to put washouts on trunk mains at every low point, the washout line being laid to discharge by gravity to the nearest watercourse. The discharge would be to a concrete pit with overflow to the watercourse in order to prevent scour from the high velocity discharge. The washout branch on the main would be a 'level invert tee'. In flat country washouts would be spaced 2 to 5 km apart depending upon pipeline gradients and valving. Where it is not possible to get a free discharge to a watercourse the washout has to discharge to a manhole from which water is pumped out to some other discharge point.

In distribution systems principal feeder mains are usually provided with washouts wherever this is convenient, regard being paid to the position of valves on the main and any branch connections and to the need to be able to empty any leg of the main in a reasonable time of one or two hours. On small mains, washouts are not normally provided because fire hydrants can be used for this purpose, but there would be a tendency to lay a specific washout to empty a part of the system where a convenient watercourse exists. Washouts are also placed at the end of every spur main; these usually comprise fire hydrants even though they may not be officially paid for by the fire authority and designated as such. They are operated regularly to sweeten the water at the end of the main.

Care is essential in the design of washouts since, under high heads, the velocity of discharge can be very high and the consequent jet discharge destructive and dangerous. Manholes receiving the jet should be of substantial construction and valves should be lockable and slow opening for high heads.

13.32 Valve chambers

A typical simple valve chamber is shown in Fig. 13.13. It should be noted that the valve is anchored on the upstream side having a Johnson flange adaptor on the downstream side which permits the valve to be removed. The valve body is therefore free to expand downstream when the valve is closed under pressure and it must therefore be specified for the open end conditions (see Section 13.22). The chamber cover can be made of precast reinforced concrete beams which can be removed for lifting out the valve. Removal of a valve should not be necessary for many years so it is frequent practice to pave or earth over the cover to the chamber, leaving only the access manhole and the spindle manhole exposed. Every effort should be made to avoid siting the chamber in a road; it should preferably be in the road verge, but even there the cover should be strong enough to take heavy vehicles which sometimes run off roads. The thrust or anchor block must be designed according to the ground conditions so that it is capable of taking the full thrust when the valve is closed.

Small valves, i.e. those of less than 300 mm diameter, are not usually installed in purpose built chambers. The valve is backfilled around the body to just below the top body flange and then a brick chamber or a glazed stoneware pipe is set round the upper part and covered with a hinged cover so that an operating tee key can be lowered on to the valve cap when it is necessary to operate the valve.

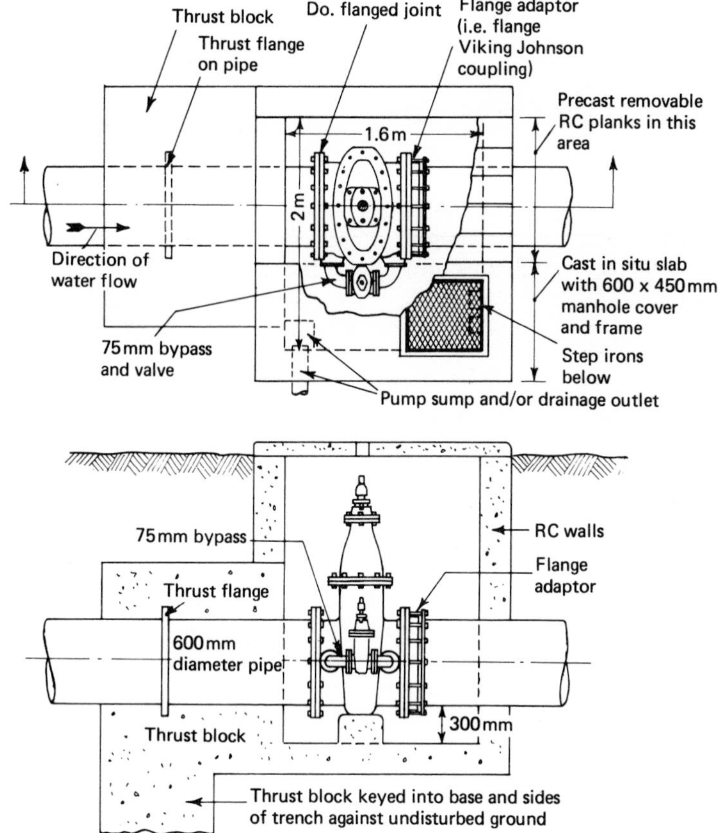

Fig. 13.13 Typical valve chamber design.

References

(1) NWC/DoE Standing Technical Committee, *The Condition of Rubber Sealing Rings in UK Water Mains*, NWC Bulletin No. 38, September 1980.
(2) Lee R A, Large Diameter Water Mains, *Proc. Pipeline Engineering Convention*, London, 1970.

14
Pipeline design

14.1 Factors in pipeline design

From the Hazen–Williams formula for the flow of water in circular pipes equation 10.5(b), Section 10.7, shows that if C and H/L are constant then

$$\text{rate of flow } Q = \frac{\pi d^2}{4} V = kd^{2.63}$$

where k is a constant for any particular case. Similarly the Manning formula gives $Q = k'd^{2.67}$. As a consequence, increasing the diameter of a pipeline by 30% doubles its flow capacity. A problem therefore arises that if a pipeline must be designed for a future demand expected to rise annually to some amount Q in x years' time, when is it economic to lay a single main now to carry the whole amount Q, and when is it economic to lay a smaller pipeline first for $0.5Q$ and to duplicate it later?

A theoretical answer to this problem can be obtained by comparing the present values of each alternative (see Section 2.8). The 'present value' of some cost £X to be paid in n years' time may be defined as that sum £A which, invested now at r% compound interest per annum, would accumulate to £X in n years, i.e. £$A(1+r)^n =$ £X where r is expressed as 0.05 for 5% interest rate, and so on. In the simple case of a main costing £X to carry $0.5Q$ the cost of a larger main to carry Q is £$1.3X$ since the diameter will need to be 30% greater and pipeline costs may be taken as roughly proportional to diameter*. The present value of the single main is therefore £$1.3X$ since all the money must be expended initially, but the present value of a dual pipeline scheme is

$$£X + \frac{£X}{(1+r)^n}$$

from which it can be seen that for the cost of the dual line not to exceed that of the single larger line

$$\frac{1}{(1+r)^n} < 0.30$$

*This rough rule applies only over the range 300 mm to about 750 mm diameter and provided that pipe wall thickness runs approximately proportional to diameter.

Referring to tables for the value of this term, for the usual rates of discount applied, it can be shown that

if $r = 8\%$ n must exceed 15 years

if $r = 10\%$ n must exceed 12 years

if $r = 12\%$ n must exceed 10 years

This theoretical calculation shows that it is seldom economic to consider laying two pipelines to meet a future demand unless the second will not be required for at least ten years. In practice, however, many other factors affect the decision. Some of the most important are as follows.

(1) Demand forecasts often prove wrong; forecasts for more than twenty years ahead are of doubtful validity.
(2) It may be difficult to lay two pipelines along a congested route.
(3) In an inflationary situation a fixed debt at a fixed rate of interest for the cost of the pipeline cheapens with time.
(4) A single pipeline creates less constructional disturbance, has larger reserve capacity, and causes less frictional head in its earlier years of use.

However, factors favouring a dual pipeline may be as follows.

(5) A dual pipeline permits supplies to be maintained when a burst main occurs and also permits shutdowns for maintenance of valves and cleaning of the pipeline.
(6) Dual lines, being of smaller diameter than a single pipeline, may offer some saving in wall thickness of pipe to meet a given pressure, thereby reducing their excess cost over a single pipeline.

14.2 Pipeline planning

It must not necessarily be assumed that the only options available consist of laying pipelines of the same diameter throughout their length. Consider the situation illustrated in Fig. 14.1. Existing sources 1, 2, and 3 feed water into the principal network shown. A new source 4 is proposed to bring, say, another 80 Ml/d into the urban area via pipeline AB. Possible areas of substantial growth of demand are labelled D1, D2 . . . etc. but, as so often occurs in practice, it is uncertain to what extent each will develop and in what order.

If D3 and D4 develop large demands then clearly point B is the right position to terminate the trunk line AB. In this case supplies from source 2 can be diverted to D4, and the new source can take over supplies formerly fed by 2. Should D1 and D5 develop into major new demand centres instead, and the main has been laid from A to B, water will then have to be transferred from B to B′, as shown in alternative (a)

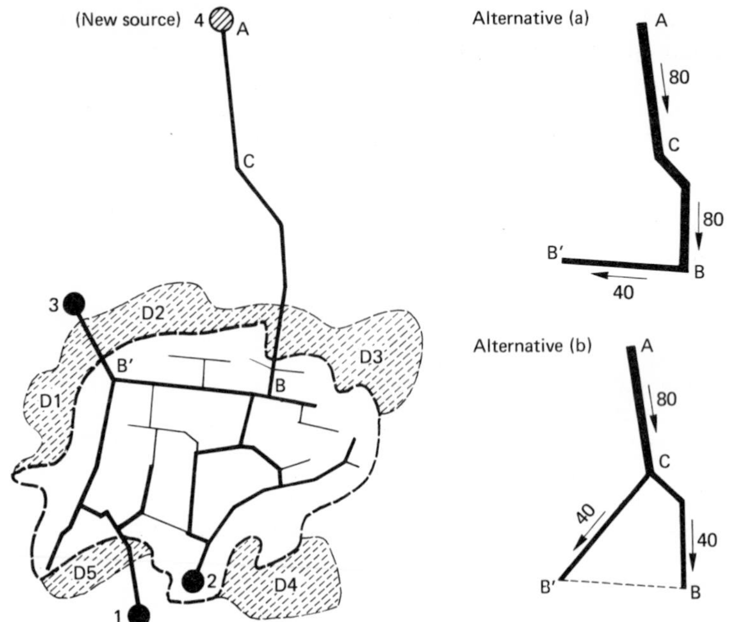

Fig. 14.1 Trunk mains alternatives.

of Fig. 14.1. If this is a possibility it would be better to lay only part of the trunk main AC for 80 Ml/d, laying CB for, say, 40 Ml/d. Then, if the extra demand arises at D1 and D5 later, the main CB′ can be laid, as shown in alternative (b). If it does not, then the length CB can be duplicated. Thus the option remains open to deal satisfactorily with the situation however it may develop. If alternative (b) is adopted a 'ring main' CBB′ is created, which permits a substantial variation of demand as between B and B′ and also permits supplies to be maintained better in the case of a pipeline burst. The phased development also permits capital expenditure to be kept more nearly in line with the rise of demand, which is financially advantageous.

In addition, it can be shown that if the length AC is laid in slightly larger diameter than would have been necessary for the whole length AB then an additional capacity can be brought forward from A to C at low cost. For example, for 80 Ml/d flow and hydraulic gradient of 2 m per km, pipeline AB would need to be 900 mm diameter ($C = 120$). For an extra 100 mm diameter on this for the length AC the capacity AC would be increased to 105 Ml/d for the same hydraulic gradient, i.e. for about 5.5% extra cost (C being taken as midway between A and B) 30% extra water is made available at C. If it is expected that source 4 will ultimately be extended to give more output, such a policy may be worth following since it may postpone or remove the need to duplicate the part AC. In any actual example other alternatives will present themselves and complicating factors will have to be taken into account, such as maximum day demands, peak output capacities of sources, service reservoir capacities and elevations, the existence of different pressure zones, and differences in the cost of production of water at the various sources.

14.3 Design of a pumping main

A pumping main usually denotes the main which conveys water from a pumped source to a distribution reservoir. In this case the flow is likely to be steady and the upper limit of flow fairly accurately known. The designer aims to find a size of main for which the total of capital loan charges and cost of power consumption against friction is a minimum. Provision for likely future developments must also be taken into account.

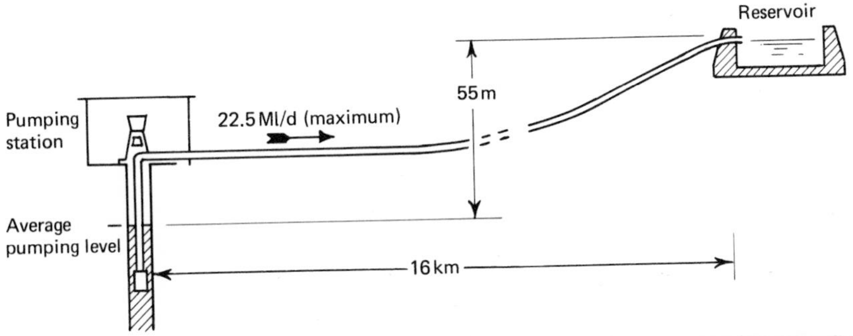

Fig. 14.2 Example for pumping main calculations.

Example Find the most economic pumping main for a maximum flow of 22.5 Ml/d and an average flow of 20.0 Ml/d, for a static lift of 55 m through 16 km of main (see Fig. 14.2). Use the following data:

cost of pipeline = £3 per m per 25 mm diameter

cost of pumps = £600 per kW installed power

cost of power (all in) = 4.5 pence per kW h

annual charges on capital = 12%

total installed power = 150% of required power (i.e. 50% standby)

Calculations for various sizes of mains can be tabulated as follows.

	Diameter of main (mm)			
	525	600	675	Notes
Method (1)				
Static lift (m)	55	55	55	
Friction (approximate) (m)	42	20	11	
Total head on pumps (m)	97	75	66	
Installed power of pumps for maximum flow rate + 50% standby (kW)	371	287	253	(1)
Power used at average rate × 0.73 efficiency × 0.85 power factor (kW)	355	275	242	(2)
	£1000	£1000	£1000	
Cost of main	1008.0	1152.0	1296.0	
Cost of pumps	222.6	172.2	151.8	
Total capital	1230.6	1324.2	1447.8	
Annual charges on capital at 12%	147.7	158.9	173.7	
Annual power charges at 4.5 p per kW h	139.9	108.4	95.4	
Total annual cost	287.6	267.3	269.1	(3)
Method (2)				
By discounting at 12% for 40 years:				
	£1000	£1000	£1000	
Cost of main	1008.0	1152.0	1296.0	
Cost of pumps	222.6	172.2	151.8	
Renewal of pumps after 20 years (discount factor = 0.1037)	23.1	17.8	15.7	
Power charges for 40 years (discount factor = 8.24)	1152.8	893.2	786.1	
Total net present cost	2406.5	2235.2	2249.6	(4)

Notes:
(1) 1 kW = $0.1135Qh$ where Q is in Ml/d and h is in metres. (Note: 1 kg wt × 1 m/s = 9.81 watts.)
(2) 1/0.73 × 1/0.85 × $0.1135Qh$ where Q is average flow of 20 Ml/d.
(3) Power against note (2) × 24 hours × 365 days × 4.5 p.
(4) Discount factors from tables.

The 600 mm diameter pipe is shown to be the most favourable by both methods. It will be noted, however, that the discounting method does not show the actual amount of the extra cost per annum in choosing one pipe in preference to another. For a 675 mm pipe the extra cost per annum is only £1800 and it might be thought that it is worth shouldering this small extra annual payment to have the larger sized main, in case a greater demand than expected occurs or in case the friction increases more than expected. However, this must be viewed in the light of the extra capital cost for the larger main.

Once a first calculation of this nature has been made, then a closer examination of the problem in detail must be carried out because there will be some assumptions which must be rectified. Amongst the factors which will need to be examined are the following.

(1) Will the different pressures on the 525, 600, and 675 mm mains require one main to be of a stronger class than another? If so, we cannot apply a rule of thumb price related to their diameter, but must price the lines individually taking into account the more expensive pipes required for the higher pressure.
(2) What will be the effect of the first 10 or 20 years' life of the main giving a higher coefficient of friction *C* in the Hazen–Williams formula and therefore a lesser friction head?
(3) What is the effect of possible increases in the cost of power?
(4) What is the effect of different discount rates?
(5) What is the effect of an output which starts at, say, 5 Ml/d and increases year by year at the rate of 3 Ml/d per annum until the maximum average output of 20 Ml/d is reached and maintained?
(6) What is the likely further development when the demand exceeds the output of the source? Will the pipeline be required to carry more water from an additional source for the whole or part of its length?

14.4 Comments on the use of discounting

Discounting permits the present value of future expenditure to be assessed so that capital projects involving high or low running costs can be compared financially. However, it must be remembered that the higher the discount rate, the less is the influence of future running and replacement costs, so that schemes which are 'cheap to buy, but expensive to run,' are put in a more favourable light when discount rates are high. High discount rates reflect, amongst other things, scarcity of capital and it is natural therefore that they direct choice towards schemes which involve least initial capital expenditure. Discount rates can also be used to put conflicting needs for capital on some equitable basis of assessment, by ensuring that projects for such things as housing, roads, schools, water supply, and so on are all devised to give some acceptable 'return on capital' which the discount rate expresses.

However, early enthusiasm for discounting has waned because the method makes assumptions about forward costs which have frequently not been borne out by events. If 'costs' were actually only money units having an invariable value they could be discounted. In practice, when future costs come to be paid they represent real values in terms of labour, materials, and fuel diverted to the project. These values can change swiftly as social and political movements force changes in the evaluation of these quantities. Thus fuel costs have risen in real terms due to political change (the 'oil crisis'). Social movements have increased the real cost of operative labour because of the pressure towards equalization of standards of living. Hence the summation of future operational expenditure as if it were a fixed cost relative to initial capital cost can be misleading. Fortunately, seldom is any project chosen by discounting alone: many other factors usually have a predominating influence on choice. Discounting has a useful part to play, particularly when it is necessary to compare schemes which have different mixtures of present and future costs, and when alternative phasings for development are possible, or when the future benefits to be derived from alternative schemes need to be compared[1], but the instinct of the engineer to build well in the first place, and to keep maintenance costs, replacement costs, and energy consumption low, may prove to be the best policy in the long run.

14.5 Design of gravity mains

In this section we consider the design of a trunk main fed by gravity from an impounding reservoir. It is best to take an example as shown in Fig. 14.3.

Example A proposed impounding reservoir 30 m deep will have a top water level of 410 m OD. From this reservoir 55 Ml/d is to pass through rapid gravity filters situated at the foot of the dam and then into a balancing reservoir beneath. From this balancing reservoir the water is to be conveyed 35 km to a service reservoir, from which a distribution system 4 km further on is fed. The ground elevations of the distribution system vary from 10 to 50 m OD, and the pipes are suitable for 100 m head.

The conditions at the headworks are therefore

minimum level in impounding reservoir		380 m OD
less loss through filters	5 m	
less loss through balancing tank	4 m	
		9 m
		371 m OD
less allowance for other possible losses		4 m
head at inlet end of trunk main		367 m OD

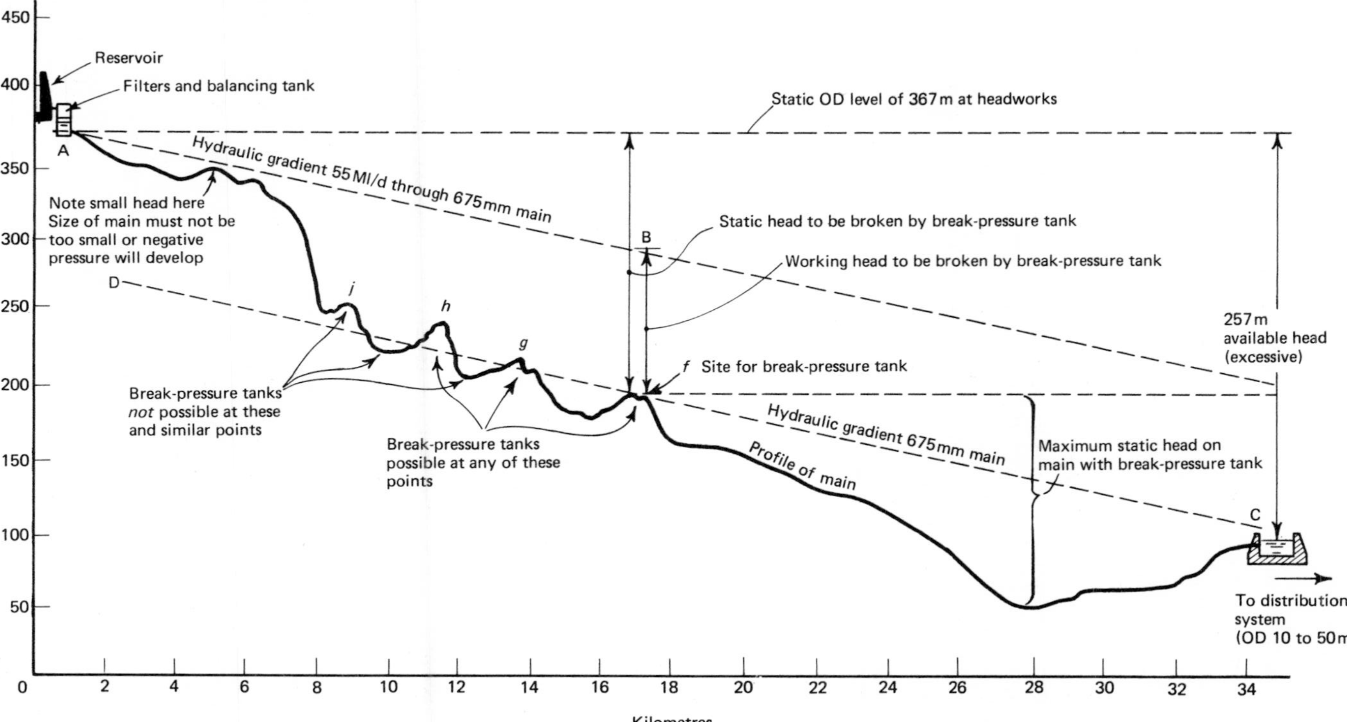

Fig. 14.3 Considerations in the design of a gravity main.

The conditions at the distribution end are

minimum elevation of distribution system	10 m OD
maximum working head for pipes	100 m
top water level of service reservoir	110 m OD

We need to check that this will give an adequate head at the high points of the distribution system. These high points have elevations of 50 m OD. Assuming 1.5 m per 1000 m friction losses in 4 km of service main from the service reservoir, the minimum pressure at the high points would be [110 – (50 + 6)] m head = 54 m head which is satisfactory. A site for a service reservoir must therefore be found at an elevation not higher than 110 m OD.

The trunk main conditions are

head at inlet	367 m OD
head at end (i.e. top water level of service reservoir)	110 m OD
allowable loss in 35 km of main	257 m head

This amount of 257 m headloss is very high. It represents a hydraulic gradient of 7.3 m per 1000 m which would imply a very high velocity through a small diameter main. High flow velocities are objectionable in water mains for three reasons: firstly, high surge pressures can possibly develop when altering the flow in the mains as, for instance, when a valve shuts off flow; secondly, they enhance any prior tendency of the water to corrode the inside of the pipe; thirdly, they indicate that a main is already at its maximum capacity. In this particular case, however, the most serious objection is that the static head on the main towards the lower end of the line will be very high indeed. Stop valves and the inlet controlling valve at the service reservoir will be very expensive if designed to work against a static head of 257 m and they may well be troublesome in operation. The pipes at low level will also be of very costly design to withstand the high pressure. The consequence of a breakage in the pipeline at the lower end under the high head might be disastrous to life and property in the area (a broken 600 mm pipe discharging under 257 m head would quickly wash away the foundations from a house). All these objections indicate that a break-pressure tank is necessary on the line.

Siting of a break-pressure tank Before we can locate the position for the break-pressure tank we must draw the hydraulic gradient for maximum flow, assuming some figure for the diameter of the pipeline that seems reasonable. A 600 mm main will dissipate all available head at 55 Ml/d. Let us therefore try a 675 mm main.

velocity of flow at 55 Ml/d in 675 mm main = 1.78 m/s

$$\text{friction loss (Hazen–Williams } C = 110\text{, worst expected)} = \frac{4.9\,\text{m}}{1000\,\text{m}}$$

$$\begin{matrix}\text{allowance for loss at inlet controller} \\ \text{to service reservoir (when wide open)}\end{matrix} = 6\,\text{m}^*$$

Therefore

$$\begin{matrix}\text{required pressure of water} \\ \text{at inlet to service reservoir}\end{matrix} = (110 + 6)\,\text{m OD} = 116\,\text{m OD}$$

Referring now to Fig. 14.3 which shows the profile of the 35 km of trunk main, we mark point C at elevation 116 m OD at the entrance to the service reservoir, and we draw back from this the line CD with a slope of 4.9 m per 1000 m which therefore represents the hydraulic gradient when the flow is 55 Ml/d. We see that this hydraulic gradient cuts the profile of the main at points *f*, *g*, *h*, *j* etc., as shown on the diagram. If we consider the matter we shall see that we can site a break-pressure tank at any of the points *f*, *g*, *h* but not at *j*. The required siting conditions for the break-pressure tank are, firstly, that it must lie on or above the hydraulic gradient CD and, secondly, that the pipeline downstream of the tank must not rise above the hydraulic gradient. (If the latter occurred the pipeline would at these points be under negative pressure at the 55 Ml/d flow.) Of the three sitings which are possible, that at *f* is the most suitable since it causes the least pressure on the pipes downstream of it: it apportions the required pressure break more equitably over the pipeline. Drawing now a gradient of 4.9 m per 1000 m from elevation 367 m OD at the inlet end of the trunk main, this will cut a vertical line through *f* at the point B, and the vertical distance B*f* will represent the pressure break that will occur at the break-pressure tank when the flow is 55 Ml/d. A typical break-pressure tank is shown in Fig. 14.4.

Having drawn such a gradient and fixed a tentative siting for the break-pressure tank, we should choose another size of main—say a 750 mm diameter main—and find a suitable site for a break-pressure tank for that size of main. It is also necessary to examine the maximum static pressures created on the pipelines, the consequent thickness of wall of pipe under these heads, and the consequent total cost of the pipeline. We shall probably find that the smaller diameter main is the cheaper, and we shall have to decide whether it is safe to adopt a high flow rate or whether we ought not to scheme out some other arrangement for breaking pressure. The most obvious matter to investigate is whether we can site the treatment works lower. The disadvantage of this is that the raw water line from the impounding reservoir to the

* This figure includes the kinetic energy loss at outlet, $v^2/2g$, which is in any case very small relative to the other losses. It is possible by using a diverging outlet to the reservoir to convert up to two-thirds of the kinetic energy into pressure energy.

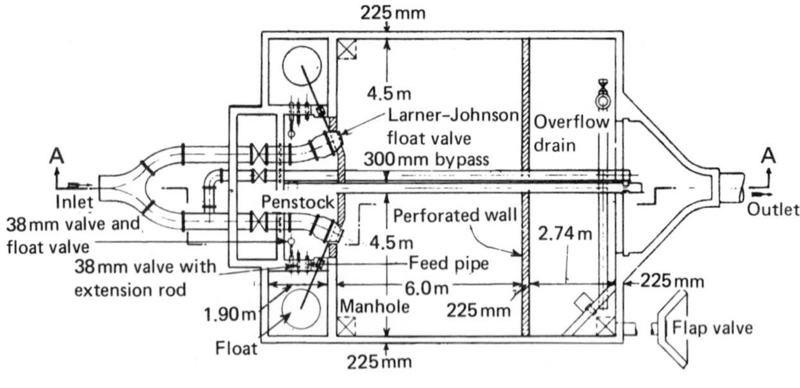

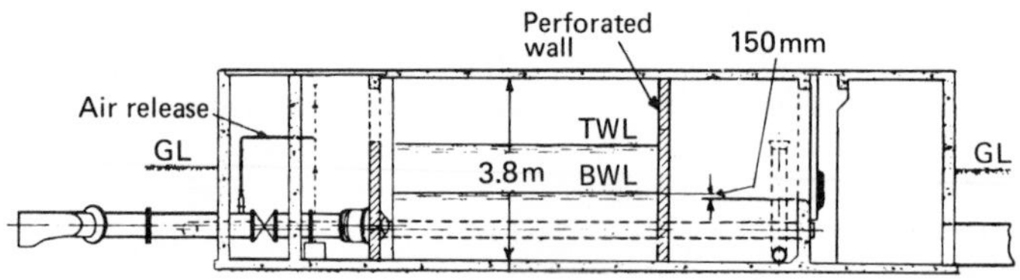

Sectional elevation A – A

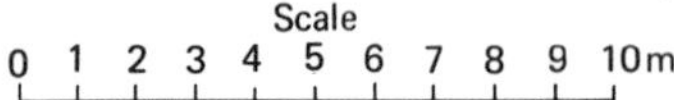

Fig. 14.4 A break-pressure tank. (*WWE.*)

treatment works must be longer, and this may be a more expensive type of pipe than the treated water line because, carrying raw water, it may have to be of slightly larger diameter because of a tendency to silt up or it may have to be concrete lined because the raw water tends to be corrosive before treatment. A point that often has a bearing on these matters is whether any provision must be made for a high level drawoff from the main.

As regards the final choice of size of main, this is one of economics and of how much provision must be made for future development. To lay a main capable of carrying 55 Ml/d at a velocity of 1.78 m/s means that the main has practically no further carrying capacity. Only another 15 Ml/d could be passed through it by using all the available 257 m head, and the resulting high static head on the pipes at lower elevation would prove more than they could stand as they would originally have been laid for a lesser pressure.

14.6 Source and pipeline layouts

One of the most desirable requirements for an economic supply system is the location of service reservoirs as near to the distribution area as possible. The reason for this lies in the fact that the service reservoir evens-out the peak demands for water, and the further the service reservoir is from the distribution area, the longer must be the lengths of main designed for peak rate of flow and, as Fig. 14.5(a) shows, the more

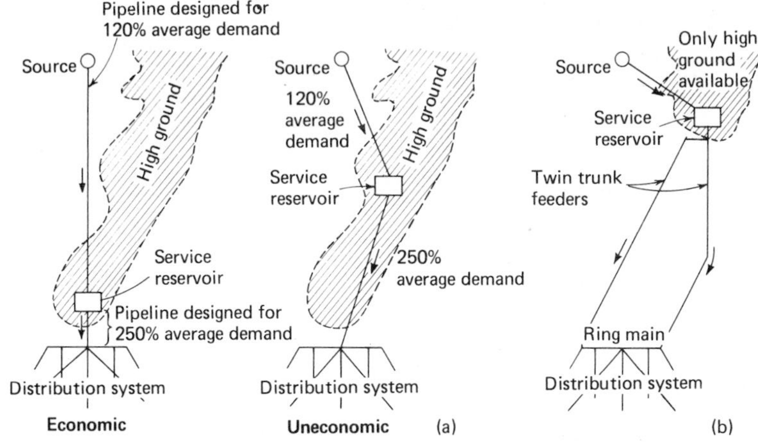

Fig. 14.5 Distribution system feeds: (a) single main feed and (b) use of ring main.

costly is the system. Further advantages are that fire-fighting protection can be better, fluctuations of pressure in the distribution system are lessened, and the development of the system is more economically achieved. Nevertheless there are occasions where the configuration of the land makes it impossible to comply with this arrangement.

If the service reservoir cannot be sited close to the distribution area then the most desirable layout in these circumstances is to have at least two major delivery mains from the reservoir, connected together at their extremities to form a ring main through the distribution area (Fig. 14.5(b)). By this means not only will the system be able to cope much more effectively with peak rates of flow but, in addition, should a repair be necessary on one main, at least some flow may be maintained to the distribution system.

A 'rise-and-fall' main can be adopted where the only possible siting of a service reservoir is further away from the distribution system than is the source of water. Figure 14.6(a) illustrates such a scheme. If the rate of output from the source is equal to the average daily supply, water flows out of the service reservoir whenever the demand rate exceeds the average and flows into the reservoir when demand is less than average.

Alternatively, if the source output is made sufficiently large to fill the reservoir during the day shifts (usually 06.00–14.00 hours; 14.00–22.00 hours), when pumping ceases the distribution area is fed only by the service reservoir. This lowers distribution pressures during the hours 22.00–06.00, thus reducing night-time losses.

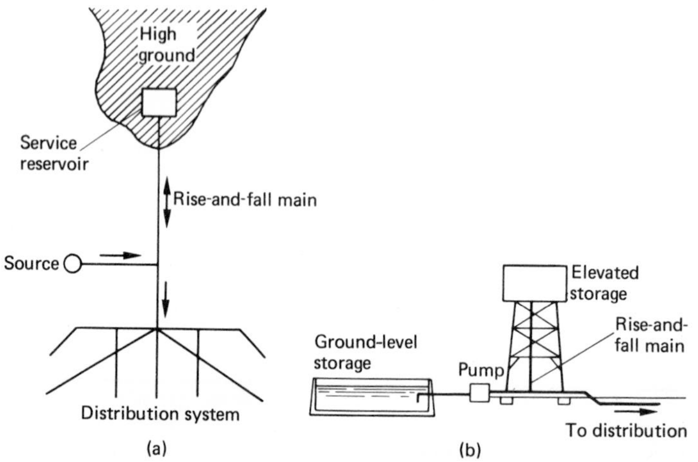

Fig. 14.6 (a) a rise-and-fall main supply. (b) use of elevated storage.

Elevated storage is often necessary on flat ground and an arrangement similar to that shown in Fig. 14.6(b) can then be adopted. As shown, the main to the elevated storage is a rise-and-fall main. If, however, the pump delivers water into the top of the tank the distribution system is at all times fed by gravity from the tank. In either case a float-operated valve can prevent the elevated tank from overflowing and water level sensors in it can control the number of pumps running, or their output, to keep the supply roughly in step with demand. The advantage of the rise-and-fall main is that, during pumping, a higher head can be maintained in the distribution system than is possible gravitationally from the elevated tank which is costly if constructed more than about 30 m high. A pressure sustaining valve could, in fact, be used to achieve the same effect without the use of an elevated tank, but even a small elevated tank is to be preferred because water level control of pumps is then possible, and this is simple and reliable in operation. Without storage, the pump output must be controlled by flow or pressure in the main feeding the distribution system, and this is more difficult to arrange and requires considerably more regular maintenance if things are not to go wrong.

Elevated storage is always expensive, hence it is seldom possible to provide the same amount of this type of storage as is usually provided when ground-level storage can be constructed. As a consequence both ground-level and elevated storage may be provided as a source in flat country, the ground-level storage providing the larger capacity. To ensure that the reserve supply in the ground-level tank can be used at all times the transfer pumps to the elevated storage usually have 50% standby capacity or more and an independent standby source of power, such as a diesel engine, may be provided as well.

14.7 Distribution system characteristics

The performance requirements for a distribution system may initially be taken as follows.

(1) *Fire demand*—the basic requirement is that any single hydrant should provide 1 m^3/min without producing negative pressures in the mains. Additionally it is desirable that a group of hydrants in an area should be capable of providing 3 to 5 m^3/min according to local fire risk (Section 15.4).
(2) *Supply pressures at peak hourly flows*—these should be not less than 20 m where dwellings do not exceed three-storey height, and 30 m where dwellings are higher.
(3) *Future demand*—the system should have reserve capacity to meet possible increases of demand in the near future.
(4) *Static pressures*—in flat areas a maximum static pressure in the range 60 to 80 m is desirable; in undulating areas a maximum static pressure of 100 to 110 m is desirable, but it may be necessary to go above this.

Higher fire flows may be necessary in high risk areas. Lower supply pressures of about 10 m head are desirable if only standpipes have to be supplied. This makes it possible to use low pressure, easily operable standpipe taps which give less maintenance problems and less wastage. However, standpipe-only systems are rare and it is not possible to supply water over large areas at very low pressure.

Occasionally static pressures have to be high because of widely fluctuating ground levels over short distances, making division of the distribution system into different pressure zones difficult. Pressure reducing valves can be adopted to reduce pressures in low lying areas, but they must be regularly maintained if they are to operate reliably and, to guard against mishaps, the low level system must be strong enough to resist the maximum possible static pressure if the valve fails.

A distribution system can be divided into three parts which have different functions:

(1) the network of small mains feeding consumers' supply pipes,
(2) the principal feeder mains, taking water from trunk mains and delivering it to focal points within the network,

(3) the trunk mains, taking water from sources or service reservoirs and supplying it to the feeder mains.

Network mains usually have diameters in the range 100 mm (4 in) to 350 mm (14 in). Above 350 mm the work of making service pipe connections to mains increases substantially, since an excavation has to be made below the pipe to accommodate the chain which holds the tapping machine on. Also a 350 mm main is of such large capacity that it will normally be used to convey water in bulk to a large area for further distribution. Notwithstanding this, in areas where there are large industrial consumers, or where the per capita consumption for domestic purposes is exceptionally high, some network mains may be 400 to 450 mm diameter. Small mains of 80 mm and below usually comprise only spur mains feeding small groups of properties so they do not form part of the network.

Principal feeder mains vary in diameter according to the size of area served: some may supply properties en route. For instance a principal main in a rural area may be as small as 150 mm diameter and this would supply properties en route, but in most urban areas 225 mm would be regarded as the smallest feeder main, and many are larger than this.

The proportions of mains laid in the different sizes in typical water undertakings are given in Table 14.1. The preponderance of mains of 80 to 150 mm size indicates that these comprise the majority of mains from which properties are served and thus form the network mains. The proportion of mains over 450 mm size usually reflects the distances between sources or reservoirs and the focal points of the area served.

Table 14.1 Percentage of mains laid in different sizes in two typical urban water undertakings in the UK.

Diameter		Undertaking A		Undertaking B	
mm	inches	%	% total	%	% total
80	(3)	(Not laid)		39.3	
100–150	(4–6)	}	86.6	43.5	
175–200	(7–8)			1.2	84.0
225–250	(9–10)	4.3		3.5	
300–350	(12–14)	3.3	7.6	4.9	8.4
375–450	(15–18)	2.8		3.8	
500–600	(20–24)	2.2	5.0	1.9	5.7
Over 600	(Over 24)	0.8	0.8	1.9	1.9
			100.0		100.0

Notes:

Some undertakings laid only 'even-inch' sizes of mains, e.g. 12″/14″/16″/18″/20″ etc. and did not lay 9″, 15″ etc.

Others laid only in the 6″/9″/12″/15″/18″/21″ sizes and did not lay 10″, 14″, 16″, or 20″.

Few undertakings laid 5″ or 7″ sizes, and none laid such sizes as 11″, 13″, 17″ etc.

14.8 Design of networks

The design of network mains is largely empirical. A main must be laid in every street along which there are properties requiring a supply. Mains most frequently used for this are 100 or 150 mm diameter, as Table 14.1 shows. Smaller mains are sometimes used, but this is not the best practice as anything less than 100 mm diameter is unlikely to have adequate capacity to meet the minimum fire demand of 1 m^3/min at any hydrant.

The diameter of street main required depends primarily upon the population density of the area served and the consequent number of persons to be served by each street main. Table 14.2 gives figures of the average number of persons served per kilometre of mains laid in typical large waterworks areas. The average number of connections per km of main is also given: this includes both domestic and trade connections. However, street mains supply only small areas in which the population density can be much higher than in large areas. Table 14.2 gives some

Table 14.2 Population served per kilometre of mains laid etc.

Large areas: UK only (50–2500 km^2)	Gross density (persons/ha)	Average no. persons per km of main	Average no. connections per km of main
Fully built-up urban areas	40–60	370–450	150–170
Predominantly residential urban areas	20–40	290–370	120–150
Well-spread urban areas with parks etc.	7–20	200–290	85–120
Mixed urban and rural areas	4–7	160–200	70–85
Well populated rural areas	1.5–4	110–160	50–70
Predominantly rural areas	0.5–1.5	65–110	25–50
Small areas (5–25 ha)	**Net density (persons/ha)**	**Persons per household**	***Persons per km of street**
Overseas			
Densely populated low income areas	800–1200	Over 6	3000–4000
Planned high occupancy dwellings	1000	5–6	3000
UK			
4-storey dwellings	500	2.75–3	2500
Residential areas with gardens	45	2.75–3	120–140
City centres (averages for 1000–1500 ha)	**Gross density (persons/ha)**	**Persons per household**	***Persons per km of street**
Densest areas	100–125	2.75–3	700–800
Next densest areas	70–100	2.75–3	600–700

Notes:
'Gross density'—on total area inclusive of open spaces etc.
'Net density'—on area reserved for housing only, including service roads.
The number of connections (including trade connections) is given for large areas only.
*The population per km of *street* is given because the population per km of main will depend upon the size of main etc.

indication of the densities that can be met in such small areas. The number of persons served by each street main will be influenced not only by the type and arrangement of the adjacent dwelling units and their occupancy rate, but also by the kind of water supply given. In densely populated low income areas of overseas cities a large proportion of the population may be served by street standpipes. In other cities similar high densities may occur due to the presence of high-rise buildings, but each such building would be fed by a single large connection.

In general, however, 100 mm diameter local street mains will supply areas having densities up to 350 persons/ha, and 150 mm mains will supply areas having up to 1000 persons/ha. This applies because in the densest areas city-block sizes are usually quite small. Table 14.3 shows the basis of the foregoing figures, the headloss in the mains being taken as 5 m/km at peak flow rate, which is a relatively modest loss rate for street mains.

Table 14.3 Feed capacity of 100 and 150 mm mains.

	100 mm diameter	150 mm diameter
Flow capacity for a main, fed from both ends, at 5 m/km loss ($C = 100$)	0.64 Ml/d	1.88 Ml/d
Peak hourly demand per person at 180 lcd × 2.5 peaking factor	450 l/d	450 l/d
Possible number of persons served	1420	4180
Average maximum size of block served per street main, 500 m × 80 m (inclusive of road area)	4 ha	4 ha
Therefore possible density of area	350 persons/ha	1040 persons/ha

To meet fire demands the whole head at a hydrant can be utilized. A single fire hydrant flow of 1 m^3/min can be relatively easily met by a 100 mm main fed from both ends. The flow rate in the main will be 1.06 m/s and whilst this will create a high rate of loss in the main (24 m/km) the length involved will usually be short. Further away the number of mains contributing to the flow will increase so that the headloss in them will be very much less.

An important characteristic of distribution networks is that because of the interconnection of mains the pressure-drop *versus* flow characteristic does not follow the single pipeline law. For a single pipeline the headloss $H \propto V^2$ (or $V^{1.85}$ in the Hazen–Williams formula) where V is the velocity of flow in the pipe. For a network $H \propto Q$ where H is the headloss to a point in the network and Q is the total flow taken by the network[2]. Hence, if the demand in a network is doubled the pressure loss is doubled, and not quadrupled as implied by the 'V^2 law' which applies to single mains only.

The computer is seldom used for the analysis of flows in small street mains, but it is useful for testing the capacity of any proposed network to meet a major fire demand (Section 14.10). Assessment of the size of street mains required is best conducted by considering the characteristics of area, block by block, and choosing the size of main needed by means of simple calculations of the kind shown in Table 14.3 above. Only a limited choice of diameters is available, and the layout must be a sensible one in which the larger mains feed the smaller ones.

14.9 Design of feeder and trunk mains

The design of feeder and trunk mains is also mainly a planning exercise for which the use of a computer is, initially, largely irrelevant or impracticable. Choices will have to be made concerning which source should supply each area, which routes are practicable for new mains and which are not, and how peak day demands are to be met when some sources have spare capacity and others have not. Oversized mains will be advisable in some areas to cater for possible additional future demand, and ring mains will be needed so that water can be transferred as necessary across the distribution system under different phases of development.

To cope with this planning it is best to start by sketching possible mains layouts on a map of the area on which have been marked the average and maximum day demands in each location. A suitable notation is thus 14.0(17.5), the lower figure being the average demand and the figure in brackets the maximum day demand.

Once realistic routes have been sketched for the mains, their sizing can proceed with reference to a summation of the flows in mains working back from the extremities of the system towards the source(s). If ring mains have been drawn in, a 'point of balance' on such mains can be found by inspection: this is the point at which the demand flow in either direction is near zero. (A closed valve can be assumed at this point.) By this means any system can be broken down into a 'tree formation' comprising main branches and sub-branches fed from some single source. If the aggregate demands have been wrongly apportioned to sources, readjustment of the layout of the feeders must occur. The maximum day demands can similarly be added back to sources and this will reveal where transfers of water may be necessary to meet peak day conditions. In this kind of work the computer is unsuitable because routes, and therefore lengths, of mains will vary from one trial to the next, and interconnections may also need to vary.

To assess quickly the required diameter of feeder and trunk mains a table can be drawn up showing the flow capacity for each diameter of main, using headlosses that are appropriate. It will be found best to draw up a table for flow capacities to meet the maximum day demand. A table of this sort is given in Table 14.4 for a maximum day of 125% the

Table 14.4 Flow capacity of mains for a maximum day demand of 125% average, and peak hourly factors as Fig. 1.4.

Pipe diameter (mm)	Velocity of flow for maximum day (125% average) (m/s)	Capacity for maximum day (Ml/d)	Headlosses (m/km)	
			Maximum day	Peak hour
100	0.60	0.41	7.7	29.8
150	0.60	0.92	4.8	17.3
200	0.60	1.63	3.4	12.3
250	0.60	2.04	2.6	8.8
300	0.65	3.97	2.1	6.4
350	0.65	5.40	1.7	5.4
400	0.70	7.60	1.7	4.8
450	0.70	9.62	1.5	4.2
500	0.75	12.7	1.5	3.9
600	0.80	19.5	1.15	3.0
700	0.85	28.3	1.08	2.8
800	0.90	39.1	1.02	2.5
900	0.90	49.5	0.89	2.1
1000	0.95	64.5	0.87	2.1
1100	1.00	82.1	0.86	2.05
1200	1.00	97.7	0.78	1.85

annual average, and for peak hourly factors in the range 2.0 to 2.5 times the average demand according to the size of population served, as suggested in Fig. 1.4 (p. 30). The figures are so chosen that, when the peak hourly demand occurs, losses through the system will be about the maximum practicable. With such a table at hand the diameter of every new main required may be directly determined by reference to the maximum day demand flow it must carry, as plotted on the layout. The capacity of any existing mains along the same routes can also easily be taken into account.

The figures in Table 14.4 can be adjusted for other peak factors. For instance, if the maximum day factor is 140% instead of the 125% taken, the velocities of flow in the table should be increased by $\sqrt{(1.4/1.25)}$, i.e. by about 6%, since the ratio of peak day : peak hour flow will decrease. Where the peak hourly flow factor is y, rather than the value 'z' taken from Fig. 1.4, then the velocities of flow in Table 14.4 should be multiplied by $\sqrt{(y/z)}$, the flow capacities being adjusted accordingly. Such adjustments will maintain nearly enough for practical purposes the same headlosses at peak hourly flows as are given in Table 14.4 and will thus ensure an economic design.

14.10 Computer analysis of network flows

A small network of mains is shown in Fig. 14.7. The 'nodes' are points of junction of mains or where a main changes diameter. The demands along each main have to be estimated and are then apportioned to the nodes at each end in a ratio which approximates to the manner in which the smaller mains (not shown on the figure) draw water. Often the demand on a main is split 50% to each end. The input to the computer comprises: the lengths, diameters, and friction coefficients of the mains,

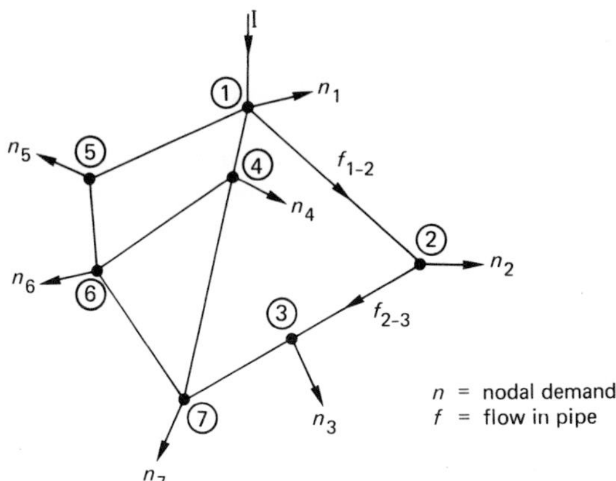

Fig. 14.7 A small mains network.

the manner in which the mains are connected, the nodal demands, the location of the input(s) to the system, the formula to be used connecting the velocity of flow V in a main to the headloss i (usually Hazen–Williams or Colebrook–White).

If there are n mains in the network the computer has n equations to solve for values of $V_1, V_2, \ldots, V_n$ subject to the following conditions.

(1) The algebraic sum of the flows entering and leaving a node must be zero.
(2) In any closed loop in the system the algebraic sum of the pressure losses must be zero.
(3) The input to the system must equal the total of the nodal demands.

The computer solves the equations by successive approximation and is usually programmed to cease calculation when the last reiteration does not change the pressure at any node by more than, say, 0.5 m. If the right program is used, no more than two or three reiterations are necessary to solve a 100 main network and the results are printed out within a few seconds of completing input of the data. Generally the cost of computer time is negligible for analysis of networks having up to 150 mains. For larger networks having up to 300 mains a main frame computer must be used.

There are a number of limitations and difficulties with respect to computer analysis of network flows which should be mentioned.

(1) The limitation with respect to the number of mains it is economic to analyse means that mains of 150 mm diameter and less are usually not included in the analysis of large systems, so their flow capacity is ignored.
(2) It is excessively time-consuming to work out the nodal demands for a large system.
(3) The nodal demands are estimates and may not represent actual demands.
(4) Losses, which commonly range from 25% to 35% of the total supply, have to be apportioned to the nodal demands in some arbitrary fashion.
(5) No diversification factor can be applied to the peak hourly demands representing reduced peaking on the larger mains since the total nodal demands must equal the input to the system.
(6) The friction coefficients have to be estimated.
(7) No account is taken of the influence of pressure at a node on the demand at that node, i.e. under high or low pressure the demand is assumed to be constant.

Variations in friction coefficients do not significantly affect flow distribution; some authorities use a uniform value of $C = 100$ (Hazen–Williams) for all pipelines when conducting a flow analysis[3].

To estimate nodal demands, plans of the smaller mains must be studied as their layout determines how the demand from an area is apportioned to nodes. Each nodal demand is based upon the estimated population (of each class) in each of the areas allocated to a node, multiplied by the per capita consumption (for each class), plus an allowance for losses, plus any specific large industrial demands in the nodal area, plus an allowance for other unidentified commercial and industrial demand. For a large city the collection of the necessary data and their translation into nodal demands may take many weeks.

The peaking factor also has to be arbitrary. Usually it is taken to be between 2.0 and 2.5 applied uniformly to all nodal demands, but in practice it may vary from 1.4 to 3.0 in different parts of the undertaking depending upon the size of the area and whether it is industrial or residential. It is possible to use different peaking factors at different nodes, provided that the sum total of the nodal demands equals the peak input to the system. However, this is not the same as allowing for a diversification factor on mains, in which the peak flow in a major main is less than the sum of the individual peak flows on the smaller mains which it supplies. If different peaking factors need to be applied, e.g. for small mains and for large mains, it is best to analyse each part separately.

Provided that the limitations are appreciated, a computer analysis is informative because of the speed with which flows in every main can be obtained. These readily show which mains are overloaded and which have capacity to spare. Proposed new mains can be included as 'dummy' mains in the initial runs, giving them a high coefficient of friction so that they virtually take no flow. To bring them into use for subsequent runs the coefficient is brought down to normal.

14.11 Other uses of the computer

In Section 14.10 the unknowns were the velocities of flow in the various mains. The values of these were computed by successive approximations to arrive at balanced flows at each node and consistent headlosses by any route through the network. The computer print-out shows the velocity of flow in each main and the resultant pressure at each node.

If field test pressures are available they can be entered as constraints into the computer program, but in consequence the computer must be permitted to adjust friction coefficients because it might not otherwise be able to find a solution. When this procedure is adopted the computer develops a solution which 'models' the behaviour of the system. It can be seen, with reference to Fig. 14.7, that when a computer is faced with this type of problem a wide range of solutions are possible, or none. For instance, if the computer first calculates the pressure difference between nodes (1) and (7) as 10 m, but is then 'told' that field tests have shown it to be a 15 m difference, alternatives exist as to which mains should have their coefficients raised by the computer. Hence, when a program of this sort is used, further instructions must be given to the computer giving the order, and possibly the maximum amount, by which mains coefficients should be altered. However, this is an arbitrary instruction which may not represent reality. The computer solution may be only one of many possible solutions. To 'prove' the model a further set of field pressures at some different rate of flow and therefore different nodal demands must be taken, and the same model must be able to simulate these pressures at the altered flow. If it does not, the model must be adjusted and retested. In practice, to model a large system may involve several months' work, including the time spent analysing demands and conducting field tests, and where field data are inadequate or of doubtful validity the procurement of a satisfactory 'model' may elude the engineer altogether. In particular, there may be conflicting data as between the first field test and the second so that 'proving' the model with any certainty is impossible.

In such circumstances a better policy is to take as many field pressures as possible during a period of average flow when conditions are steadiest. From these pressures a pressure-contour map of the system is drawn from which pressures at all nodes can be estimated. The contour map can reveal areas of exceptionally low pressure. These can arise from any of the following: valves left shut, a large leak, a washout left open, errors in the assumed sizing or interconnection of mains, and errors in estimating nodal demands. Each of these can be tried in turn on the computer to see which is most likely and field investigations can be directed accordingly. If the computer is programmed to adjust friction coefficients, a high value in any main may indicate a closed valve or an undersized main. Some authorities find it helpful to calculate nodal demands from the observed pressures to see whether or not the computed values are realistic[4].

In fact, a great deal of knowledge about a distribution system can be obtained by regular measurement and recording of pressures. This is the most easily measurable phenomenon by which a distribution system exhibits its behaviour. The measurement of flows in distribution mains is fraught with many difficulties, and rarely is it possible to measure flows on many mains simultaneously: the instantaneous demands on a network are usually unknown, the nodal demands are largely theoretical, and at peak demand times pressures may be by no means static throughout the system, but may be continuously changing as the flows themselves alter. However, given average and more steady conditions, pressures must be consistent with flows and are often repeated from one day to the next. Also coefficients of friction for mains are well known, being seldom below 100 (Hazen–Williams) for 300 mm diameter mains, and rarely above 130 for larger mains[5].

References

(1) See, for example, *An Introduction to Engineering Economics*, Chapter 5, ICE, 1969.
(2) Shamir U and Howard C D D, Engineering Analysis of Water Distribution and Systems, *JAWWA*, September 1977, p. 510.
(3) Becher A E, Bizjack G J and Schulz J W, Computer Techniques for Water Distribution Analysis, *JAWWA*, July 1972, p. 410.
(4) Lawson W R, Some Practical Aspects of Water Distribution Network Analysis, *Proc. IPHE/IWES Meeting*, January 1970. See Discussion by Harrison J D.
(5) Reference 4, Discussion by Coe A L.

15

Distribution practice

15.1 Distribution organization

The normal functions of a distribution department are as follows:

(1) maintenance of supplies throughout the system,
(2) extension of mains and laying of new services,
(3) repair and maintenance of mains, booster stations, and service reservoirs,
(4) metering of supplies,
(5) leak detection and repair and reduction of waste,
(6) inspection of plumbing systems and enforcement of water byelaws,
(7) training.

In large undertakings serving several million people or covering very large areas a three-tier organization would be adopted comprising: (1) the Head Office (Departmental) Organization, (2) Divisions, (3) Districts. The size of the Divisions and Districts will depend upon the characteristics of the water authority area, Divisional size often being determined by the existence of geographical barriers which naturally divide the area, or by the magnitude of population served or area covered*. District sizes will vary also, tending to be larger when there is no Divisional organization; generally speaking they will cover from 50 000 to 100 000 population in urban areas or perhaps one-tenth of that in rural areas. Many exceptions to these sizes will be found.

Figure 15.1 shows the type of Head Office organization adopted for a Distribution Department and Fig. 15.2 shows a large Divisional organization and a small District organization. Where Districts are large they would be strengthened by the addition of a local sub-depot and stores so that they are competent to carry out all service laying and small mains repairs in their area.

Most large undertakings utilize several sources of water, and the output of these to the various service reservoirs, and thence into the distribution systems, will need to come under continuous control. Often sources will be remote from distribution districts, and each source will have its own characteristics, some being expensive to run and others cheaper, some having strictly limited outputs and others having capacity

* For instance the Istanbul Water Authority is divided by the Bosphorous into Asian and European sides, and is further divided on the European side by the Golden Horn.

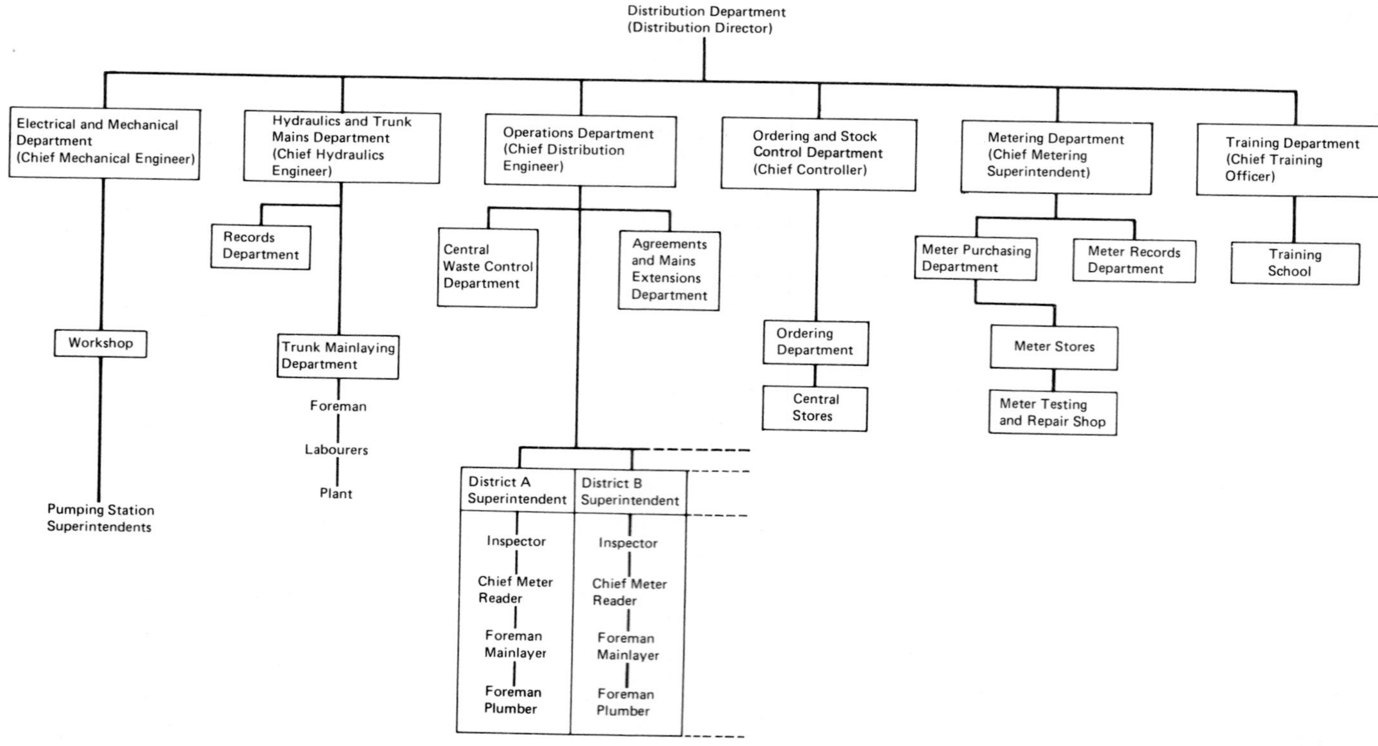

Fig. 15.1 The distribution department for a large waterworks undertaking.

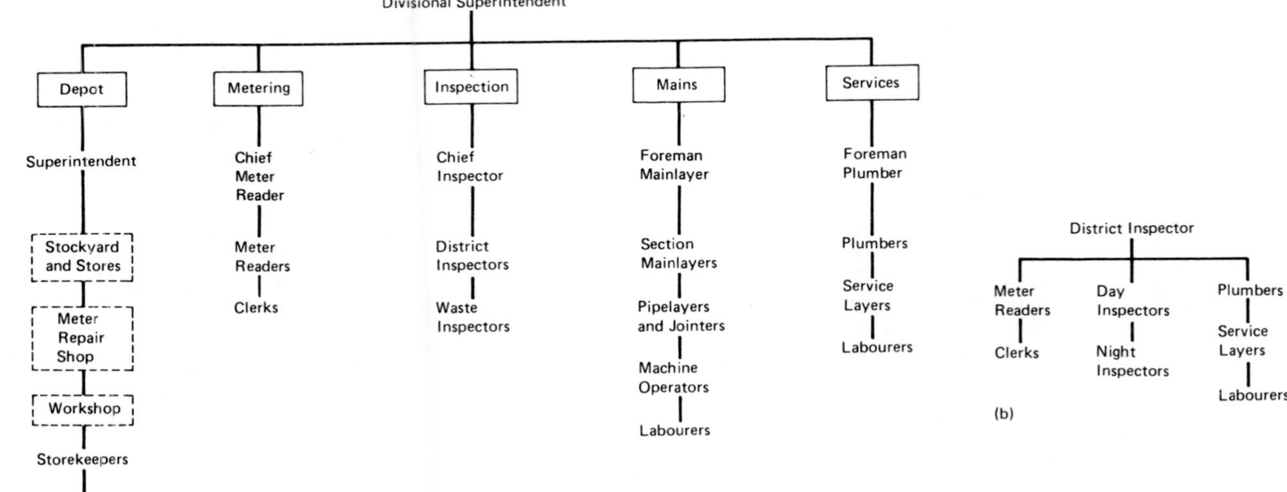

Fig. 15.2 (a) divisional organization for a large distribution area. (b) district organization for a small distribution area.

to spare. On the distribution side water demands will fluctuate both regularly, according to the time of day and the day of the week, and irregularly, as burst mains occur, water is required for fire-fighting, and large industrial consumers suddenly change their demand. These varying demands and the source outputs have to be coordinated. In large undertakings this is normally done through a Central Operations Room at Headquarters into which all messages flow. Source outputs can be directed from this centre, and standby mains repair gangs and plumbers can be called out to deal with distribution problems.

In the smaller water undertakings with no central operations room communication between demand and supply sections is by telephone from the Distribution Superintendent to the appropriate sourceworks, either routinely reporting rates of demand at certain fixed hours of the day (usually at 09.00 hours to report the previous day's demand and at 16.00 hours to report the current day's likely demand)*, or on occasion notifying special needs calling for amendment of output from sources.

15.2 Operation of districts

The key operative in the distribution system is the Waterworks Inspector (or 'District Inspector' or 'Turnkey'). It is he who liaises with the public and investigates their complaints, watches that supplies are maintained, inspects properties for waste, and keeps the area under supervision for signs of leakage, taking part in waste detection measures. It is his intimate knowledge of the local mains system and its consumers that is invaluable in diagnosing the cause of troubles and directing where remedial measures are necessary. When burst mains occur or a big fire demand for water arises it is he who initiates the necessary valve operations. His principal task is to maintain supplies in his area with the minimum of waste.

In undertakings where supplies must be intermittent through lack of source capacity valves will have to be operated daily to apportion water in rotation to different areas. Even when a 24-hour supply is possible a consistent pattern of flows must usually be maintained through the distribution network. Reversal of flows in mains must not be haphazard or 'red' water is produced, and consumer complaints arise as the unaccustomed flow direction causes mains deposits to be stirred up and taken into suspension in the water. As these are mainly iron corrosion products from cast iron mains the water is usually coloured dull red. Regular daily reversal of flow is acceptable once the deposits initially raised have been cleared by flushing out the main. 'Boundary valves' on mains which connect one Distribution District to another are often kept

* The Superintendent will discuss water levels in service reservoirs and, with the Divisional Engineer's help as necessary, will decide the appropriate source output for the next period, having regard to the expected demand characteristics of the day and the need to cater for mains shutdowns for repairs.

closed in large urban areas if there is little normal flow through them and the Districts are fed from different sources. This gives better monitoring of consumption, and such valves can always be opened to permit one District to feed another when required. Division of the distribution system into separate supply Districts forms the basis for the assessment of consumption and losses by the Districts, and this is the first step towards efficient control of wastage (see Section 15.16).

15.3 Mains extensions

The design of mains extensions and trunk mains is normally undertaken by the engineering department at Head Office or Divisional Office. When the extension is authorized the necessary materials are drawn from the stores and the main is laid by a foreman mainlayer with a jointer and labourers to assist. Generally a gang of six to seven men is required to lay pipes up to 300 mm diameter, but for the laying of large trunk mains of 750 mm diameter and over through city streets or open country, a larger gang of 20 to 24 men is required. Progress is frequently measured by the number of pipes laid per gang per day. It should average six pipes per gang per day in the smaller diameters (100 to 200 mm), and five pipes per gang per day in the larger diameters. Minimum progress when difficulties are encountered may be only one-third these figures, and best progress perhaps 50% more.

The extension of distribution mains to serve new consumers may be by 'mains guarantee' arrangement in England and Wales. Under this system the consumers who wish to take the public supply from the proposed extension are asked to guarantee to pay 12.5% of the cost of the main annually, if the normal water rate they would pay annually is less than this 12.5%. When additional consumers connect to the extension the water rates they pay are credited against the 12.5%, and when the total water rates payable by all consumers on the extension equal the 12.5% the guarantee agreement finishes. Often the local authority within whose area the extension is laid acts as the guarantor, and grants were also made in the UK towards the cost of the extension by the Government under successive Rural Water Supplies Acts. By these means a high percentage of rural inhabitants were provided with public water supply in England and Wales during the 1950s and 1960s.

15.4 Fire-fighting requirements

Fire hydrants must be fitted on all distribution mains in Britain as required by the Fire Authority. The cost of the hydrant installation is borne by the Fire Authority, but the actual installation is undertaken by the water undertaking. In Britain standard fire hydrants are placed below ground in a chamber, but in the warmer parts of the USA and some other countries pillar hydrants are used. These are not suitable for cold climates, are more expensive than underground hydrants, and are more easily damaged by traffic.

There are no nationally agreed standards for the discharge capacity of fire hydrants, or for their spacing, but the figures given in Table 15.1 show a general consistency.

From Table 15.1 it will be seen that the minimum requirement is 1 m^3/min from any hydrant. Several authorities report that this flow will be sufficient to put out the great majority of fires. When larger fires occur, suction hose lines will be put out to additional hydrants further afield, thus spreading the water demand over more mains. In the USA the National Board of Fire Underwriters stipulates minimum fire flow capacity of mains according to size of area and nature of property. These requirements often exceed the peak domestic rate of demand and therefore dominate the design of distribution mains, but some engineers question whether such high rates are necessary.

Fire appliances in common use have built-in pumps of capacities 2.3 and 4.5 m^3/min. Nozzle sizes are commonly 13 and 19 mm, using flows

Table 15.1 Fire hydrant and fire flow requirements[1].

	England and Wales	USA	Continent
Hydrant characteristics	2 m^3/min at 1.7 bar	2 m^3/min	1.0 to 1.5 m^3/min minimum
Initial flow in close proximity to fire	1.4 m^3/min upwards	Not stated	Generally 1.0 m^3/min
Subsequent rates of flow (as required) for a single fire	9 m^3/min or more, up to 25 m^3/min for major fires	4 m^3/min for 1000 population, rising to 45 m^3/min for 200 000 population[2]	3.6 to 6.0 m^3/min in high risk areas, or more
Minimum residual pressure in mains	Preferably 0.7 bar, but not less than zero	Not stated	Not stated
Possible quantity of water used in a fire	Maximum flow times 'several hours', say 6 hours	Maximum flow for 4 to 10 hours	Maximum not stated; average '75% incidents last less than 2 hours' (France)
Spacing of hydrants	100 to 150 m generally, but 30 m in high risk areas	100 to 150 m	80 to 100 m in urban areas; 120 to 140 m in residential areas; wider in rural areas

Notes:

(1) Information for USA and continental requirements as reported by J Bernis in Water Requirements for Firefighting, Special Subject 7, *Proc. IWSA Congress*, 1976, and for UK requirements from a private communication from the Home Office.

(2) These flows are National Board of Fire Underwriters requirements; actual fire requirements according to records are 6.3 to 12.5 m^3/min in built-up areas and up to 25 m^3/min in high risk areas.

of 0.16 and 0.45 m^3/min respectively. Larger nozzles of 25 mm diameter requiring 1.1 m^3/min are occasionally required, but this is the largest practicable size for hand-held branches. Fixed but portable monitors may use up to 2.7 m^3/min or more.

15.5 Service pipes

A service pipe connection from a main to a property is usually laid as shown in Fig. 15.3. The ferrule is normally inserted into the main by

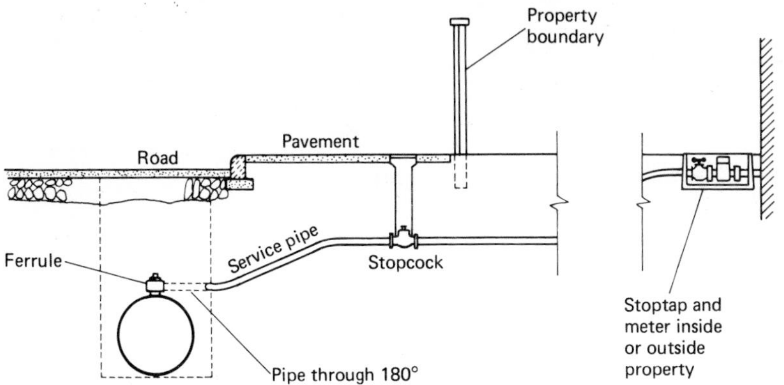

Fig. 15.3 Typical service pipe connection.

means of an 'under-pressure' tapping machine as shown in Fig. 15.4. This bores a hole into the main and taps a screw-thread in the hole, and then by rotating the head of the machine the ferrule is brought into position over the hole and screwed into it. The ferrule has a plug in it which can be screwed down, so cutting off the supply.

Service pipe connections for houses are usually to a standard size according to the practice of the water authority, 13 mm being the normal size for a one-family house or flat which is the usual distance (a little under 30 m) from the distribution main. The pressure in the main would usually be about 30 m. Materials used for service pipes are as follows:

(1) lead and lead alloys,
(2) copper,
(3) steel,
(4) uPVC,
(5) polyethylene.

Lead and lead alloy pipes are now no longer used, but many older pipes in the UK are of lead and still continue in service. Provided that they convey water which is not plumbosolvent, i.e. the water conveyed does not take lead into solution, such pipes can continue in use.

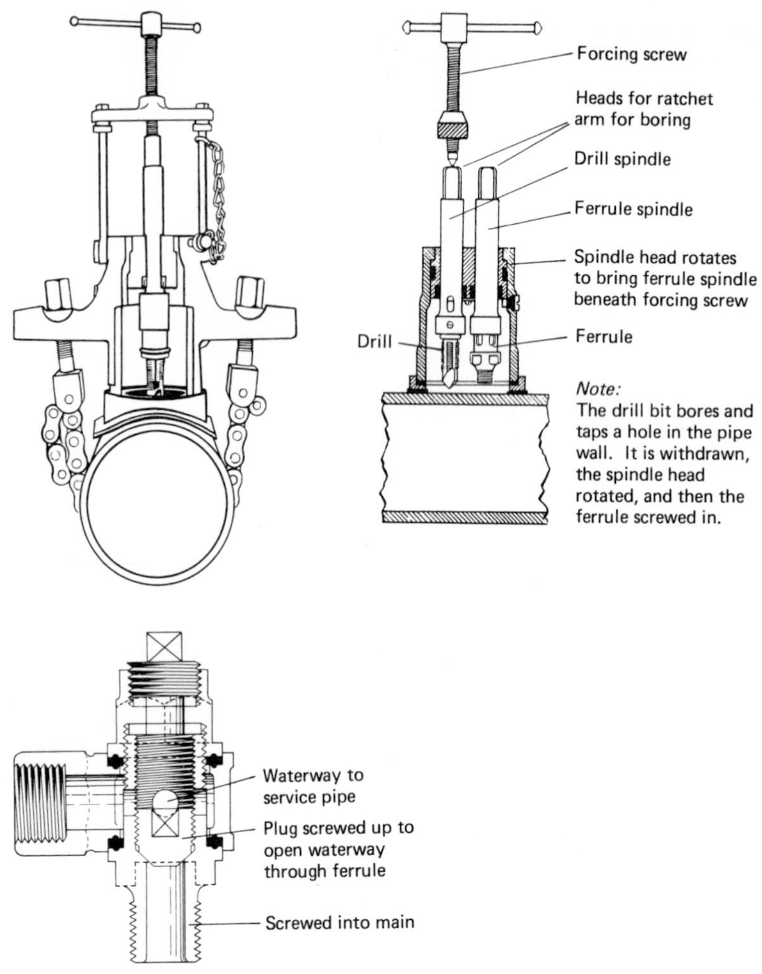

Fig. 15.4 Under-pressure tapping in a service main showing the type of ferrule inserted. (F W Talbot & Co. Ltd.)

15.6 Copper pipes

Copper pipes are expensive, but they are strong, durable, resistant to corrosion, easily jointed, and capable of withstanding high internal pressures. The pipes can be obtained in straight lengths of semi-rigid pipe 4 to 6 m long, or in coils of longer length of fully annealed copper. BS 2871 Part 1 Table X designates copper tubing for above ground work and Table Y designates tubing for laying underground. The standard diameters and wall thicknesses are as given in Table 15.2. Fittings to suit

Table 15.2 Copper tubing to BS 2871.

Minimum outside diameter to nearest mm	Wall thicknesses	
	Table X for above ground work (mm)	Table Y for laying underground (mm)
12	0.6	0.8
15	0.7	1.0
18	0.8	1.0
22	0.9	1.2
28	0.9	1.2
35	1.2	1.5
42	1.2	1.5

Note: Tolerance on wall thickness ± 10%.

the foregoing are specified in BS 864. Maximum working pressures decrease with increasing diameter, but all sizes are amply strong enough to meet the highest distribution pressures likely to occur in practice. Joints (see Fig. 15.5) are either compression joints or capillary joints.

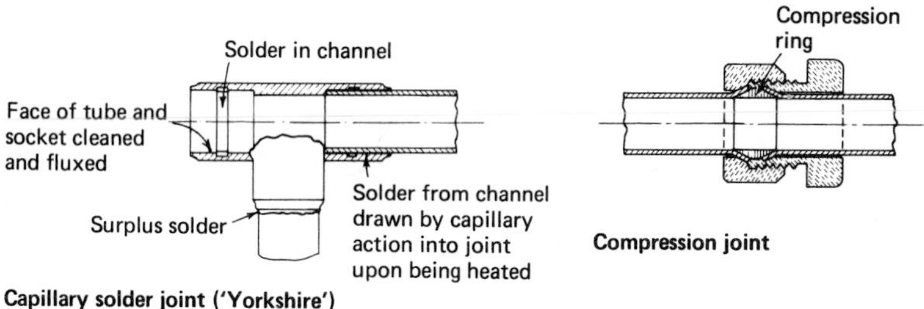

Fig. 15.5 Joints for copper tubing.

For compression joints the two ends of the pipe are belled out and forced on to a cone-shaped copper ring by screwed collars. The capillary joints are close fitting sleeves of copper within which two rings of solder have been cast. These sleeves are slipped over the ends of the pipes to be jointed and are warmed with a blow lamp. This causes the solder to melt and run out to fill the annular space between the pipe and sleeve by capillary action. These joints are very neat and can be quickly made. The pipe ends must be cleaned and fluxed before insertion into the sleeve. Fittings of all kinds are made.

15.7 Steel pipes

Steel pipes are widely used because they are one of the cheapest forms of service pipe and can sustain high pressures. They are supplied in straight lengths about 9 to 10 m long and can be bent to curves on a portable bending machine. They may be supplied 'black' (i.e. untreated), galvanized, or bitumen-coated inside and out, or additionally sheathed on the exterior with glass fibre cloth and a further coating of bituminous compound. Steel pipes have screwed ends to BS 21 taper thread and are connected by steel couplings. A wide variety of specials are made, including flanges which are screwed on to the pipe ends. Most steel service pipes laid by water undertakings are galvanized, and where they are to be laid in corrosive soils they are additionally sheathed with bitumen and glass fibre cloth wrapping. Pipes must be to BS 1387 which stipulates three classes of steel pipe, of which only Heavy Grade should be laid underground. Diameters and thicknesses of heavy grade galvanized steel tube (colour banded red for identification) to BS 1387 (metric) are as set out in Table 15.3. Permissible working pressures are ample for the highest waterworks distribution pressures likely to be met in practice.

15.8 uPVC pipes

uPVC or unplasticized polyvinyl chloride piping is described in Section 13.13 above. It is used for cold water service piping in temperate climates; it is not wholly suitable for hot climates because of the reduction of maximum working pressures at temperatures above 20 °C which is a 2% reduction in working pressure per 1 °C above 20 °C. The pipes should be stored under cover to protect them from the ultraviolet rays of sunlight. The pipes are corrosion-resistant, light to handle, and easy to joint using either solvent cement or compression joints. Fittings of all types may be obtained in uPVC. Diameters and bores of the smaller sizes of tubing are given in Table 15.4. Further details are given in Section 13.13 and it should be noted that some problems accompany the use of uPVC so that there is a current tendency to use modified or medium density polyethylene tubing instead.

15.9 Polyethylene pipes

Polyethylene piping is described in Section 13.15 and pressure ratings and sizes for the larger diameter pipes are given in Table 13.4. Bore sizes of the smaller diameter low density and high density polyethylene tubing are given in Table 15.5. The reduction of working pressure with rise of temperature above 20 °C given in Table 13.4 should be noted. The pipes are not suitable for hot water, nor for internal plumbing as they need continuous support. Low density polyethylene tubing is frequently used for long service pipes, such as to farms, in temperate climates, and it can be mole ploughed in lengths into soft ground. Unlike uPVC piping it can be laid at temperatures below freezing if necessary. Modified or medium density polyethylene tubing is now coming into greater use and is dealt with in Section 13.15.

Table 15.3 Heavy grade steel tube to BS 1387—suitable for underground work.

Nominal diameter (mm)	Outside diameter (mm)	Minimum wall thickness (mm)	Mean bore (mm)
15	21.1–21.7	3.25	14.9
20	26.6–27.2	3.25	20.4
25	33.4–34.2	4.05	25.7
32	42.1–42.9	4.05	34.4
40	48.0–48.8	4.05	40.3
50	59.8–60.8	4.50	51.3

Table 15.4 Unplasticized (rigid) PVC pipes to BS 3505.

		Bore to nearest mm	
Nominal diameter (mm)	Minimum OD to nearest mm	Class D 120 m working head	Class E 150 m working head
15	21	—	17
20	27	—	22
25	33	—	28
30	42	37	36
40	48	42	41

Notes:
Bores are based upon the minimum outside diameter and the maximum average wall thickness measured at six points on a circumference.
Tolerances: outside diameter +0.3 mm; wall thickness −0.6 to −0.4 mm according to diameter.

Table 15.5 Polyethylene tubing to BS 1972 and 3284.

			Bore to nearest mm					
			Class B 60 m working head		Class C 90 m working head		Class D 120 m working head	
Nominal bore mm	Nominal bore in	Outside diameter (mm)	Low density	High density	Low density	High density	Low density	High density
13	(½)	21	—	—	16	17	14	16
19	(¾)	27	22	—	19	22	18	20
25	(1)	33	27	29	25	27	—	26
38	(1½)	48	39	42	35	39	—	37
50	(2)	60	49	53	44	49	—	46

Notes:
Bores are based upon the minimum OD less the mean of the minimum and maximum wall thicknesses.
Tolerances are +0.3 mm on wall thicknesses up to 25 mm diameter and thereafter about +0.5 mm.

15.10 Domestic flow requirements

The maximum demand rate from a house will depend upon the amount of storage, if any, provided on the premises. In most UK plumbing systems the only storage provided is that on the hot water system and in WC flushing cisterns. In such cases all cold water supplies to taps, showers, WC ball valves, and the cold water feed tank to the hot water system will be fed directly from the mains. In other cases only the cold water taps at the kitchen and at washbasins will be direct off the mains, the WC cisterns and showers being fed via the cold water storage tank. The maximum rate of flow required to a property therefore depends upon the type of plumbing arrangement adopted, but it can be estimated using the figures given in Table 15.6. However, a householder does not expect to get maximum flow to all his water consuming facilities simultaneously: he knows that full opening of the kitchen tap will tend to reduce the flow to washbasin and bath taps on the floor above. If the pressure in the mains is high the effect may be scarcely noticeable; it is only when pressures are below 30 m that an observable effect will be caused. The last two columns of Table 15.6 show the likely peak demand for a typical one-family household in the UK.

Where 'low-rise' flats occur the same sort of plumbing systems will be adopted, but for 'high-rise' flats the cold water supply from the mains will be boosted by a small pump to a roof tank which then feeds all the supplies to the flats below. The peak demand from the group of flats will therefore depend upon the amount of storage provided and the pump characteristics.

Table 15.6 Typical one-family in-house domestic flow requirements (in litres per minute).

					Likely maximum drawoff rate	
	Fast flow	Good flow	Reason-able flow	Just ade-quate	High pressure	Low pressure
Kitchen tap	20	15	12	7	15	12
Bath tap (cold)	20	15	12	10	—	—
WC cistern	—	7	6	4	7	4
Washbasin (cold)	15	10	7	6	7	5
Cold water storage tank	—	10	7	6	10	5
Totals					39	26

Notes:
Bath tap assumed to be on cold water storage.
WC cistern assumed to be direct on mains supply.
WC cistern and cold water storage will be ball valve controlled and hence will have an upper flow limit according to size of valve.

15.11 Design of service pipes

For the design of a service pipe the peak demand rate must be estimated and the minimum mains pressure must be known. The flow and headloss through the piping can be derived from the Hazen–Williams formula which, expressed in convenient units for small pipes, is

$$\text{flow (litres per minute)} = 0.0918CD_{\text{cm}}^{2.63}i^{0.54}$$

where C is the Hazen–Williams coefficient, D is the actual internal bore of pipe in cm, and i is the hydraulic gradient (i.e. the headloss per unit length).

Figure 15.6 is based upon the above formula taking $C = 110$. Actual bore diameters can be obtained from Sections 15.6 to 15.9 above.

The Hazen–Williams formula is used in preference to any other because of the facility it gives for adjustment of the C value as may be thought necessary for any particular pipe. Blair's formula[1], although well checked against test results, is not so practical. Blair quotes a different formula for each type of pipe but, unfortunately, only his formula relating to new galvanized steel tubing is appropriate; he does not quote a formula specifically for copper and uPVC pipe, only a formula for 'smooth' pipe which would not be appropriate for design purposes. It must be remembered that the actual headloss through a service pipe may depart substantially from the theoretical loss due to age of pipe, state of joints, quality of water conveyed etc. A Hazen–Williams C value of 110 will suffice for new galvanized iron piping (it gives practically the same values as the Blair formula), except that for flows over 1.0 m/s in 15 mm pipe and over 1.5 m/s in 25 mm pipe the C value should be reduced to 105. For old piping no particular value of C can be quoted and the engineer must use judgement having regard to the probable state of the pipe when fixing an appropriate C value. It should be noted, however, that instead of using C values below about 90 it is preferable to assume some reduction in the internal diameter of the pipe by an amount equal to its estimated roughness (see Section 10.6). For uPVC or copper tubing it is suggested a design value of $C = 120$ should be taken. Whilst theoretically a higher value may apply, it is prudent not to overestimate the capacity of a pipe when designing it. For values of C other than $C = 110$ the flow read from Fig. 15.6 for a given friction loss should be multiplied by $C'/110$ where C' is the revised value. To obtain the headloss with revised C' the flow should first be adjusted by multiplying it by $110/C'$ before using Fig. 15.6 to read the headloss.

It must be appreciated that no formula gives consistently identical results to those obtained even in laboratory tests and, in practical terms, discrepancies will be not less than ±5% and frequently ±10%. Hence for design purposes flows should be taken as something less than the strict theoretical value and headlosses as something more.

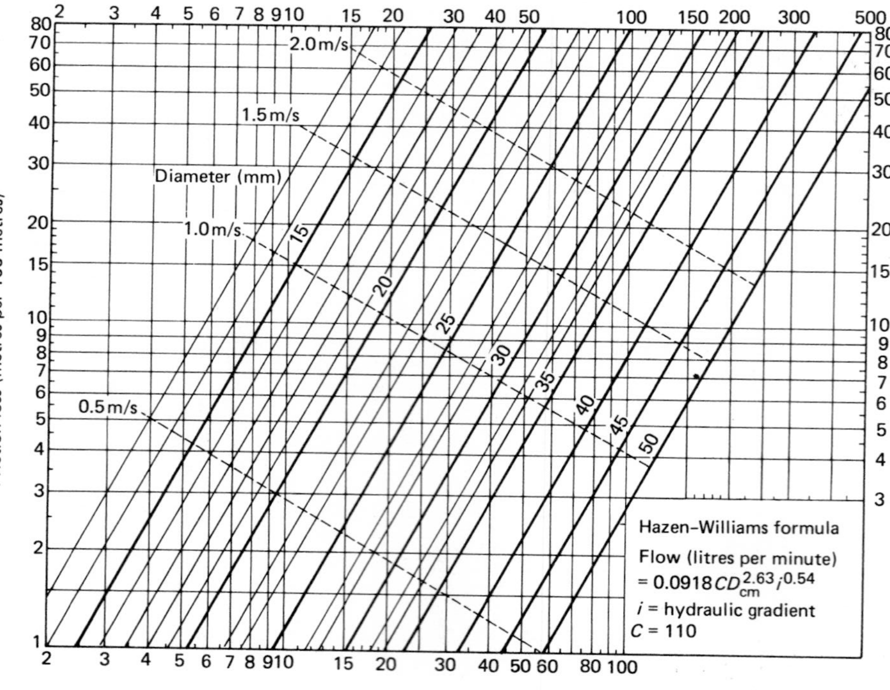

Fig. 15.6 Flow versus friction loss in service pipes ($C = 110$).

15.12 Pressure loss in fittings

Pressure losses in the tapping to the main, in stopcocks, and at ball valves and bibtaps may represent the major pressure losses on the delivery line. Losses on fittings which have been in use for a number of years may be very high. Connections to the distribution main are usually made under pressure, the size of the tapping being 6 mm less in diameter than the service pipe diameter, except in the case of a 13 mm service pipe diameter (the minimum size) which has a tapping of the same diameter as itself. There are usually two stopcocks on the service pipeline: one at the boundary to the consumer's property, which may be operated only by the water authority, and one just inside the consumer's property for his own operation. Pressure losses through these and other fittings may be taken as shown in Tables 15.7 and 15.8, although it must be borne in mind that there is considerable variation of pressure loss according to the design of the fitting and its age. Where losses through

Table 15.7 Pressure losses through service pipe fittings.

Fitting	Value of k in $kv^2/2g$
Elbows	1.0
Short radius bends	0.75
Long radius bends	0.7
Fully open gate valve	0.15
Fully open screwdown valve (Fig. 13.6)	8.2
Tee-flow to branch (same size)	1.3
Sudden contraction 2:1	0.3
Sudden contraction 4:1	0.4
Sudden enlargement 2:1	0.4
Sudden enlargement 4:1	0.7
Joints in 13 mm diameter pipe	1.2
Joints in 25 mm diameter pipe	0.8

Note: The figures for joint losses are for service piping as laid without special attention to jointing.

Table 15.8 Headlosses (in m) at various flows.

	Flow (in litres per minute)		
	7.5	15	25
13 mm ferrule	0.3	0.7	2.4
19 mm ferrule	0.15	0.4	1.1
13 mm bibtap open	0.7	1.7	4.9
19 mm bibtap open	0.3	0.7	2.4
13 mm ball valve open (6 mm orifice)	1.7	4.7	—
19 mm ball valve open (9 mm orifice)	0.7	1.8	4.0

Note: Losses for ferrules, bibtaps, and ball valves are for average conditions. Losses greatly in excess of these can be experienced with old fittings.

fittings are not individually estimated the total pipeline loss should be increased by 5%.

Losses through meters will vary according to type and age of meter installed. For semi-positive meters (see Section 15.18) the loss through a new meter will be about 1.0 m at normal maximum recommended continuous rating, which is 45–50 l/min for 13 and 20 mm sizes and about 75 l/min for 25 mm meters. Inferential meters, which would be used for larger flows, cause less headloss.

15.13 Leakage rates on mains and services

Repair and maintenance of mains and services is a continuous necessity in all water undertakings. The likely incidence of defects is given in Table 15.9 based upon figures reported previously from annual reports of UK undertakings 1970–71, as supported by later international data reported by Reed[2], Lackington[3] and Clark[4]. Mains fractures found and repaired by a water authority do not necessarily represent all fractures occurring since a low level of reported fractures may represent less resources applied to tracing them. Table 15.9 figures allow for this. Newport[5], in an analysis of data from the Severn–Trent area, shows that small diameter mains suffer a higher fracturing rate than do the large mains, and that an unduly high rate of fracturing was experienced in the

Table 15.9 Incidence of defects on mains and services.

Type of defect	Usual average number per annum
Mains	
Trunk main fracture rate	
mains in good condition	1.5–2 per 100 km
below average condition	6–7 per 100 km
Distribution main fracture rate	
lowest rate	10–15 per 100 km
average rate	20–35 per 100 km
above average rate	40–55 per 100 km
Valve, hydrant, and joint defects	1–1.5 times number of distribution mains fractures
Services	
Leakages and other defects on service pipes and stoptaps	1–2 per 100 services
Defects on consumers' premises causing wastage of water	About 2 per 100 services

Midlands with spun (grey) cast iron pipes laid since 1940. Ground movement is considered to be the principal cause of mains fractures: hence the number of bursts experienced during severely cold periods may be ten times the average. Prolonged hot dry weather may also cause an increase of fractures.

It is important to keep a record of all mains bursts because, after a period, analysis of the data may reveal which mains are particularly prone to fracturing. The causes may be various, such as fracturing due to mining settlement, poor original laying, age of pipes, faulty original manufacture, inadequate cover for increased traffic or protection against frost, or original class strength too low for current pressures. Where evidence accumulates that a main is frequently fracturing, it will eventually need to be relaid. Since many undertakings have systems initiated a century ago, many are continuously involved in renovating or relaying mains.

Defects giving rise to leaks from joints, hydrants, valves, service pipes and their connections to mains, and from consumers' plumbing systems, are much more numerous than burst mains. They will total several thousand per annum for even a moderately large undertaking: hence a continuous routine inspection of all service pipes and properties for signs of leakage is necessary. Some authorities manage to inspect 100% of all services annually; others manage only 33% or thereabouts per annum.

15.14 Cleaning and relining of mains

It is not normally possible to clean mains by flushing them. A flow of 1.5 to 2.0 m/s is required to remove silt from 80 or 100 mm diameter mains, and higher scouring velocities are required for larger mains. Such flows cannot normally be achieved by opening fire hydrants or washouts, and this type of 'flushing' is principally adopted for removing dirty water after a mains repair.

Foam swabs can be used for removing soft or loose material, such as organic debris, iron and manganese deposits, sand, and stones. The swabs are of polyurethene foam. It is usual to start with the softest grade first: if the main is rough this swab will tear and be ejected in pieces. The next grade is used and so on until the swab comes out whole, which indicates the hardest grade suitable for that main. The swab should have a diameter of 25 to 75 mm larger than the bore of the pipe and a length of 1.5 to 2.0 times the bore. Pipes up to a large size can be cleaned in this manner. For 80 and 100 mm pipes the swab may be inserted into the main via a hydrant branch, but in other cases the main must be opened. Speed of travel of the swab can be controlled by controlling the outflow at the discharge end, the speed of the swab being usually one-half to three-quarters of the water velocity because of flow past the swab. An electrical transmitter incorporated in the swab will assist in locating it. Swabbing can greatly improve the flow characteristics of a slimed main. It is also useful to swab a newly laid main before putting it into service.

For the removal of scale a scraper must be used. There are various devices, from the traditional disc with trailing metal scrapers to a straw rope ball tightly wound with barbed wire. Several passes of such scrapers may be necessary and it is essential to have a good flow of water past the device to prevent debris accumulating in front of it. Quite long lengths of main—up to several kilometres—can be scraped in 'one go' if the run is fairly straight and no obstacles to the passage of the device exist. The speed of the scraper can be about a fast walking speed and, again, a sound-making device or electrical transmitter should be incorporated in it so that, if it gets stuck anywhere, it can be located*. If the main is a distribution main the service pipe connections must be shut off and, after the passage of the scraper, must be blown back with freshwater.

Unless the interior of a main is still in a protected condition after scraping, it should be relined to avoid exposure of bare metal. Mortar lining in situ is normal practice in this case, although the lengths which can be so treated at a time are considerably shorter than the lengths which can be scraped at one time. It is also necessary to get a cable through the pipeline. This cable pulls a machine through the pipe, the machine being fed continuously with mortar of the right consistency which is thrown out on to the wall of the pipe by rotating ejector nozzles. A circular 'squeegee' behind the nozzles smoothes the mortar to an even thickness upon the pipe. Despite the slight reduction of pipe bore that results the flow capacity of the main can be greatly increased, the amount depending upon the initial condition of the main; the pipe is also considerably strengthened. Even quite small pipes can be treated in this manner. A lining thickness of 3 to 6 mm is usual. Relining with bitumen is an earlier technique sometimes still used for small diameter mains, but it does not have a long life. Epoxy resins have been used for relining mains: the thickness should be a minimum of 1.0 mm as it has been found that if it falls below 0.55 mm significant pinholing can occur. Care has to be taken in the application to avoid ridging. Also it is preferable to use resins which are free of benzene alcohol in order to reduce the amount of microbiological aftergrowth commonly experienced[6].

The relining of mains often substantially reduces losses from the local distribution system because of the preparatory work involved in ensuring that all service pipe connections can be shut off.

15.15 Waste detection

Methods of leakage and waste detection will be described first since this leads to a consideration of measures for reducing wastage as a whole. Four methods are available for locating leaks from mains and services:

(1) visual observation,
(2) 'sounding',
(3) waste metering,
(4) cutting off mains.

* One engineer reported that a dog as his 'assistant' was more effective.

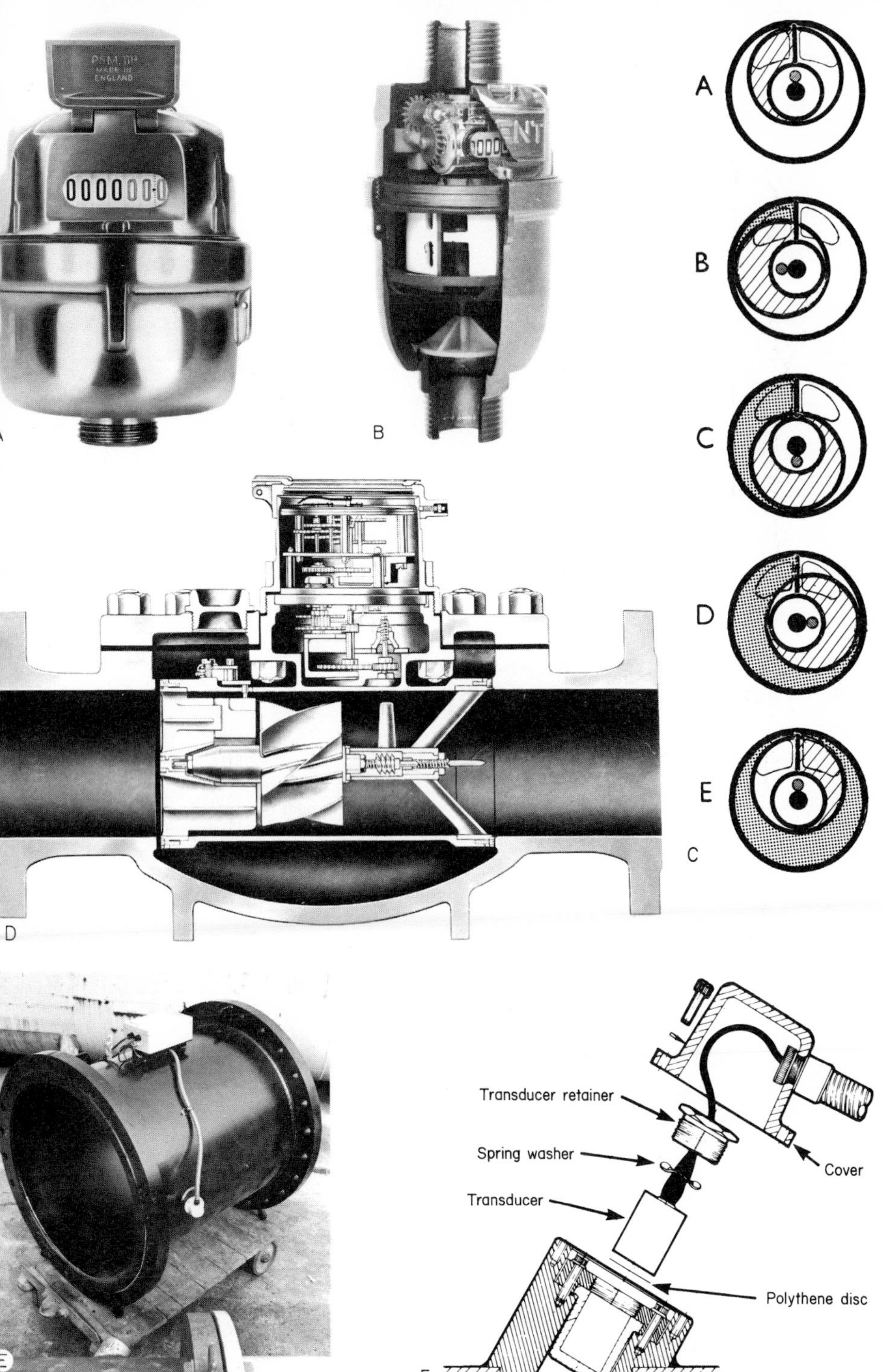

Plate 21 (A) and (B) Semi-positive rotary piston meters of 15 and 20 mm size respectively, and (C) sequence of operation of the meter. (Kent Meters Ltd.) (D) The 'Helix' inferential meter. (Kent Meters Ltd.) (E) An ultrasonic transit time meter showing the nearer transducer head inserted at 45° through the wall of the pipe, and the transmitter fixed to the crown of the pipe. (Bestobell Sparling Ltd.) (F) Details of the transducer head of an ultrasonic meter. (Bestobell Sparling Ltd.)

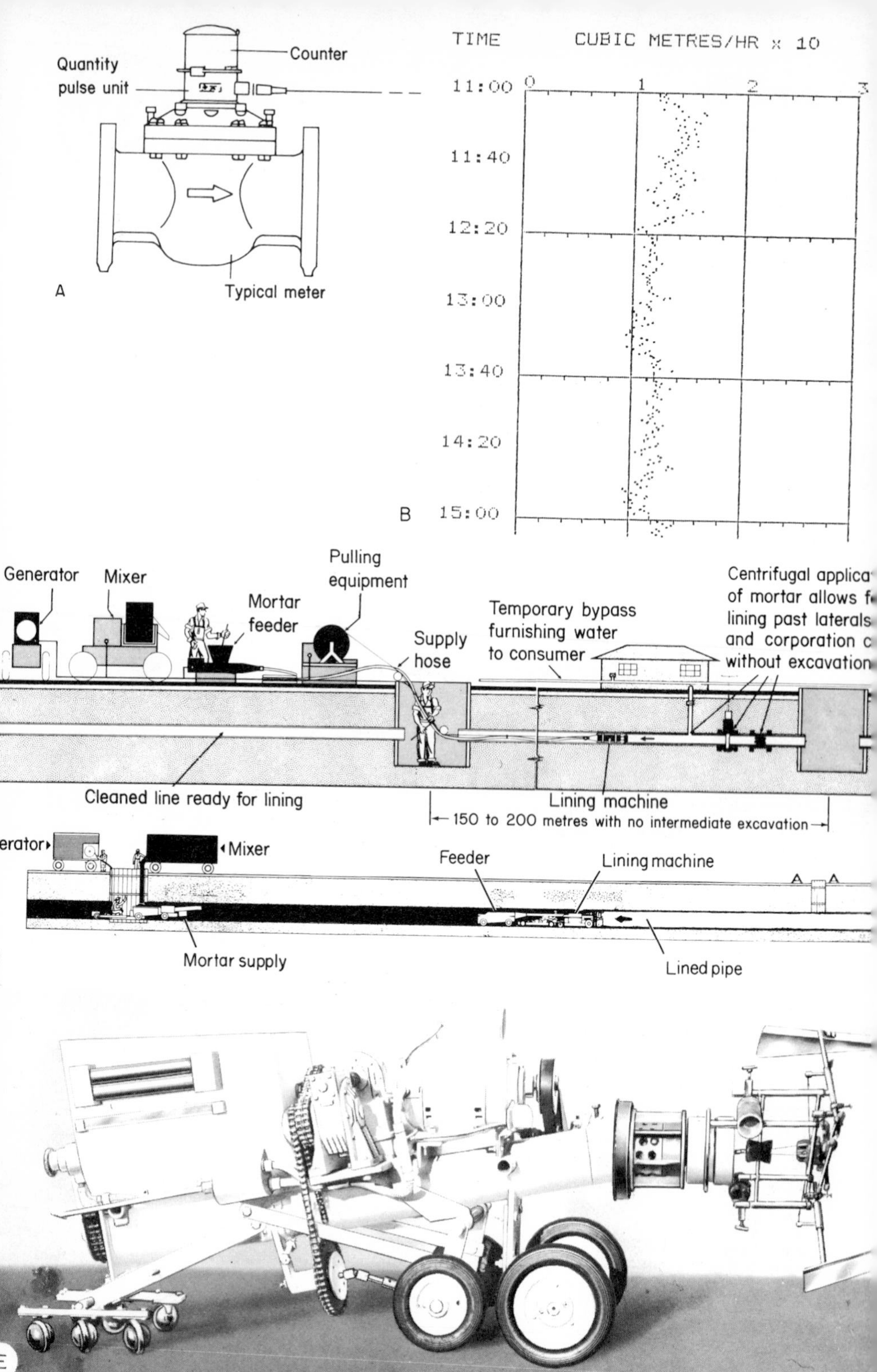

Plate 22 (A) 'Helix' meter fitted with an electrical pulse unit, and (B) the print-out of flow at one-minute intervals which may be obtained to a range of scales, using a data-logger, microcomputer and printer. (Kent AFA System: Kent Meters Ltd.) (C) and (D) In-situ mortar lining of a small diameter main. (Centriline Ltd.) (E) Mortar lining machine for a large diameter main. (Centriline Ltd. and Tate Pipe Lining Processes Ltd.)

With each method the final act is to dig down on the pipe to find the actual leak coming from it: hence all the several methods do is locate where a leak might be*.

Visual observation can detect many leaks. The District Inspector must continually be on the watch for signs of leakage and should, as a matter of routine, inspect the whole of his District on a regular basis. Trunk main routes should be walked monthly. It is particularly important to make sure that all washouts are kept closed. However, in many kinds of ground a leak from a main or service pipe may not cause a damp patch on the ground so that visual examination may not by itself be sufficient.

Sounding establishes the presence of a leak by its sound, using a 'listening stick' which need be no more than a light bar of solid metal about 1.5 m long. One end is placed on an exposed part of the main or service pipe, e.g. on a valve, hydrant, or stopcock, and the other (which is sometimes equipped with an earpiece) is placed on to the ear. The sound emitted by a leak, if it emits a sound at all (which is not always the case), may be variously described as 'a small low drumming noise', 'a continuous buzzing sound' etc. It has two useful identifying characteristics: it is continuous without any change of audibility or quality and it stops abruptly when, and if, the water can be turned off. An experienced Waste Inspector using a sounding rod can detect even a small leak at a distance of 10 to 15 m, *if* it is making a sound and *if* the pipeline is of metal. The principal problem is the difficulty of hearing the noise in the daytime when the noise of traffic is about, but it is quite possible for an experienced man to do so provided that the main is not on a major traffic route where there are no lulls in the traffic. Daytime sounding is preferred by some water authorities since, at night, although it may be easier to listen for a leak, the presence of parked cars may prevent access to valves and stopcock boxes and it is not possible to call on private premises at night to see whether the leak is on the service pipe to the premises. Small leaks from glands of valves or pipe joints and also leaks from plastic pipes are seldom detectable by sounding.

Various electronic devices exist called 'leak detectors' working on the principle of amplifying sounds heard in a pipe, but as they amplify all sounds from a pipe the particular noise made by a leak may be no easier to identify than with a listening stick. This criticism applies even if the sound vibration patterns are displayed visually on a cathode ray tube screen. Most water authorities regard such electronic apparatus as an aid to leak detection, rather than as a substitute for the listening stick which is very much cheaper and which, in the hands of an experienced man, can be just as effective. Both methods can only roughly show the

* This point is apparently not appreciated by those water authorities who let contracts for 'leak finding' by use of some (magic) apparatus and who do not make it a condition of contract that the existence of all leaks shall be confirmed by excavations taken down to locate them.

position of a leak by judging the extent to which the sound is more audible from one point of contact with a main than it is from another.

The 'leak noise correlator' is more effective in pinpointing the possible position for a leak whose sound has already been identified. In this system two probes are used simultaneously to make contact with the main several metres apart. The pattern of sound received from one probe (displayed on a cathode ray tube screen) is time-delayed by a few milliseconds until it matches a similar pattern received from the other probe. From the time difference applied the position of the leak with reference to the probes can be estimated. The apparatus, which is expensive and is contained in a mobile van, is not intended for routine sounding, but tests have shown that it can locate the position of a sound-emitting leak within 1 m, thus reducing the time spent digging holes to no effect.

Waste metering does not locate specific leaks, but can identify mains on which there is probably leakage. A small part of the distribution system of about 1000 properties or so (a 'waste district') must be valved off so that the supply to it is provided through one main only. This single supply main must have a valve on it controlling the supply to the district. A bypass containing a recording flow meter, i.e. a 'waste meter', must be constructed around the valve. On a given night the control valve on the single supply main is shut so that all night-time flow passes through the waste meter. The mains within the district are then shut down progressively, one by one, according to a pre-arranged pattern at 15- or 30-minute intervals. With each shutdown the drop in flow is registered on the waste meter chart. If a disproportionately large drop in flow occurs when a particular main is shut down, this may mean that such a main is causing a substantial amount of the flow and hence may be leaking. The results of the test are interpreted and passed to the District Day Inspector who then has to arrange for sounding of the mains which appear to have a high leakage. When it is thought that all leaks have been found and repaired the night-time test is repeated to find out if there has been an improvement.

The cost of the exercise is high in both labour and materials. All boundary valves to each district must shut off properly: if they are faulty they must be repaired or renewed. The making of the bypass for insertion of the waste meter will involve putting valved tees into the supply main either side of the control valve if anything larger than a 100 mm waste meter is used. It takes about a month to prepare each waste district and the cost may run into several thousand pounds per district. The waste meter used can be of the 'gate' type, but it is usually not practicable to use gate meters larger than 80 or 100 mm which record up to 1100 and 1500 m^3/d respectively. Provided that the flows to be registered are not too large, these meters can be trailer mounted and connected via fire hoses to hydrants upstream and downstream of the control valve on the supply main to the district, thus saving the cost of a bypass. With such small meters the waste districts have to be quite small, covering 600 to 1200 connections. A recent development is to use the electrical pulse-recording inferential meter of the vane type shown in Plates 22A and 22B. This can give more accurate results than the gate type waste meter as it can print out flow rates at intervals of 1 minute to 30 minutes to any scale desired. With this meter it is possible to adopt larger waste districts, thus reducing the number of waste metering positions required. However, if larger waste districts are adopted it may be necessary to shut down several mains at a time in the 'step-test' procedure described above in order to cause an accurately measurable flow drop, so that individual leaking mains are not so clearly identified.

Cutting off mains means isolating a particular main or group of mains, shutting off all (or most) of the service pipe connections from it, and putting it under a pressure or flow test. This is a large exercise which may have to be resorted to when leakage is known to be large but the points of leakage cannot be found by other methods. It is the sort of procedure that has to be followed when a system is not under 24-hour supply, in which case many consumers may not have turned off their taps, and standpipes may be left running etc. Further trouble and expense is incurred if stopcocks must first be inserted on service pipes and valves on branch connections do not shut off properly. Often the first attempt results in additional bursts occurring on the main under test because the system has long been on intermittent supply at low pressures.

15.16 Waste reduction as a whole

Total waste comprises both distribution leakages from mains and service pipes and consumer wastage. The latter includes all leaks and overflows on consumers' premises as well as abstractions by illegal unmetered connections. For a start all source meters have to be checked for accuracy: unless regularly serviced they will almost certainly be in error[7]. The general accuracy of the supply meters must also be assessed by the methods described in Section 1.3. A survey can reveal large quantities of water under-recorded by meters (or meter readings). Figures for acceptable levels of total waste were discussed in Section 1.16, whilst consumer wastage was dealt with in Section 1.9 and probable metering errors in Section 1.15.

As a permanent measure the setting up of waste districts as described in the preceding section is desirable. Thus regular waste metering followed up by sounding is the best practice. If no waste districts have been set up progressively as an undertaking has expanded it can be an expensive and long job creating them in an existing large undertaking. For a large undertaking serving one million population some 200 to 400 waste detection districts are required*. The cost will run into several thousand pounds sterling per district. If this cost cannot be met then it is better to use 'zonal metering'. These zones are much bigger than waste districts, containing 10 000 to 20 000 connections each, or possibly more. For economy they should use as many natural divisions of the distribution system as exist already which are fed by one main on which a permanent meter can be installed. Again boundary valves to each zone must be tested and repaired or replaced as necessary to ensure that they shut down properly. The meter has to be arranged on a bypass and can be an appropriate large waste meter, a Dall tube, or a typical inferential meter. The recording instrument may be made transportable to save expenditure. By such means the night-time flow to each zone can be regularly recorded and the problem of assessing levels of waste can then be broken down into zones so that those zones which appear to have the highest losses can come under more detailed inspection for leaks.

Irrespective of whether or not district waste metering or zonal monitoring can be adopted, *sounding* comprises the basic means whereby leaks and wastage by consumers are detected. The cheapest method by far is to use the sounding stick and properly trained men should routinely apply it every four to six months to every service pipe connection at night. The accuracy of all large metered supplies should also be regularly checked. By this means every main and service pipe will be checked for leakage at least twice per annum†.

Where an undertaking is in a bad state and has high losses the problem can only be tackled by setting up temporary zonal metering so that the areas suffering highest losses can be identified and then put under more detailed inspection and testing. When there are only intermittent supplies the difficulties of leak and waste detection are greatly exacerbated and cutting off mains, as described in Section 15.15, may have to be resorted to. However, along with these measures, once a main can be put under a 24-hour supply for a week or ten days, visual observation and sounding can reveal many leaks. Checking distribution pressures can also be helpful once the supply on a 24-hour basis can be sufficient to meet the demand. However, waste reduction must not be

* The number depends upon the average number of persons per connection in the area. A district should reasonably cover between 600 and 1000 connections. At three persons per connection the number of districts required for one million population is therefore in the range 333 to 555 according to the size of the districts. However, only one-half to two-thirds of this number of waste meters will be required because it is usually possible to make one meter serve two districts.

† Giles (see Reference 22 of Chapter 1) considers that an Inspector should be able to sound 80–120 connections per day, on average. This would be for overseas conditions. Ingham (Reference 36 of Chapter 1) reports that the rate of progress on a typical UK system is 200 soundings per man per day, but also states that 'a very high percentage' of connections would have no leakage, which is not a typical overseas experience.

thought of as an exercise to be tried now and then: it has to be organized as a permanent activity of the undertaking to which an adequate number of properly trained men must be allocated under the charge of a responsible engineer.

15.17 Field measurement of pipeline flows

A need in waste detection and network analysis is to have temporary means of measuring and recording flow in mains. A number of devices are available, but all have limitations as described below.

(1) Gate type waste meters are only convenient in 80 or 100 mm size for measuring rates of flow up to 1500 m^3/d. When trailer mounted they can be connected by fire hoses to hydrants upstream and downstream of some valve that will shut off properly. Larger sizes are available, but these have to be connected by branch tees inserted into the main to form a bypass around the control valve: this becomes expensive.
(2) Inferential meters of the rotating vane type can measure large flows, and are especially valuable if equipped with electrical pulse recording (see Section 15.15 and Plates 22A and 22B). One meter mechanism can serve several meter bodies of the same size, which greatly reduces the cost if only temporary flow recording is required.
(3) Venturi meters to various designs, e.g. the Dall tube, need permanent installation. On large mains the cost is several thousand pounds (see Section 10.19).
(4) The pitot tube can be used. This is inserted into a main via a screwed gland tapping of about 50 mm diameter. The apparatus is cumbersome, easily broken, needs skilled technical supervision for calibration of pipe flow, and must be attended and read every half hour to obtain a record of flow. Procurement of the mercury required may present a problem in some countries because it is a listed poison and may be difficult to obtain or import.
(5) The insertion vane meter comprises a small current meter on the end of a rod which is inserted into the main and used in the same manner as a pitot tube. It is more robust than a pitot tube and easier to read. However, if the vane is damaged the meter then possesses an unknown error and little confidence can be placed on its accuracy unless the vane is replaced by a new one or sent back to the manufacturers for repair and recalibration.
(6) The ultrasonic flow meter working on the measurement of transit times of ultrasonic beams passing diagonally across the pipe (*not* the Doppler effect meter—see Section 10.20) has considerable promise as a field flow-measuring device now that it is produced in kit form for insertion by means of tappings into an existing main. An accuracy of ±2.5% is quoted if the pipe diameter is accurately known and the meter is properly positioned. A portable transmitter and calibration unit

means that flows can be recorded on a number of installations on mains of different sizes.

(7) The magnetic flow meter is expensive and comprises a permanent installation which must be designed and set up by a skilled electrician. It is as accurate as the Venturi meter and has the advantage of being able to measure flow in either direction.

(8) The heat pulse meter has the special purpose of measuring flows at very low velocity. It has not been developed for ordinary flow velocities, and is not yet commercially available*.

(9) Chemical dilution flow methods have not yet been developed for continuous flow measurement and recording.

Of the meters listed above, numbers (1), (2) and (5) are those most used in practice for temporary flow recording. Of the others, the use of the transit time ultrasonic meter, which can be inserted through tappings on a main, promises to be of considerable value.

15.18 Supply meters

Meters in common use on service pipes are of two types: semi-positive or inferential†. These terms describe the manner in which they measure flow.

The semi-positive meter is almost universally used for metering domestic supplies and small trade supplies. Several different designs are available, the most usual type in the UK being the rotary piston (or rotary cylinder) type. In this an eccentrically pivoted, light weight, and freely moving cylinder inside the cylindrical body of the meter is pushed around by the water, opening and closing inlet and outlet ports as it turns. This movement operates a revolution counter summating the total flow. A new meter will register flows down to 4 l/h in the smaller size and is eminently suitable for domestic purposes. Another type of semi-positive meter uses a nutating disc which wobbles around in a circle as water is drawn through the meter and endeavours to pass above or below the disc. This meter will not measure low flows to the same degree of accuracy as the rotary piston meter, but is widely used in the USA where domestic flows are generally higher. All semi-positive meters should incorporate a strainer upstream as the meter is only suitable for measuring water which is free from grit or sand.

The inferential meter comprises a propeller vane which is turned by the flow of water, i.e. it 'infers' the quantity of water passing by counting the revolutions of the vane. Any particular design must therefore be

* Pioneering work on the heat pulse meter and chemical dilution has been carried out by the Water Research Centre in the UK which has also investigated other types of metering. A number of commercial firms must also receive credit for developing meters and measuring kits into their most practical form.

† The positive meter is not used in water supply: it works on a piston displacement method.

calibrated. It is primarily used on industrial supplies, being suitable for large flows. For the measurement of widely fluctuating flows, beyond the range of any single meter, two meters of different sizes may be connected in parallel, an automatic arrangement ensuring that the smaller flows go through the smaller meter and vice versa. This arrangement is called a 'combination meter'. Inferential meters can measure water containing some suspended matter.

15.19 Meter testing

All meters must be brought in regularly for maintenance, repair, and testing before re-issue. Stocks of spare parts for meters must be held. A meter testing bench is shown in Fig. 15.7.

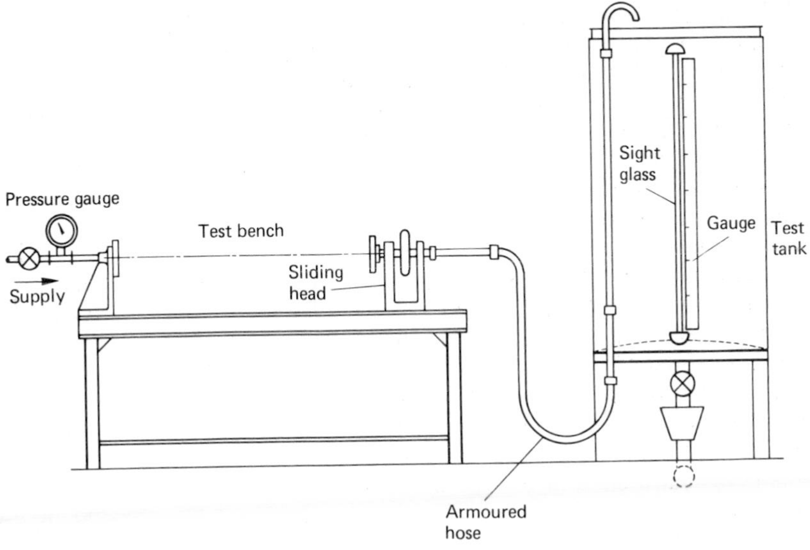

Fig. 15.7 Meter testing bench.

15.20 Waterworks byelaws

Nearly all water authorities have byelaws or regulations setting out a variety of requirements with respect to consumers' use of water and the design of plumbing systems. In the UK Model Water Byelaws have been produced by the Ministry of Housing and Local Government, now the Department of the Environment. Among the matters laid down by such byelaws are the following.

(1) The consumer shall not cause 'waste, undue consumption, misuse, erroneous measurement or contamination of water, or reverberation in pipes'.

(2) The consumer must repair or replace pipes, fittings, and apparatus causing or likely to cause waste of water.
(3) Only approved quality pipes, fittings, and apparatus shall be used.
(4) Pipes and fittings must conform to certain layout, protection, and laying requirements.
(5) Stoptaps, overflows, and cisterns etc. must be of approved quality and comply with certain conditions with regard to their size, design, and positioning.
(6) Cross-connections from pipes under mains pressure to other systems are forbidden and provisions with regard to the prevention of back-siphonage must be complied with.
(7) The use of water for motive or power purposes is forbidden.

By the foregoing means the water authority can ensure that fittings, such as ball valves, are of the right type having regard to the pressure in the mains and that plumbing systems are generally durable and not liable to give rise to mounting waste of water. Provisions with regard to the prevention of back-siphonage have become increasingly important since contamination of drinking water can occur if back-siphonage is not prevented. A simple measure is to ensure that every tap is above the maximum overflow level that can occur in washbasins, baths etc. If this is not done, an open tap in some basin or bath could become drowned and, if the mains pressure should drop or a lower tap should be opened, water in the basin or bath could be drawn off by siphonic action through the tap. Particular danger exists in hospitals which have complex plumbing systems.

For the same reason the WC flushing cistern is preferred in the UK to the hand-operated flushing valve. The use of the latter brings the pressurized supply down to the toilet pan and thereby increases the danger of back-siphonage. The ball valve cistern, however, is at higher level and acts as a barrier to back-siphonage and is the traditional method whereby the pressurized mains supply is divorced from the local plumbing system feeding receptacles that can contain contaminated water. Such provisions need not be regarded as unduly rigorous since cases of cross-contamination do occur despite all precautions.

To facilitate compliance with byelaws in the UK fittings made by manufacturers can be approved by a central organization acting on behalf of all water authorities, before such materials are put on the market. Many requirements as to quality of materials and strength and thickness of pipes and fittings are specified by reference to British Standards which are national. A similar procedure is adopted in many other countries and makes a major contribution to the reduction of waste.

References

(1) Blair J S, New Formulae for Water Flow in Pipes, *Proc. IME*, February 1951.
(2) Reed E C, Report on Water Losses, *Aqua.: JIWSA*, 1980, p. 178.
(3) Lackington D W, Survey of Renovation of Water Mains, Paper No. 4, *Proc. Symposium on the Deterioration of Underground Assets*, IWES, 1983.
(4) Clark P G, Factors Relating to the Deterioration of Water Mains, Paper No. 2, Symposium referred to in Reference 3.
(5) Newport R, Factors Influencing the Occurrence of Bursts in Iron Water Mains, *Aqua.: JIWSA*, 1980, p. 274.
(6) Parkinson R W and Warren I C, Recent Investigations of the Epoxy Resin Lining of Water Mains, *JIWES*, June 1983, p. 257.
(7) Baker G P *et al.*, Calibrating Large Water Mains in Situ, Paper to the Conference on Drinking Water Distribution Techniques, WRC, June 1974.

Conversion factors

Metric to British units	British to metric units
Length:	
1 m = 39.37 in	1 in = 25.40 mm
1 m = 3.2808 ft	1 ft = 304.80 mm
1 m = 1.0936 yd	1 yd = 0.914 40 m
1 km = 0.6214 mile	1 mile = 1.609 34 km

Note: In nominal sizing 300 mm is taken as equivalent to 1 ft and 25 mm as equivalent to 1 in.

Area:

1 m^2 = 1.196 yd^2	1 ft^2 = 0.0929 m^2
1 ha = 2.471 acres	1 yd^2 = 0.8361 m^2
1 km^2 = 0.386 square mile	1 acre = 0.4047 ha
	1 square mile = 2.5900 km^2

Notes:
1 km^2 = 100 ha (hectares) and 1 ha = 10 000 m^2.
1 square mile = 640 acres and 1 acre = 4840 yd^2.

Volume:

1 m^3 = 35.314 ft^3	1 ft^3 = 0.028 32 m^3
1 m^3 = 1.3079 yd^3	1 ft^3 = 28.32 litres
1 m^3 = 219.97 gallons (British)	1 yd^3 = 0.764 56 m^3
1 litre = 0.219 97 gallon	1 gallon = 4.546 litres
1 Ml = 0.219 97 Mg	1 Mg = 4.546 × 10^3 m^3 = 4.546 Ml

Notes:
1 m^3 = 1000 l (litres) and 1 Ml = 1000 m^3.
1 US gallon = 0.832 67 British gallon; also 1 ft^3 = 6.2288 British gallons.

Mass:

1 kg = 2.2046 lb	1 lb = 0.453 59 kg
50 kg = 0.9842 cwt	1 cwt = 50.802 kg
1 Mg = 19.684 cwt	1 ton = 1.016 05 Mg
1 tonne = 0.9842 ton	

Metric to British units	**British to metric units**
Pressure:	
1 metre head of water $= 1.422\ \mathrm{lb/in^2}$	1 ft head of water $= 0.030\,48\ \mathrm{kgf/cm^2}$ $= 0.002\,99\ \mathrm{N/mm^2}$
$1\ \mathrm{kgf/cm^2} = 14.223\ \mathrm{lb/in^2}$	$1\ \mathrm{lb/in^2} = 0.0703\ \mathrm{kgf/cm^2}$ $= 0.006\,895\ \mathrm{N/mm^2}$
$1\ \mathrm{N/mm^2} = 145.038\ \mathrm{lb/in^2}$	

Notes:
$1\ \mathrm{N/mm^2} = 10.197\ \mathrm{kgf/cm^2}$, $1\ \mathrm{kgf/cm^2} = 10$ metres head of water, and 1 bar = 10.197 metres head of water.
$1\ \mathrm{lb/in^2} = 2.3067$ ft head of water.

Density:

$1\ \mathrm{kg/m^3} = 0.062\,43\ \mathrm{lb/ft^3}$	$1\ \mathrm{lb/ft^3} = 16.018\ \mathrm{kg/m^3}$

Flow rates:

$1\ \mathrm{m^3/s} = 35.31\ \mathrm{ft^3/s}$	$1\ \mathrm{ft^3/s} = 0.0283\ \mathrm{m^3/s}$
$1\ \mathrm{m^3/s} = 19.00$ mgd	1 mgd $= 0.052\,62\ \mathrm{m^3/s}$
1 litre/s = 13.20 gpm = 0.019 mgd	1 gpm = 0.0758 litre/s

Notes:
mgd = million gallons per day; gpm = gallons per minute.
$1\ \mathrm{m^3/s} = 86.4 \times 10^3\ \mathrm{m^3/d} = 86.4$ Ml/d.
$1\ \mathrm{ft^3/s} = 86\,400\ \mathrm{ft^3/d} = 0.538\,17$ mgd.

Hydrological units:

1 litre/s per $\mathrm{km^2}$ $= 0.091\,46\ \mathrm{ft^3/s}$ per 1000 acres	$1\ \mathrm{ft^3/s}$ per 1000 acres $= 6.997$ litres/s per $\mathrm{km^2}$
1 mm rainfall per $\mathrm{km^2}$ $= 1000\ \mathrm{m^3}$ $= 0.220$ Mg	$1\ \mathrm{ft^3/s}$ per square mile $= 10.933$ litres/s per $\mathrm{km^2}$
	1 in rainfall per square mile $= 65\,786\ \mathrm{m^3}$

Note: 1 in of rainfall per square mile = 14.47 million gallons.

Filtration rate:

Note: 100 gallons per $\mathrm{ft^2}$ per hour $= 117.44\ \mathrm{m^3}$ per $\mathrm{m^2}$ per day
$= 4.89$ m/h
$= 1.36$ mm/s.

Power:

1 joule (J) = 0.737 56 ft lb	1 horsepower (hp) = 0.745 70 kW
1 kW = 1.3410 hp	

Notes:
1 J/s = 1 watt (W).
1 Ml/d of water raised through 8.81 m = 1 kW (at 100% efficiency).
1 hp = 550 ft lb/s.

Index